# Mechanics of Continuous Media: an Introduction

# Mechanics of Continuous Media: an Introduction

John Botsis
Michel Deville

Translated from french by
Ray Snyder

EPFL Press

Cover artwork: © royyimzy, fotolia.com

*Original version*
**Mécanique des milieux continus : une introduction**
John Botsis, Michel Deville
Second edition, ISBN 978-2-88915-154-7


EPFL Press

is an imprint owned by Presses polytechniques et universitaires romandes, a Swiss academic publishing company whose main purpose is to publish the teaching and research works of the Ecole polytechnique fédérale de Lausanne (EPFL) and other universities and institutions of higher learning.

Presses polytechniques et universitaires romandes
EPFL – Rolex Learning Center
Post office box 119
CH-1015 Lausanne, Switzerland
E-mail: ppur@epfl.ch
Phone: 021 /693 21 30
Fax: 021 /693 40 27

www.epflpress.org


ISBN 978-2-88915-281-0

Printed in France

To our wives Maria and Christina
and our families

# Preface

The Mechanics of Continuous Media (MCM) offers a unified approach to the concepts and general principles for modeling the motion or deformation of diverse materials. By considering an appropriate length scale, based on the molecular arrangements or the microstructure, any material can be approximated by a continuum where smoothly varying properties are assigned to every point in space occupied by the investigated body at a given time during its motion. This topic of mechanics was considerably developed in the second half of the 20th century, and today is considered an important subject in all branches of continuum physics.

The reasons for this development are the modeling of new materials with unusual mechanical behaviors in solid or fluid form including rubber, polymers and gels, living tissues, and soft matter. If we consider classical continuum like air or water, the phenomena of turbulent flow are far from being understood in their totality and this area is still a subject of intense research. Also, modern engineering analysis uses sophisticated tools for design of advanced structures and devices in transportation, energy, life science, sports, etc. These tools involve simulation software that performs numerical integration of partial differential equations that describe the continuous models. In order to be able to analyze the results obtained, the engineer must have a sound theoretical foundation.

This book presents the basic concepts for the mathematical modeling of classical solid and fluid continuous media. In the first chapter the indicial notation is defined and discussed with several examples to illustrate its use in demonstrations of vector identities. The theory of Cartesian tensors and their properties are presented afterwards. Next, the important relations of polar decomposition of a tensor are given. The different vector operators are defined with several illustrative examples, and the mathematical tools and theorems necessary to develop the continuous mechanics model are described.

In chapter 2 the kinematics of a continuum is developed using the Lagrangian (material) and Eulerian (spatial) descriptions. The material derivative is defined, and velocity and acceleration are introduced. The motion of a rigid body is described afterwards, followed by the definitions of the deformation gradient tensor and the various deformation tensors, the properties of which are illustrated in detail with representative examples. The transformation of line, surface, and volume elements between material and spatial coordinates are explained. Also developed in this chapter is the important topic of linearization to obtain the infinitesimal strain tensor used in linear elasticity followed by the definition of the deformation gradient tensor. In the last part of this chapter, the objectivity with respect to two different observers is defined and the relevant definitions of objective tensors are given.

In chapter 3, entitled dynamics of continuous media, the laws of conservation of mass, momentum, and angular momentum are presented. Cauchy's stress tensor, Cauchy's theorem, and the equations of motion of a continuum using spatial description are introduced. The properties of the stress tensor and the equation of equilibrium are analyzed to illustrate their significance in continuum mechanics. Representative simple examples are given to illustrate the importance of stress and the properties of Cauchy's stress tensor. The Piola-Kirchhoff stress tensors with respect to the material description are also defined. The effect of linearization is highlighted, and its ramifications for the Cauchy and Piola-Kirchhoff stress tensors are illustrated.

Chapter 4 is devoted to the thermodynamics of continuous media. The first principle, which deals with the conservation of total energy, is developed in detail using the spatial description followed by the principle of conservation of mechanical energy in the material description to arrive at the set of conjugate tensor parameters that define energy. The notion of entropy is discussed afterwards, followed by the second principle of thermodynamics and the Clausius-Duhem inequality to account for the irreversibility of the phenomena in a continuous medium. The general principles for establishing constitutive equations are described in chapter 5.

In Chapter 6 classical Newtonian fluids are introduced, followed by the theory of hyperelasticity from which the constitutive laws of hyperelastic materials are deduced. The description is illustrated with simple representative cases of hyperelastic material behavior. The infinitesimal linear isotropic elasticity theory is defined next. After introducing Fourier's law for heat conduction, the chapter ends with considerations of the second principle of thermodynamics applied to viscous fluids, ideal gases, and linear elastic materials.

The seventh chapter deals with linear, isotropic elasticity. In the first part the general theory is defined for solids in static equilibrium, followed by the definitions of plane strain and plane stress. Afterwards, the methods of solutions of Navier's equations using the method of potentials are presented with applications to selected problems. The solution to some advanced problems including those of Kelvin, Cerutti, and Boussinesq are discussed next. The important solution method based on Airy's stress function is developed for two-dimensional problems with examples of solutions for representative problems. In the second part of the chapter, the wave propagation equation in linear elastic solids is deduced from Navier's equations, and the solution for Rayleigh surface waves is discussed in some detail. Lastly d'Alembert's solution to the one-dimensional wave equation is provided with representative examples.

The last chapter deals with the mechanics of Newtonian fluids. Some physical observations are presented for laminar and turbulent flow of an incompressible fluid and then, for subsonic and supersonic flow of a compressible fluid. The Navier-Stokes equations are derived in compressible and incompressible cases.

Analytical solutions are proposed for simple cases. The dynamics of vorticity is described. The equation for the fluid circulation is presented and the Bernoulli equation is obtained. The chapter ends with acoustic waves and simple solutions for steady state, irrotational, and isentropic flow of a perfect compressible fluid.

Appendices give the necessary additions to follow the work and represent the field equations in cylindrical and spherical coordinate systems. A list of symbols and suggestions for solutions to the exercises in each chapter are also provided.

This book is the result of our teaching of the Mechanics of Continuous Media to second-year students in mechanical engineering at the EPFL and contributes modestly to the knowledge base in this direction. The subject is developed in a simple-to-follow pedagogical manner that a reader can work through on her own.

*Audience*: This book is intended for engineering and physics students who want to learn the basic principles of continuum mechanics. They will find in this work a complete modern introduction that opens the door to this vast territory of knowledge. For an introductory course in continuum mechanics in engineering or physics curricula, we recommend covering the first four chapters, i.e., Cartesian tensors, kinematics, dynamics, and energetics of continuous media as well as selected sections in chapters 6, 7 and 8 on constitutive equations for solids and fluids.

*Prerequisites*: We assume that potential readers have taken courses in Newtonian mechanics, linear algebra, calculus, and an introductory course in structural mechanics. The curious reader will also be able to further explore this area by referring to the many more advanced texts quoted in the bibliography.

*Acknowledgements*: M. Deville thanks Professor Marcel Crochet, who was his PhD supervisor at the Catholic University of Louvain in Belgium, for having opened the doors of this large intellectual area represented by MCM. Many hours of discussion with his deceased colleague François Dupret deeply influenced the sections on thermodynamics and viscous fluids. Some developments were inspired by lecture notes on MCM by M. Crochet, UCL, 1992 and on Fluid Mechanics by M. Deville, F. Dupret and P. Wauters, UCL, 1992.

J. Botsis is thankful to Professor Alexander Chudnovsky for his mentorship and friendship, especially during the early years of his career.

We are grateful for the help of a number of people through their remarks, their interest for the subject, and the drawing up of figures. These persons are Roland Bouffanais, Matteo Galli, Qi-Chang Hi, Laurent Humbert, David LoJacono, and Aïssa Mellal.

Special thanks go to Georgios Pappas for his relevant remarks on the figures and his skillful drawing of the new figures.

The authors also thank Dr. Ray Snyder for the translation of the original French text into English.

*General Rules for the Notations*: In this monograph, the scalars are in italic characters, such as $p$ and $T$. Vectors and tensors are in bold italic characters, such as $\boldsymbol{v}$, $\boldsymbol{\sigma}$ and $\boldsymbol{L}$. Matrices are in italics in square brackets, for example, $[C]$. When tensors are written in index notation, the indices are chosen in most cases from the letters $i$, $j$, $k$, $l$, $m$, $n$. Therefore the vector $\boldsymbol{v}$ will have as components $v_i$ and the second order tensor $\boldsymbol{\sigma}$ will have components $\sigma_{ij}$.

*John Botsis*
*Michel Deville*
Lausanne
September 2018

# Contents

Preface vii

CHAPTER 1 **Cartesian Tensors** **1**

1.1 Introduction ... 1
1.2 Vector Algebra ... 3
1.2.1 Generalities ... 3
1.2.2 Orthogonal Transformation of Coordinate Systems ... 6
1.2.3 Scalars ... 9
1.2.4 Vectors ... 10
1.2.5 Permutation Symbol and Vector Product ... 12
1.3 Tensor Algebra ... 14
1.3.1 Definition of Tensors of Order 2 ... 14
1.3.2 Tensor Product or Dyadic Product of Two Vectors ... 16
1.3.3 The Transformation Rule for Cartesian Tensor Components ... 17
1.3.4 Tensors of Any Order ... 18
1.3.5 Algebra of Tensors of Order 2 ... 20
1.3.6 Tensor Properties ... 21
1.3.7 Dual Vector of a Tensor of Order 2 ... 25
1.3.8 Eigenvalues and Eigenvectors of a Tensor ... 26
1.3.9 Square Root of a Tensor ... 30
1.3.10 Polar Decomposition ... 32
1.3.11 Isotropic Tensor Function of a Symmetric Tensor ... 33
1.3.12 Scalar Function of a Tensor ... 34
1.4 Tensor Analysis ... 34
1.4.1 Derivative of Vector, Tensor or Scalar Function ... 35
1.4.2 Gradient of a Scalar Field ... 36
1.4.3 Gradient of a Vector Field ... 37
1.4.4 Gradient of a Scalar Valued Tensor Function ... 38
1.4.5 Gradient of a Tensor Valued Tensor Function ... 38
1.4.6 Divergence of Vectors and Tensors ... 38
1.4.7 Curl of a Vector Field ... 39
1.4.8 Laplacian Operator ... 40
1.4.9 Curvilinear Coordinates ... 42

1.4.10 Scalars, Vectors, and Tensors in Orthogonal Curvilinear Coordinates ... 46
1.4.11 Gradient of Scalar and Vector Fields in Orthogonal Curvilinear Coordinates ... 46
1.4.12 Definition of the Notion of Flux ... 48
1.4.13 Integral Theorems of Gauss and Stokes ... 49
1.5 Exercises ... 51

CHAPTER 2 **Kinematics of Continuous Media** **55**
2.1 Introduction ... 55
2.2 Bodies, Configurations and Motion ... 55
2.3 Material and Spatial Descriptions ... 58
2.4 Velocity, Material Derivative and Acceleration ... 64
2.4.1 Velocity ... 64
2.4.2 Material Derivative ... 64
2.4.3 Acceleration ... 66
2.5 Pathlines, Streamlines and Streaklines ... 67
2.6 Kinematic Equations for Rigid Body Motion ... 69
2.6.1 Translation of a Rigid Body ... 69
2.6.2 Rotation of a Rigid Body Around a Fixed Point ... 70
2.6.3 General Rigid Body Motion ... 71
2.7 Gradient and Deformation Tensors ... 72
2.7.1 Definition ... 72
2.7.2 Deformation Tensors ... 74
2.7.3 Geometric Interpretation ... 78
2.8 Homogeneous Deformations ... 81
2.9 Small Displacements and Infinitesimal Strain Tensors ... 85
2.9.1 Small Displacements ... 85
2.9.2 Infinitesimal Strain Tensor ... 87
2.9.3 Compatibility Equations for the Infinitesimal Strain Tensor ... 90
2.10 Velocity Gradient and Associated Tensors ... 91
2.11 Objectivity of the Kinematic Quantities ... 94
2.12 Exercises ... 99

CHAPTER 3 **Dynamics of Continuous Media** **103**
3.1 Introduction ... 103
3.2 Reynolds Transport Theorem ... 103
3.2.1 Background ... 103
3.3 Conservation of Mass ... 107
3.3.1 Material Form ... 107
3.3.2 Spatial Form ... 110
3.4 Volume Forces, Contact Forces and Cauchy's Postulate ... 112
3.5 Conservation of Momentum and Angular Momentum ... 115

3.6 Cauchy's Theorem and Equation of Motion ........... 117
3.7 Properties of the Cauchy Stress Tensor ................ 124
3.8 Simplified Stress States ................................ 128
3.9 Piola-Kirchhoff Stress Tensors ........................ 131
3.9.1 General Considerations ........................ 131
3.9.2 First and Second Piola-Kirchhoff Tensors ...... 132
3.9.3 Linearization of the Stress Tensors ............. 135
3.10 Exercises ............................................. 138

CHAPTER 4 **Energy** **141**
4.1 Introduction ........................................... 141
4.2 Conservation of Energy .............................. 141
4.3 Conservation of Mechanical Energy in the Material Representation ......................................... 147
4.4 Interpretation of the Conservation Laws by the First Principle ............................................... 149
4.4.1 First Case: Uniform Translation ............... 150
4.4.2 Second Case: Rigid Body Rotation ............. 151
4.5 The Notion of Entropy ................................ 151
4.6 Second Principle of Thermodynamics ................. 154
4.7 Exercises ............................................... 156

CHAPTER 5 **Constitutive Equations: Basic Principles** **159**
5.1 Introduction ........................................... 159
5.2 General Principles .................................... 162
5.2.1 Hypothesis of Causality or Determinism ........ 162
5.2.2 Local Action Principle ........................... 162
5.2.3 Principle of Objectivity ........................ 163
5.2.4 Principle of Material Invariance ................ 165
5.2.5 Principle of Memory ............................. 166
5.2.6 Principle of Admissibility ..................... 166
5.3 Consequence of the Principle of Local Action .......... 166
5.4 Thermomechanical Constitutive Equations ............ 168
5.4.1 Principle of Determinism ....................... 168
5.4.2 Principle of Equipresence ...................... 168
5.4.3 Principle of Local Action ....................... 169
5.4.4 Principle of Objectivity ........................ 169
5.5 Definition of a Fluid and a Solid ...................... 170
5.6 Principle of Regular Memory .......................... 170
5.7 Exercises ............................................... 171

CHAPTER 6 **Classical Constitutive Equations** **173**
6.1 Introduction ........................................... 173
6.2 Simple Fluids .......................................... 173
6.3 Classical Fluids or Newtonian Viscous Fluids .......... 175
6.4 Isothermal Isotropic Elastic Media .................... 177

6.5 Hyperelastic Materials ... 179
6.5.1 Isotropic Hyperelastic Materials ... 181
6.5.2 Forms of the Strain Energy Function ... 186
6.5.3 Reduction to Simple Stress States ... 188
6.6 Linear Infinitesimal Elasticity ... 191
6.7 Heat Conduction ... 195
6.8 Second Principle of Thermodynamics for Viscous Fluids ... 195
6.9 Ideal Gas Thermodynamics ... 198
6.10 Second Principle of Thermodynamics for Classical Elastic Media ... 200
6.11 Thermoelasticity ... 201
6.12 Exercises ... 203

CHAPTER 7 **Introduction to Solid Mechanics** **207**

7.1 Introduction ... 207
7.2 Fundamental Equations of Static Linear Elasticity ... 207
7.2.1 Static Linear Elastic Field Equations ... 208
7.2.2 Boundary Conditions ... 210
7.2.3 Superposition Principle ... 211
7.3 Plane Isotropic Linear Elasticity ... 211
7.3.1 Plane Strain States ... 211
7.3.2 Plane Stress States ... 215
7.4 Solution Methods in Linear Elasticity ... 217
7.4.1 Displacement Functions ... 219
7.4.2 Stress Functions and Airy Solutions for Plane Problems ... 232
7.5 Wave Propagation in a Linear Elastic Medium ... 242
7.5.1 Shear and Dilatation Waves ... 242
7.5.2 Rayleigh Surface Waves ... 244
7.5.3 One-Dimensional Elastic Plane Waves ... 248
7.5.4 Propagation of a Wave in an Elastic Cord ... 251
7.6 Exercises ... 263

CHAPTER 8 **Introduction to Newtonian Fluid Mechanics** **267**

8.1 Introduction ... 267
8.2 Physical Considerations for Laminar and Turbulent Incompressible Flows ... 268
8.3 Physical Considerations for Compressible Fluid Flows ... 272
8.3.1 Subsonic, Supersonic and Hypersonic Flows ... 272
8.3.2 Shock Waves ... 273
8.4 Navier-Stokes Equations ... 276
8.4.1 Navier-Stokes Equations for an Ideal Gas with Constant Heat Capacity ... 276
8.4.2 Navier-Stokes Equations for an Incompressible Fluid in Isothermal Flow ... 279

8.5 Non-Dimensional Form of the Navier-Stokes Equations ... 279
8.5.1 Compressible Fluid Case ... 279
8.5.2 Case of an Incompressible Fluid in Isothermal Flow ... 282
8.6 Boundary and Initial Conditions ... 283
8.6.1 Viscous Fluid ... 283
8.6.2 Perfect Fluid ... 284
8.7 Exact Solutions of the Navier-Stokes Equations ... 284
8.7.1 Plane Stationary Flows ... 284
8.7.2 Stationary Axisymmetric Flows ... 293
8.7.3 Plane Non-Stationary Flows ... 297
8.8 Stokes Flow ... 300
8.8.1 Plane Creeping Flows ... 301
8.8.2 Parallel Flow Around a Sphere ... 303
8.9 Vorticity and Vortex Kinematics ... 307
8.10 Dynamic Vorticity Equation ... 310
8.10.1 General Equation ... 310
8.10.2 Physical Interpretation of Vorticity Dynamics ... 311
8.11 Vorticity Equation for a Newtonian Viscous Fluid ... 313
8.12 Circulation Equation ... 314
8.13 Vorticity Equation for a Perfect Fluid ... 315
8.14 Bernoulli's Equation ... 316
8.15 Acoustic Waves ... 317
8.16 Stationary, Irrotational, Isentropic Flow of a Compressible Perfect Fluid ... 320
8.16.1 Small Perturbation Theory ... 321
8.16.2 Two-Dimensional Flow of a Compressible Fluid in the Neighborhood of a Sinusoidal Wavy Wall ... 322
8.17 Exercises ... 325

APPENDIX A **Cylindrical Coordinates** **327**

APPENDIX B **Spherical Coordinates** **333**

List of Symbols 339

Suggestions for Solutions to the Exercises 345

Bibliography 349

Index 353

CHAPTER 1

# Cartesian Tensors

## 1.1 Introduction

The Mechanics of Continuous Media (MCM) is the branch of mechanics for modeling the behavior of solid or fluid materials. We know from physics that matter is composed of elementary particles. At the atomic scale (on the order of a nanometer or less) one would typically use statistical mechanics to describe the physical phenomena. In fact, every constituent "particle" could be described with Newtonian laws, but the value of Avogadro's number, $N_A = 6 \times 10^{23}$ per mole, renders any solution of the resulting set of equations impossible, either analytically or numerically. In the case of compressible fluids, the kinetic theory of gases is a good example of a statistical approach using a probability density function to describe the particle in a phase space of velocity and position. These particles undergo random Brownian motion characterized by the mean free path $\lambda$. This leads to the definition of the Knudsen number $Kn = \lambda/L$, the ratio of the mean free path to a reference length, $L$, for the problem under consideration. If $Kn < 1$, the medium is sufficiently dense to allow the behavior of individual particles to be ignored; the continuous medium hypothesis at the scale $L$ is valid. If, on the contrary, $Kn \sim 1$ or $Kn > 1$, then the continuous medium model is no longer appropriate. It is thus seen that the notion of a continuous medium depends directly on the observation scale.

Another way to define the notion of a continuous medium consists of studying the evolution of the mass density or simply the density of a cube as a function of its size. For simplicity, we consider water as the physical system in a cube centered on a point P with sides of length $h$. A certain number of molecules are found in the cube with an average density $\rho_h$ defined as $\rho_h = M_h/h^3$, where $M_h$ is the mass of water in the cube.

Now consider the variation of $\rho_h$ as a function of $h$ at a given time. When $h$ is very small, the cube contains just a few molecules and a small change in $h$ causes a large change in $\rho_h$, as molecules may be excluded when $h$ is reduced. Note that 1 cm$^3$ contains around $3 \times 10^{22}$ water molecules and when $h$ is around $10^{-7}$ cm there are about 30 water molecules in the cube. Thus, large variations of $\rho_h$ are expected for values less than $h^* \sim 10^{-7} cm$ (fig. 1.1).

Above this value of $h$, corresponding to the characteristic value $\rho_h = \rho^*$, the density is constant, assuming uniform temperature in the medium, and it

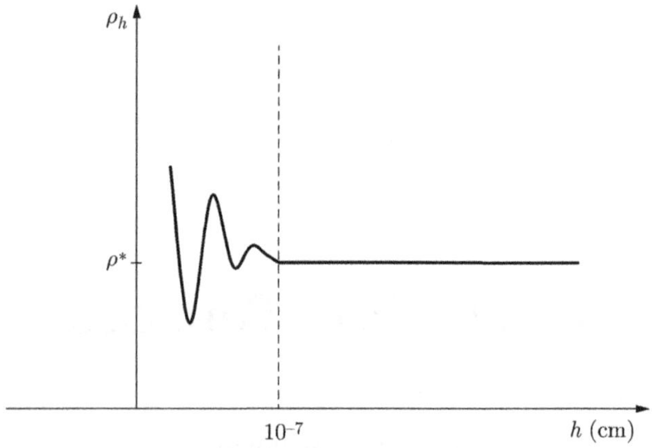


**Fig. 1.1** Qualitative variation of the average mass density inside a cube of water of size $h$

is considered to be the density of water at point P under the approximation of continuous media. Continuous velocity, acceleration, etc., can be defined using similar reasoning. Another example where the continuous medium hypothesis can easily be invalid and where the size of the cube must be precisely defined is for a system with a porous structure. When $h$ is very small, small variations in size can result in large variations in density (fig. 1.2).

The density becomes independent of $h$, i.e., $\rho_h = \rho^*$, (and thus of the size of the cube) above a certain limiting value $h^*$, and remains so as long as the microstructure of the solid stays the same. In general, large variations of the density or other physical variables are observed when $h$ is of the order of the dimension of the microstructure (that is, the distance between molecules or the characteristic length in a material with a complex microstructure found in porous, polycrystalline, or composite materials).

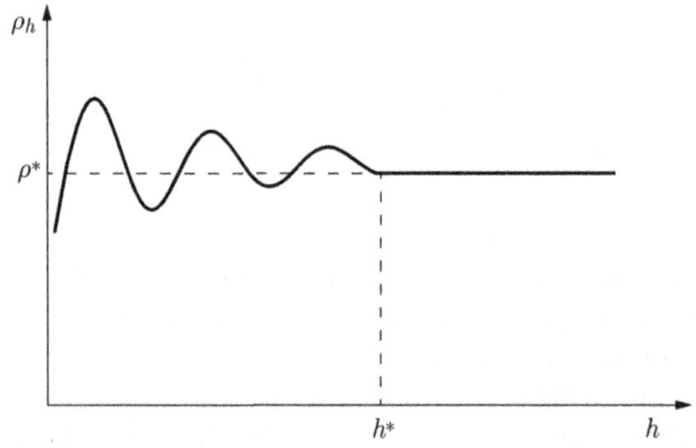


**Fig. 1.2** Qualitative variation of the average mass density of a porous material inside a cube of size $h$

Therefore, for engineering analysis or design it is not necessary to go down to the atomic or microstructure scale. For these purposes we consider the materials to be continuous so that they can be modeled at the macroscopic scale. This simplification of the reality ignores the discrete nature of the material and considers that its properties such as viscosity, density, elastic modulus, etc., attributed to a point in the medium, are spatially continuous functions. These quantities are the averages obtained over a large number of particles inside a small volume of material containing the point. The specific dimension $h$ of the volume element depends on the structure of the medium and needs to be defined with mathematical techniques and physical arguments.

The theory that we will develop is a phenomenological theory, which represents the generalization of the rational laws of mechanics (point mechanics) to continuous media. With respect to statistical mechanics, the models that we will establish are mathematically satisfactory. We will consider that the transformation between two regions of space that the material can occupy at different times is a continuous transformation. With this abstraction we may speak of the velocity at a point in a more suitable way than for the same notion based on a molecular model. For the latter we would need to take the average velocity of the molecules in the neighborhood of the point under consideration. Thus, the question of the definition of a neighborhood becomes crucial. If it is too large, its relation to the point is lost; if it is too small, the notion of an average is in doubt. To establish a valid link between molecular and continuous models, it is necessary to employ more sophisticated notions of average which are beyond the scope of this book.

The beginner in MCM may wonder why the first concepts introduced are related to ***vectors*** and ***tensors***. The reason is that tensors and the associated algebra are the natural tools for the theory of fields or continuous media. Specifically, we would like the physical quantities that describe a continuous medium to be independent of the coordinate system in which we work. This objective can only be attained by using tensors.

Numerous publications dedicated to MCM deal with the vector and tensor tools used in MCM. Not intending to be exhaustive, we refer the reader to the following for complementary reading: [2, 8, 20, 24, 36, 45].

## 1.2 Vector Algebra

### 1.2.1 Generalities

In mechanics of continuous media, motion and the associated physical quantities are described in ***Euclidean space*** $\mathbb{R}^3$ (physical space) with which a three-dimensional vector space $E^3$ is associated. The elements of $\mathbb{R}^3$ and $E^3$ are called the points and vectors, respectively. The scalars, vectors, and tensors that describe the physical quantities which will be introduced later on are also attached (most of them) to a space (typically $\mathbb{R}^3$) and form what are called scalar, vector, or tensor fields.

First recall that a ***vector space*** is defined uniquely from the properties of operations on its elements and assumes the existence of an arbitrary field (typically the field of real numbers $\mathbb{R}$) whose elements are called scalars. The ***vector space*** $E^3$ is then the set of elements denoted $\boldsymbol{u}$, $\boldsymbol{v}$, $\boldsymbol{w}$, ... such that

$$\begin{array}{ll} \boldsymbol{u}+\boldsymbol{v} \in E^3 & a\,\boldsymbol{u} \in E^3 \\ (\boldsymbol{u}+\boldsymbol{v})+\boldsymbol{w} = \boldsymbol{u}+(\boldsymbol{v}+\boldsymbol{w}) & 1\boldsymbol{u} = \boldsymbol{u} \\ \exists\,\boldsymbol{0} \in E^3 \mid \boldsymbol{u}+\boldsymbol{0} = \boldsymbol{u} & a(b\boldsymbol{u}) = (ab)\boldsymbol{u} \\ \exists\,-\boldsymbol{u} \in E^3 \mid \boldsymbol{u}+(-\boldsymbol{u}) = \boldsymbol{0} & (a+b)\boldsymbol{u} = a\boldsymbol{u}+b\boldsymbol{u} \\ \boldsymbol{u}+\boldsymbol{v} = \boldsymbol{v}+\boldsymbol{u} & a(\boldsymbol{u}+\boldsymbol{v}) = a\boldsymbol{u}+a\boldsymbol{v} \end{array} \tag{1.1}$$

for every $\boldsymbol{u}, \boldsymbol{v}, \boldsymbol{w} \in E^3$ and $a, b \in \mathbb{R}$. We note that the axioms (1.1) can be divided in two. Those in the first part concern the additive structure of the vector space and show that $E^3$ is a commutative group with respect to addition. The other axioms translate action of the body $\mathbb{R}$ on the vector space (distributive with respect to vector and scalar addition).

By providing $E^3$ with a scalar product, in order to later be able to calculate lengths and angles, it takes the name Euclidean space. The scalar product associates with every pair of vectors $\boldsymbol{u}, \boldsymbol{v} \in E^3$ a scalar denoted $\boldsymbol{u} \cdot \boldsymbol{v}$ with the following properties:

$$\begin{aligned} &\boldsymbol{u} \cdot \boldsymbol{v} = \boldsymbol{v} \cdot \boldsymbol{u} \\ &\boldsymbol{u} \cdot (\alpha\boldsymbol{v} + \beta\boldsymbol{w}) = \alpha(\boldsymbol{u} \cdot \boldsymbol{v}) + \beta(\boldsymbol{u} \cdot \boldsymbol{w}) \\ &\boldsymbol{u} \cdot \boldsymbol{u} \geq 0 \end{aligned} \tag{1.2}$$

for every $\boldsymbol{u}, \boldsymbol{v}, \boldsymbol{w} \in E^3$ and $\alpha, \beta \in \mathbb{R}$. The ***scalar product*** is consequently an application of $E^3 \times E^3$ in $\mathbb{R}$ that is linear with respect to each of its arguments. It is also called a positive definite bilinear form as shown in the last relation in (1.2). The scalar product permits the definition of the ***vector norm*** $\boldsymbol{u}$, denoted $\|\boldsymbol{u}\|$ by the relation

$$\|\boldsymbol{u}\| = \sqrt{\boldsymbol{u} \cdot \boldsymbol{u}}\,. \tag{1.3}$$

The vector $\boldsymbol{u}$ is called a unit vector when $\|\boldsymbol{u}\| = 1$, and two vectors $\boldsymbol{u}$ and $\boldsymbol{v}$ are orthogonal if and only if $\boldsymbol{u} \cdot \boldsymbol{v} = 0$. Every vector in $E^3$ can be decomposed uniquely according to a basis formed of three linearly independent vectors of $E^3$. The choice of a basis is arbitrary but generally one uses the ***canonical basis*** $(\boldsymbol{e}_1, \boldsymbol{e}_2, \boldsymbol{e}_3)$ defined by

$$\boldsymbol{e}_i \cdot \boldsymbol{e}_j = \begin{cases} 1 & \text{if } i = j \\ 0 & \text{if } i \neq j\,, \quad i, j = 1, 2, 3. \end{cases} \tag{1.4}$$

As (1.4) shows, the vectors of the canonical basis are unit vectors and each pair is orthogonal (that is, orthonormal). The basis is called ***orthogonal*** when the basis vectors are not unit vectors but are still orthogonal.

By choosing a fixed point $O$ (arbitrary) in the space $\mathbb{R}^3$, we have a correspondence between each vector $\boldsymbol{x}$ of $E^3$ and one and only one point P in

$\mathbb{R}^3$ (different than $O$) such that $\boldsymbol{OP} = \boldsymbol{x}$. The Cartesian coordinate system $(O, \boldsymbol{e}_1, \boldsymbol{e}_2, \boldsymbol{e}_3)$ for the space $\mathbb{R}^3$ is, by definition, the set formed by the point $O$ taken as the origin and the three orthonormal basis vectors $(\boldsymbol{e}_1, \boldsymbol{e}_2, \boldsymbol{e}_3)$ at this origin. The Cartesian coordinates of the point P in the system, as well as the components of the associated vector $\boldsymbol{x}$ with respect to $(\boldsymbol{e}_1, \boldsymbol{e}_2, \boldsymbol{e}_3)$, are thus represented by a unique triplet of numbers $(x_1, x_2, x_3)$ such that $\boldsymbol{OP} = \boldsymbol{x} = \sum_{i=1}^{3} x_i \boldsymbol{e}_i$ (in this context $\boldsymbol{x}$ is the vector position of the point P). Using the properties (1.2) and (1.4), the scalar product of two vectors $\boldsymbol{x} = \sum_{i=1}^{3} x_i \boldsymbol{e}_i$ and $\boldsymbol{y} = \sum_{j=1}^{3} y_j \boldsymbol{e}_j$ is given by

$$\boldsymbol{x} \cdot \boldsymbol{y} = \sum_{i=1}^{3} x_i y_i \,. \tag{1.5}$$

The simplicity of the result (1.5) clearly shows the value of using the canonical basis (or another orthonormal basis) to perform operations on vectors.

There are many physical quantities for which only one value can be associated. For example, one of them is the mass ***density*** of a material. We denote it $\rho$ and it has for SI [(1)] units kg/m$^3$ or dimensions ML$^{-3}$ where M is the mass and L a length. For water, at standard sea level temperature, $\rho$ is $1,000$ kg/m$^3$. In the neighborhood of a point, this density is practically constant and, in addition, there is no direction associated with its value. It is thus a scalar quantity.

Other quantities have not only a value but also a direction. A force of one Newton is that which, applied to a point, gives it an ***acceleration*** of 1 m s$^{-2}$ per kg. Since this force has a direction, it is a vector. We know that vectors are expressed in terms of coordinates for the system in use. In a given coordinate system, this vector is specified by its components. Going from one set of axes to another, the vector remains invariant and only the components of the vector change by a transformation rule.

Finally, we introduce the concept of a tensor in a rudimentary way as follows. For example, a stress is a force per unit surface. As we have seen. a force is a vector. But an element of a surface is also a vector as we must specify both its size and its orientation, that is, the direction of the normal vector. If $\boldsymbol{f}$ describes the force vector and $\boldsymbol{s}$ the normal vector of the surface $S$, then we might think that the stress $\boldsymbol{T}$ could be expressed by $\boldsymbol{f}/\boldsymbol{s}$. But, as the division of two vectors is an undefined operation, we get around the difficulty by saying that given $\boldsymbol{s}$, we can find $\boldsymbol{f}$ by multiplying $\boldsymbol{s}$ by a new entity $\boldsymbol{T}$ such that

$$\boldsymbol{f}(\boldsymbol{s}) = \boldsymbol{T}\boldsymbol{s} \,.$$

This new mathematical object is a tensor which yields the stress at a point. In this case we have a tensor of order 2. One gathers intuitively that this quantity is associated with two spatial directions, not just one as for vectors or none for

[(1)] International System of Units, Système International d'Unités, designated SI in all languages.

scalars. In fact, this tensor can be represented by a matrix with two indices, each index corresponding to one direction in Euclidean space. It is thus an entity with nine components. Again, we would like the physical quantity (the tensor) to remain invariant when we change the coordinate system. For this, the components of a tensor will follow a transformation rule when coordinate systems are changed.

### 1.2.2 Orthogonal Transformation of Coordinate Systems

In the physical Euclidean space $\mathbb{R}^3$, let there be a Cartesian orthonormal coordinate system $(O, \boldsymbol{e}_1, \boldsymbol{e}_2, \boldsymbol{e}_3)$, that we denote also as $Ox_i$ $(i = 1, 2, 3)$, with origin at $O$ and the unit vectors $\boldsymbol{e}_i$ $(i = 1, 2, 3)$ directed along the axes $Ox_i$ (fig. 1.3). Another system $Ox'_i$ $(i = 1, 2, 3)$ with unit vectors $\boldsymbol{e}'_i$ defines a Cartesian coordinate system with the same origin $O$. The ***direction cosines*** of the axes $x'_p$ with respect to the axes $x_i$, denoted by $c_{pi}$, are given by the scalar products of the basis vectors

$$c_{pi} = \cos(x'_p, x_i) = \boldsymbol{e}'_p \cdot \boldsymbol{e}_i \qquad i, p = 1, 2, 3. \tag{1.6}$$

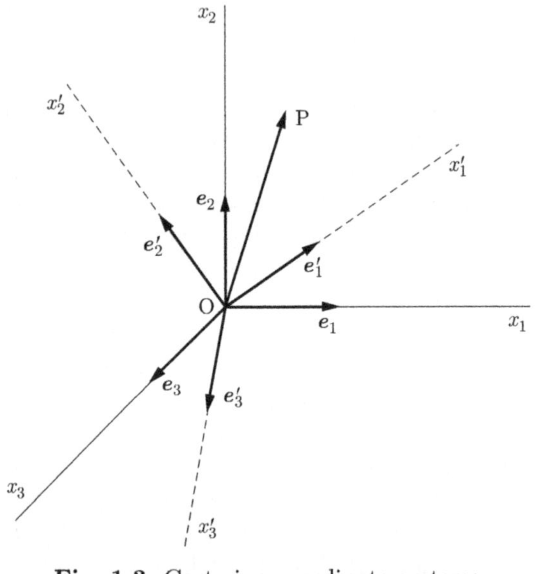


**Fig. 1.3** Cartesian coordinate systems

Similarly, the direction cosines of the first system with respect to the second are given by

$$c'_{pi} = \boldsymbol{e}_p \cdot \boldsymbol{e}'_i = c_{ip}, \tag{1.7}$$

the last equality being obtained from (1.6).

Let P be a point with coordinates $x_i$ in the first system and $x'_i$ in the second. From equation (1.6), the coordinates $x'_i$ are related to those of $x_i$ by the equations

$$\begin{aligned} x_1' &= c_{11}x_1 + c_{12}x_2 + c_{13}x_3 \\ x_2' &= c_{21}x_1 + c_{22}x_2 + c_{23}x_3 \\ x_3' &= c_{31}x_1 + c_{32}x_2 + c_{33}x_3 \,. \end{aligned} \tag{1.8}$$

We can write (1.8) in the form

$$x_i' = \sum_{j=1}^{3} c_{ij}x_j \qquad i = 1, 2, 3 \,. \tag{1.9}$$

We can easily see that the inverse of (1.8) is given by

$$\begin{aligned} x_1 &= c_{11}x_1' + c_{21}x_2' + c_{31}x_3' \\ x_2 &= c_{12}x_1' + c_{22}x_2' + c_{32}x_3' \\ x_3 &= c_{13}x_1' + c_{23}x_2' + c_{33}x_3' \end{aligned} \tag{1.10}$$

or

$$x_i = \sum_{j=1}^{3} c_{ji}x_j' = \sum_{j=1}^{3} c_{ij}'x_j' \,. \tag{1.11}$$

We can suppress the symbol $\sum$ by adopting, from here on, the Einstein ***summation convention*** for repeated indices, that is when an index appears twice in a product, a sum with respect to that index is implied by taking successively all its possible values (in this case, $i = 1, 2, 3$). In this way equations (1.9) and (1.11) are written in the compact form

$$x_i' = c_{ij}x_j \qquad x_i = c_{ji}x_j' \qquad j = 1, 2, 3 \,. \tag{1.12}$$

To illustrate the summation convention, we can write

$$\sigma_{ij}n_j = \sum_{j=1}^{3} \sigma_{ij}n_j = \sigma_{i1}n_1 + \sigma_{i2}n_2 + \sigma_{i3}n_3$$

$$\begin{aligned} \sigma_{ij}n_jn_i = \sum_{j=1}^{3}\sum_{i=1}^{3} \sigma_{ij}n_jn_i &= \sigma_{11}n_1^2 + \sigma_{22}n_2^2 + \sigma_{33}n_3^2 + (\sigma_{12} + \sigma_{21})n_1n_2 \\ &\quad + (\sigma_{23} + \sigma_{32})n_2n_3 + (\sigma_{31} + \sigma_{13})n_3n_1 \,. \end{aligned}$$

In the expression $\sigma_{ij}n_j$, the index $i$ is fixed and has a value among $1, 2, 3$. We call it the free index.

The symbol $u_i$ will designate the set of the $3^1$ quantities $u_1, u_2, u_3$ (3 for space and 1 for the free index). Similarly, the symbol $L_{ij}$ signifies the set of the $3^2$ quantities $L_{11}, L_{12}, L_{13}, L_{21}, L_{22}, L_{23}, L_{31}, L_{32}, L_{33}$ (3 for space and 2 for the free indices). For the parameter $L_{ii}$, we have $3^0 = 1$ quantity, namely a scalar. Thus, we can write

$$
\begin{aligned}
L_{ii} &= \sum_{i=1}^{3} L_{ii} = L_{11} + L_{22} + L_{33} \\
A_i B_k C_i &= \sum_{i=1}^{3} A_i B_k C_i = B_k \sum_{i=1}^{3} A_i C_i = B_k (A_1 C_1 + A_2 C_2 + A_3 C_3) \\
ds^2 &= dx_1^2 + dx_2^2 + dx_3^2 = \sum_{i=1}^{3} dx_i\, dx_i = dx_i\, dx_i\ .
\end{aligned}
$$

Note that the index over which we sum is a dummy index; we can change the notation of this index without changing the significance of the sum. Thus

$$
\begin{aligned}
\sigma_{ij} n_j &= \sigma_{ik} n_k = \sigma_{il} n_l \\
M_{ijk} u_i v_j w_k &= M_{jik} u_j v_i w_k = M_{ikj} u_i v_k w_j = \ldots
\end{aligned}
$$

A dummy index is not allowed to appear more than twice in an expression. Therefore, to insert the second equation (1.12) in the first we need to rewrite it, for example, in the form

$$x_j = c_{qj} x'_q\ .$$

From which

$$x'_i = c_{ij} c_{qj} x'_q \qquad \text{and similarly} \qquad x_i = c_{ji} c_{jq} x_q\ . \tag{1.13}$$

It is clear that the coefficient of $x'_q$ in the first equation (1.13) should be equal to unity for $i = q$ and 0 for $i \neq q$. This is also true for the second equation (1.13). If we introduce the ***Kronecker delta***

$$\delta_{ij} = \begin{cases} 1 & \text{if } i = j \\ 0 & \text{if } i \neq j\,, \end{cases} \tag{1.14}$$

then we obtain

$$c_{ij} c_{qj} = \delta_{iq}\ . \tag{1.15}$$

The transformation (1.12) which satisfies (1.15) is called orthogonal and (1.15) are the orthogonality conditions. The components $c_{ij}$ form an ***orthogonal matrix*** $[C]$ such that its transpose is equal to its inverse

$$c_{ik} c_{kj}^{-1} = c_{ik} c_{jk} = \delta_{ij} \qquad \text{or} \qquad [C][C]^{-1} = [C][C]^T = [I]\,, \tag{1.16}$$

with $[I]$ denoting the unit or identity matrix. The matrix of $c_{ij}$ is such that

$$\det [C] = \boldsymbol{e}_1 \cdot (\boldsymbol{e}_2 \times \boldsymbol{e}_3)\,, \tag{1.17}$$

that is, $\det [C] = \pm 1$. The sign + or − characterizes a direct (rotation) or retrograde (reflection) change of the coordinate system. Here, the symbol $\times$ represents the vector product.

Note that by using the Kronecker delta, the index of a component can be changed as follows:

$$\begin{aligned}
L_{ik} &= \delta_{ij} L_{jk} , \\
A_i B_k C_i &= \delta_{ij} A_i B_k C_j , \\
\frac{\partial u_j}{\partial x_i} &= u_{j,i} = \delta_{kj} \frac{\partial u_k}{\partial x_i} = \delta_{kj} u_{k,i} \\
\frac{\partial^2 u_i}{\partial x_j \partial x_k} &= u_{i,jk} = \delta_{jl} \frac{\partial^2 u_i}{\partial x_l \partial x_k} = \delta_{jl} u_{i,lk} .
\end{aligned}$$

### 1.2.3 Scalars

Let P be a point in a continuous medium and $F(\mathrm{P})$ the real value of a continuous function at P. If the value $F(\mathrm{P})$ does not depend on the coordinate system, then the function $F$ is called a scalar function, or scalar, or a tensor of order 0. This is the case, for example, for temperature, pressure, kinetic energy, etc. It is not the case for the components of a vector which depend on the coordinate system. Of course, this does not mean that the form of the function that yields the value $F(\mathrm{P})$ is independent of the chosen coordinate system. If the point P has coordinates $x_i$ and if $F(\mathrm{P})$ has a value $f(x_i)$, the change of coordinate systems in the second equation in (1.12) for the scalar $F(\mathrm{P})$ leads to

$$F(P) = f(x_i) = f(c_{ji} x'_j) = f'(x'_j) . \tag{1.18}$$

Take, for example, a linear temperature field $T(P)$ given in a coordinate system $x_i$ by

$$T(x_i) = T_0 + \frac{T_1 - T_0}{L} x_1 ,$$

such that $T(0) = T_0$ and $T(L) = T_1$. A rotation of 45° about the axis $x_3$, moves us to the coordinate system $x'_i$ with the transformation

$$[C] = \begin{pmatrix} \frac{\sqrt{2}}{2} & \frac{\sqrt{2}}{2} & 0 \\ -\frac{\sqrt{2}}{2} & \frac{\sqrt{2}}{2} & 0 \\ 0 & 0 & 1 \end{pmatrix} .$$

More generally, the matrix that expresses the change of coordinate systems by rotation around the axis $\boldsymbol{e}_3$ by an angle $\theta$, is given by

$$[C] = \begin{pmatrix} \cos\theta & \sin\theta & 0 \\ -\sin\theta & \cos\theta & 0 \\ 0 & 0 & 1 \end{pmatrix} .$$

The temperature in the new coordinate system becomes

$$T'(x'_i) = T_0 + \frac{T_1 - T_0}{L} \frac{\sqrt{2}}{2} (x'_1 - x'_2) ,$$

as from (1.11)

$$\begin{aligned} x_1 &= c_{11}x'_1 + c_{21}x'_2 + c_{31}x'_3 \\ &= \frac{\sqrt{2}}{2}\,x'_1 - \frac{\sqrt{2}}{2}\,x'_2 \,. \end{aligned}$$

### 1.2.4 Vectors

Let $\boldsymbol{v} = \boldsymbol{PQ}$ be a vector having its origin at point P and its extremity at point Q (fig. 1.4). This vector has a direction and three components $v_i$. The vector itself is independent of the coordinate system.

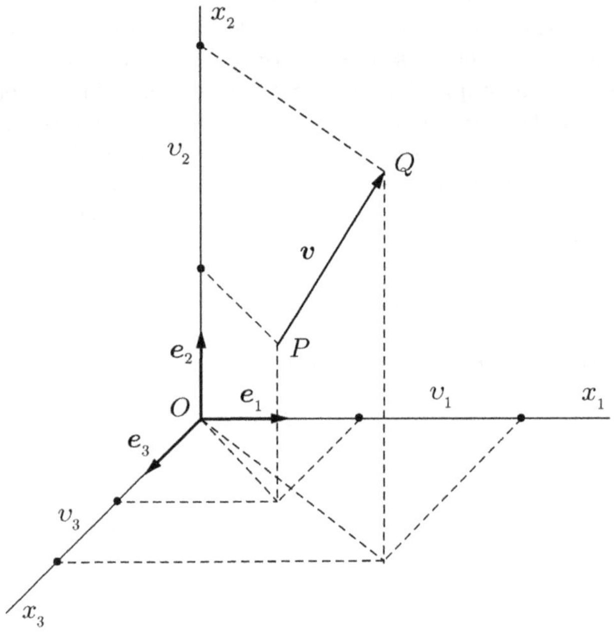


**Fig. 1.4** Vector in a Cartesian coordinate system

We examine the representations of $\boldsymbol{v}$ in the two coordinate systems linked by the relations

$$x'_i = c_{ij}x_j \qquad \text{and} \qquad x_i = c_{ji}x'_j \,. \tag{1.19}$$

Let $x_i$, $y_i$ be the coordinates of the points P and Q in the first coordinate system and $x'_i$, $y'_i$ in the second. The components of $\boldsymbol{v}$ in the first system are written as

$$v_i = y_i - x_i$$

and

$$v'_i = y'_i - x'_i$$

in the second. We relate the components $v'_i$ to those of $v_i$ by the transformation rule

$$v'_i = y'_i - x'_i = c_{ij}(y_j - x_j) = c_{ij}v_j \,. \tag{1.20}$$

Since in a Cartesian coordinate system, the direction cosines $c_{ij}$ are independent of the coordinates of P, we can write

$$\frac{\partial x'_i}{\partial x_j} = c_{ij} \qquad \frac{\partial x_i}{\partial x'_j} = c_{ji} \tag{1.21}$$

and thus

$$\frac{\partial x_i}{\partial x'_j} = \frac{\partial x'_j}{\partial x_i} \,.$$

Combining (1.20) and (1.21), we obtain

$$v'_i = \frac{\partial x'_i}{\partial x_j} v_j \qquad \text{or} \qquad v'_i = \frac{\partial x_j}{\partial x'_i} v_j \,. \tag{1.22}$$

By definition, we will say that a mathematical object $\boldsymbol{v}$, characterized by the three components $v_i$ in a Cartesian coordinate system, is a vector or a tensor of order 1 if its components are transformed according to the rule (1.22) during a coordinate system change that is an orthogonal transformation according to (1.16). Consequently, a triplet of numbers in a coordinate system does not necessarily give a vector. It is the transformation rule (1.22) associated with the invariant character of the vector that determines its nature.

The index notation allows us to implement the standard algebra of vectors and their scalar components. For example, if $a$ is a scalar, the $i^{\text{th}}$ component of $a\boldsymbol{v}$ is $av_i$. With (1.22), we can show that the multiplication $a\boldsymbol{v}'$ is a vector since

$$(av'_i) = c_{ij}(av_j) = \frac{\partial x'_i}{\partial x_j}(av_j) \,.$$

The addition of two vectors is done by the addition of their respective components, that is,

$$w_i = u_i + v_i \,. \tag{1.23}$$

In vector notation, we have

$$\boldsymbol{w} = \boldsymbol{u} + \boldsymbol{v} \,.$$

Now we look at the ***scalar product of two vectors***. We denote the product $b$ as the sum $u_i v_i$. In symbolic form, $b = \boldsymbol{u} \cdot \boldsymbol{v} = \boldsymbol{v} \cdot \boldsymbol{u}$ and

$$b = u_i v_i = u_1 v_1 + u_2 v_2 + u_3 v_3 \,. \tag{1.24}$$

Let us see if the product $b$ is affected by an orthogonal change of coordinate system. Relation (1.20) can be expressed in the form

$$v_i = y_i - x_i = c_{ji}(y'_j - x'_j) = c_{ji}v'_j \,.$$

And then we have

$$\begin{aligned} b = u_i v_i &= c_{ji} u'_j c_{ki} v'_k \\ &= c_{ji} c_{ki} u'_j v'_k \\ &= \delta_{jk} u'_j v'_k \\ &= u'_j v'_j \,. \end{aligned}$$

Thus the ***product*** $u_i v_i$ is a ***scalar*** as its value does not change during a coordinate system change. With vector analysis it can be shown that

$$b = \|\boldsymbol{u}\| \, \|\boldsymbol{v}\| \cos\theta \,, \tag{1.25}$$

where $\|\boldsymbol{u}\|$ and $\|\boldsymbol{v}\|$ are the norms of the vectors $\boldsymbol{u}$ and $\boldsymbol{v}$ respectively, and $\theta$ is the angle between the two vectors.

Recall that an important consequence of (1.25) is that two (non-zero) orthogonal vectors have a zero scalar product. Another interesting result is that the scalar product of a vector with itself yields the square of its norm:

$$v_i v_i = \|\boldsymbol{v}\|^2 \,. \tag{1.26}$$

If we consider a vector from the origin of the coordinate system to a point P of the Euclidean space, this vector is the position vector $\boldsymbol{x}$ such that $\boldsymbol{x} = \boldsymbol{x}(x_1, x_2, x_3) = \boldsymbol{OP}$. The position vector is a fundamental concept of continuous media kinematics that we will develop more completely in chapter 2. Its components depend on the spatial coordinates. Recalling (1.14) we can write

$$\frac{\partial x_i}{\partial x_j} = \delta_{ij} \,. \tag{1.27}$$

### 1.2.5 Permutation Symbol and Vector Product

The permutation symbol is defined as follows:

$$\varepsilon_{ijk} = \begin{cases} 1 & \text{if } ijk \text{ is an even permutation of } 1\,2\,3 \\ -1 & \text{if } ijk \text{ is an odd permutation of } 1\,2\,3 \\ 0 & \text{all other cases} \end{cases} \tag{1.28}$$

or as

$$\varepsilon_{ijk} = \frac{1}{2}(i-j)(j-k)(k-i) \,. \tag{1.29}$$

From the definition (1.28), we can move an index toward the back and inversely

$$\varepsilon_{ijk} = \varepsilon_{jki} = \varepsilon_{kij} \,.$$

The permutation of two adjacent indices results in a sign change

$$\begin{aligned} \varepsilon_{ijk} &= -\varepsilon_{jik} \\ \varepsilon_{ijk} &= -\varepsilon_{ikj} \,. \end{aligned}$$

From (1.28) and the definition of the Kronecker delta (1.14), we can prove the very useful identity

$$\varepsilon_{ijk}\varepsilon_{ilm} = \delta_{jl}\delta_{km} - \delta_{jm}\delta_{kl}\,. \tag{1.30}$$

In an orthonormal basis of $\mathbb{R}^3$, the vector product of two vectors $\boldsymbol{w} = \boldsymbol{u} \times \boldsymbol{v}$, sometimes denoted $\boldsymbol{u} \wedge \boldsymbol{v}$, is defined by the equality

$$w_i = \varepsilon_{ijk} u_j v_k\,. \tag{1.31}$$

For example we can evaluate

$$w_1 = \varepsilon_{123} u_2 v_3 + \varepsilon_{132} u_3 v_2 = u_2 v_3 - u_3 v_2\,.$$

We could do similarly for $w_2$ and $w_3$.

An important point of the notation (1.31) is that the first index of $\varepsilon_{ijk}$ must be that of the component of the vector $\boldsymbol{w}$; the second, the same as that of the first vector of the product $\boldsymbol{u} \times \boldsymbol{v}$; and the last, that of the last vector of the product.

The norm of a vector product is equal to the product of the norms of the vectors multiplied by the sine of the angle $\theta$ that these vectors form

$$\|\boldsymbol{w}\| = \|\boldsymbol{u}\|\,\|\boldsymbol{v}\| \sin\theta\,. \tag{1.32}$$

A few formulas in symbolic notation present diverse combinations of scalar and vector products. For example

$$\boldsymbol{u} \times \boldsymbol{v} = -\boldsymbol{v} \times \boldsymbol{u} \tag{1.33}$$

$$\boldsymbol{u} \times (\boldsymbol{v} + \boldsymbol{w}) = \boldsymbol{u} \times \boldsymbol{v} + \boldsymbol{u} \times \boldsymbol{w} \tag{1.34}$$

$$(\boldsymbol{u} \times \boldsymbol{v}) \cdot \boldsymbol{w} = (\boldsymbol{v} \times \boldsymbol{w}) \cdot \boldsymbol{u} = (\boldsymbol{w} \times \boldsymbol{u}) \cdot \boldsymbol{v} \tag{1.35}$$

$$(\boldsymbol{a} \times \boldsymbol{b}) \cdot (\boldsymbol{c} \times \boldsymbol{d}) = (\boldsymbol{a} \cdot \boldsymbol{c})(\boldsymbol{b} \cdot \boldsymbol{d}) - (\boldsymbol{a} \cdot \boldsymbol{d})(\boldsymbol{b} \cdot \boldsymbol{c})\,. \tag{1.36}$$

EXAMPLE 1.1

The vector product $\boldsymbol{u} \times \boldsymbol{v}$ generates a vector $\boldsymbol{w}$ perpendicular to the plane of the two vectors, and the three vectors $\boldsymbol{u}$, $\boldsymbol{v}$, $\boldsymbol{w}$ form a direct system. It can be shown that $\boldsymbol{w}$ is orthogonal to $\boldsymbol{v}$ since $\boldsymbol{v} \cdot \boldsymbol{w}$ is zero. Using (1.28), we have

$$\begin{aligned} v_i w_i &= v_i \varepsilon_{ijk} u_j v_k = \varepsilon_{ijk} v_i v_k u_j \\ &= \frac{1}{2}\left(\varepsilon_{ijk} v_i v_k u_j + \varepsilon_{ijk} v_i v_k u_j\right) = \frac{1}{2}\left(\varepsilon_{ijk} v_i v_k u_j + \varepsilon_{kji} v_k v_i u_j\right) \\ &= \frac{1}{2}\left(\varepsilon_{ijk} v_i v_k u_j - \varepsilon_{ijk} v_i v_k u_j\right) = 0\,. \end{aligned}$$

EXAMPLE 1.2
We use index notation algebra to verify the identity (1.36). The term on the left $\mathfrak{L}$ is written as

$$\mathfrak{L} = (\boldsymbol{a} \times \boldsymbol{b}) \cdot (\boldsymbol{c} \times \boldsymbol{d}) = \varepsilon_{ijk} a_j b_k \varepsilon_{ilm} c_l d_m \,.$$

With (1.30), we obtain

$$\begin{aligned} \mathfrak{L} &= (\delta_{jl}\delta_{km} - \delta_{jm}\delta_{kl})\, a_j b_k c_l d_m \\ &= \delta_{jl} a_j c_l \delta_{km} b_k d_m - \delta_{jm} a_j d_m \delta_{kl} b_k c_l \,. \end{aligned}$$

Using the properties of the Kronecker delta $\delta_{ij}$, we set $l = j$ and $m = k$ in the first term and $m = j$ and $l = k$ in the second. Then

$$\mathfrak{L} = a_j c_j b_k d_k - a_j d_j b_k c_k \,.$$

The right-hand term of this relation is none other than the index notation representation of the right-hand term of (1.36).

EXAMPLE 1.3
If $P_i = \varepsilon_{ijk} u_{k,j}$ where $u_i$ are continuous functions of $x_i$ with continuous partial derivatives, show that $P_{i,i} = 0$.

We use the properties of the permutation symbol to modify the expression as follows:

$$\begin{aligned} P_i &= \varepsilon_{ijk} u_{k,j} \quad \Rightarrow \quad P_{i,i} = \varepsilon_{ijk} u_{k,ji} = \varepsilon_{jik} u_{k,ij} \\ \varepsilon_{jik} &= -\varepsilon_{ijk}, \quad u_{k,ij} = u_{k,ji} \\ \Rightarrow P_{i,i} &= \varepsilon_{ijk} u_{k,ji} = -\varepsilon_{ijk} u_{k,ji} \\ 2P_{i,i} &= \varepsilon_{ijk} u_{k,ji} - \varepsilon_{ijk} u_{k,ji} = 0 \quad \Rightarrow \quad P_{i,i} = 0 \,. \end{aligned}$$

## 1.3 Tensor Algebra

### 1.3.1 Definition of Tensors of Order 2

The notion of a tensor of order 2 is introduced by respecting the representation of an invariant object. Let $E^3$ be the Euclidean vector space of vectors associated with $\mathbb{R}^3$, and $\boldsymbol{L}$ a linear mapping on $E^3$ that transforms a vector to another, that is,

$$\boldsymbol{L} : E^3 \to E^3 \qquad \text{such that} \qquad \boldsymbol{u} \mapsto \boldsymbol{L}\boldsymbol{u} \,. \tag{1.37}$$

If $\boldsymbol{L}$ transforms $\boldsymbol{u}_1$ to $\boldsymbol{v}_1$ and $\boldsymbol{u}_2$ to $\boldsymbol{v}_2$ with the relations

$$\begin{aligned} \boldsymbol{L}\boldsymbol{u}_1 &= \boldsymbol{v}_1 \\ \boldsymbol{L}\boldsymbol{u}_2 &= \boldsymbol{v}_2 \ , \end{aligned}$$

and if $\boldsymbol{L}$ has the following properties:

$$\begin{aligned} \boldsymbol{L}(\boldsymbol{u}_1 + \boldsymbol{u}_2) &= \boldsymbol{L}\boldsymbol{u}_1 + \boldsymbol{L}\boldsymbol{u}_2 \\ \boldsymbol{L}(\alpha \boldsymbol{u}_1) &= \alpha \boldsymbol{L}\boldsymbol{u}_1 \ , \end{aligned} \tag{1.38}$$

where $\boldsymbol{u}_1$ and $\boldsymbol{u}_2$ are two arbitrary vectors of $E^3$ and $\alpha \in \mathbb{R}$, then we say that $\boldsymbol{L}$ is a linear transformation. It is also a tensor of order 2, or simply, a tensor. The ***unit tensor*** $\boldsymbol{I}$ and the ***zero tensor*** $\boldsymbol{O}$ are defined by the relations $\boldsymbol{u} = \boldsymbol{I}\boldsymbol{u}$ and $\boldsymbol{0} = \boldsymbol{O}\boldsymbol{u}$, respectively.

For every vector $\boldsymbol{u}$, the vector $\boldsymbol{v}$ is such that

$$\boldsymbol{v} = \boldsymbol{L}\boldsymbol{u} = \boldsymbol{L}u_i\boldsymbol{e}_i = u_i\boldsymbol{L}\boldsymbol{e}_i \ . \tag{1.39}$$

The components of $\boldsymbol{v}$ are obtained by taking the scalar product

$$v_i = \boldsymbol{e}_i \cdot \boldsymbol{v} \ . \tag{1.40}$$

Combining (1.39) and (1.40), we have

$$v_i = \boldsymbol{e}_i \cdot (u_j\boldsymbol{L}\boldsymbol{e}_j) = u_j\boldsymbol{e}_i \cdot \boldsymbol{L}\boldsymbol{e}_j \ . \tag{1.41}$$

The terms of this equation, such as, for example, $\boldsymbol{e}_1 \cdot \boldsymbol{L}\boldsymbol{e}_1$ and $\boldsymbol{e}_2 \cdot \boldsymbol{L}\boldsymbol{e}_1$, are the components along $\boldsymbol{e}_1$ and $\boldsymbol{e}_2$ of $\boldsymbol{L}\boldsymbol{e}_1$. We agree to write the components as $L_{11} = \boldsymbol{e}_1 \cdot \boldsymbol{L}\boldsymbol{e}_1$, $L_{21} = \boldsymbol{e}_2 \cdot \boldsymbol{L}\boldsymbol{e}_1$, etc. In general we will have

$$L_{ij} = \boldsymbol{e}_i \cdot \boldsymbol{L}\boldsymbol{e}_j \ . \tag{1.42}$$

These elements $L_{ij}$ are the components of the tensor $\boldsymbol{L}$. From (1.41) and (1.42), we obtain

$$v_i = L_{ij}u_j \ . \tag{1.43}$$

This last relation can be written in matrix form

$$\begin{pmatrix} v_1 \\ v_2 \\ v_3 \end{pmatrix} = \begin{pmatrix} L_{11} & L_{12} & L_{13} \\ L_{21} & L_{22} & L_{23} \\ L_{31} & L_{32} & L_{33} \end{pmatrix} \begin{pmatrix} u_1 \\ u_2 \\ u_3 \end{pmatrix} . \tag{1.44}$$

The matrix

$$\begin{pmatrix} L_{11} & L_{12} & L_{13} \\ L_{21} & L_{22} & L_{23} \\ L_{31} & L_{32} & L_{33} \end{pmatrix}$$

is the matrix of the tensor $\boldsymbol{L}$ with respect to the basis vectors $\{\boldsymbol{e}_i\}$. Note that the components of the first column are the components of the vector $\boldsymbol{L}\boldsymbol{e}_1$, those of the second are the components of the vector $\boldsymbol{L}\boldsymbol{e}_2$ and so on. Thus we have

$$\boldsymbol{L}\boldsymbol{e}_1 = L_{11}\boldsymbol{e}_1 + L_{21}\boldsymbol{e}_2 + L_{31}\boldsymbol{e}_3 = L_{j1}\boldsymbol{e}_j \ ,$$

that is,

$$\boldsymbol{L}\boldsymbol{e}_i = L_{ji}\boldsymbol{e}_j \,. \tag{1.45}$$

We see then that the components of a tensor depend on the coordinate system defined by the basis $\{\boldsymbol{e}_i\}$ in the same way as the components of a vector depend on the system.

Nevertheless, a tensor that is an invariant ***linear operator*** has an intrinsic character, as, for example, a force per unit surface. Only its components will be affected by changing the basis. Note the ***matrix associated with the tensor*** $\boldsymbol{L}$

$$[L] = \begin{pmatrix} L_{11} & L_{12} & L_{13} \\ L_{21} & L_{22} & L_{23} \\ L_{31} & L_{32} & L_{33} \end{pmatrix} \tag{1.46}$$

and its ***determinant***

$$\det \boldsymbol{L} = \det[L] = \det \begin{pmatrix} L_{11} & L_{12} & L_{13} \\ L_{21} & L_{22} & L_{23} \\ L_{31} & L_{32} & L_{33} \end{pmatrix} . \tag{1.47}$$

A tensor is called ***singular*** if and only if $\det \boldsymbol{L} = 0$.

### 1.3.2 Tensor Product or Dyadic Product of Two Vectors

The tensor product or dyadic product $\boldsymbol{a} \otimes \boldsymbol{b}$ of two vectors $\boldsymbol{a}$ and $\boldsymbol{b}$ is defined as the tensor which, for any vector $\boldsymbol{v}$, yields the vector $(\boldsymbol{b} \cdot \boldsymbol{v})\boldsymbol{a}$

$$(\boldsymbol{a} \otimes \boldsymbol{b})\boldsymbol{v} = (\boldsymbol{b} \cdot \boldsymbol{v})\boldsymbol{a} = \boldsymbol{a}(\boldsymbol{b} \cdot \boldsymbol{v}) \,. \tag{1.48}$$

For every vector $\boldsymbol{v}$ and $\boldsymbol{w}$ and for $\alpha$, $\beta \in \mathbb{R}$, we have

$$\begin{aligned} (\boldsymbol{a} \otimes \boldsymbol{b})(\alpha\boldsymbol{v} + \beta\boldsymbol{w}) &= \big(\boldsymbol{b} \cdot (\alpha\boldsymbol{v} + \beta\boldsymbol{w})\big)\boldsymbol{a} \\ &= \big(\alpha(\boldsymbol{b} \cdot \boldsymbol{v}) + \beta(\boldsymbol{b} \cdot \boldsymbol{w})\big)\boldsymbol{a} \\ &= \alpha(\boldsymbol{b} \cdot \boldsymbol{v})\boldsymbol{a} + \beta(\boldsymbol{b} \cdot \boldsymbol{w})\boldsymbol{a} \\ &= \alpha(\boldsymbol{a} \otimes \boldsymbol{b})\boldsymbol{v} + \beta(\boldsymbol{a} \otimes \boldsymbol{b})\boldsymbol{w} \,. \end{aligned}$$

This shows that $(\boldsymbol{a} \otimes \boldsymbol{b})$ is a tensor. Its components with respect to the basis $\{\boldsymbol{e}_i\}$ $(i = 1, 2, 3)$ are

$$(a \otimes b)_{ij} = \boldsymbol{e}_i \cdot (\boldsymbol{a} \otimes \boldsymbol{b})\boldsymbol{e}_j = \boldsymbol{e}_i \cdot \big((\boldsymbol{b} \cdot \boldsymbol{e}_j)\boldsymbol{a}\big) = \boldsymbol{e}_i \cdot (\boldsymbol{a} b_j) = (\boldsymbol{e}_i \cdot \boldsymbol{a}) b_j = a_i b_j \,.$$

Thus, we have

$$(\boldsymbol{a} \otimes \boldsymbol{b})_{ij} = a_i b_j \,. \tag{1.49}$$

The corresponding matrix will be given by

$$[\boldsymbol{a} \otimes \boldsymbol{b}] = \begin{pmatrix} a_1 \\ a_2 \\ a_3 \end{pmatrix} \begin{pmatrix} b_1 & b_2 & b_3 \end{pmatrix} = \begin{pmatrix} a_1 b_1 & a_1 b_2 & a_1 b_3 \\ a_2 b_1 & a_2 b_2 & a_2 b_3 \\ a_3 b_1 & a_3 b_2 & a_3 b_3 \end{pmatrix} .$$

In general $(\boldsymbol{u} \otimes \boldsymbol{v}) \neq (\boldsymbol{v} \otimes \boldsymbol{u})$. The tensor product of the base vectors $\boldsymbol{e}_i \otimes \boldsymbol{e}_j$ is expressed as

$$(\boldsymbol{e}_i \otimes \boldsymbol{e}_j)\boldsymbol{u} = (\boldsymbol{e}_j \cdot \boldsymbol{u})\boldsymbol{e}_i = u_j \boldsymbol{e}_i \,. \tag{1.50}$$

From (1.50) and (1.43), we can write

$$\boldsymbol{v} = v_i \boldsymbol{e}_i = \boldsymbol{L}\boldsymbol{u} = L_{ij} u_j \boldsymbol{e}_i = L_{ij}(\boldsymbol{e}_i \otimes \boldsymbol{e}_j)\boldsymbol{u} \,.$$

Consequently, we obtain

$$\boldsymbol{L} = L_{ij}(\boldsymbol{e}_i \otimes \boldsymbol{e}_j) \,. \tag{1.51}$$

EXAMPLE 1.4

The identity $\boldsymbol{I}$ and the dyadic product tensors can be expressed as

$$\begin{aligned} \boldsymbol{I} &= \delta_{ij}(\boldsymbol{e}_i \otimes \boldsymbol{e}_j) \\ \boldsymbol{a} \otimes \boldsymbol{b} &= a_i b_j (\boldsymbol{e}_i \otimes \boldsymbol{e}_j) \,. \end{aligned}$$

### 1.3.3 The Transformation Rule for Cartesian Tensor Components

The representation in the Cartesian coordinate system $x_i$ of the linear operator $\boldsymbol{L}$, which is invariant, is given by its components $L_{ij}$ (1.42). In the coordinate system $x'_i$, the components of $\boldsymbol{L}$ are expressed as

$$L'_{ij} = \boldsymbol{e}'_i \cdot \boldsymbol{L}\boldsymbol{e}'_j \,.$$

We can easily evaluate the relation between the components $L_{ij}$ and $L'_{ij}$. From (1.20), the previous relation becomes

$$L'_{ij} = (c_{ik}\boldsymbol{e}_k) \cdot \boldsymbol{L}(c_{jl}\boldsymbol{e}_l) = c_{ik}c_{jl}\boldsymbol{e}_k \cdot \boldsymbol{L}\boldsymbol{e}_l = c_{ik}c_{jl}L_{kl} \,. \tag{1.52}$$

Invoking (1.21), we have

$$L'_{ij} = \frac{\partial x'_i}{\partial x_k}\frac{\partial x'_j}{\partial x_l}L_{kl} \qquad \text{or} \qquad L'_{ij} = \frac{\partial x_k}{\partial x'_i}\frac{\partial x_l}{\partial x'_j}L_{kl} \,. \tag{1.53}$$

In matrix notation, equation (1.52) is written as

$$[L'] = [C][L][C]^T \,. \tag{1.54}$$

By definition, a matrix $[L]$ with nine components corresponds to a tensor of order 2 if its components are transformed according to (1.53) during a coordinate change that obeys (1.20) and that is an orthogonal transformation according to (1.16). By extension, we will also refer to the tensor components $L_{ij}$. The transformation rules (1.53) guarantee the invariance of $\boldsymbol{L}$ with respect to the choice of coordinates.

As an example, let us verify that the tensor product $\boldsymbol{T} = \boldsymbol{a} \otimes \boldsymbol{b}$ is a tensor of order 2. We have

$$T'_{ij} = a'_i b'_j = \frac{\partial x'_i}{\partial x_k}\frac{\partial x'_j}{\partial x_l}a_k b_l = \frac{\partial x'_i}{\partial x_k}\frac{\partial x'_j}{\partial x_l}T_{kl} \,.$$

### 1.3.4 Tensors of Any Order

Recall now the transformation rules for vectors (tensors of order 1) and tensors of order 2 between the systems $x_i$ and $x'_i$

$$v'_i = \frac{\partial x'_i}{\partial x_j} v_j \qquad\qquad L'_{ij} = \frac{\partial x'_i}{\partial x_k}\frac{\partial x'_j}{\partial x_l} L_{kl}\,.$$

By generalizing these rules, we can define a tensor of order $n$. Let $\mathcal{T}$ be an object whose value in a coordinate system $x_i$ is given by $3^n$ components $\mathcal{T}_{i_1 i_2 \cdots i_n}$, where the indices $i_j$ take the values 1, 2, or 3. By definition, $\mathcal{T}$ is a tensor of order $n$ if, during a coordinate transformation, its components are transformed according to the rule

$$\mathcal{T}'_{i_1 i_2 \cdots i_n} = \frac{\partial x'_{i_1}}{\partial x_{j_1}}\frac{\partial x'_{i_2}}{\partial x_{j_2}} \cdots \frac{\partial x'_{i_n}}{\partial x_{j_n}} \mathcal{T}_{j_1 j_2 \cdots j_n}\,. \tag{1.55}$$

Note that for $n = 1$ and 2, we obtain the transformation rules for vectors and tensors of order 2.

We can also verify that the permutation symbol introduced in equation (1.28) is a tensor of order 3. To show this, apply to $\varepsilon_{ijk}$ the tensor transformation (1.55) and verify that the components $\varepsilon'_{ijk}$ satisfy the relations (1.28). Using (1.21), we write

$$\varepsilon'_{ijk} = c_{im} c_{jn} c_{kp} \varepsilon_{mnp}\,.$$

The second term can be developed taking into account (1.28) to obtain

$$\varepsilon'_{ijk} = c_{i1}c_{j2}c_{k3} + c_{i2}c_{j3}c_{k1} + c_{i3}c_{j1}c_{k2} - c_{i2}c_{j1}c_{k3} - c_{i1}c_{j3}c_{k2} - c_{i3}c_{j2}c_{k1}\,.$$

The right-hand side of the equation is none other than the determinant of the matrix

$$\begin{pmatrix} c_{i1} & c_{i2} & c_{i3} \\ c_{j1} & c_{j2} & c_{j3} \\ c_{k1} & c_{k2} & c_{k3} \end{pmatrix},$$

which is the orthogonal matrix of direction cosines between $x_i$ and $x'_i$. When $(i, j, k)$ is an even permutation of $(1, 2, 3)$, $\varepsilon'_{ijk}$ is 1 if the coordinate system is right-handed; when $(i, j, k)$ is an odd permutation of $(1, 2, 3)$, the lines are permuted an odd number of times, and the determinant is $-1$. Finally, in every other case, at least two lines are equal and the determinant is zero.

Tensors of order $n$ satisfy the following rules.

*Multiplication by a Scalar:*
: the multiplication of a tensor of order $n$ by a scalar is carried out by multiplying each component of the tensor by the scalar. The result is a tensor of order $n$.

*Linear Combination:*
: the linear combination of two tensors of order $n$ is by linear combination of the corresponding components. A tensor of the same order is obtained.

*Zero Tensor:*
is the tensor for which all the components are equal to zero.

*Equivalent Tensors:*
when the components of two tensors of the same order are equal term by term in a coordinate system, then they are equal in every other system; the tensors are equivalent. Consequently, if a tensor relation is verified in one coordinate system, it is true in all coordinate systems.

*Exterior Product of Tensors:*
let $\mathcal{A}_{i_1\cdots i_n}$ and $\mathcal{B}_{j_1\cdots j_m}$ be the respective components of a tensor of order $n$ and a tensor of order $m$ in a coordinate system. The $3^{n+m}$ quantities obtained by

$$\mathcal{C}_{i_1\cdots i_n j_1\cdots j_m} = \mathcal{A}_{i_1\cdots i_n}\mathcal{B}_{j_1\cdots j_m}$$

form a tensor $\boldsymbol{\mathcal{C}}$ of order $n+m$. As an example, we have previously shown that the dyadic product of two vectors yields a tensor of order 2.

*Tensor Contraction:*
let $\boldsymbol{\mathcal{A}}$ be a tensor of order $n$ whose components in a coordinate system are $\mathcal{A}_{i_1\cdots i_n}$. The contraction operation consists of equating two indices of the tensor, for example, the $j^{\text{th}}$ and the $k^{\text{th}}$ with $j$ and $k \leq n$, and of summing over these indices $(j, k = 1, 2, 3)$ to form a tensor of order $n-2$ thus having $3^{n-2}$ components. This tensor is obtained by contraction of the indices $j$ and $k$.
For example, $L_{ii}$ is the only contraction possible of $L_{ij}$. It is then no longer a tensor of order 2, but a scalar (tensor of order 0).

Now consider two tensors $\boldsymbol{S}$ and $\boldsymbol{T}$ of order 2. Their exterior product results in a tensor of order 4 whose components are

$$\mathcal{R}_{ijkl} = S_{ij}T_{kl}\,.$$

The components obtained by contraction of the second and third indices of $\boldsymbol{\mathcal{R}}$ are

$$\mathcal{R}_{imml} = S_{im}T_{ml}\,.$$

We can show that this is a tensor of order 2. From the transformation rule (1.55), we have

$$\mathcal{R}'_{ijkl} = c_{ip}c_{jq}c_{kr}c_{ls}\mathcal{R}_{pqrs}$$

and thus

$$\mathcal{R}'_{imml} = c_{ip}c_{mq}c_{mr}c_{ls}\mathcal{R}_{pqrs}\,.$$

From (1.15), we obtain

$$\begin{aligned}\mathcal{R}'_{imml} &= c_{ip}c_{ls}\,\delta_{qr}\mathcal{R}_{pqrs} = c_{ip}c_{ls}\mathcal{R}_{prrs}\\ &= \frac{\partial x'_i}{\partial x_p}\frac{\partial x'_l}{\partial x_s}\mathcal{R}_{prrs}\,.\end{aligned}$$

This last equation proves that we do indeed have a tensor of order 2 $\mathcal{R}_{imml}$.

Looking back, we can see that the left-hand term of equation (1.30), $\varepsilon_{ijk}\varepsilon_{ilm}$ is a tensor of order 4 obtained by contraction. In the same way, the contraction

$$\varepsilon_{ijk}\varepsilon_{ijl} = 2\,\delta_{kl} \tag{1.56}$$

yields a tensor of order 2, and the relation

$$\varepsilon_{ijk}\varepsilon_{ijk} = 6$$

yields a scalar.

### 1.3.5 Algebra of Tensors of Order 2

**Sum of Tensors**

Let $\boldsymbol{L}$ and $\boldsymbol{T}$ be two tensors of order 2. Their sum $(\boldsymbol{T}+\boldsymbol{L})$ is such that for every vector $\boldsymbol{a}$, we have

$$(\boldsymbol{T}+\boldsymbol{L})\boldsymbol{a} = \boldsymbol{T}\boldsymbol{a} + \boldsymbol{L}\boldsymbol{a}\,. \tag{1.57}$$

The components of $(\boldsymbol{T}+\boldsymbol{L})$ are then

$$(\boldsymbol{T}+\boldsymbol{L})_{ij} = \boldsymbol{e}_i \cdot (\boldsymbol{T}+\boldsymbol{L})\boldsymbol{e}_j = \boldsymbol{e}_i \cdot \boldsymbol{T}\boldsymbol{e}_j + \boldsymbol{e}_i \cdot \boldsymbol{L}\boldsymbol{e}_j$$

or

$$(\boldsymbol{T}+\boldsymbol{L})_{ij} = T_{ij} + L_{ij}\,. \tag{1.58}$$

**Interior Product of Two Tensors**

If $\boldsymbol{L}$ and $\boldsymbol{T}$ are two tensors of order 2, then $\boldsymbol{LT}$ and $\boldsymbol{TL}$ are defined by the equations

$$(\boldsymbol{LT})\boldsymbol{a} = \boldsymbol{L}(\boldsymbol{T}\boldsymbol{a}) \tag{1.59}$$

and

$$(\boldsymbol{TL})\boldsymbol{a} = \boldsymbol{T}(\boldsymbol{L}\boldsymbol{a}). \tag{1.60}$$

The components of $\boldsymbol{LT}$ are obtained from (1.42) and taking into account (1.45),

$$\begin{aligned}(\boldsymbol{LT})_{ij} &= \boldsymbol{e}_i \cdot (\boldsymbol{LT})\boldsymbol{e}_j = \boldsymbol{e}_i \cdot \boldsymbol{L}(\boldsymbol{T}\boldsymbol{e}_j)\\ &= \boldsymbol{e}_i \cdot \boldsymbol{L}T_{mj}\boldsymbol{e}_m = T_{mj}\boldsymbol{e}_i \cdot \boldsymbol{L}\boldsymbol{e}_m\,.\end{aligned}$$

Thus

$$(\boldsymbol{LT})_{ij} = L_{im}T_{mj}\,. \tag{1.61}$$

Similarly we have

$$(\boldsymbol{TL})_{ij} = T_{im}L_{mj}\,. \tag{1.62}$$

Note that the interior product of tensors $\boldsymbol{L}$ and $\boldsymbol{T}$ is a contracted multiplication of tensors of order 2, with the contraction over the last index of the first tensor and the first index of the second tensor. In matrix form, we can write the matrix of the interior product as equal to the product of the matrices of the two tensors

$$\big[(LT)\big] = [L][T] \qquad \text{and} \qquad \big[(TL)\big] = [T][L]\,.$$

Note that the interior product of tensors is not commutative in general, that is $\boldsymbol{LT} \neq \boldsymbol{TL}$.

If $\boldsymbol{L}$, $\boldsymbol{T}$, and $\boldsymbol{S}$ are three tensors, then we can evaluate

$$\big(\boldsymbol{L}(\boldsymbol{ST})\big)\boldsymbol{a} = \big(\boldsymbol{L}(\boldsymbol{ST})\boldsymbol{a}\big) = \boldsymbol{L}\big(\boldsymbol{S}(\boldsymbol{Ta})\big)$$

and

$$(\boldsymbol{LS})(\boldsymbol{Ta}) = \boldsymbol{L}\big(\boldsymbol{S}(\boldsymbol{Ta})\big)\,.$$

From which we obtain

$$\boldsymbol{L}(\boldsymbol{ST}) = (\boldsymbol{LS})\,\boldsymbol{T}\,. \tag{1.63}$$

The interior product of tensors is associative. For the case $\boldsymbol{L} = \boldsymbol{T}$, we can introduce the following notations $\boldsymbol{TT} = \boldsymbol{T}^2$, $\boldsymbol{TT}^2 = \boldsymbol{T}^3$, etc.

We also have the property

$$\det(\boldsymbol{ST}) = \det \boldsymbol{S}\, \det \boldsymbol{T}\,. \tag{1.64}$$

Note also the relations

$$\boldsymbol{L}(\boldsymbol{a}\otimes\boldsymbol{b}) = (\boldsymbol{La})\otimes\boldsymbol{b} \tag{1.65}$$

$$(\boldsymbol{u}\otimes\boldsymbol{v})(\boldsymbol{a}\otimes\boldsymbol{b}) = (\boldsymbol{v}\cdot\boldsymbol{a})\boldsymbol{u}\otimes\boldsymbol{b} = \boldsymbol{u}\otimes\boldsymbol{b}(\boldsymbol{v}\cdot\boldsymbol{a})\,. \tag{1.66}$$

EXAMPLE 1.5

We use index notation algebra to verify identity (1.65)

$$\begin{aligned}\big(\boldsymbol{L}(\boldsymbol{a}\otimes\boldsymbol{b})\big)_{ij} &= L_{im}(\boldsymbol{a}\otimes\boldsymbol{b})_{mj} = L_{im}a_m b_j \\ &= (\boldsymbol{La})_i b_j = \big((\boldsymbol{La})\otimes\boldsymbol{b}\big)_{ij}\,.\end{aligned}$$

### 1.3.6 Tensor Properties

**Transpose of a Tensor**

The transpose of a tensor is obtained by exchanging two indices: the transpose of $L_{ij}$ is $L_{ji}$. We denote it $\boldsymbol{L}^T$, and have the relation

$$\left(\boldsymbol{L}^T\right)_{ij} = L_{ji}\,. \tag{1.67}$$

Consequently, we can easily show that

$$(\boldsymbol{LS})^T = \boldsymbol{S}^T\boldsymbol{L}^T \tag{1.68}$$

and

$$\boldsymbol{u}\cdot\boldsymbol{L}^T\boldsymbol{v} = \boldsymbol{Lu}\cdot\boldsymbol{v} = \boldsymbol{v}\cdot\boldsymbol{Lu}\,. \tag{1.69}$$

For a dyad

$$(\boldsymbol{a}\otimes\boldsymbol{b})^T = \boldsymbol{b}\otimes\boldsymbol{a}\,, \tag{1.70}$$

$$(\boldsymbol{a}\otimes\boldsymbol{b})\boldsymbol{L} = \boldsymbol{a}\otimes\boldsymbol{L}^T\boldsymbol{b}\,. \tag{1.71}$$

And finally note the property

$$\det \boldsymbol{L}^T = \det \boldsymbol{L}\,. \tag{1.72}$$

### Inverse of a Tensor

For a non-singular tensor $\boldsymbol{L}$, that is $\det \boldsymbol{L} \neq 0$, there exists a unique tensor called the inverse tensor $\boldsymbol{L}^{-1}$ of $\boldsymbol{L}$ that satisfies the relation

$$\boldsymbol{L}\boldsymbol{L}^{-1} = \boldsymbol{L}^{-1}\boldsymbol{L} = \boldsymbol{I}\,. \tag{1.73}$$

By definition of the inverse, we can show that

$$\left(\boldsymbol{L}^{-1}\right)^{-1} = \boldsymbol{L} \tag{1.74}$$

$$(\alpha\boldsymbol{L})^{-1} = \frac{1}{\alpha}\,\boldsymbol{L}^{-1} \tag{1.75}$$

$$\det\left(\boldsymbol{L}^{-1}\right) = \left(\det \boldsymbol{L}\right)^{-1}. \tag{1.76}$$

For two invertible tensors $\boldsymbol{S}$ and $\boldsymbol{T}$, we have

$$(\boldsymbol{S}\boldsymbol{T})^{-1} = \boldsymbol{T}^{-1}\boldsymbol{S}^{-1}\,. \tag{1.77}$$

### Symmetric Tensors

A tensor is said to be symmetric when it is equal to its transpose. The tensor $\boldsymbol{L}$ is symmetric if

$$\boldsymbol{L} = \boldsymbol{L}^T\,, \tag{1.78}$$

that is, if

$$L_{ij} = L_{ji}\,. \tag{1.79}$$

Consequently, a symmetric tensor has six independent components.

### Antisymmetric Tensors

A tensor $\boldsymbol{L}$ is said to be antisymmetric if it is equal to the opposite of its transpose, or

$$\boldsymbol{L} = -\boldsymbol{L}^T\,, \tag{1.80}$$

that is,

$$L_{ij} = -L_{ji}\,. \tag{1.81}$$

In this case the diagonal components of $\boldsymbol{L}$ are zero and only three components are independent.

Now we can prove that all tensors $\boldsymbol{L}$ of order 2 can be uniquely decomposed into the sum of a symmetric tensor $\boldsymbol{L}^S$ and an antisymmetric tensor $\boldsymbol{L}^A$. To demonstrate that this decomposition is possible, we write

$$\begin{aligned} L_{ij} &= L^S_{ij} + L^A_{ij} \\ L^S_{ij} &= \frac{1}{2}\left(L_{ij} + L_{ji}\right) \\ L^A_{ij} &= \frac{1}{2}\left(L_{ij} - L_{ji}\right). \end{aligned} \tag{1.82}$$

To show the uniqueness of the decomposition, suppose that two decompositions exist, that is,

$$L_{ij} = L_{ij}^{S} + L_{ij}^{A} = L_{ij}'^{S} + L_{ij}'^{A} .$$

The last equation can be rearranged as

$$L_{ij}^{S} - L_{ij}'^{S} = L_{ij}'^{A} - L_{ij}^{A} . \tag{1.83}$$

Here the left-hand side (1.83) is symmetric and the right-hand side is antisymmetric. Only the zero tensor (all components zero) satisfies condition (1.83). We have then

$$L_{ij}^{S} = L_{ij}'^{S} \qquad L_{ij}^{A} = L_{ij}'^{A} .$$

**Trace of a Tensor**

The trace of a tensor $\boldsymbol{L}$ of order 2, denoted tr, is the sum of its diagonal components

$$\operatorname{tr}(\boldsymbol{L}) = \operatorname{tr}\left(L_{ij}\left(\boldsymbol{e}_i \otimes \boldsymbol{e}_j\right)\right) = L_{ij} \operatorname{tr}\left(\boldsymbol{e}_i \otimes \boldsymbol{e}_j\right) = L_{ij}\, \delta_{ij} = L_{ii} . \tag{1.84}$$

The trace of the tensor product of two vectors reduces to the scalar product of the vectors

$$\operatorname{tr}(\boldsymbol{a} \otimes \boldsymbol{b}) = \boldsymbol{a} \cdot \boldsymbol{b} . \tag{1.85}$$

The properties of the trace are

$$\operatorname{tr} \boldsymbol{L} = \operatorname{tr} \boldsymbol{L}^T \tag{1.86}$$

$$\operatorname{tr}(\boldsymbol{S} + \boldsymbol{T}) = \operatorname{tr} \boldsymbol{S} + \operatorname{tr} \boldsymbol{T} \tag{1.87}$$

$$\operatorname{tr}(\alpha \boldsymbol{L}) = \alpha \operatorname{tr} \boldsymbol{L} \tag{1.88}$$

$$\operatorname{tr}(\boldsymbol{A}\boldsymbol{L}) = \operatorname{tr}(\boldsymbol{L}\boldsymbol{A}) , \tag{1.89}$$

where $\alpha \in \mathbb{R}$.

**Deviatoric Tensors**

A tensor $\boldsymbol{L}$ can be decomposed as the sum of a spherical tensor $\boldsymbol{L}^s$ and a tensor with a zero trace $\boldsymbol{L}^d$, called the deviatoric tensor, so that

$$\boldsymbol{L} = \boldsymbol{L}^s + \boldsymbol{L}^d . \tag{1.90}$$

The spherical component of $\boldsymbol{L}^s$ is one third of its trace: $L_{ij}^s = \frac{1}{3} L_{kk} \delta_{ij}$ and the deviatoric components of $\boldsymbol{L}^d$ are defined by

$$L_{ij}^d = L_{ij} - \frac{1}{3} \operatorname{tr}(\boldsymbol{L})\, \delta_{ij} . \tag{1.91}$$

These components $L_{ij}^d$ are not independent since the trace of $\boldsymbol{L}^d$ is zero.

### Orthogonal Tensors

A tensor $\boldsymbol{Q}$ is orthogonal if it satisfies the condition

$$\boldsymbol{Qu} \cdot \boldsymbol{Qv} = \boldsymbol{u} \cdot \boldsymbol{v} \tag{1.92}$$

for every vector $\boldsymbol{u}$ and $\boldsymbol{v}$. Using (1.69), the condition (1.92) is written as

$$\boldsymbol{u} \cdot \boldsymbol{Q}^T \boldsymbol{Qv} = \boldsymbol{u} \cdot \boldsymbol{v}. \tag{1.93}$$

Thus an orthogonal tensor satisfies the equation $\boldsymbol{Q}^T\boldsymbol{Q} = \boldsymbol{I}$. Since $\boldsymbol{u} \cdot \boldsymbol{v}$ is preserved in this transformation, the angle between the vectors and their norms $\|\boldsymbol{u}\|$, $\|\boldsymbol{v}\|$ are also preserved. The tensor $\boldsymbol{Q}$ has the property that $\det \boldsymbol{Q} = \pm 1$. If $\det \boldsymbol{Q} = +1$, the tensor corresponds to a rotation; the tensor is said to be a proper rotation. When $\det \boldsymbol{Q} = -1$, it is said to be an improper rotation as it corresponds to a reflection. It is important to notice the difference between the orthogonal tensor $\boldsymbol{Q}$, mapping two vectors, and the coordinate change resulting from the rotation of the axes presented in section 1.2.2 that describe the same vector. There we defined a new vector basis $\boldsymbol{e}'_i$ $(i = 1, 2, 3)$ by rotation of the vector basis $\boldsymbol{e}_i$ $(i = 1, 2, 3)$, both systems of axes being orthogonal Cartesian systems describing the same vector. Since by (1.92), the angle between vectors is conserved, the coordinate change can be considered to be an orthogonal transformation which rotates the basis vectors $\boldsymbol{e}_i$ to $\boldsymbol{e}'_i$. Consequently, the matrix of the tensor $\boldsymbol{Q}$ is equal to the matrix $[C]$.

### Scalar Product of two Tensors

The scalar product of two tensors of order 2 is the scalar defined by the double sum

$$a = S_{ij} T_{ij} \,. \tag{1.94}$$

Symbolically, we denote it $a = \boldsymbol{S} : \boldsymbol{T}$. We can see that it is a multiplication with double contraction. Then, we can write successively

$$\boldsymbol{S} : \boldsymbol{T} = \operatorname{tr}(\boldsymbol{S}^T\boldsymbol{T}) = \operatorname{tr}(\boldsymbol{T}^T\boldsymbol{S}) = \operatorname{tr}(\boldsymbol{S}\boldsymbol{T}^T) = \operatorname{tr}(\boldsymbol{T}\boldsymbol{S}^T) = \boldsymbol{T} : \boldsymbol{S} \,. \tag{1.95}$$

The norm of a tensor $\|\boldsymbol{L}\|$ is defined by the relation

$$\|\boldsymbol{L}\| = (\boldsymbol{L} : \boldsymbol{L})^{1/2} = (L_{ij} L_{ij})^{1/2} \geq 0 \,. \tag{1.96}$$

The scalar product also satisfies the following properties:

$$\boldsymbol{L} : (\boldsymbol{ST}) = \left(\boldsymbol{S}^T\boldsymbol{L}\right) : \boldsymbol{T} = \left(\boldsymbol{L}\boldsymbol{T}^T\right) : \boldsymbol{S} \tag{1.97}$$

$$(\boldsymbol{u} \otimes \boldsymbol{v}) : (\boldsymbol{a} \otimes \boldsymbol{b}) = (\boldsymbol{u} \cdot \boldsymbol{a})(\boldsymbol{v} \cdot \boldsymbol{b}) \tag{1.98}$$

$$\boldsymbol{L} : (\boldsymbol{a} \otimes \boldsymbol{b}) = \boldsymbol{a} \cdot \boldsymbol{Lb} = (\boldsymbol{a} \otimes \boldsymbol{b}) : \boldsymbol{L} \,. \tag{1.99}$$

EXAMPLE 1.6
To verify identity (1.99) we use index notation algebra. With the definition (1.94), we have

$$(\boldsymbol{L} : (\boldsymbol{a} \otimes \boldsymbol{b})) = L_{ij}(\boldsymbol{a} \otimes \boldsymbol{b})_{ij} = L_{ij}a_i b_j = a_i L_{ij} b_j = (\boldsymbol{a} \cdot \boldsymbol{L}\boldsymbol{b}) \; . \qquad (1.100)$$

EXAMPLE 1.7
Consider the tensors $\boldsymbol{A}$ and $\boldsymbol{B}$ such that $A_{ij} = A_{ji}$ is symmetric and $B_{ij} = -B_{ji}$ is antisymmetric. The scalar product of these two tensors is zero. (Note that $A_{ji}B_{ij} = A_{ij}B_{ji}$ since the result is a scalar). Using definitions (1.78), (1.80), and (1.94), we have

$$\begin{aligned}(\boldsymbol{A} : \boldsymbol{B}) &= A_{ij}B_{ij} = \frac{1}{2}(A_{ij}B_{ij} + A_{ij}B_{ij}) = \frac{1}{2}(A_{ij}B_{ij} - A_{ij}B_{ji}) \\ &= \frac{1}{2}(A_{ij}B_{ij} - A_{ji}B_{ij}) = \frac{1}{2}(A_{ij}B_{ij} - A_{ij}B_{ij}) = 0 \; . \qquad (1.101)\end{aligned}$$

**Right Product of a Tensor and a Vector**

The right product of a tensor $\boldsymbol{L}$ and a vector $\boldsymbol{u}$ is defined as

$$v_j = u_i L_{ij} = L_{ij} u_i \; . \qquad (1.102)$$

We write it $\boldsymbol{uL}$. Here, the order of the symbols is important, which is not the case for index notation. The symbolic form $\boldsymbol{Lu}$ represents a different vector, which in index notation is written as

$$w_i = L_{ij} u_j = u_j L_{ij} \; . \qquad (1.103)$$

Note that this last relation is none other than (1.43).

### 1.3.7 Dual Vector of a Tensor of Order 2

The components $d_i$ of the dual (or axial) vector of a tensor $\boldsymbol{L}$ are definined by the product

$$d_i = \frac{1}{2}\varepsilon_{ikj}L_{jk} = -\frac{1}{2}\varepsilon_{ijk}L_{jk} \; . \qquad (1.104)$$

Or explicitly

$$\begin{aligned} d_1 &= -\frac{1}{2}(\varepsilon_{123}L_{23} + \varepsilon_{132}L_{32}) = -\frac{1}{2}(L_{23} - L_{32}) \\ d_2 &= -\frac{1}{2}(\varepsilon_{231}L_{31} + \varepsilon_{213}L_{13}) = -\frac{1}{2}(L_{31} - L_{13}) \\ d_3 &= -\frac{1}{2}(\varepsilon_{312}L_{12} + \varepsilon_{321}L_{21}) = -\frac{1}{2}(L_{12} - L_{21}) \; . \end{aligned}$$

Note in passing that if tensor $\boldsymbol{L}$ is symmetric, the dual vector $\boldsymbol{d}$ is zero. Decomposing $\boldsymbol{L}$ into its symmetric and antisymmetric parts, we have from (1.82)

$$d_i = -\frac{1}{2}\left(\varepsilon_{ijk}L^S_{jk} + \varepsilon_{ijk}L^A_{jk}\right). \tag{1.105}$$

As $\varepsilon_{ijk}$ is by definition antisymmetric with respect to any pair of its indices, the first term on the right-hand side is zero as it is the interior product of a symmetric and an antisymmetric tensor (1.101). Thus the dual vector depends uniquely on the antisymmetric part of a tensor

$$d_i = -\frac{1}{2}\,\varepsilon_{ijk}L^A_{jk}\,. \tag{1.106}$$

The inverse of the relation (1.104) is obtained by multiplying the two sides by $\varepsilon_{ilm}$, that is,

$$\varepsilon_{ilm}d_i = \varepsilon_{lmi}d_i = -\frac{1}{2}\,\varepsilon_{ilm}\varepsilon_{ijk}L_{jk}\,.$$

With (1.30), we obtain successively

$$\begin{aligned} \varepsilon_{ilm}d_i &= -\frac{1}{2}\left(\delta_{lj}\,\delta_{mk} - \delta_{lk}\,\delta_{mj}\right)L_{jk} \\ &= -\frac{1}{2}\left(L_{lm} - L_{ml}\right) = -L^A_{lm} \end{aligned}$$

or

$$-L^A_{lm} = \varepsilon_{lmi}d_i\,. \tag{1.107}$$

The three independent components of an antisymmetric tensor (recall that the diagonal components are zero) are equivalent to the three components of the dual vector in that they yield the same information as $d_1 = -L^A_{23}$, $d_2 = -L^A_{31}$, $d_3 = -L^A_{12}$.

### 1.3.8 Eigenvalues and Eigenvectors of a Tensor

Let $\boldsymbol{L}$ be a second order tensor. If $\boldsymbol{u}$ is a vector that, when $\boldsymbol{L}$ is applied, is transformed into a vector parallel to itself, that is,

$$\boldsymbol{L}\boldsymbol{u} = \lambda\boldsymbol{u}\,, \tag{1.108}$$

then the vector $\boldsymbol{u}$ is an eigenvector of $\boldsymbol{L}$, and $\lambda$ is the corresponding eigenvalue. We know the eigenvectors can always be of arbitrary length. However, for simplicity, we will normalize them to unit length. Let $\boldsymbol{n}$ be a ***unit eigenvector***. Then, if we introduce the unit tensor $\boldsymbol{I}$, we can write

$$\boldsymbol{L}\boldsymbol{n} = \lambda\boldsymbol{n} = \lambda\boldsymbol{I}\boldsymbol{n}\,, \tag{1.109}$$

which gives

$$(\boldsymbol{L} - \lambda\boldsymbol{I})\boldsymbol{n} = \boldsymbol{0} \qquad \text{with} \qquad \boldsymbol{n}\cdot\boldsymbol{n} = 1\,. \tag{1.110}$$

In index notation, with $\boldsymbol{n} = n_i\boldsymbol{e}_i$, we have

$$(L_{ij} - \lambda\,\delta_{ij})n_j = 0 \qquad\qquad n_jn_j = 1\,. \tag{1.111}$$

Since equation (1.111) is valid for any $\boldsymbol{n} \neq \boldsymbol{0}$, we must solve the equation

$$\det([L] - \lambda[I]) = 0 \tag{1.112}$$

to obtain a solution.

If the tensor $\boldsymbol{L}$ is symmetric, then the characteristic equation (1.112) allows us to invoke a theorem from linear algebra (see [67] for a proof) which is stated as follows.

**Theorem**

*The eigenvalues of a real $n \times n$ symmetric matrix are all real. The corresponding eigenvectors are orthogonal.*

It will be the same for the eigenvalues corresponding to the tensor $\boldsymbol{L}$, which we will call principal values, as is common in mechanics for stress and strain tensors. Similarly, the associated eigenvectors define the ***principal directions***.

We can show that for a symmetric tensor, there are always three principal directions which are orthogonal. Let $\boldsymbol{n}_1$ and $\boldsymbol{n}_2$ be two eigenvectors corresponding to the respective eigenvalues $\lambda_1$ and $\lambda_2$ of the tensor $\boldsymbol{L}$. Then,

$$\boldsymbol{L}\boldsymbol{n}_1 = \lambda_1 \boldsymbol{n}_1 \tag{1.113}$$

and

$$\boldsymbol{L}\boldsymbol{n}_2 = \lambda_2 \boldsymbol{n}_2 \,. \tag{1.114}$$

Let us take the scalar product of (1.113) with $\boldsymbol{n}_2$ and of (1.114) with $\boldsymbol{n}_1$. Switching to index notation we obtain

$$L_{ij}(n_1)_j(n_2)_i = \lambda_1(n_1)_i(n_2)_i \tag{1.115}$$

and

$$L_{ij}(n_2)_j(n_1)_i = \lambda_2(n_2)_i(n_1)_i \,. \tag{1.116}$$

Because of the symmetric property of $\boldsymbol{L}$ the first term of (1.116) can be transformed as follows:

$$L_{ji}(n_1)_j(n_2)_i = \lambda_2(n_2)_i(n_1)_i \,. \tag{1.117}$$

Subtracting (1.117) from (1.115), we have the result

$$(\lambda_1 - \lambda_2)(\boldsymbol{n}_1 \cdot \boldsymbol{n}_2) = 0 \,. \tag{1.118}$$

As $\lambda_1 \neq \lambda_2$, $\boldsymbol{n}_1 \cdot \boldsymbol{n}_2 = 0$ and so $\boldsymbol{n}_1$ and $\boldsymbol{n}_2$ are orthogonal. Thus, we can conclude that if the principal values are all distinct, then the three principal directions are mutually orthogonal.

If $\lambda_1 = \lambda_2 \neq \lambda_3$, we have $\boldsymbol{n}_1 \cdot \boldsymbol{n}_3 = \boldsymbol{n}_2 \cdot \boldsymbol{n}_3 = 0$. The directions $\boldsymbol{n}_1$ and $\boldsymbol{n}_2$ are then chosen mutually orthogonal and normal to $\boldsymbol{n}_3$.

If $\lambda_1 = \lambda_2 = \lambda_3$, the directions $\boldsymbol{n}_1$, $\boldsymbol{n}_2$, and $\boldsymbol{n}_3$ are chosen mutually orthogonal and without any restriction.

Let us examine now the form of the matrix of a tensor with respect to its principal directions. We denote by $\boldsymbol{n}_1$, $\boldsymbol{n}_2$, and $\boldsymbol{n}_3$ the unit vectors in these directions. If we use these vectors as basis vectors, and (1.42), then the tensor components become

$$\begin{aligned} L_{11} &= \boldsymbol{n}_1 \cdot \boldsymbol{L}\boldsymbol{n}_1 = \boldsymbol{n}_1 \cdot (\lambda_1 \boldsymbol{n}_1) = \lambda_1 \\ L_{22} &= \lambda_2 \\ L_{33} &= \lambda_3 \\ L_{12} &= \boldsymbol{n}_1 \cdot \boldsymbol{L}\boldsymbol{n}_2 = (\boldsymbol{n}_1 \cdot \lambda_2 \boldsymbol{n}_2) = 0 = L_{21} \\ L_{13} &= L_{31} = 0 \\ L_{23} &= L_{32} = 0\,. \end{aligned} \tag{1.119}$$

The matrix is thus diagonal and its diagonal elements are its eigenvalues.

**Scalar Invariants of a Tensor and the Cayley-Hamilton Theorem**

Developing (1.112) the characteristic equation of a tensor is cubic in $\lambda$; we can write

$$\lambda^3 - I_1(\boldsymbol{L})\lambda^2 + I_2(\boldsymbol{L})\lambda - I_3(\boldsymbol{L}) = 0\,, \tag{1.120}$$

where $I_1(\boldsymbol{L})$, $I_2(\boldsymbol{L})$, and $I_3(\boldsymbol{L})$ are the invariants of the tensor $\boldsymbol{L}$ given by the expressions

$$\begin{aligned} I_1(\boldsymbol{L}) &= L_{ii} = \operatorname{tr}\boldsymbol{L} \\ I_2(\boldsymbol{L}) &= \begin{vmatrix} L_{11} & L_{12} \\ L_{21} & L_{22} \end{vmatrix} + \begin{vmatrix} L_{22} & L_{23} \\ L_{32} & L_{33} \end{vmatrix} + \begin{vmatrix} L_{11} & L_{13} \\ L_{31} & L_{33} \end{vmatrix} \\ &= \frac{1}{2}\left(L_{ii}L_{jj} - L_{ij}L_{ji}\right) \\ &= \frac{1}{2}\left((\operatorname{tr}\boldsymbol{L})^2 - \operatorname{tr}(\boldsymbol{L}\boldsymbol{L})\right) = \frac{1}{2}\left((\operatorname{tr}\boldsymbol{L})^2 - \operatorname{tr}(\boldsymbol{L}^2)\right) \\ I_3(\boldsymbol{L}) &= \varepsilon_{ijk}L_{i1}L_{j2}L_{k3} = \det\boldsymbol{L}\,. \end{aligned} \tag{1.121}$$

As, by definition, the eigenvalues $\lambda_1, \lambda_2, \lambda_3$ of $\boldsymbol{L}$ are independent of the basis vectors $\{\boldsymbol{e}_i\}$, then the coefficients of the cubic equation (1.120) should be the same for all $\{\boldsymbol{e}_i\}$. This is why they are called 'invariants'.

Example 1.8

Find the principal values (eigenvalues) and the corresponding unit vectors (eigenvectors) of the symmetric tensor with matrix

$$[L] = \begin{pmatrix} 2 & 1 & -2 \\ 1 & 4 & -3 \\ -2 & -3 & -2 \end{pmatrix}. \tag{1.122}$$

Using the expressions (1.121), the corresponding invariants are

$$I_1(\boldsymbol{L}) = 4, \quad I_2(\boldsymbol{L}) = -18, \quad I_3(\boldsymbol{L}) = -36,$$

and the corresponding characteristic equation (1.112) is given by

$$\lambda^3 - 4\lambda^2 - 18\lambda + 36 = 0.$$

This equation has three distinct roots which are the principal values of $\boldsymbol{L}$

$$\lambda_1 = 6, \quad \lambda_2 = 1.65, \quad \lambda_3 = -3.65.$$

Let us assume that the eigenvector $\boldsymbol{n}_1$ corresponding to $\lambda_1$ has components $(n_1)_1$, $(n_1)_2$, and $(n_1)_3$. For this particular eigenvalue, equation (1.111) gives the following system of equations:

$$\begin{aligned} (2-6)(n_1)_1 + (n_1)_2 - 2(n_1)_3 &= 0 \\ (n_1)_1 + (4-6)(n_1)_2 - 3(n_1)_3 &= 0 \\ -2(n_1)_1 - 3(n_1)_2 - (2+6)(n_1)_3 &= 0. \end{aligned}$$

The first two equations give $(n_1)_2 = -2(n_1)_3$. Inserting into the third equation, we obtain $(n_1)_1 = -(n_1)_3$. By the orthogonality condition we have

$$(n_1)_1^2 + (n_1)_2^2 + (n_1)_3^2 = 1 .$$

With this condition and the results for $(n_1)_1$ and $(n_1)_2$, we obtain $(n_1)_3 = 0.4082$. Then with this value, the other components are $(n_1)_1 = -0.4082$ and $(n_1)_2 = -0.8165$. Proceeding similarly we calculate the unit vectors for the two other principal values. Finally, we have

$$\begin{aligned} \lambda_1 = 6 \quad &: (n_1)_1 = -0.4082, (n_1)_2 = -0.8165, (n_1)_3 = +0.4082 \\ \lambda_2 = 1.65 \quad &: (n_2)_1 = +0.8736, (n_2)_2 = -0.4792, (n_2)_3 = -0.0849 \\ \lambda_3 = -3.65 \quad &: (n_3)_1 = +0.2650, (n_3)_2 = +0.3220, (n_3)_3 = +0.9089 . \end{aligned}$$

EXAMPLE 1.9

Show that the expression $a = \frac{1}{2}\varepsilon_{ijk}\varepsilon_{ist}L_{js}L_{kt}$ is an invariant of the symmetric tensor $\boldsymbol{L}$.

We use identity (1.30) to modify the expression as follows:

$$\begin{aligned} 2a = \varepsilon_{ijk}\varepsilon_{ist}L_{js}L_{kt} &= (\delta_{js}\delta_{kt} - \delta_{jt}\delta_{ks})\, L_{js}L_{kt} \\ &= \delta_{js}L_{js}\delta_{kt}L_{kt} - \delta_{jt}L_{js}\delta_{ks}L_{kt} \\ &= L_{jj}L_{kk} - L_{ts}L_{st} = L_{jj}L_{kk} - L_{ts}L_{ts} . \end{aligned}$$

**Cayley-Hamilton Theorem**

*Any tensor* $\boldsymbol{L}$ *satisfies its own characteristic equation*

$$\boldsymbol{L}^3 - I_1(\boldsymbol{L})\boldsymbol{L}^2 + I_2(\boldsymbol{L})\boldsymbol{L} - I_3(\boldsymbol{L})\boldsymbol{I} = \boldsymbol{0} \,. \tag{1.123}$$

*(See exercise 1.16.)*

**Positive Definite Tensors**

We introduce now the notion of a positive definite tensor. A tensor $\boldsymbol{L}$ is said to be positive definite if for any non-zero vector $\boldsymbol{v}$ it satisfies the inequality

$$\forall \boldsymbol{v} \in E^3, \quad \boldsymbol{v} \cdot \boldsymbol{L}\boldsymbol{v} > 0 \,. \tag{1.124}$$

We can easily show that the eigenvalues of a positive definite tensor are strictly positive. Let $\lambda$ be an eigenvalue of the positive definite tensor $\boldsymbol{L}$ and let $\boldsymbol{n}$ be the corresponding unit eigenvector. As $\boldsymbol{L}\boldsymbol{n} = \lambda\boldsymbol{n}$ and $\|\boldsymbol{n}\| = 1$, then

$$\boldsymbol{n} \cdot \boldsymbol{L}\boldsymbol{n} = \lambda > 0 \,.$$

**Spectral Decomposition of a Tensor**

Let $\boldsymbol{L}$ be a symmetric tensor which has three real eigenvalues $\lambda_i$ and three real, orthogonal eigenvectors $\boldsymbol{n}_i$. These eigenvectors form the basis of the spectral decomposition that is written as the sum of three principal self-dyads $\boldsymbol{n}_i \otimes \boldsymbol{n}_i$

$$\boldsymbol{L} = \sum_{i=1}^{3} \lambda_i \, \boldsymbol{n}_i \otimes \boldsymbol{n}_i \,. \tag{1.125}$$

This expression is known as the spectral decomposition or the spectral representation of the tensor $\boldsymbol{L}$. It is easily established by expressing $\boldsymbol{L} = \boldsymbol{L}\boldsymbol{I}$, $\boldsymbol{I} = \boldsymbol{n}_i \otimes \boldsymbol{n}_i$ and using (1.65) and (1.109).

### 1.3.9 Square Root of a Tensor

**Theorem**

*Let* $\boldsymbol{C}$ *be a symmetric, positive definite tensor whose eigenvalues are* $\lambda_i^2$ *with the corresponding eigenvectors* $\boldsymbol{n}_i$*. Then, there exists a unique, symmetric, positive definite tensor* $\boldsymbol{U}$ *such that*

$$\boldsymbol{U}^2 = \boldsymbol{C} \,. \tag{1.126}$$

*We denote* $\sqrt{\boldsymbol{C}} = \boldsymbol{U}$.

PROOF.
1) *Existence.* By (1.125), let

$$C = \sum_{i=1}^{3} \lambda_i^2 \, \boldsymbol{n}_i \otimes \boldsymbol{n}_i \,. \tag{1.127}$$

Define $\boldsymbol{U}$ by the relation

$$\boldsymbol{U} = \sum_{i=1}^{3} \lambda_i \, \boldsymbol{n}_i \otimes \boldsymbol{n}_i \,. \tag{1.128}$$

Then equation (1.126) is the direct consequence of

$$\boldsymbol{U}^2 = \boldsymbol{U}\boldsymbol{U} = \sum_{i=1}^{3} \lambda_i^2 (\boldsymbol{n}_i \otimes \boldsymbol{n}_i)(\boldsymbol{n}_i \otimes \boldsymbol{n}_i) = \sum_{i=1}^{3} \lambda_i^2 (\boldsymbol{n}_i \otimes \boldsymbol{n}_i) = \boldsymbol{C}$$

as, for any set of eigenvectors,

$$(\boldsymbol{n}_i \otimes \boldsymbol{n}_i)(\boldsymbol{n}_j \otimes \boldsymbol{n}_j) = \begin{cases} 0 & \text{if } i \neq j \\ (\boldsymbol{n}_i \otimes \boldsymbol{n}_i) & \text{if } i = j \,. \end{cases}$$

2) *Uniqueness.* Suppose that there are two tensors $\boldsymbol{U}$ and $\boldsymbol{V}$ such that

$$\boldsymbol{U}^2 = \boldsymbol{V}^2 = \boldsymbol{C} \,.$$

Let $\boldsymbol{n}$ be the eigenvector of $\boldsymbol{C}$ corresponding to the eigenvalue $\lambda > 0$. Then,

$$0 = (\boldsymbol{U}^2 - \lambda^2 \boldsymbol{I})\boldsymbol{n} = (\boldsymbol{U} + \lambda \boldsymbol{I})(\boldsymbol{U} - \lambda \boldsymbol{I})\boldsymbol{n} \,. \tag{1.129}$$

Let us set

$$\boldsymbol{v} = (\boldsymbol{U} - \lambda \boldsymbol{I})\boldsymbol{n} \,. \tag{1.130}$$

Then from (1.129)

$$\boldsymbol{U}\boldsymbol{v} = -\lambda \boldsymbol{v} \,.$$

But the vector $\boldsymbol{v}$ must vanish, otherwise $-\lambda$ would be an eigenvalue of $\boldsymbol{U}$, which is impossible as $\boldsymbol{U}$ is positive definite and $\lambda > 0$. Thus, by (1.130), we obtain

$$\boldsymbol{U}\boldsymbol{n} = \lambda \boldsymbol{n} \,. \tag{1.131}$$

In the same way, we have $\boldsymbol{V}\boldsymbol{n} = \lambda \boldsymbol{n}$, thus $\boldsymbol{U}\boldsymbol{n} = \boldsymbol{V}\boldsymbol{n}$ for all eigenvectors $\boldsymbol{n}$ of $\boldsymbol{C}$. And, as we can form a basis from the eigenvectors of $\boldsymbol{C}$, then $\boldsymbol{U}$ and $\boldsymbol{V}$ must coincide.

### 1.3.10 Polar Decomposition

**Theorem**

*Let $\boldsymbol{F}$ be a tensor belonging to the set of tensors $\boldsymbol{F}$ with determinant* $\det \boldsymbol{F} > 0$*. Then, there exist symmetric positive definite tensors $\boldsymbol{U}$ and $\boldsymbol{V}$, and a rotation, that is, an orthogonal tensor with a positive determinant (equal to 1), $\boldsymbol{R}$ such that*

$$\boldsymbol{F} = \boldsymbol{R}\boldsymbol{U} = \boldsymbol{V}\boldsymbol{R}. \tag{1.132}$$

*Each of these decompositions is unique. We have*

$$\boldsymbol{U} = \sqrt{\boldsymbol{F}^T\boldsymbol{F}}\,, \qquad \boldsymbol{V} = \sqrt{\boldsymbol{F}\boldsymbol{F}^T}\,. \tag{1.133}$$

*The representation $\boldsymbol{F} = \boldsymbol{R}\boldsymbol{U}$ (and the representation $\boldsymbol{F} = \boldsymbol{V}\boldsymbol{R}$) is the right polar decomposition (and the left polar decomposition) of $\boldsymbol{F}$.*

PROOF.
1) We show that $\boldsymbol{F}^T\boldsymbol{F}$ and $\boldsymbol{F}\boldsymbol{F}^T$ belong to the set of all symmetric positive definite tensors. The two tensors are clearly symmetric. In addition,

$$\boldsymbol{v} \cdot \boldsymbol{F}^T\,\boldsymbol{F}\boldsymbol{v} = v_j F_{mi} F_{mj} v_i = F_{mi} v_i\, F_{mj} v_j = (\boldsymbol{F}\boldsymbol{v}) \cdot (\boldsymbol{F}\boldsymbol{v}) \geq 0\,.$$

This last scalar product cannot be equal to zero unless $\boldsymbol{F}\boldsymbol{v} = 0$; consequently it can only be zero when $\boldsymbol{v} = 0$. Therefore, $\boldsymbol{F}^T\boldsymbol{F}$ belongs to the set of symmetric positive definite tensors. Similar reasoning applies for $\boldsymbol{F}\boldsymbol{F}^T$.

2) *Uniqueness.* Let $\boldsymbol{F} = \boldsymbol{R}\boldsymbol{U}$ be the right polar decomposition of $\boldsymbol{F}$. As $\boldsymbol{R}$ is a rotation, we have

$$\boldsymbol{F}^T\boldsymbol{F} = \boldsymbol{U}\boldsymbol{R}^T\boldsymbol{R}\boldsymbol{U} = \boldsymbol{U}^2\,.$$

But by the square root theorem, there can only be one tensor $\boldsymbol{U}$ belonging to the set of symmetric positive definite tensors for which the square is $\boldsymbol{F}^T\boldsymbol{F}$. Thus, the first relation of (1.133) is valid and $\boldsymbol{U}$ is unique. Because $\boldsymbol{R} = \boldsymbol{F}\boldsymbol{U}^{-1}$, $\boldsymbol{R}$ is also unique. In addition, if $\boldsymbol{F} = \boldsymbol{V}\boldsymbol{R}$ is the left polar decomposition, then

$$\boldsymbol{F}\boldsymbol{F}^T = \boldsymbol{V}^2\,, \tag{1.134}$$

and $\boldsymbol{V}$ is determined by (1.133) with $\boldsymbol{R} = \boldsymbol{V}^{-1}\boldsymbol{F}$.

3) *Existence.* Let $\boldsymbol{U}$ be a tensor belonging to the set of symmetric positive definite tensors given by (1.133) and let

$$\boldsymbol{R} = \boldsymbol{F}\boldsymbol{U}^{-1}\,. \tag{1.135}$$

To verify that $\boldsymbol{F} = \boldsymbol{R}\boldsymbol{U}$ is in fact a ***polar decomposition***, we show that $\boldsymbol{R}$ belongs to the set of all rotations.

As $\det \boldsymbol{F} > 0$ and $\det \boldsymbol{U} > 0$, then $\det \boldsymbol{R} > 0$ ($\det \boldsymbol{U} > 0$ because all the eigenvalues of $\boldsymbol{U}$ are strictly positive). We only need to show that $\boldsymbol{R}$ belongs to the set of orthogonal tensors.

We proceed as follows:

$$\boldsymbol{R}^T\boldsymbol{R} = \boldsymbol{U}^{-1}\boldsymbol{F}^T\boldsymbol{F}\boldsymbol{U}^{-1} = \boldsymbol{U}^{-1}\boldsymbol{U}^2\boldsymbol{U}^{-1} = \boldsymbol{I}\,. \tag{1.136}$$

Now show that $\det \boldsymbol{R} = 1$. We know that $\det \boldsymbol{F} = J > 0$. Thus, $\det \boldsymbol{U}^2 = J^2$ and $\det \boldsymbol{U}^{-1} = 1/J > 0$. From (1.136), we have $\det \boldsymbol{R} = \pm 1$. Relation (1.135) allows us to deduce $\det \boldsymbol{R} = +1$, which corresponds to a rotation. If we define

$$\boldsymbol{V} = \boldsymbol{R}\boldsymbol{U}\boldsymbol{R}^T\,, \tag{1.137}$$

then $\boldsymbol{V}$ belongs to the set of symmetric positive definite tensors and

$$\boldsymbol{V}\boldsymbol{R} = \boldsymbol{R}\boldsymbol{U}\boldsymbol{R}^T\boldsymbol{R} = \boldsymbol{R}\boldsymbol{U} = \boldsymbol{F}\,.$$

### 1.3.11 Isotropic Tensor Function of a Symmetric Tensor

An isotropic tensor function $\boldsymbol{f}$, for which the variable $\boldsymbol{T}$ is a symmetric tensor of order 2, satisfies by definition the identity

$$\boldsymbol{Q}\boldsymbol{f}(\boldsymbol{T})\boldsymbol{Q}^T = \boldsymbol{f}(\boldsymbol{Q}\boldsymbol{T}\boldsymbol{Q}^T) \tag{1.138}$$

for any orthogonal tensor $\boldsymbol{Q}$. Then the application of the vector space of symmetric tensors in itself yields a symmetric tensor such that

$$\boldsymbol{L} = \boldsymbol{f}(\boldsymbol{T})\,. \tag{1.139}$$

In the following we present without demonstration the representation theorem. For a proof, the reader is referred to the appendix of [49].

**Rivlin-Ericksen Theorem**

*Expression (1.139) can be written in the form*

$$\begin{aligned}\boldsymbol{L} = {} & \varphi_0\big(I_1(\boldsymbol{T}), I_2(\boldsymbol{T}), I_3(\boldsymbol{T})\big)\boldsymbol{I} + \varphi_1\big(I_1(\boldsymbol{T}), I_2(\boldsymbol{T}), I_3(\boldsymbol{T})\big)\boldsymbol{T} \\ & + \varphi_2\big(I_1(\boldsymbol{T}), I_2(\boldsymbol{T}), I_3(\boldsymbol{T})\big)\boldsymbol{T}^2\,,\end{aligned} \tag{1.140}$$

*where the $\varphi_i$ ($i = 0, 1, 2$) are scalar functions of the invariants of $\boldsymbol{T}$.*

For an isotropic function (1.140), the principal directions of $\boldsymbol{T}$ and of $\boldsymbol{f}(\boldsymbol{T})$ coincide; $\boldsymbol{T}$ and $\boldsymbol{f}(\boldsymbol{T})$ are said to be coaxial.

### 1.3.12 Scalar Function of a Tensor

A function $\mathcal{W}(\boldsymbol{T})$ is defined as a scalar function of the tensor $\boldsymbol{T}$ and yields a scalar value. When $\boldsymbol{T}$ is symmetric and the condition

$$\mathcal{W}(\boldsymbol{T}) = \mathcal{W}(\boldsymbol{Q}\boldsymbol{T}\boldsymbol{Q}^T) \tag{1.141}$$

is satisfied, the function $\mathcal{W}(\boldsymbol{T})$ is an isotropic tensor function of $\boldsymbol{T}$. It can be represented by the relation

$$\mathcal{W}(\boldsymbol{T}) = \Phi\big(I_1(\boldsymbol{T}), I_2(\boldsymbol{T}), I_3(\boldsymbol{T})\big)\,. \tag{1.142}$$

This function is also a scalar invariant of the tensor $\boldsymbol{T}$. The preceding representation is equivalent to

$$\mathcal{W}(\boldsymbol{T}) = \phi(\lambda_1, \lambda_2, \lambda_3)\,, \tag{1.143}$$

where the $\lambda_i$ $(i = 1, 2, 3)$ are the eigenvalues of $\boldsymbol{T}$.

We can show that for an isotropic function $\mathcal{W}(\boldsymbol{T})$, its derivative with respect to $\boldsymbol{T}$ is expressed in the form

$$\frac{\partial \mathcal{W}}{\partial \boldsymbol{T}} = \sum_{i=1}^{3} \frac{\partial \mathcal{W}}{\partial \lambda_i} \boldsymbol{n}_i \otimes \boldsymbol{n}_i\,, \tag{1.144}$$

where $\boldsymbol{n}_i$ $(i = 1, 2, 3)$ are the principal directions corresponding to $\lambda_i$. It follows that

$$\boldsymbol{T}\frac{\partial \mathcal{W}}{\partial \boldsymbol{T}} = \frac{\partial \mathcal{W}}{\partial \boldsymbol{T}}\boldsymbol{T}\,. \tag{1.145}$$

This last relation indicates that the two tensors $\boldsymbol{T}$ and $\partial\mathcal{W}/\partial\boldsymbol{T}$ are coaxial or have the same eigenvectors or principal directions.

## 1.4 Tensor Analysis

In this section we will introduce the concepts related to the differentiation and differentials of tensors. We will denote, for example, $F$ a scalar, $v_i$ a vector component, and $L_{ij}$ a tensor component, all of which are functions of position $x_i$ in space and time $t$. The notation $F(\boldsymbol{x}, t)$ signifies $F(x_1, x_2, x_3, t)$, while the notation $v_i(x_m, t)$ covers the three functions $v_1(x_1, x_2, x_3, t)$, $v_2(x_1, x_2, x_3, t)$, and $v_3(x_1, x_2, x_3, t)$. Also $L_{ij}(x_i, t)$ covers all components of a tensor.

Thus when $x_i$ is between parentheses, to indicate a function, the summation rules do not apply to the independent space variable: $F(x_i, t)$ is not a vector; it is a scalar field, while $v_i(x_m, t)$ is a vector field, etc.

### 1.4.1 Derivative of Vector, Tensor or Scalar Function

Time derivatives of several field parameters play important roles in continuum mechanics based modeling, i.e., velocity, acceleration, strain rate, power, etc. In section 1.4.1 we define the essential time derivatives of second order tensors and vectors before presenting the spatial derivatives in the following sections 1.4.2 to 1.4.8. Afterwards, the definition of curvilinear coordinates (sec. 1.4.9) and flux (sec. 1.4.12) are given, followed by two important integral theorems of calculus (sec. 1.4.13).

Let $\boldsymbol{L} = \boldsymbol{L}(t)$ be a tensor function of a scalar $t$ (time, for example). The derivative of $\boldsymbol{L}$ with respect to $t$ is the tensor of order 2 given by

$$\dot{\boldsymbol{L}} = \frac{d\boldsymbol{L}}{dt} = \lim_{\Delta t \to 0} \frac{\boldsymbol{L}(t + \Delta t) - \boldsymbol{L}(t)}{\Delta t}. \tag{1.146}$$

In terms of its components,

$$\dot{\boldsymbol{L}} = \frac{dL_{ij}(t)}{dt} \boldsymbol{e}_i \otimes \boldsymbol{e}_j = \dot{L}_{ij} \boldsymbol{e}_i \otimes \boldsymbol{e}_j . \tag{1.147}$$

The first derivative of a vector function $\boldsymbol{v}(t)$ with respect to $t$ is defined similarly, $\dot{\boldsymbol{v}} = d\boldsymbol{v}/dt$. In terms of its components,

$$\frac{d\boldsymbol{v}}{dt} = \dot{\boldsymbol{v}} = \dot{v}_i(t)\boldsymbol{e}_i . \tag{1.148}$$

And the second derivative is

$$\frac{d^2\boldsymbol{v}}{dt^2} = \ddot{\boldsymbol{v}} = \ddot{v}_i(t)\boldsymbol{e}_i . \tag{1.149}$$

One can easily establish the following identities:

$$\frac{d}{dt}(\boldsymbol{u} \pm \boldsymbol{v}) = \frac{d\boldsymbol{u}}{dt} \pm \frac{d\boldsymbol{v}}{dt} \tag{1.150}$$

$$\frac{d}{dt}(\boldsymbol{u} \otimes \boldsymbol{v}) = \frac{d\boldsymbol{u}}{dt} \otimes \boldsymbol{v} + \boldsymbol{u} \otimes \frac{d\boldsymbol{v}}{dt} \tag{1.151}$$

$$\frac{d}{dt}(\boldsymbol{L} \pm \boldsymbol{T}) = \frac{d\boldsymbol{L}}{dt} \pm \frac{d\boldsymbol{T}}{dt} \tag{1.152}$$

$$\frac{d}{dt}(\alpha(t)\boldsymbol{L}) = \frac{d\alpha(t)}{dt}\boldsymbol{L} + \alpha(t)\frac{d\boldsymbol{L}}{dt} \tag{1.153}$$

$$\frac{d}{dt}(\boldsymbol{L}\boldsymbol{T}) = \frac{d\boldsymbol{L}}{dt}\boldsymbol{T} + \boldsymbol{L}\frac{d\boldsymbol{T}}{dt} \tag{1.154}$$

$$\frac{d}{dt}(\boldsymbol{L}\boldsymbol{a}) = \frac{d\boldsymbol{L}}{dt}\boldsymbol{a} + \boldsymbol{L}\frac{d\boldsymbol{a}}{dt} \tag{1.155}$$

$$\frac{d}{dt}(\boldsymbol{L}^T) = \left(\frac{d\boldsymbol{L}}{dt}\right)^T . \tag{1.156}$$

We can demonstrate (1.155) for example. By definition (1.146), we obtain

$$\begin{aligned}\frac{d}{dt}(\boldsymbol{La}) &= \lim_{\Delta t\to 0}\frac{\boldsymbol{L}(t+\Delta t)\boldsymbol{a}(t+\Delta t)-\boldsymbol{L}(t)\boldsymbol{a}(t)}{\Delta t}\\ &= \lim_{\Delta t\to 0}\frac{1}{\Delta t}\Big(\boldsymbol{L}(t+\Delta t)\boldsymbol{a}(t+\Delta t)-\boldsymbol{L}(t)\boldsymbol{a}(t)\\ &\qquad +\boldsymbol{L}(t)\boldsymbol{a}(t+\Delta t)-\boldsymbol{L}(t)\boldsymbol{a}(t+\Delta t)\Big)\\ &= \lim_{\Delta t\to 0}\frac{\big(\boldsymbol{L}(t+\Delta t)-\boldsymbol{L}(t)\big)\boldsymbol{a}(t+\Delta t)}{\Delta t}\\ &\qquad +\lim_{\Delta t\to 0}\frac{\boldsymbol{L}(t)\big(\boldsymbol{a}(t+\Delta t)-\boldsymbol{a}(t)\big)}{\Delta t}\\ &= \frac{d\boldsymbol{L}}{dt}\boldsymbol{a}+\boldsymbol{L}\frac{d\boldsymbol{a}}{dt}\,.\end{aligned}$$

### 1.4.2 Gradient of a Scalar Field

Associated with a scalar field $F(\boldsymbol{x})$ is a vector field called the gradient of $F$. This gradient is denoted $\boldsymbol{\nabla} F$ or $\mathbf{grad}\, F$. It is such that the scalar product with $d\boldsymbol{x}$ gives the difference between the values of $F$ evaluated at $\boldsymbol{x}+d\boldsymbol{x}$ and at $\boldsymbol{x}$. We obtain

$$dF = F(\boldsymbol{x}+d\boldsymbol{x}) - F(\boldsymbol{x}) = \boldsymbol{\nabla} F\cdot d\boldsymbol{x}\,. \tag{1.157}$$

If we denote by $dx$ the norm of $d\boldsymbol{x}$, and by $\boldsymbol{e}$ the unit vector in the direction of $d\boldsymbol{x}$ ($\boldsymbol{e} = d\boldsymbol{x}/dx$), then equation (1.157) yields

$$\left(\frac{dF}{dx}\right)_{\text{in direction }\boldsymbol{e}} = \boldsymbol{\nabla} F\cdot\boldsymbol{e}\,. \tag{1.158}$$

This last relation shows that the component of $\boldsymbol{\nabla} F$ in the direction $\boldsymbol{e}$ gives the variation of $F$ in that direction (directional derivative). Since we have

$$\left(\frac{dF}{dx}\right)_{\text{in direction }\boldsymbol{e}_1} = \frac{\partial F}{\partial x_1} = \boldsymbol{\nabla} F\cdot\boldsymbol{e}_1 = (\boldsymbol{\nabla} F)_1\ ,$$

with similar relations in directions 2 and 3, the Cartesian components of $\boldsymbol{\nabla} F$ are $\partial F/\partial x_i$, or

$$\boldsymbol{\nabla} F = \frac{\partial F}{\partial x_1}\boldsymbol{e}_1 + \frac{\partial F}{\partial x_2}\boldsymbol{e}_2 + \frac{\partial F}{\partial x_3}\boldsymbol{e}_3 = \frac{\partial F}{\partial x_i}\boldsymbol{e}_i\,. \tag{1.159}$$

We can define the gradient operator as

$$\boldsymbol{\nabla}(\bullet) = \frac{\partial(\bullet)}{\partial x_1}\boldsymbol{e}_1 + \frac{\partial(\bullet)}{\partial x_2}\boldsymbol{e}_2 + \frac{\partial(\bullet)}{\partial x_3}\boldsymbol{e}_3 = \frac{\partial(\bullet)}{\partial x_i}\boldsymbol{e}_i\ , \tag{1.160}$$

where $(\bullet)$ indicates an arbitrary function.

It is easy to show that the gradient of a scalar field is a vector. From relation (1.18), we have

$$\begin{aligned}\frac{\partial F(\boldsymbol{x})}{\partial x'_i} &= \frac{\partial}{\partial x'_i} f'(x'_j) = \frac{\partial}{\partial x'_i} f\big(x_m(x'_j)\big) \\ &= \frac{\partial}{\partial x_k} f(x_m) \frac{\partial x_k}{\partial x'_i} = \frac{\partial x_k}{\partial x'_i} \frac{\partial}{\partial x_k} F(\boldsymbol{x}) \,.\end{aligned}$$

This equation is a transformation of the form (1.22).

### 1.4.3 Gradient of a Vector Field

With a vector field $\boldsymbol{v}(\boldsymbol{x})$, we associate a tensor, called the gradient of $\boldsymbol{v}$, and denote it $\boldsymbol{\nabla v}$. It is a tensor of order 2 which, applied to $d\boldsymbol{x}$, gives the difference of $\boldsymbol{v}$ between $\boldsymbol{x} + d\boldsymbol{x}$ and $\boldsymbol{x}$. We have

$$d\boldsymbol{v} = \boldsymbol{v}(\boldsymbol{x} + d\boldsymbol{x}) - \boldsymbol{v}(\boldsymbol{x}) = (\boldsymbol{\nabla v})\, d\boldsymbol{x} \,. \tag{1.161}$$

Again, let us define $dx = \|d\boldsymbol{x}\|$ and $\boldsymbol{e} = d\boldsymbol{x}/dx$. We obtain

$$\left(\frac{d\boldsymbol{v}}{dx}\right)_{\text{in direction } \boldsymbol{e}} = (\boldsymbol{\nabla v})\, \boldsymbol{e} \,. \tag{1.162}$$

The tensor $(\boldsymbol{\nabla v})$ of order 2 (it is left to the reader to prove that it is a tensor of order 2) transforms a unit vector $\boldsymbol{e}$ to a vector describing the variation of $\boldsymbol{v}$ in that direction. Since

$$\left(\frac{d\boldsymbol{v}}{dx}\right)_{\text{in direction } \boldsymbol{e}_1} = \frac{\partial \boldsymbol{v}}{\partial x_1} = (\boldsymbol{\nabla v})\, \boldsymbol{e}_1 \,,$$

in a Cartesian coordinate system we have

$$(\boldsymbol{\nabla v})_{11} = \boldsymbol{e}_1 \cdot (\boldsymbol{\nabla v})\, \boldsymbol{e}_1 = \boldsymbol{e}_1 \cdot \frac{\partial \boldsymbol{v}}{\partial x_1} = \frac{\partial}{\partial x_1} (\boldsymbol{e}_1 \cdot \boldsymbol{v}) = \frac{\partial v_1}{\partial x_1} \,.$$

Similar reasoning leads us to write

$$(\boldsymbol{\nabla v})_{ij} = \frac{\partial v_i}{\partial x_j} \,. \tag{1.163}$$

The tensor $\boldsymbol{\nabla v}$ can be expressed as (see (1.49) and (1.51))

$$\boldsymbol{\nabla v} = \boldsymbol{\nabla} \otimes \boldsymbol{v} = \frac{\partial v_i}{\partial x_j} (\boldsymbol{e}_i \otimes \boldsymbol{e}_j) \,. \tag{1.164}$$

### 1.4.4 Gradient of a Scalar Valued Tensor Function

For a regular, scalar valued, smooth function $\mathcal{W}(\boldsymbol{T})$ of a tensor $\boldsymbol{T}$ of order 2, the first two terms of a Taylor series expansion around $\boldsymbol{T}$ are

$$\mathcal{W}(\boldsymbol{T}+d\boldsymbol{T}) = \mathcal{W}(\boldsymbol{T}) + d\mathcal{W}(\boldsymbol{T}) + \mathrm{o}(d\boldsymbol{T})\,, \tag{1.165}$$

where $\mathrm{o}(d\boldsymbol{T})$ is the remainder of the expansion which tends to zero as $d\boldsymbol{T} \to 0$, as expressed in the relation

$$\lim_{d\boldsymbol{T}\to 0} \frac{\mathrm{o}(d\boldsymbol{T})}{\|d\boldsymbol{T}\|} = 0\,. \tag{1.166}$$

The total differential is expressed as follows:

$$d\mathcal{W}(\boldsymbol{T}) = \frac{\partial\mathcal{W}(\boldsymbol{T})}{\partial\boldsymbol{T}} : d\boldsymbol{T} = \mathrm{tr}\left(\left(\frac{\partial\mathcal{W}(\boldsymbol{T})}{\partial\boldsymbol{T}}\right)^T d\boldsymbol{T}\right). \tag{1.167}$$

In index notation,

$$d\mathcal{W}(\boldsymbol{T}) = \frac{\partial\mathcal{W}(\boldsymbol{T})}{\partial T_{ij}} dT_{ij}. \tag{1.168}$$

The tensor $\partial\mathcal{W}(\boldsymbol{T})/\partial\boldsymbol{T}$ of order 2 is called the gradient of $\mathcal{W}(\boldsymbol{T})$ in $\boldsymbol{T}$.

### 1.4.5 Gradient of a Tensor Valued Tensor Function

For a regular, tensor valued, smooth function $\boldsymbol{S}(\boldsymbol{T})$ of a tensor $\boldsymbol{T}$ of order 2, the first two terms of a Taylor expansion around $\boldsymbol{T}$ are

$$\boldsymbol{S}(\boldsymbol{T}+d\boldsymbol{T}) = \boldsymbol{S}(\boldsymbol{T}) + d\boldsymbol{S}(\boldsymbol{T}) + o(d\boldsymbol{T})\,. \tag{1.169}$$

When $d\boldsymbol{T} \to 0$, we have

$$d\boldsymbol{S}(\boldsymbol{T}) = \frac{\partial\boldsymbol{S}(\boldsymbol{T})}{\partial\boldsymbol{T}} : d\boldsymbol{T}\,. \tag{1.170}$$

In index notation

$$dS_{ij} = \frac{\partial S_{ij}}{\partial T_{kl}} dT_{kl}. \tag{1.171}$$

The tensor $\partial\boldsymbol{S}(\boldsymbol{T})/\partial\boldsymbol{T}$ of order 4 is the gradient of $\boldsymbol{S}(\boldsymbol{T})$ in $\boldsymbol{T}$.

### 1.4.6 Divergence of Vectors and Tensors

Let $\boldsymbol{v}(\boldsymbol{x})$ be a vector field. The divergence of $\boldsymbol{v}(\boldsymbol{x})$ is the scalar obtained by a contraction. Thus

$$\mathrm{div}\,\boldsymbol{v} = \frac{\partial v_i}{\partial x_i} = \mathrm{tr}\,(\boldsymbol{\nabla}\boldsymbol{v})\,. \tag{1.172}$$

Note that when the divergence of a vector field $\boldsymbol{v}$ is zero, that is, $\mathrm{div}\,\boldsymbol{v} = 0$, the field $\boldsymbol{v}$ is called a **solenoidal** field.

We can also express the divergence of the field $\boldsymbol{v}$ as the following scalar product:

$$\boldsymbol{\nabla}\cdot\boldsymbol{v} = \frac{\partial v_i}{\partial x_i}\,. \tag{1.173}$$

For a tensor, the divergence of $\boldsymbol{L}$ is the vector field denoted $\mathbf{div}\,\boldsymbol{L}$, defined by

$$(\mathbf{div}\,\boldsymbol{L})_i = \frac{\partial L_{ij}}{\partial x_j} = L_{ij,j} \tag{1.174}$$

or

$$\begin{aligned}\mathbf{div}\,\boldsymbol{L} &= \frac{\partial L_{ik}}{\partial x_j}\left(\boldsymbol{e}_i\otimes\boldsymbol{e}_k\right)\boldsymbol{e}_j = \frac{\partial L_{ik}}{\partial x_j}\left(\boldsymbol{e}_k\cdot\boldsymbol{e}_j\right)\boldsymbol{e}_i\\ &= \frac{\partial L_{ik}}{\partial x_j}\delta_{kj}\boldsymbol{e}_i = \frac{\partial L_{ij}}{\partial x_j}\,\boldsymbol{e}_i\,.\end{aligned} \tag{1.175}$$

### 1.4.7 Curl of a Vector Field

Let $\boldsymbol{v}(\boldsymbol{x})$ be a vector field. The curl of $\boldsymbol{v}$ is a vector field defined by the equation

$$\mathbf{curl}\,\boldsymbol{v} = \boldsymbol{\nabla}\times\boldsymbol{v}\,. \tag{1.176}$$

In index notation, we have

$$(\mathbf{curl}\,\boldsymbol{v})_i = \varepsilon_{ijk}\frac{\partial v_k}{\partial x_j}\,. \tag{1.177}$$

With the properties of the permutation symbol, we can easily verify that

$$\begin{aligned}(\mathbf{curl}\,\boldsymbol{v})_1 &= \frac{\partial v_3}{\partial x_2} - \frac{\partial v_2}{\partial x_3}\\ (\mathbf{curl}\,\boldsymbol{v})_2 &= \frac{\partial v_1}{\partial x_3} - \frac{\partial v_3}{\partial x_1}\\ (\mathbf{curl}\,\boldsymbol{v})_3 &= \frac{\partial v_2}{\partial x_1} - \frac{\partial v_1}{\partial x_2}\,.\end{aligned} \tag{1.178}$$

If the curl of the field $\boldsymbol{v}$ is zero, that is, $\boldsymbol{\nabla}\times\boldsymbol{v} = 0$, the field is called **irrotational**.

In the following examples $\Phi$ and $\boldsymbol{a}$ are continuously differentiable scalar and vector functions, respectively.

EXAMPLE 1.10
Prove that

$$\mathbf{curl}\,(\nabla\,\Phi) = \boldsymbol{\nabla} \times \nabla\Phi = \boldsymbol{\nabla} \times (\nabla\Phi) = 0\ . \tag{1.179}$$

We have successively

$$(\boldsymbol{\nabla} \times \nabla\Phi)_i = \varepsilon_{ijk}\frac{\partial}{\partial x_j}(\frac{\partial\Phi}{\partial x_k}) = \varepsilon_{ijk}\frac{\partial^2\Phi}{\partial x_j \partial x_k} = \varepsilon_{ikj}\frac{\partial^2\Phi}{\partial x_k \partial x_j} = -\varepsilon_{ijk}\frac{\partial^2\Phi}{\partial x_j \partial x_k}.$$

Thus, $2\varepsilon_{ijk}\dfrac{\partial^2\Phi}{\partial x_j \partial x_k} = 0$ .

EXAMPLE 1.11
Demonstrate that

$$\mathrm{div}(\boldsymbol{\nabla} \times \boldsymbol{a}) = \mathrm{div}\,\mathbf{curl}\,\boldsymbol{a} = 0\ . \tag{1.180}$$

Taking into account (1.177), we obtain

$$\mathrm{div}(\boldsymbol{\nabla} \times \boldsymbol{a}) = \frac{\partial}{\partial x_i}\left(\varepsilon_{ijk}\frac{\partial a_k}{\partial x_j}\right) = \varepsilon_{ijk}\frac{\partial^2 a_k}{\partial x_i \partial x_j}\ .$$

By following the same steps as in the previous example, we show that (1.180) is satisfied.

EXAMPLE 1.12
Show that

$$\mathrm{div}(\Phi\boldsymbol{a}) = \Phi\ \mathrm{div}\,\boldsymbol{a} + \boldsymbol{a}\cdot(\nabla\Phi)\ . \tag{1.181}$$

Taking into account (1.172), we have

$$\mathrm{div}(\Phi\boldsymbol{a}) = \frac{\partial(\Phi a_i)}{\partial x_i} = \frac{\partial\Phi}{\partial x_i}a_i + \Phi\frac{\partial a_i}{\partial x_i} = (\nabla\Phi)\cdot\boldsymbol{a} + \Phi\ \mathrm{div}\,\boldsymbol{a}\ .$$

### 1.4.8 Laplacian Operator

**Laplacian of a Scalar Field**

We also encounter second order derivatives in expressions of physical quantities in continuum mechanics. For example, the divergence of the gradient of a scalar function is

$$\frac{\partial^2 F}{\partial x_i \partial x_i} \quad \text{or} \quad \nabla\cdot(\boldsymbol{\nabla} F) \text{ or } \mathrm{div}\,(\mathbf{grad}\,F)\ , \tag{1.182}$$

which is also the Laplacian of $F$, symbolically denoted $\nabla^2 F$, or sometimes as $\triangle F$,

$$\frac{\partial^2 F}{\partial x_i \partial x_i} = \frac{\partial^2 F}{\partial {x_1}^2} + \frac{\partial^2 F}{\partial {x_2}^2} + \frac{\partial^2 F}{\partial {x_3}^2}\ . \tag{1.183}$$

A function $F$ that satisfies the equation

$$\nabla^2 F = 0 \tag{1.184}$$

is said to be harmonic. Equation (1.184) is known as Laplace's equation.

If the relation reads

$$\nabla^2 F = f \,, \tag{1.185}$$

where $f$ is a scalar function, then it is called Poisson's equation.

**Laplacian of a Vector Field**

We can also treat a vector function in the same way. The divergence of the gradient of a vector is written as

$$\frac{\partial^2 v_j}{\partial x_i \partial x_i} \qquad \text{or} \qquad \boldsymbol{\nabla} \cdot (\boldsymbol{\nabla} \boldsymbol{v}) \text{ or } \mathbf{div}\,(\boldsymbol{\nabla} \boldsymbol{v}) \,. \tag{1.186}$$

The result of these operations is a vector. We also denote the operation as $\nabla^2$, that is,

$$\boldsymbol{\nabla} \cdot (\boldsymbol{\nabla} \boldsymbol{v}) = \nabla^2 \boldsymbol{v} \,. \tag{1.187}$$

The symbol $\nabla^2 \boldsymbol{v}$ does not pose a problem in rectangular coordinates since $\nabla^2 \boldsymbol{v}$ has three components so that

$$\left(\nabla^2 \boldsymbol{v}\right)_i = \nabla^2 v_i \,.$$

But, a difficulty appears in curvilinear coordinates where

$$\left(\nabla^2 \boldsymbol{v}\right)_i \neq \nabla^2 (v_i) \,.$$

EXAMPLE 1.13

Prove that

$$\text{div}(\boldsymbol{\nabla} \Phi) = \nabla^2 \Phi. \tag{1.188}$$

We have, successively

$$\text{div}(\boldsymbol{\nabla} \Phi) = \frac{\partial}{\partial x_i} \left( \frac{\partial \Phi}{\partial x_i} \right) = \frac{\partial^2 \Phi}{\partial x_i \partial x_i} = \nabla^2 \Phi \,. \tag{1.189}$$

EXAMPLE 1.14

Show that

$$\boldsymbol{\nabla} \times (\boldsymbol{\nabla} \times \boldsymbol{a}) = \nabla(\text{div}\, \boldsymbol{a}) - \nabla^2 \boldsymbol{a} \,. \tag{1.190}$$

In index notation, we have

$$(\boldsymbol{\nabla} \times (\boldsymbol{\nabla} \times \boldsymbol{a}))_l = \varepsilon_{lmi} \frac{\partial}{\partial x_m} \left( \varepsilon_{ijk} \frac{\partial a_k}{\partial x_j} \right) = \varepsilon_{lmi} \varepsilon_{ijk} \frac{\partial^2 a_k}{\partial x_m \partial x_j} \,.$$

Using (1.30) and the properties of the permutation symbol, we have

$$\begin{aligned}
\varepsilon_{lmi}\varepsilon_{ijk}\frac{\partial^2 a_k}{\partial x_m \partial x_j} &= (\delta_{lj}\delta_{mk} - \delta_{lk}\delta_{mj})\frac{\partial^2 a_k}{\partial x_m \partial x_j} \\
&= \delta_{lj}\delta_{mk}\frac{\partial^2 a_k}{\partial x_m \partial x_j} - \delta_{lk}\delta_{mj}\frac{\partial^2 a_k}{\partial x_m \partial x_j} \\
&= \frac{\partial^2 a_m}{\partial x_m \partial x_l} - \frac{\partial^2 a_l}{\partial x_j \partial x_j} \\
&= \frac{\partial}{\partial x_l}\left(\frac{\partial a_m}{\partial x_m}\right) - \frac{\partial^2 a_l}{\partial x_j \partial x_j} \\
&= \frac{\partial}{\partial x_l}(\operatorname{div}\boldsymbol{a}) - \nabla^2 a_l = \left(\nabla(\operatorname{div}\boldsymbol{a}) - \nabla^2\boldsymbol{a}\right)_l ,
\end{aligned}$$

and thus

$$(\boldsymbol{\nabla}\times(\boldsymbol{\nabla}\times\boldsymbol{a}))_l = \left(\nabla(\operatorname{div}\boldsymbol{a}) - \nabla^2\boldsymbol{a}\right)_l ,$$

which is valid for the three components of the vector $\boldsymbol{a}$.

EXAMPLE 1.15
Demonstrate that

$$\operatorname{div}\left(\nabla^2\boldsymbol{a}\right) = \nabla^2\left(\operatorname{div}\boldsymbol{a}\right) . \qquad (1.191)$$

In index notation, we have

$$\left(\nabla^2\boldsymbol{a}\right)_i = \frac{\partial^2 a_i}{\partial x_j \partial x_j} .$$

Thus

$$\left(\operatorname{div}\left(\nabla^2\boldsymbol{a}\right)_i\right) = \frac{\partial}{\partial x_i}\left(\frac{\partial^2 a_i}{\partial x_j \partial x_j}\right) = \frac{\partial^2}{\partial x_j \partial x_j}\left(\frac{\partial a_i}{\partial x_i}\right) = \nabla^2\left(\operatorname{div}\boldsymbol{a}\right) .$$

### 1.4.9 Curvilinear Coordinates

Let $x_i$ be the Cartesian coordinates of a point P. In a curvilinear coordinate system $\theta_i$ (for example, cylindrical or spherical coordinates), the position of the point P is given by the three numbers $\theta_i$ which represent the coordinate curves that pass through P (fig. 1.5), that is, by the curves on which two of the three coordinates $\theta_i$ are constant. The curvilinear coordinates can be considered as functions of Cartesian coordinates

$$\theta_i = \theta_i(x_j) , \qquad (1.192)$$

and if the condition that the Jacobian $J = \det(\partial\theta_i/\partial x_j)$, not be zero is met, then the transformation (1.192) is invertible.

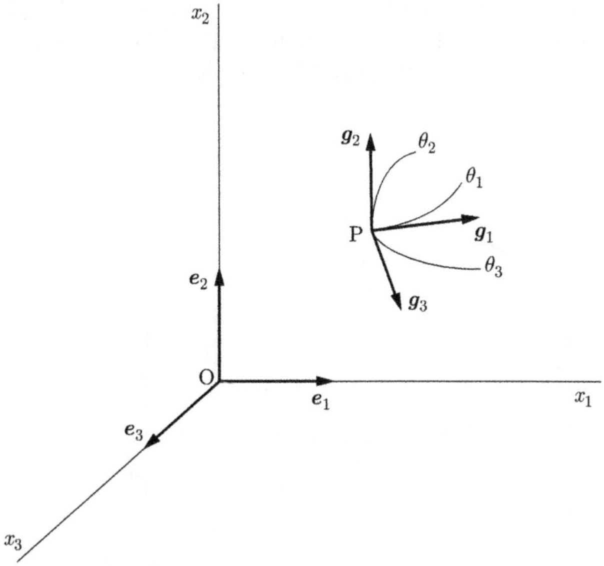


**Fig. 1.5** Curvilinear coordinate system

The set of points for which the curvilinear coordinate $\theta_i$ is a constant, represents a surface given by the equation

$$\theta_i(x_j) = \text{cnst}\,. \tag{1.193}$$

Consider the cylindrical and spherical coordinate systems. Referring to figures 1.6 and 1.7, we go, respectively, from Cartesian to cylindrical or spherical coordinates with the relations

$$\begin{aligned}\theta_1 &= r = \sqrt{x_1^2 + x_2^2}\\ \theta_2 &= \theta = \tan^{-1}\frac{x_2}{x_1}\\ \theta_3 &= z = x_3\end{aligned} \tag{1.194}$$

and

$$\begin{aligned}\theta_1 &= r = \sqrt{x_1^2 + x_2^2 + x_3^2}\\ \theta_2 &= \theta = \tan^{-1}\frac{\sqrt{x_1^2 + x_2^2}}{x_3}\\ \theta_3 &= \varphi = \tan^{-1}\frac{x_2}{x_1}\,.\end{aligned} \tag{1.195}$$

As seen in figure 1.7, the angle $\varphi$ is the longitude and $\theta$ is the colatitude (latitude with 0 at the North Pole) or polar angle. The $(r, \theta, \varphi)$ coordinates are commonly used in physics as suggested by an ISO convention. Mathematicians swap the two angles.

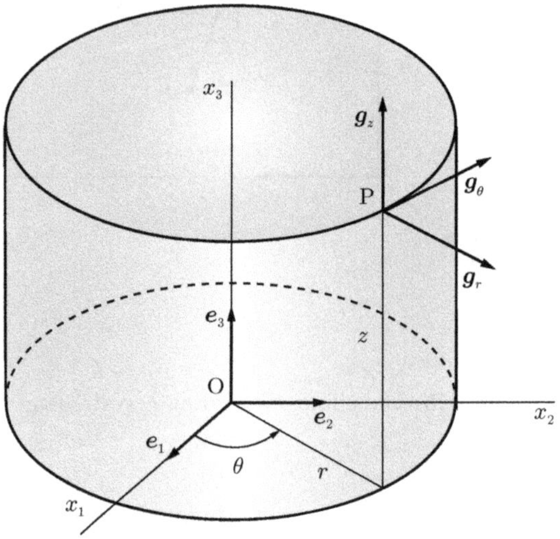


**Fig. 1.6** Cylindrical coordinate system

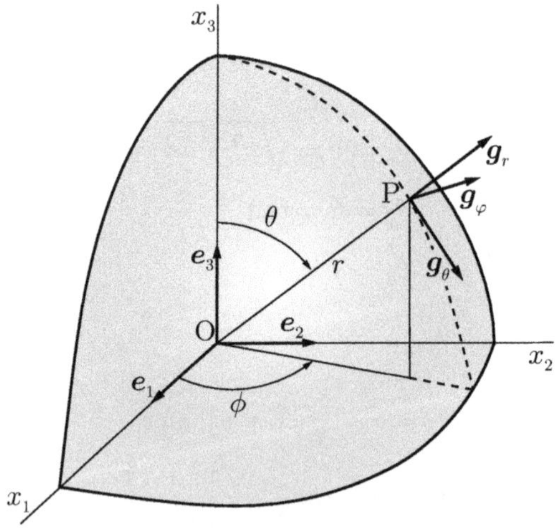


**Fig. 1.7** Spherical coordinate system

The inverse relations are easily obtained.

*Cylindrical coordinates:*

$$
\begin{aligned}
x_1 &= r\cos\theta \\
x_2 &= r\sin\theta \\
x_3 &= z\,.
\end{aligned}
$$

*Spherical coordinates:*

$$
\begin{aligned}
x_1 &= r\sin\theta\cos\varphi \\
x_2 &= r\sin\theta\sin\varphi \\
x_3 &= r\cos\theta\,.
\end{aligned}
$$

For later calculations of volume integrals, the element $dV$ is written in the three coordinate systems:

- Cartesian $dV = dx\,dy\,dz$,
- cylindrical $dV = r\,dr\,d\theta\,dz$,
- spherical $dV = r^2\sin\theta\,dr\,d\theta\,d\varphi$.

The position vector of a point P in Cartesian coordinates is written as

$$
\boldsymbol{OP} = \boldsymbol{r} = x_i\boldsymbol{e}_i\,. \tag{1.196}
$$

In curvilinear coordinates, the basis vectors at P are the three unit vectors tangent to the coordinate lines passing through P. We define them with the relations

$$
\boldsymbol{g}_i = \frac{\dfrac{\partial\boldsymbol{r}}{\partial\theta_i}}{\left\|\dfrac{\partial\boldsymbol{r}}{\partial\theta_i}\right\|}\,. \tag{1.197}
$$

These vectors $\boldsymbol{g}_i$ are shown in the two figures 1.6 and 1.7 for the two contexts.

In cylindrical coordinates, we write

$$
\begin{aligned}
&\boldsymbol{r} = r\cos\theta\boldsymbol{e}_1 + r\sin\theta\boldsymbol{e}_2 + z\boldsymbol{e}_3 \\
&\frac{\partial\boldsymbol{r}}{\partial r} = \cos\theta\boldsymbol{e}_1 + \sin\theta\boldsymbol{e}_2 \qquad \left\|\frac{\partial\boldsymbol{r}}{\partial r}\right\| = 1 \qquad \boldsymbol{g}_r = \frac{\partial\boldsymbol{r}}{\partial r} \\
&\frac{\partial\boldsymbol{r}}{\partial\theta} = -r\sin\theta\boldsymbol{e}_1 + r\cos\theta\boldsymbol{e}_2 \qquad \left\|\frac{\partial\boldsymbol{r}}{\partial\theta}\right\| = r \qquad \boldsymbol{g}_\theta = \frac{1}{r}\frac{\partial\boldsymbol{r}}{\partial\theta} \\
&\frac{\partial\boldsymbol{r}}{\partial z} = \boldsymbol{e}_3 \qquad \boldsymbol{g}_z = \boldsymbol{e}_3\,.
\end{aligned} \tag{1.198}
$$

With the same reasoning, we obtain for spherical coordinates

$$
\boldsymbol{g}_r = \frac{\partial\boldsymbol{r}}{\partial r} \qquad \boldsymbol{g}_\theta = \frac{1}{r}\frac{\partial\boldsymbol{r}}{\partial\theta} \qquad \boldsymbol{g}_\varphi = \frac{1}{r\sin\theta}\frac{\partial\boldsymbol{r}}{\partial\varphi}\,. \tag{1.199}
$$

Note that the basis vectors in cylindrical and spherical coordinates are orthogonal and the corresponding curvilinear coordinates are orthogonal.

The direction cosines of the basis vectors $\boldsymbol{g}_i$ with respect to the basis $\boldsymbol{e}_j$ are obtained using (1.6) as follows:

$$c_{ij} = \boldsymbol{g}_i \cdot \boldsymbol{e}_j \,. \tag{1.200}$$

In cylindrical coordinates, by (1.198) and (1.200), we have

$$[C] = \begin{pmatrix} \cos\theta & \sin\theta & 0 \\ -\sin\theta & \cos\theta & 0 \\ 0 & 0 & 1 \end{pmatrix} . \tag{1.201}$$

Note that this matrix is orthogonal.

### 1.4.10 Scalars, Vectors, and Tensors in Orthogonal Curvilinear Coordinates

The value of a scalar field is given in either Cartesian or curvilinear coordinates by the transformation

$$F(\mathrm{P}) = f(x_i) = f(x_i(\theta_j)) = f'(\theta_j) \,. \tag{1.202}$$

Let $\boldsymbol{v}$ be a vector at a point P that has coordinates $x_i, \theta_i$. The ***physical components*** $v'_i$ of the vector $\boldsymbol{v}$ in curvilinear coordinates at P are its components following the basis vectors $\boldsymbol{g}_i$, that is,

$$\boldsymbol{v} = v_i \boldsymbol{e}_i = v'_i \boldsymbol{g}_i \,. \tag{1.203}$$

Multiplying by $g_i$ we obtain

$$v'_i = \boldsymbol{g}_i \cdot \boldsymbol{e}_j v_j = c_{ij} v_j \,. \tag{1.204}$$

If a tensor $\boldsymbol{L}$ applied to a vector $\boldsymbol{u}$ produces the vector $\boldsymbol{v}$, $\boldsymbol{v} = \boldsymbol{L}\boldsymbol{u}$, the physical components of $\boldsymbol{L}$ in curvilinear coordinates at P are such that

$$v'_i = L'_{ij} u'_j \,. \tag{1.205}$$

We can easily verify that

$$L'_{ij} = c_{ik} c_{jl} L_{kl} \,. \tag{1.206}$$

### 1.4.11 Gradient of Scalar and Vector Fields in Orthogonal Curvilinear Coordinates

The gradient of the property $f$ given by (1.159) is a vector $\boldsymbol{h}$ with components $\partial f / \partial x_j$ in Cartesian coordinates. In curvilinear coordinates, we have for the physical components of the vector

$$h'_i = c_{ij} h_j = c_{ij} \frac{\partial f'}{\partial \theta_k} \frac{\partial \theta_k}{\partial x_j} \,. \tag{1.207}$$

In cylindrical coordinates, with relations (1.194), we have

$$\begin{aligned}\frac{\partial r}{\partial x_1} &= \frac{x_1}{r} = \cos\theta & \frac{\partial r}{\partial x_2} &= \frac{x_2}{r} = \sin\theta \\ \frac{\partial \theta}{\partial x_1} &= -\frac{x_2}{r^2} = -\frac{\sin\theta}{r} & \frac{\partial \theta}{\partial x_2} &= \frac{x_1}{r^2} = \frac{\cos\theta}{r}\,.\end{aligned} \tag{1.208}$$

Combining equations (1.207) and (1.208), we easily obtain

$$h'_r = \frac{\partial f'}{\partial r} \qquad h'_\theta = \frac{1}{r}\frac{\partial f'}{\partial \theta} \qquad h'_z = \frac{\partial f'}{\partial z}\,. \tag{1.209}$$

If we consider a scalar function $f$ defined in the two coordinate systems, Cartesian and cylindrical, since $x_i$ and $\theta_i$ are related by (1.192), we have

$$\begin{aligned}\frac{\partial}{\partial x_i} &= \frac{\partial \theta_j}{\partial x_i}\frac{\partial}{\partial \theta_j} \\ \frac{\partial}{\partial x_m}\left(\frac{\partial}{\partial x_i}\right) &= \frac{\partial^2}{\partial x_m \partial x_i} = \frac{\partial}{\partial x_m}\left(\frac{\partial \theta_j}{\partial x_i}\frac{\partial}{\partial \theta_j}\right).\end{aligned}$$

The last relation becomes

$$\frac{\partial^2}{\partial x_m \partial x_i} = \frac{\partial^2 \theta_j}{\partial x_m \partial x_i}\frac{\partial}{\partial \theta_j} + \frac{\partial \theta_k}{\partial x_m}\frac{\partial \theta_j}{\partial x_i}\frac{\partial^2}{\partial \theta_k \partial \theta_j}\,. \tag{1.210}$$

We can now calculate the first derivatives

$$\frac{\partial}{\partial x_1} = \cos\theta\frac{\partial}{\partial r} - \frac{\sin\theta}{r}\frac{\partial}{\partial \theta} \tag{1.211}$$

$$\frac{\partial}{\partial x_2} = \sin\theta\frac{\partial}{\partial r} + \frac{\cos\theta}{r}\frac{\partial}{\partial \theta} \tag{1.212}$$

$$\frac{\partial}{\partial x_3} = \frac{\partial}{\partial z} \tag{1.213}$$

and the second derivatives

$$\begin{aligned}\frac{\partial^2}{\partial x_1^2} = \cos^2\theta\frac{\partial^2}{\partial r^2} &- \frac{\sin 2\theta}{r}\frac{\partial^2}{\partial r \partial \theta} + \frac{\sin 2\theta}{r^2}\frac{\partial}{\partial \theta} \\ &+ \frac{\sin^2\theta}{r}\frac{\partial}{\partial r} + \frac{\sin^2\theta}{r^2}\frac{\partial^2}{\partial \theta^2}\end{aligned} \tag{1.214}$$

$$\begin{aligned}\frac{\partial^2}{\partial x_2^2} = \sin^2\theta\frac{\partial^2}{\partial r^2} &+ \frac{\sin 2\theta}{r}\frac{\partial^2}{\partial r \partial \theta} - \frac{\sin 2\theta}{r^2}\frac{\partial}{\partial \theta} \\ &+ \frac{\cos^2\theta}{r}\frac{\partial}{\partial r} + \frac{\cos^2\theta}{r^2}\frac{\partial^2}{\partial \theta^2}\end{aligned} \tag{1.215}$$

$$\frac{\partial^2}{\partial x_3^2} = \frac{\partial^2}{\partial z^2}\,. \tag{1.216}$$

Combining relations (1.214)–(1.216) and taking (1.201) into account, we obtain for the Laplacian operator

$$\nabla^2 f = \frac{1}{r}\frac{\partial f}{\partial r} + \frac{\partial^2 f}{\partial r^2} + \frac{1}{r^2}\frac{\partial^2 f}{\partial \theta^2} + \frac{\partial^2 f}{\partial z^2}\,. \tag{1.217}$$

To express the gradient of a vector in cylindrical and spherical coordinates, we must define the metric tensors of these coordinate systems and resort to an analysis beyond the scope of this book. The reader can find complementary reading in [2, 8, 47]. As an example, for the gradient of a vector, we will have

$$\boldsymbol{L} = \boldsymbol{\nabla v}$$

$$\boldsymbol{L} = \begin{pmatrix} \dfrac{\partial v_r}{\partial r} & \dfrac{1}{r}\dfrac{\partial v_r}{\partial \theta} - \dfrac{v_\theta}{r} & \dfrac{\partial v_r}{\partial z} \\ \dfrac{\partial v_\theta}{\partial r} & \dfrac{1}{r}\dfrac{\partial v_\theta}{\partial \theta} + \dfrac{v_r}{r} & \dfrac{\partial v_\theta}{\partial z} \\ \dfrac{\partial v_z}{\partial r} & \dfrac{1}{r}\dfrac{\partial v_z}{\partial \theta} & \dfrac{\partial v_z}{\partial z} \end{pmatrix} . \tag{1.218}$$

Other operations, including the divergence of a vector, the components of the curl, and the Laplacian of a vector for cylindrical coordinates, are given in appendix A.

All the corresponding expressions for spherical coordinates are presented in appendix B.

### 1.4.12 Definition of the Notion of Flux

Consider in figure 1.8 a body in a three-dimensional space and a surface element of area $ds$. The volume of material that passes through $ds$ in the time interval $dt$ is given by

$$v_i n_i \, dt \, ds \qquad \text{or} \qquad \boldsymbol{v} \cdot \boldsymbol{n} \, dt \, ds \, , \tag{1.219}$$

where $v_i n_i \, dt$ corresponds to the length of the cylinder of material that has passed through the surface element during the interval $dt$. Then, adopting the notation $\partial\omega$ for the surface that encloses the volume $\omega$, we have the following definitions:

- the flux of a vector

$$\int_{\partial\omega} v_i n_i \, ds \qquad \text{or} \qquad \int_{\partial\omega} \boldsymbol{v} \cdot \boldsymbol{n} \, ds \, ; \tag{1.220}$$

- the flux of an arbitrary scalar quantity $\rho$ through the surface of a body (scalar quantity)

$$\int_{\partial\omega} \rho v_i n_i \, ds \qquad \text{or} \qquad \int_{\partial\omega} \rho \boldsymbol{v} \cdot \boldsymbol{n} \, ds \, ; \tag{1.221}$$

- the flux of kinetic energy (scalar quantity)

$$\int_{\partial\omega} \frac{1}{2} \rho v_i v_i v_j n_j \, ds \qquad \text{or} \qquad \int_{\partial\omega} \frac{1}{2} \rho (\boldsymbol{v} \cdot \boldsymbol{v})(\boldsymbol{v} \cdot \boldsymbol{n}) \, ds \, ; \tag{1.222}$$

- the flux of a property $Q$

$$\int_{\partial\omega} Q v_i n_i \, ds \qquad \text{or} \qquad \int_{\partial\omega} Q (\boldsymbol{v} \cdot \boldsymbol{n}) \, ds \, . \tag{1.223}$$

In this case, $Q$ can be a scalar, a vector, or a tensor.

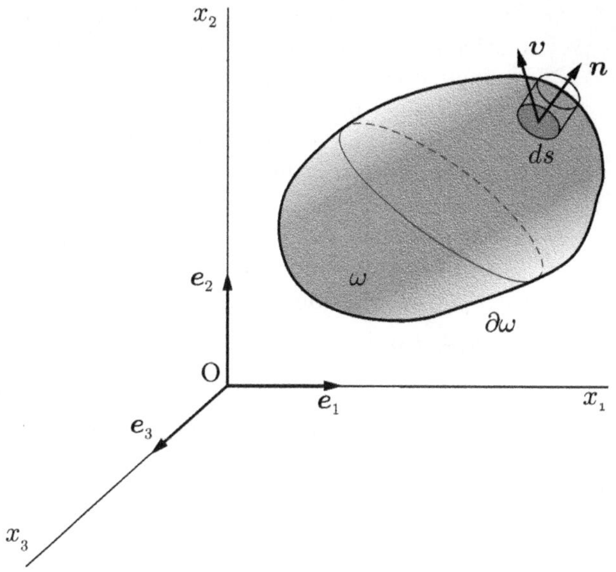


**Fig. 1.8** The notion of flux

### 1.4.13 Integral Theorems of Gauss and Stokes

The basic theory of analysis establishes the relation between the integral and the derivative of the integrand. If we have for the integrand $f = dF/dx$, then

$$\int_a^b f\,dx = \int_a^b \frac{dF}{dx}\,dx = F(b) - F(a)\,.$$

**Gauss' Theorem**

*The equivalent theorem for a volume integral is called Gauss' theorem or* ***divergence theorem****, which is written for an arbitrary component $T_{jk\cdots}(x_i)$ as*

$$\int_\omega \frac{\partial T_{jk\cdots}}{\partial x_i}\,dv = \int_{\partial\omega} n_i T_{jk\cdots}\,ds\,. \tag{1.224}$$

*Gauss' theorem is most often used in the form of the divergence theorem. This theorem transforms the volume integral of the divergence of a property of a continuous medium into a surface integral and plays an important role in the mechanics of continuous media [67].*

Carl Friedrich Gauss (1777–1855) was a talented mathematician, so much so that he was crowned the "prince of mathematics" at the age of 24 by experts from all over Europe. Named professor of mathematics at the University of Göttingen, he was responsible for major contributions to number theory, geodesy, geometry, statistics (including the least squares method), and physics. He took as a motto *Pauca sed matura* (few, but ripe) and declined to publish important manuscripts because they did not satisfy this criterion.

**Fig. 1.9** Carl Friedrich Gauss

In (1.224), $T_{jk\cdots}$ can be a scalar or a component of a vector or tensor of arbitrary rank, and the symbol $\boldsymbol{n}$ represents the unit normal vector to the element $ds$ (fig. 1.10).

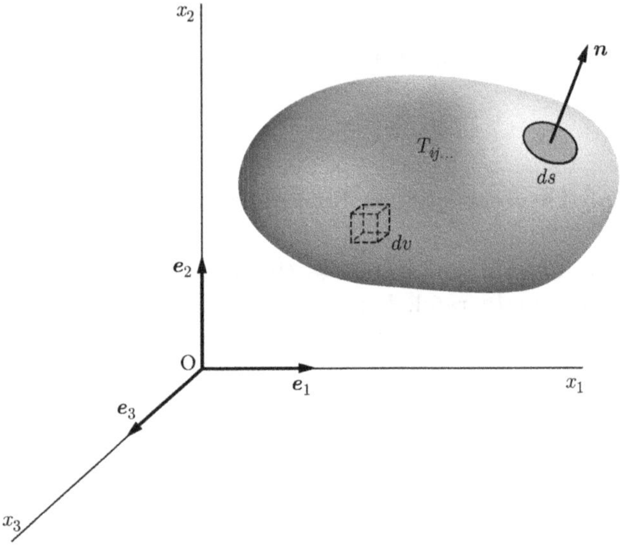


**Fig. 1.10** Elements of surface and volume used in Gauss' theorem

Take the case where $T_{jk\cdots} = F$ (scalar function). Then the statement of the theorem becomes

$$\int_\omega \frac{\partial F}{\partial x_i}\, dv = \int_\omega (\boldsymbol{\nabla} F)_i \, dv = \int_{\partial\omega} n_i F \, ds\,. \tag{1.225}$$

If $T_{jk\cdots}$ is a component of a vector function $v_i$, we have

$$\begin{aligned}\int_\omega \frac{\partial v_i}{\partial x_i}\, dv &= \int_\omega \operatorname{div} \boldsymbol{v}\, dv = \int_{\partial\omega} n_i v_i \, ds \\ &= \int_{\partial\omega} \boldsymbol{n}\cdot\boldsymbol{v}\, ds.\end{aligned} \tag{1.226}$$

The surface integral on the right-hand side expresses the flux of vector $\boldsymbol{v}$ passing through the surface $\partial\omega$. For a tensor quantity $\boldsymbol{L}$ the theorem is written as

$$\int_\omega \frac{\partial L_{ji}}{\partial x_i}\,dv = \int_{\partial\omega} L_{ji}\, n_i\, ds \quad \text{or} \quad \int_\omega \mathbf{div}\, \boldsymbol{L}\, dv = \int_{\partial\omega} \boldsymbol{L}\boldsymbol{n}\, ds\,. \tag{1.227}$$

For an arbitrary property $Q$, the theorem gives

$$\int_\omega \frac{\partial Q v_i}{\partial x_i}\,dv = \int_{\partial\omega} Q v_i\, n_i\, ds \text{ or} \int_\omega \mathrm{div}(\boldsymbol{v}Q)\, dv = \int_{\partial\omega} Q\boldsymbol{v}\cdot\boldsymbol{n}\, ds\,. \tag{1.228}$$

Another important theorem used very often in fluid mechanics is Stokes' theorem.

**Stokes' Theorem**

*Stokes' theorem equates the integral on an open surface to a contour integral along the curve $C$ that bounds the surface.*

$$\int_{\partial\omega} \mathbf{curl}\, \boldsymbol{v}\cdot\boldsymbol{n}\, ds = \oint_C \boldsymbol{v}\cdot d\boldsymbol{l}\,, \tag{1.229}$$

*where the vector $d\boldsymbol{l}$ is tangent to $C$ (see [67]).*

George Gabriel Stokes (1819–1903) was an Irish mathematician (born in Skreen in County Sligo). Named professor of mathematics at Cambridge University at the age of 30, he did much work in hydrodynamics, elasticity, and optics. "Stokes' law" for the motion of particles in a viscous fluid was developed by him. He precisely wrote the equations of viscous fluid dynamics now known as the Navier-Stokes equations.

**Fig. 1.11** George Gabriel Stokes

## 1.5 Exercises

**1.1** Prove that the Kronecker delta function is a tensor of order 2.

**1.2** Show that $\delta_{ij}\delta_{ik}\delta_{jk} = 3$.

**1.3** Prove that $\varepsilon_{ijk}u_iu_j = 0$ and that $\delta_{ij}\varepsilon_{ijk} = 0$.

**1.4** Demonstrate that $\boldsymbol{t}\times(\boldsymbol{u}\times\boldsymbol{v}) = (\boldsymbol{t}\cdot\boldsymbol{v})\boldsymbol{u} - (\boldsymbol{t}\cdot\boldsymbol{u})\boldsymbol{v}$ .

**1.5** Calculate the expression equivalent to

$$(\boldsymbol{a} \times \boldsymbol{b}) \times (\boldsymbol{c} \times \boldsymbol{d})$$

where all vectors $\boldsymbol{a}, \boldsymbol{b}, \boldsymbol{c}$, and $\boldsymbol{d}$ are non zero. Express the result in vector notation.

**1.6** Let $\boldsymbol{a}$ and $\boldsymbol{b}$ be two arbitrary vectors. Using the following notation,

$$[(\boldsymbol{a} \cdot \boldsymbol{\nabla})\boldsymbol{b}]_i = a_j \frac{\partial b_i}{\partial x_j} ,$$

demonstrate the following identities:

$$\begin{aligned}
\boldsymbol{\nabla} \cdot (\boldsymbol{a} \times \boldsymbol{b}) &= (\boldsymbol{\nabla} \times \boldsymbol{a}) \cdot \boldsymbol{b} - \boldsymbol{a} \cdot (\boldsymbol{\nabla} \times \boldsymbol{b}) && (1.230)\\
\boldsymbol{\nabla}(\boldsymbol{a} \cdot \boldsymbol{b}) &= (\boldsymbol{a} \cdot \boldsymbol{\nabla})\boldsymbol{b} + (\boldsymbol{b} \cdot \boldsymbol{\nabla})\boldsymbol{a} \\
&\quad + \boldsymbol{a} \times (\boldsymbol{\nabla} \times \boldsymbol{b}) + \boldsymbol{b} \times (\boldsymbol{\nabla} \times \boldsymbol{a}) && (1.231)\\
\mathbf{curl}(\boldsymbol{a} \times \boldsymbol{b}) &= (\boldsymbol{b} \cdot \boldsymbol{\nabla})\boldsymbol{a} - (\boldsymbol{a} \cdot \boldsymbol{\nabla})\boldsymbol{b} + \boldsymbol{a} \operatorname{div} \boldsymbol{b} - \boldsymbol{b} \operatorname{div} \boldsymbol{a} && (1.232)\\
\mathbf{div}(\boldsymbol{a} \otimes \boldsymbol{b}) &= (\boldsymbol{\nabla} \boldsymbol{a})\boldsymbol{b} + \boldsymbol{a} \operatorname{div} \boldsymbol{b} \, . && (1.233)
\end{aligned}$$

**1.7** Let $\boldsymbol{a}$ be an arbitrary vector function and $\Phi$ a continuously differentiable scalar function.

Prove the following identities:

$$\begin{aligned}
\mathbf{curl}(\Phi \boldsymbol{a}) &= \Phi \, \mathbf{curl}\boldsymbol{a} - \boldsymbol{a} \times \boldsymbol{\nabla} \Phi && (1.234)\\
\boldsymbol{\nabla}(\Phi \boldsymbol{a}) &= \Phi \boldsymbol{\nabla} \boldsymbol{a} + \boldsymbol{a} \otimes \boldsymbol{\nabla} \Phi && (1.235)\\
\nabla^2 (\boldsymbol{\nabla} \Phi) &= \boldsymbol{\nabla} \left(\nabla^2 \Phi\right) && (1.236)\\
\boldsymbol{\nabla} \times \left(\nabla^2 \boldsymbol{a}\right) &= \nabla^2 (\boldsymbol{\nabla} \times \boldsymbol{a}) && (1.237)\\
\Delta \boldsymbol{a} &= \boldsymbol{\nabla} \cdot \boldsymbol{\nabla} \boldsymbol{a} = \boldsymbol{\nabla}(\boldsymbol{\nabla} \cdot \boldsymbol{a}) - \mathbf{curl} \; \mathbf{curl} \; \boldsymbol{a} \, . && (1.238)
\end{aligned}$$

**1.8** Let $\boldsymbol{a}$ be an arbitrary vector function, $\Phi$ a continuously differentiable scalar function, and $\boldsymbol{x}$ the position vector. Show that

$$\begin{aligned}
\nabla (\boldsymbol{a} \cdot \boldsymbol{x}) &= \boldsymbol{a} + (\nabla \boldsymbol{a})^T \boldsymbol{x} && (1.239)\\
\nabla^2 (\boldsymbol{a} \cdot \boldsymbol{x}) &= 2 \operatorname{div} \boldsymbol{a} + \boldsymbol{x} \cdot \left(\nabla^2 \boldsymbol{a}\right) && (1.240)\\
\nabla^2 (\Phi \boldsymbol{x}) &= 2 \nabla \Phi + \boldsymbol{x} \nabla^2 \Phi \, . && (1.241)
\end{aligned}$$

**1.9** Verify that $\operatorname{tr}(\boldsymbol{L})$ is an invariant.

**1.10** Prove (1.69) and (1.71).

**1.11** Derive relations (1.121).

**1.12** Let $[A]$ be a matrix with constant coefficients. Verify the following relation:

$$\boldsymbol{\nabla}(A_{jk} \, x_j x_k) = (A_{ij} + A_{ji}) x_j \boldsymbol{e}_i \, . \qquad (1.242)$$

**1.13** Show that the quadratic form $D_{ij}\, x_i x_j$ remains unchanged if we replace $D_{ij}$ by its symmetric part.

**1.14** Show that for an orthogonal tensor $\boldsymbol{Q}$, the following conditions are satisfied:

$$\boldsymbol{Q}^T\boldsymbol{Q} = \boldsymbol{Q}\boldsymbol{Q}^T = \boldsymbol{I}, \quad \text{ou} \quad \boldsymbol{Q}^T = \boldsymbol{Q}^{-1} \tag{1.243}$$

$$\det\big(\boldsymbol{Q}^T\boldsymbol{Q}\big) = (\det\boldsymbol{Q})^2 \,. \tag{1.244}$$

**1.15** Find the invariants of an antisymmetric tensor and its eigenvectors.

**1.16** Using the characteristic equation of a tensor (1.120), relation (1.109), and the property (1.59), prove (1.123).

**1.17** Using the Cayley-Hamilton theorem (1.123), prove that the representation theorem (1.140) can be written in the form

$$\begin{aligned} \boldsymbol{L} = {} & \alpha_0\big(I_1(\boldsymbol{T}), I_2(\boldsymbol{T}), I_3(\boldsymbol{T})\big)\boldsymbol{I} + \alpha_1\big(I_1(\boldsymbol{T}), I_2(\boldsymbol{T}), I_3(\boldsymbol{T})\big)\,\boldsymbol{T} \\ & + \alpha_2\big(I_1(\boldsymbol{T}), I_2(\boldsymbol{T}), I_3(\boldsymbol{T})\big)\boldsymbol{T}^{-1}\,, \end{aligned} \tag{1.245}$$

where $\alpha_i$ $(i = 1, 2, 3)$ are scalar functions of the invariants of $\boldsymbol{T}$.

**1.18** For a matrix $[A]$ of order 3, prove the following relations:

$$\det[A] = \frac{1}{6}\,\varepsilon_{ijk}\varepsilon_{lmn}A_{il}A_{jm}A_{kn} \tag{1.246}$$

$$\big([A]^{-1}\big)_{ij} = \frac{1}{2\det[A]}\,\varepsilon_{ikl}\varepsilon_{jmn}A_{km}A_{ln}\,. \tag{1.247}$$

**1.19** Prove that for the scalar functions $f$ and $g$, we have

$$\nabla^2(fg) = f\nabla^2 g + g\nabla^2 f + 2\boldsymbol{\nabla} f \cdot \boldsymbol{\nabla} g\,. \tag{1.248}$$

CHAPTER 2

# Kinematics of Continuous Media

## 2.1 Introduction

Kinematics is the study of the motion of solid or fluid bodies. This motion is described by the successive positions of each point of the body as a function of time. We are not interested in the cause of the motion. However, we impose the condition that the body remains a continuous medium in the sense that its density and all the other parameters describing the motion be continuous spatial and temporal functions. In this chapter we will define the various parameters that characterize the motion of the body and present in detail how its deformation is defined; the properties and significance of these descriptions will be highlighted. In addition, the invariance of these various parameters with respect to a reference frame associated with an observer will be discussed. This description will be general and not limited to a particular form of motion or the composition of the body.

The kinematics of continuous media is covered in the following works: [15, 20, 24, 29, 31, 33, 34, 36, 49, 59].

## 2.2 Bodies, Configurations and Motion

A ***body*** $\mathcal{B}$ is a set of ***particles*** or ***material points***. These particles correspond to infinitesimal volumes surrounding the points. At any time $t$, each particle occupies a point in a three-dimensional Euclidean space. The volume $V$ occupied by all particles of $\mathcal{B}$ at time $t$ is called the ***configuration*** $\mathcal{R}_t$ or $\mathcal{R}$. In particular, the configuration of $\mathcal{B}$ at time $t = 0$ is defined as the ***initial configuration*** and will be named $\mathcal{R}_0$ (fig. 2.1). In addition, the boundary of the body is indicated by $\partial\mathcal{R}_0$ or $\partial\mathcal{R}$.

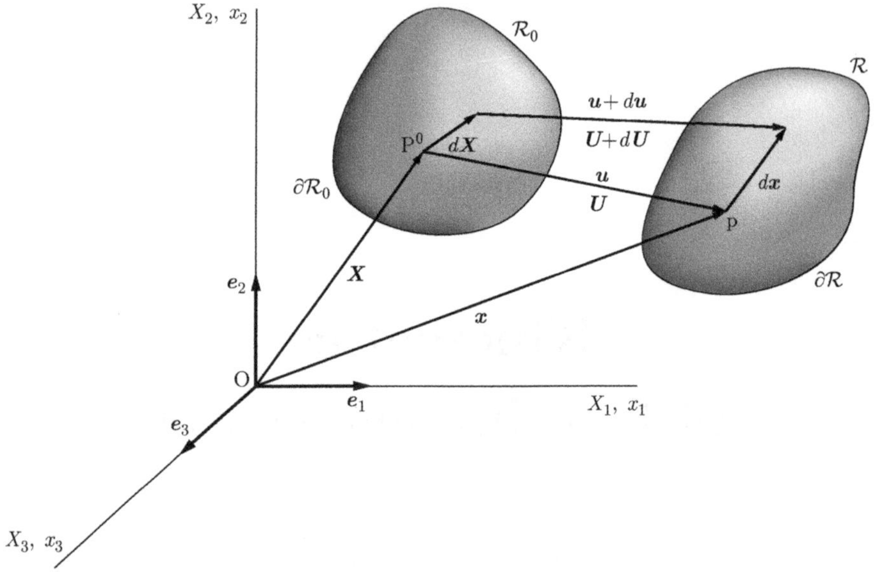


**Fig. 2.1** Initial configuration at $t = 0$ and configuration at the time $t$ of $\mathcal{B}$

A ***motion*** of $\mathcal{B}$ is a continuous sequence of configurations of $\mathcal{B}$ as seen by an ***observer***. The notion of motion is clearly linked to that of the reference frame. In rational mechanics, a reference frame is a set of $N$ points ($N \geq 4$), that are not coplanar, that are immobile with respect to each other, and with respect to which motion can be studied. The choice of the reference frame being arbitrary, an inertial, or Galilean, reference frame is often used in classical mechanics. Reference frames and coordinate systems should not be confused; for a given reference frame an infinite number of coordinate systems exist. As in classical mechanics, we will define the concept of an observer by

$$\textit{\textbf{an observer}} = \textit{\textbf{a clock}} + \textit{\textbf{a coordinate system}}\,.$$

In the observer's fixed Cartesian coordinate system, having $O$ as origin, the position $\mathrm{P}^0$ of a particle of $\mathcal{B}$ at $t = 0$ is represented by the ***initial position vector*** $\boldsymbol{X}$ and its current position p at time $t \geq 0$ by the ***current position vector*** $\boldsymbol{x}$. Thus, the motion of $\mathcal{B}$ is described by a vector function $\boldsymbol{\chi}$ defined over time $t$ that depends on $\boldsymbol{X}$:

$$\boldsymbol{x} = \boldsymbol{\chi}(\boldsymbol{X}, t)\,. \tag{2.1}$$

The vector function $\boldsymbol{\chi}$ is called the ***motion*** or ***deformation*** of the body. Furthermore, if the reference configuration ($t = 0$) coincides with the current form, the function $\boldsymbol{\chi}$ must satisfy the condition

$$\boldsymbol{X} = \boldsymbol{\chi}(\boldsymbol{X}, 0)\,. \tag{2.2}$$

The motion $\boldsymbol{\chi}$ is a bijection between the configurations $\mathcal{R}_0$ and $\mathcal{R}$; that is, there is a one-to-one correspondence between the initial and current positions of the particles of $\mathcal{B}$. The existence of the function $\chi : \mathcal{R}_0 \to \mathcal{R}$ and its ***inverse*** $\chi^{-1} : \mathcal{R} \to \mathcal{R}_0$

$$\boldsymbol{X} = \chi^{-1}(\boldsymbol{x}, t) \tag{2.3}$$

with

$$\boldsymbol{X} = \chi^{-1}(\boldsymbol{X}, 0) \tag{2.4}$$

guarantees the integrality and the unity of the body. The continuity of the motion function and its inverse rule out any separation (fragmentation) or local fusion (junction, superposition) of the material. Continuity implies that a connected domain remains connected throughout the motion, that two points infinitesimally close in the initial configuration remain so in the deformed configurations ($t > 0$). We assume that the functions $\chi$ and $\chi^{-1}$ are continuously differentiable at least two times with respect to spatial and temporal variables. These hypotheses then allow, without difficulty, the preservation of the regularity of the body, the definition and calculation of velocities and accelerations, deformation tensors, as well as of the equilibrium and compatibility equations.

The equation above says that $\boldsymbol{X}$ is the initial position of a particle that is currently found at $\boldsymbol{x}$. Equations (2.3) and (2.4) were obtained from (2.1) by calculating $\boldsymbol{X}$ as a function of $\boldsymbol{x}$. By definition, this means that

$$\chi(\chi^{-1}(\boldsymbol{x}, t), t) = \boldsymbol{x}$$
$$\chi^{-1}(\chi(\boldsymbol{X}, t), t) = \boldsymbol{X} .$$

Also by definition, referring to figure 2.1, the vector ***displacement*** $\boldsymbol{u}$ is the vector difference

$$\boldsymbol{u} = \boldsymbol{x} - \boldsymbol{X} = \chi(\boldsymbol{X}, t) - \boldsymbol{X} = \boldsymbol{x} - \chi^{-1}(\boldsymbol{x}, t) , \tag{2.5}$$

where (2.1) and (2.3) are used in the second and third equations.

Joseph-Louis de Lagrange (1736–1813) was a French mathematician (born in Turin). After teaching at the Academy of Artillery in Turin, he succeeded Euler as director of mathematics at the Berlin Academy. Later he taught at the École Normale and École Polytechnique in Paris. His book *Mécanique analytique* was a mathematical presentation of mechanics; its publication in 1788 was approved by a committee including Laplace and Legendre.

**Fig. 2.2** Joseph-Louis de Lagrange.

## 2.3 Material and Spatial Descriptions

In mechanics of continuous media the ***material description***, also called the ***Lagrangian description***, signifies the study of physical or mechanical phenomena under consideration by observing what happens to ***a particle*** P ***of the body***. Alternatively the ***spatial description***, known as the ***Eulerian description***, consists of observing the events ***occurring at a fixed point in space***. Thus, when the events at all fixed points in space are recorded, we obtain the spatial description. In what follows we will consider the same coordinate system to describe the motion of a body in material and spatial description and that the origins of the basis vectors in the two descriptions coincide and are indicated as $\boldsymbol{e}_i$, $(i = 1, 2, 3)$ (fig. 2.1).

Figure 2.3 shows the two representations for the case of a fluid flowing in a tube with a varying section.

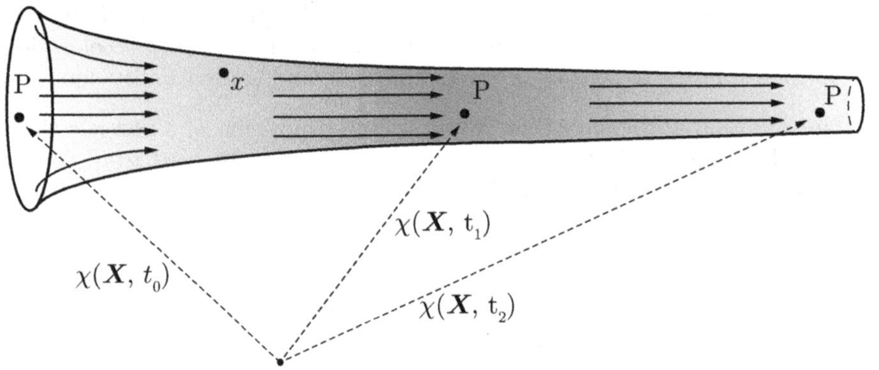


**Fig. 2.3** Schematic of the material and spatial descriptions for a fluid flow represented by the arrows

From a practical point of view, problems in solid mechanics are often easiest to formulate and solve with a material description while those in fluid mechanics are easier in a spatial description.

In order to exactly define the material and spatial descriptions, we first introduce the notion of the ***reference configuration***. By definition, it is a particular configuration $\mathcal{R}_r$ used to identify each particle of $\mathcal{B}$. In the following, the configuration $\mathcal{R}_0$ of $\mathcal{B}$ at $t = 0$ will be chosen as the reference configuration. Thus, we have the following definitions.

*Material description:*
: description in which the components of the initial vector position $\boldsymbol{X}$ are independent spatial variables.

*Spatial description:*
: description in which the components of the vector position at later times $\boldsymbol{x}$ are independent spatial variables.

When we calculate the partial derivatives of the various functions that are pertinent for the mechanics of continuous media, we can use either the material coordinates or the spatial coordinates as independent variables, that is $\boldsymbol{X}$ or $\boldsymbol{x}$. Since $\boldsymbol{X}$ and $\boldsymbol{x}$ are related, we also need to relate the derivatives of the function with respect to these variables. This is done with the application of the rule for derivatives of composite functions (chain rule).

**Fig. 2.4** Leonhard Euler (1707–1783) was a famous Swiss mathematician born in Basel. He succeeded Daniel Bernoulli at the Saint Petersburg Academy of Sciences. Later he taught at the Berlin Academy. Besides being one of the best mathematicians of all time, in mechanics Euler conceived the "Euler equations" which describe inviscid fluid dynamics. For a long time he was honored by appearing on the Swiss 10 franc note with his hydraulic turbine project. Euler was the prolific author of many books and nearly 900 memoirs (articles, as we say today). With the basics of the calculus of variations, he opened the door to modern methods of scientific calculation. In his *Lettres à une Princesse d'Allemagne* [14], he described, in French and without equations, the knowledge of physics of his time. There are also numerous religious reflections in them, as Euler was a fervent believer.

To simplify the derivation, we introduce the following convention:

- functions written with ***small letters*** refer to functions of ***spatial variables***, for example, $f(\boldsymbol{x}, t)$;
- functions written with ***capital letters*** refer to functions of ***material variables***, for example, $F(\boldsymbol{X}, t)$.

With this convention, we can write

$$f(\boldsymbol{x}, t) \longrightarrow f\big(\chi(\boldsymbol{X}, t), t\big) = F(\boldsymbol{X}, t) \tag{2.6}$$

$$F(\boldsymbol{X}, t) \longrightarrow F\big(\chi^{-1}(\boldsymbol{x}, t), t\big) = f(\boldsymbol{x}, t)\,. \tag{2.7}$$

In (2.6) we substituted $\boldsymbol{x}$ as a function of $\boldsymbol{X}$, while in (2.7) we substituted $\boldsymbol{X}$ as a function of $\boldsymbol{x}$. We note that the functions $f$ and $F$ are different although they represent the same physical phenomena. However, their values at the corresponding points $\boldsymbol{X}$ and $\boldsymbol{x}$ are equal, as shown in equations (2.6) and (2.7).

Consider the following example: let $\theta(\boldsymbol{x}, t)$ be the temperature at time $t$ and position $\boldsymbol{x}$, and $\Theta(\boldsymbol{X}, t)$ be the temperature at time $t$ of a particle that was initially located at $\boldsymbol{X}$. We have, according to (2.6) and (2.7),

$$\theta(\boldsymbol{x}, t) \longrightarrow \theta\big(\chi(\boldsymbol{X}, t), t\big) = \Theta(\boldsymbol{X}, t)$$

$$\Theta(\boldsymbol{X}, t) \longrightarrow \Theta\big(\chi^{-1}(\boldsymbol{x}, t), t\big) = \theta(\boldsymbol{x}, t)\,.$$

Equations (2.6) and (2.7) illustrate the transformation between the spatial description (Eulerian) and the material description (Lagrangian).

These terms are more precisely defined in the following way.

*Spatial description:*
$\boldsymbol{x}$, $t$ are the independant variables.

*Material description:*
$\boldsymbol{X}$, $t$ are the independent variables.

An observer can measure the velocity, the density, etc., at a given point in space. If these measurements are made for each point in the region of interest, we have a spatial description. To obtain the material description, the observer performs the measurements while travelling with a particle, at its velocity.

In most cases we assume that the reference condition is the configuration at time $t = 0$ (that is, the Lagrangian configuration). The material and spatial coordinates are generally measured with respect to the same coordinate axes. Note that for fluids as well as solids, the choice of the reference configuration is arbitrary.

Now consider a particle initially ($t = 0$) at $\boldsymbol{X}$ which, after a certain time $t$, is found at the position $\boldsymbol{x}$. In figure 2.1, we see that

$$\boldsymbol{x} = \boldsymbol{\chi}(\boldsymbol{X}, t) = \boldsymbol{X} + \boldsymbol{U}(\boldsymbol{X}, t) , \tag{2.8}$$

where $\boldsymbol{U}(\boldsymbol{X}, t)$ is the displacement in material coordinates. In spatial coordinates the displacement is given by

$$\boldsymbol{u}(\boldsymbol{x}, t) = \boldsymbol{U}\big(\boldsymbol{\chi}^{-1}(\boldsymbol{x}, t), t\big) = \boldsymbol{U}(\boldsymbol{X}, t) . \tag{2.9}$$

It is seen that the two vector functions $\boldsymbol{u}$ and $\boldsymbol{U}$ are equal, as they represent the same physical reality. This equation, while taking into account (2.3), allows us to write (2.8) in the form

$$\boldsymbol{x} = \boldsymbol{\chi}^{-1}(\boldsymbol{x}, t) + \boldsymbol{u}(\boldsymbol{x}, t) . \tag{2.10}$$

This equation says that the initial position of a particle located at the present time at $\boldsymbol{x}$ plus the displacement of the particle evaluated at $\boldsymbol{x}$ is equal to its position at that time.

The reader should not be confused by the use of $\boldsymbol{U}$ as a material displacement vector in (2.8) and the tensor $\boldsymbol{U}$ in (1.126) and in polar decomposition (1.132). The distinction should be clear from the context of the expressions.

EXAMPLE 2.1
The transformation of a body is described by

$$\begin{aligned} x_1 &= X_1 + aX_2 \\ x_2 &= X_2 + aX_1 \\ x_3 &= X_3 , \end{aligned} \tag{2.11}$$

where $a$ is a constant.

1) Express the motion and the displacements of the body in material and spatial coordinates.

2) For a cubic body defined by

$$\Omega = \big\{\boldsymbol{X} \in E^3 \mid 0 \leq X_1 \leq 1,\ 0 \leq X_2 \leq 1,\ 0 \leq X_3 \leq 1\big\} , \tag{2.12}$$

sketch $\Omega$ in its deformed configuration after the transformation (2.11) for $a = 1/3$.

3) After deformation, what is the form of the surface defined by

$$\Omega = \big\{\boldsymbol{X} \in E^3 \mid X_1^2 + X_2^2 \leq 1/(1 - a^2), X_3 = 0\big\}? \tag{2.13}$$

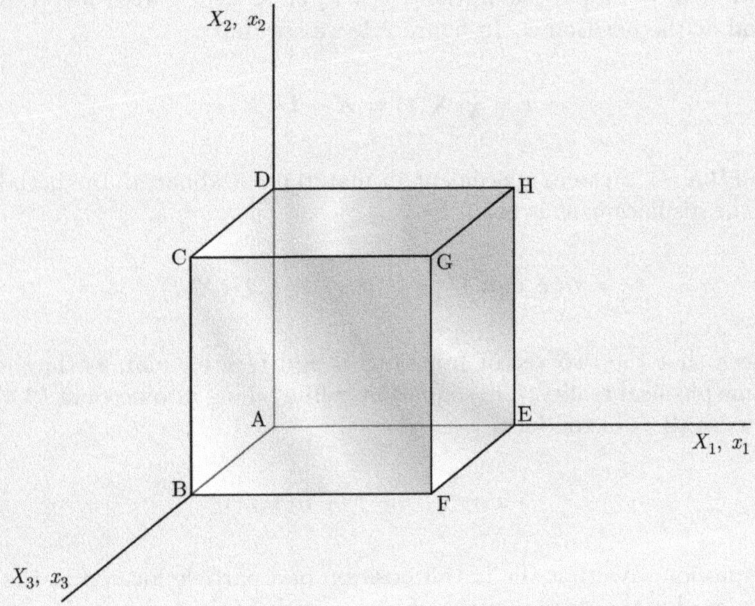


**Fig. 2.5** Solid unit cube: initial configuration

1) The motion in the spatial description is obtained by inverting relations (2.11)

$$\begin{aligned} X_1 &= (x_1 - a x_2)/(1 - a^2) \\ X_2 &= (x_2 - a x_1)/(1 - a^2) \\ X_3 &= x_3 \, . \end{aligned} \tag{2.14}$$

Using (2.8), the components of the displacement vector in material coordinates are

$$\begin{aligned} U_1 &= x_1 - X_1 = X_1 + aX_2 - X_1 = aX_2 \\ U_2 &= x_2 - X_2 = X_2 + aX_1 - X_2 = aX_1 \\ U_3 &= x_3 - X_3 = 0 \, . \end{aligned} \tag{2.15}$$

Using (2.10), the components of the displacement vector in spatial coordinates are

$$\begin{aligned} u_1 &= x_1 - X_1 = a(x_2 - a x_1)/(1 - a^2) \\ u_2 &= x_2 - X_2 = a(x_1 - a x_2)/(1 - a^2) \\ u_3 &= x_3 - X_3 = 0 \, . \end{aligned} \tag{2.16}$$

Note that $\boldsymbol{u}(\boldsymbol{x}, t) = \boldsymbol{U}\big(\chi^{-1}(\boldsymbol{x}, t), t\big)$.

2) By giving values to the variables $X_i$ in equation (2.15), we can construct the deformed body. For example, consider the following cases:

For the edge $X_1 = X_1, X_2 = X_3 = 0$, we have $U_1 = U_3 = 0, U_2 = aX_1$.
For the edge $X_1 = 0, X_2 = X_2, X_3 = 0$, we have $U_1 = aX_2, U_2 = U_3 = 0$.
For the edge $X_1 = X_2 = 0, X_3 = X_3$, one obtains $U_1 = U_2 = U_3 = 0$.

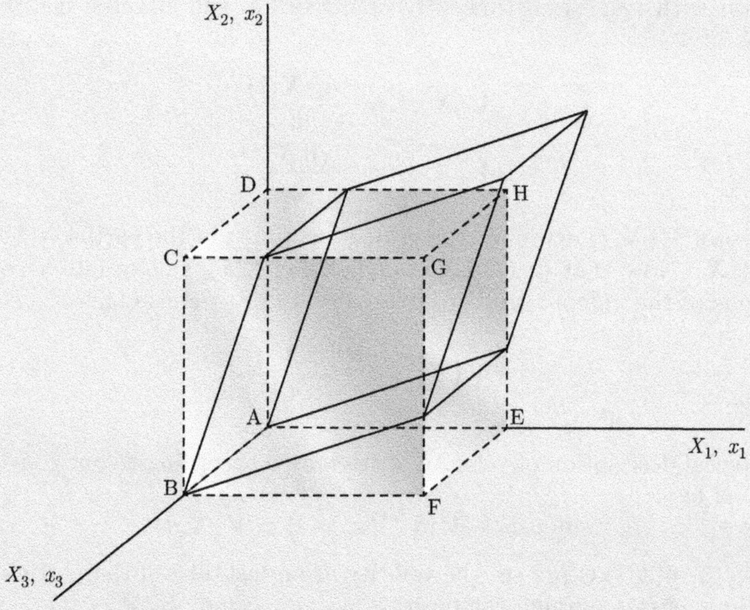


**Fig. 2.6** Solid unit cube: deformed configuration

3) The surface given is a cylinder described by

$$X_1^2 + X_2^2 \leq 1/(1 - a^2) .$$

Inserting (2.14) in the last expression, we obtain

$$\left(\frac{x_1 - ax_2}{1 - a^2}\right)^2 + \left(\frac{x_2 - ax_1}{1 - a^2}\right)^2 \leq 1/(1 - a^2) ,$$

that we can also write as

$$x_1^2(1 + a^2) + x_2^2(1 + a^2) - 4ax_1x_2 \leq (1 - a^2) ,$$

which is an elliptic surface.

## 2.4 Velocity, Material Derivative and Acceleration

### 2.4.1 Velocity

The ***velocity*** of a material particle at time $t$ is the derivative of the motion function with respect to time. By definition, in the material description we have

$$\boldsymbol{V}(\boldsymbol{X},t) = \frac{\partial \boldsymbol{\chi}(\boldsymbol{X},t)}{\partial t} \tag{2.17}$$

$$V_i(\boldsymbol{X},t) = \frac{\partial \chi_i(\boldsymbol{X},t)}{\partial t}\,. \tag{2.18}$$

The vector $\boldsymbol{V}(\boldsymbol{X},t)$ expresses the velocity at time $t$ of the particle that initially was at $\boldsymbol{X}$. Note that (2.17) is obtained using (2.1), taking into account that $\boldsymbol{X}$ is one of the independent variables. From (2.8) we also have

$$\boldsymbol{V}(\boldsymbol{X},t) = \frac{\partial \boldsymbol{U}(\boldsymbol{X},t)}{\partial t}\,. \tag{2.19}$$

The spatial description of velocity, written as $\boldsymbol{v}$ according to our convention, is obtained by

$$\boldsymbol{v}(\boldsymbol{x},t) = \boldsymbol{V}\big(\boldsymbol{\chi}^{-1}(\boldsymbol{x},t),t\big) = \boldsymbol{V}(\boldsymbol{X},t)\,. \tag{2.20}$$

The vector $\boldsymbol{v}(\boldsymbol{x},t)$ expresses the velocity at an instant $t$ of the particle that, at that time, passes through the position $\boldsymbol{x}$.

### 2.4.2 Material Derivative

Now we introduce the notion of the ***material derivative*** for a spatial field. Let $\varphi$ be a scalar field of $\mathcal{B}$. [1] During a motion $\boldsymbol{\chi}$ of $\mathcal{B}$, the material derivative of $\varphi(\boldsymbol{x},t)$, written $\dot{\varphi}$ or $D\varphi/Dt$, is the rate of change of $\varphi(\boldsymbol{x},t)$ with time (the derivative with respect to time) for a ***single*** particle of $\mathcal{B}$. To obtain the material derivative of the field $\varphi(\boldsymbol{x},t)$, we use the chain rule. In the material description, that is $\varphi\big(\boldsymbol{\chi}(\boldsymbol{X},t),t\big) = \Phi(\boldsymbol{X},t)$ (refer to equations (2.6) and (2.7)), we simply have

$$\frac{D\varphi(\boldsymbol{x},t)}{Dt} = \dot{\varphi} = \left.\frac{\partial \Phi(\boldsymbol{X},t)}{\partial t}\right|_{\boldsymbol{X}=\boldsymbol{\chi}^{-1}(\boldsymbol{x},t)}\,. \tag{2.21}$$

The last equation shows that the material derivative is applied to the same particle. For that reason, some authors call it the particle derivative. Since we can write $\Phi(\boldsymbol{X},t) = \Phi\big(\boldsymbol{\chi}^{-1}(\boldsymbol{x},t),t\big) = \varphi(\boldsymbol{x},t)$, we obtain

$$\frac{\partial \Phi(\boldsymbol{X},t)}{\partial t} = \left.\frac{\partial \varphi}{\partial x_1}\frac{\partial \chi_1}{\partial t} + \frac{\partial \varphi}{\partial x_2}\frac{\partial \chi_2}{\partial t} + \frac{\partial \varphi}{\partial x_3}\frac{\partial \chi_3}{\partial t} + \frac{\partial \varphi}{\partial t}\right|_{\boldsymbol{x}=\boldsymbol{\chi}(\boldsymbol{X},t)}\,. \tag{2.22}$$

[1] We will sometimes omit the arguments of the vector and scalar functions in order to simplify the equations.

Using the definition of the velocity (2.17), the preceding equation takes the following form:

$$\frac{\partial \Phi(\boldsymbol{X},t)}{\partial t} = \left.\frac{\partial \varphi}{\partial t}\right|_{\boldsymbol{x}=\boldsymbol{\chi}(\boldsymbol{X},t)} + V_i(\boldsymbol{X},t)\left.\frac{\partial \varphi}{\partial x_i}\right|_{\boldsymbol{x}=\boldsymbol{\chi}(\boldsymbol{X},t)} . \tag{2.23}$$

Since the goal is to express the rightmost term of (2.23) in spatial coordinates, we make the substitution $\boldsymbol{X} = \boldsymbol{\chi}^{-1}(\boldsymbol{x},t)$ in the last equation which gives

$$\left.\frac{\partial \Phi(\boldsymbol{X},t)}{\partial t}\right|_{\boldsymbol{X}=\boldsymbol{\chi}^{-1}(\boldsymbol{x},t)} = \frac{\partial \varphi}{\partial t} + v_i(\boldsymbol{x},t)\frac{\partial \varphi}{\partial x_i} , \tag{2.24}$$

where we used

$$V_i(\boldsymbol{X},t)\Big|_{\boldsymbol{X}=\boldsymbol{\chi}^{-1}(\boldsymbol{x},t)} = v_i(\boldsymbol{x},t) . \tag{2.25}$$

Now we can define the following derivative:

$$\dot{\varphi}(\boldsymbol{x},t) = \frac{D\varphi(\boldsymbol{x},t)}{Dt} \equiv \left.\frac{\partial \Phi(\boldsymbol{X},t)}{\partial t}\right|_{\boldsymbol{X}=\boldsymbol{\chi}^{-1}(\boldsymbol{x},t)} , \tag{2.26}$$

where, from (2.24),

$$\frac{D\varphi(\boldsymbol{x},t)}{Dt} = \frac{\partial \varphi(\boldsymbol{x},t)}{\partial t} + \boldsymbol{v}(\boldsymbol{x},t)\cdot\nabla\varphi(\boldsymbol{x},t) , \tag{2.27}$$

$$= \frac{\partial \varphi(\boldsymbol{x},t)}{\partial t} + v_j\frac{\partial \varphi(\boldsymbol{x},t)}{\partial x_j} . \tag{2.28}$$

The derivative $D\varphi(\boldsymbol{x},t)/Dt$ is called the ***material derivative*** and represents the rate of change of the function $\varphi$ following the same particle whose velocity is $\boldsymbol{v}(\boldsymbol{x},t)$. Alternatively, this derivative can be considered as giving the change of $\varphi$ over time, as seen by an observer moving with the particle that is at $\boldsymbol{x}$.

As an example, consider the function $\varphi$ that represents the temperature of a fluid particle in a river, which we will denote here $\theta$. We would like to know the variation $D\theta/Dt$, and for that we have a thermometer. In the first phase of the experiment, we get in a boat and immerse the thermometer in the water. Riding with the flow, we measure the variation of $\theta$ of the fluid particle that we are travelling with, that is, we measure $\partial\Theta/\partial t$ corresponding to the last term of equation (2.26). Thus the name particle derivative for that quantity. In the second phase, we attach the thermometer to a bridge pillar. The thermometer is located at point $\boldsymbol{x}$. There we measure $D\theta/Dt$, which is composed of $\partial\theta/\partial t$, that is, the time variation of the temperature at that point, and of the variation due to the local advection $\boldsymbol{v}\cdot\boldsymbol{\nabla}\theta$ induced by the changing velocity field (eqn. (2.27)).

For a vector field $\boldsymbol{w}$, we have a similar formula for its material derivatives:

$$\frac{D\boldsymbol{w}}{Dt} = \dot{\boldsymbol{w}} = \left.\frac{\partial \boldsymbol{W}(\boldsymbol{X},t)}{\partial t}\right|_{\boldsymbol{X}=\boldsymbol{\chi}^{-1}(\boldsymbol{x},t)} \tag{2.29}$$

$$\frac{Dw_i}{Dt} = \dot{w}_i = \left.\frac{\partial W_i(\boldsymbol{X},t)}{\partial t}\right|_{\boldsymbol{X}=\boldsymbol{\chi}^{-1}(\boldsymbol{x},t)}$$

$$\dot{\boldsymbol{w}} = \frac{\partial \boldsymbol{w}(\boldsymbol{x},t)}{\partial t} + \big(\nabla \boldsymbol{w}(\boldsymbol{x},t)\big)\left.\frac{\partial \boldsymbol{\chi}(\boldsymbol{X},t)}{\partial t}\right|_{\boldsymbol{X}=\boldsymbol{\chi}^{-1}(\boldsymbol{x},t)} \tag{2.30}$$

$$\dot{w}_i = \frac{\partial w_i(\boldsymbol{x},t)}{\partial t} + \frac{\partial w_i(\boldsymbol{x},t)}{\partial x_j} v_j \,.$$

Note that the material derivative of a field $\Phi(\boldsymbol{X},t)$ is the standard partial derivative

$$\dot{\Phi}(\boldsymbol{X},t) = \frac{D\Phi(\boldsymbol{X},t)}{Dt} = \frac{\partial \Phi(\boldsymbol{X},t)}{\partial t} \,. \tag{2.31}$$

### 2.4.3 Acceleration

The acceleration $\boldsymbol{A}$ of a material particle at time $t$ is the derivative of its velocity $\boldsymbol{V}$ with respect to time, that is, the material derivative of $\boldsymbol{V}$. In the material description we have

$$\boldsymbol{A}(\boldsymbol{X},t) = \frac{\partial \boldsymbol{V}(\boldsymbol{X},t)}{\partial t} = \frac{\partial^2 \boldsymbol{\chi}(\boldsymbol{X},t)}{\partial t^2} \tag{2.32}$$

$$A_i = \dot{V}_i = \frac{\partial^2 \chi_i(\boldsymbol{X},t)}{\partial t^2} \,,$$

and in the spatial description, we have

$$\boldsymbol{a} = \dot{\boldsymbol{v}} = \frac{\partial \boldsymbol{v}(\boldsymbol{x},t)}{\partial t} + \big(\boldsymbol{\nabla} \boldsymbol{v}(\boldsymbol{x},t)\big)\boldsymbol{v}(\boldsymbol{x},t) \tag{2.33}$$

$$a_i = \dot{v}_i = \frac{\partial v_i(\boldsymbol{x},t)}{\partial t} + \frac{\partial v_i(\boldsymbol{x},t)}{\partial x_j} v_j(\boldsymbol{x},t) \,.$$

The first term on the right-hand side of (2.33) can be considered as the acceleration due to the time dependence of the velocity at a fixed point in space. The second term can be interpreted as the contribution to the acceleration of the material particle due to the heterogeneity of the velocity field. These terms are sometimes called the local and convective (or advective) parts, respectively, of the acceleration. The advection corresponds to the transport of the velocity field by itself.

Note that equation (2.33) can also be written

$$\boldsymbol{a} = \frac{\partial \boldsymbol{v}(\boldsymbol{x},t)}{\partial t} + \big(\boldsymbol{v}(\boldsymbol{x},t)\cdot \boldsymbol{\nabla}\big)\boldsymbol{v}(\boldsymbol{x},t) \,,$$

with the definition

$$((\boldsymbol{v}\cdot\boldsymbol{\nabla})\boldsymbol{v})_i = v_j \frac{\partial v_i}{\partial x_j} \,.$$

This expression is not to be confused with $\boldsymbol{\nabla}\cdot\boldsymbol{v} = \partial v_j/\partial x_j$.

## 2.5 Pathlines, Streamlines and Streaklines

Equation (2.1) gives the successive positions $\boldsymbol{x}$ of a particle $\boldsymbol{X}$ with time $t$ as a parameter and thus describes a curve in space. This curve is called the ***pathline*** or ***trajectory*** of the particle $\boldsymbol{X}$. In differential form it is given by one of the following sets of equations:

$$d\boldsymbol{x} = \boldsymbol{V}(\boldsymbol{X},t)\,dt \quad \text{or} \quad dx_i = V_i(\boldsymbol{X},t)\,dt \tag{2.34}$$

$$d\boldsymbol{x} = \boldsymbol{v}(\boldsymbol{x},t)\,dt \quad \text{or} \quad dx_i = v_i(\boldsymbol{x},t)\,dt\ , \tag{2.35}$$

with the initial condition $\boldsymbol{x}(0) = \boldsymbol{X}$.

A ***streamline*** at a given instant $\bar{t}$ is the curve in space that is everywhere tangent to the velocity vector. It is thus determined in terms of a parameter $s$ by the differential equation

$$d\boldsymbol{x}(s) = \boldsymbol{v}\big(\boldsymbol{x}(s),\bar{t}\big)\,ds \tag{2.36}$$

$$dx_i(s) = v_i\big(\boldsymbol{x}(s),\bar{t}\big)\,ds\,. \tag{2.37}$$

The motion is said to be ***stationary*** if the velocity for all points $\boldsymbol{x}$ is time independent, that is $\boldsymbol{v} = \boldsymbol{v}(\boldsymbol{x},t) = \boldsymbol{v}(\boldsymbol{x},0) = \boldsymbol{v}_0(\boldsymbol{x})$, in which case, equation (2.36) is identical to (2.35). Thus, ***for stationary motion, the streamlines and the pathlines coincide***.

The ***streakline*** through a given point in space $\bar{\boldsymbol{x}}$ and at a given instant $\bar{t}$ is the curve composed of all the particles that previously occupied $\bar{\boldsymbol{x}}$; stated differently, all the particles that have passed through $\bar{\boldsymbol{x}}$ for time between 0 and $\bar{t}$. This curve can be parameterized in $t$, as follows:

$$\boldsymbol{x} = \boldsymbol{x}\big(\chi^{-1}(\bar{\boldsymbol{x}},t),\bar{t}\big) \qquad 0 \le t \le \bar{t}\,. \tag{2.38}$$

EXAMPLE 2.2

Consider the following example where a plane flow is given by the velocity field:

$$v_1 = \frac{x_1}{1+t} \qquad v_2 = x_2 \qquad v_3 = 0\,. \tag{2.39}$$

First we calculate the streamlines. Equation (2.37) applied to (2.39) gives

$$dx_1 = \frac{x_1}{1+t}\,ds \qquad dx_2 = x_2\,ds \qquad dx_3 = 0\,. \tag{2.40}$$

Setting $t = \bar{t}$ and integrating yields

$$x_1 = c_1\,e^{\frac{s}{1+\bar{t}}} \qquad x_2 = c_2\,e^{s} \qquad x_3 = c_3\,. \tag{2.41}$$

These are the equations of the streamline that passes through the point $\boldsymbol{c}$.

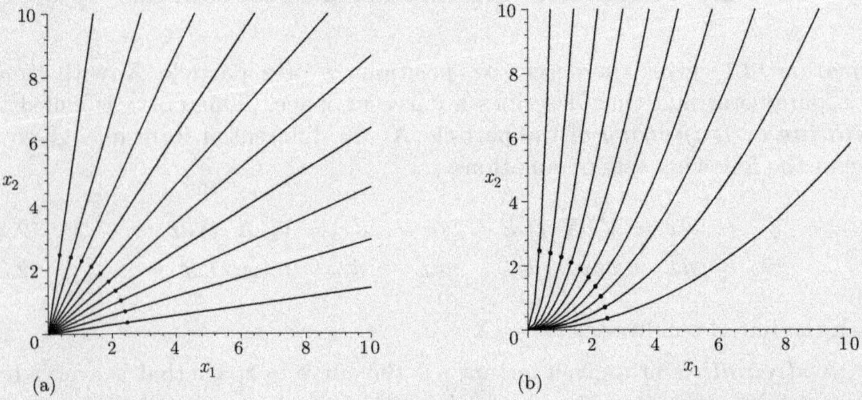


**Fig. 2.7** Streamlines: (a) for $\bar{t} = 0$, (b) for $\bar{t} = 1$. The points shown on the lines correspond to values of $\boldsymbol{c}$ uniformly distributed on a circle of radius $R = 2.5$

Figure 2.7 shows the streamlines that are curves in the plane $x_3 = c_3$ such that

$$\frac{x_2}{c_2} = \left(\frac{x_1}{c_1}\right)^{(1+\bar{t})} . \tag{2.42}$$

The calculation of the pathlines is achieved by combining (2.35) and (2.39). The result is

$$\int_{X_1}^{x_1} \frac{dx_1'}{x_1'} = \int_0^t \frac{dt'}{1+t'} \qquad \int_{X_2}^{x_2} \frac{dx_2'}{x_2'} = \int_0^t dt' \qquad \int_{X_3}^{x_3} \frac{dx_3'}{x_3'} = 0 . \tag{2.43}$$

The first integral becomes $\ln x_1' \big|_{X_1}^{x_1} = \ln(1+t')\big|_0^t$ and thus $\ln x_1 - \ln X_1 = \ln(1+t)$ as $x_1 = X_1$ at time $t = 0$. We obtain, finally,

$$x_1 = X_1(1+t) \qquad x_2 = X_2\, e^t \qquad x_3 = X_3 . \tag{2.44}$$

The pathlines are curves in the plane $x_3 = c_3$ given by

$$x_2 = X_2\, e^{(x_1 - X_1)/X_1} \tag{2.45}$$

and are drawn in figure 2.8.

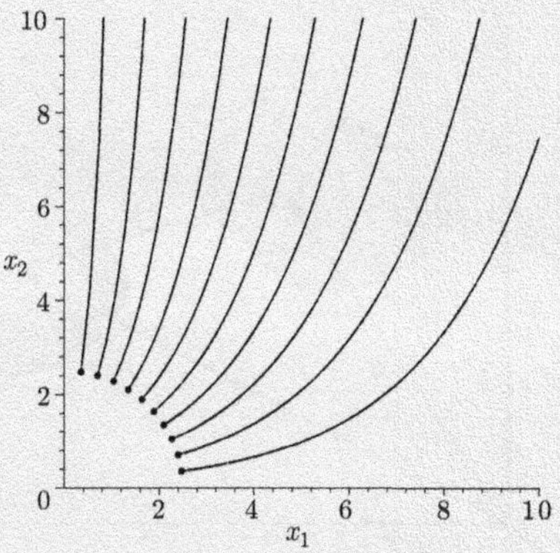


**Fig. 2.8** Pathlines; the points corresponding to $\boldsymbol{X}$ are the same as those in figure 2.7

To calculate the streaklines, we first invert the pathline relations

$$X_1 = \frac{x_1}{1+t} \qquad X_2 = x_2\, e^{-t} \qquad X_3 = x_3\,. \tag{2.46}$$

Since the particle passes through $\overline{\boldsymbol{x}}$ at time $t \leq \bar{t}$, we have

$$X_1 = \frac{\overline{x}_1}{1+t} \qquad X_2 = \overline{x}_2\, e^{-t} \qquad X_3 = \overline{x}_3\,. \tag{2.47}$$

Substituting (2.47) in (2.44) evaluated at time $\bar{t}$, we obtain the parametric equations of the streaklines

$$x_1 = \overline{x}_1 \frac{1+\bar{t}}{1+t} \qquad x_2 = \overline{x}_2\, e^{\bar{t}-t} \qquad x_3 = \overline{x}_3\,. \tag{2.48}$$

## 2.6 Kinematic Equations for Rigid Body Motion

When referring to objectivity while discussing the general principles governing the constitutive equations of continuous media, we will primarily examine the rotation of a rigid body. To prepare for that, we will study a few relations of rigid body kinematics. Rigid body motion is that for which lengths and angles are conserved.

### 2.6.1 Translation of a Rigid Body

For the case illustrated in figure 2.9, the equation of motion is given by

$$\boldsymbol{x} = \boldsymbol{\chi}(\boldsymbol{X}, t) = \boldsymbol{X} + \boldsymbol{c}(t)\ , \tag{2.49}$$

with $\boldsymbol{c}(0) = \boldsymbol{0}$.

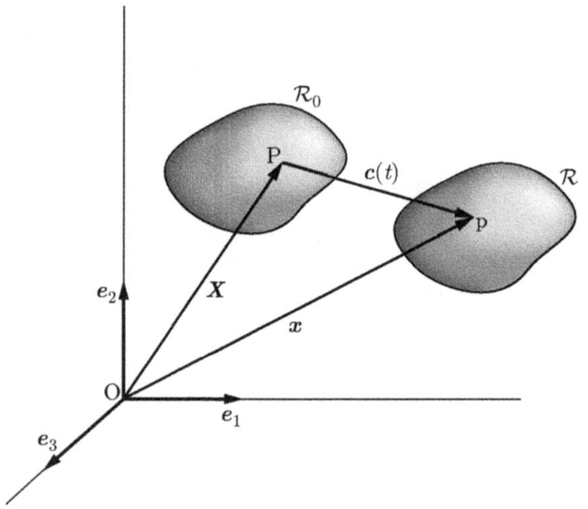


**Fig. 2.9** Translation of a rigid body

We notice that the displacement vector $\boldsymbol{U}$ is independent of $\boldsymbol{X}$, as we have

$$\boldsymbol{U} = \boldsymbol{x} - \boldsymbol{X} = \boldsymbol{c}(t)\,. \tag{2.50}$$

Thus every material point is displaced identically with the same magnitude and direction at time $t$.

### 2.6.2 Rotation of a Rigid Body Around a Fixed Point

The motion illustrated in figure 2.10 is described by the equation

$$\chi(\boldsymbol{X}, t) = \boldsymbol{x} = \boldsymbol{b} + \boldsymbol{Q}(t)(\boldsymbol{X} - \boldsymbol{b})\,, \tag{2.51}$$

where $\boldsymbol{Q}$ is an orthogonal tensor such that $\boldsymbol{Q}(0) = \boldsymbol{I}$ and $\boldsymbol{b}$ is a constant vector.

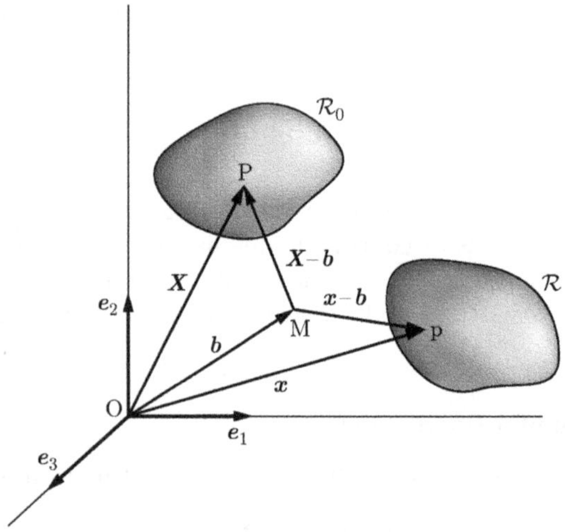


**Fig. 2.10** Rotation of a rigid body around a fixed point

We note that when the material point $\boldsymbol{X} = \boldsymbol{b}$ coincides with the spatial point $\boldsymbol{x} = \boldsymbol{b}$, the rotation is always about a fixed point $\boldsymbol{x} = \boldsymbol{b}$. If the center of rotation is placed at the origin, then $\boldsymbol{b} = \boldsymbol{0}$. The rotational motion is given by $\boldsymbol{x} = \boldsymbol{Q}(t)\boldsymbol{X}$.

### 2.6.3 General Rigid Body Motion

The motion illustrated in figure 2.11 is expressed by

$$\boldsymbol{x} = \chi(\boldsymbol{X}, t) = \boldsymbol{Q}(t)\boldsymbol{X} + \boldsymbol{d}(t), \tag{2.52}$$

where $\boldsymbol{Q}$ is a rotation tensor as before and $\boldsymbol{d}(t) = -\boldsymbol{Q}(t)\boldsymbol{b}+\boldsymbol{c}(t)$. The vector $\boldsymbol{c}(t)$ is such that $\boldsymbol{c}(0) = \boldsymbol{b}$. Equation (2.52) indicates that the motion is composed of a rotation $\boldsymbol{Q}(t)$ and a translation $\boldsymbol{c}(t)$ of a material point $\boldsymbol{X} = \boldsymbol{b}$.

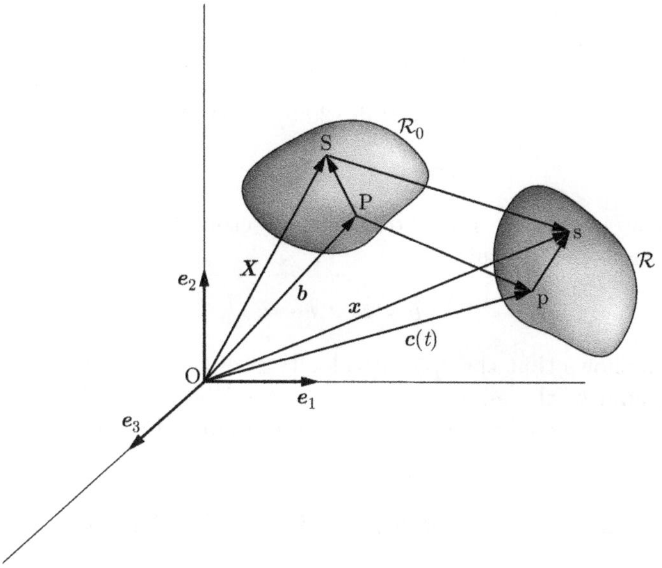


**Fig. 2.11** General rigid body motion

The velocity is obtained by taking the derivative with respect to time of equation (2.52)

$$\boldsymbol{V} = \dot{\boldsymbol{Q}}(\boldsymbol{X} - \boldsymbol{b}) + \dot{\boldsymbol{c}}. \tag{2.53}$$

Now from (2.52), we have

$$\boldsymbol{X} - \boldsymbol{b} = \boldsymbol{Q}^T(\boldsymbol{x} - \boldsymbol{c}). \tag{2.54}$$

Taking into account (2.20), we can write

$$\boldsymbol{v} = \dot{\boldsymbol{Q}}\boldsymbol{Q}^T(\boldsymbol{x} - \boldsymbol{c}) + \dot{\boldsymbol{c}}, \tag{2.55}$$

since the tensor $\boldsymbol{Q}$ is orthogonal, $\boldsymbol{Q}\boldsymbol{Q}^T = \boldsymbol{I}$ (eqn. (1.243)). Taking the derivative of this last expression with respect to time, we obtain

$$\dot{\boldsymbol{Q}}\boldsymbol{Q}^T + \boldsymbol{Q}\dot{\boldsymbol{Q}}^T = 0\,, \tag{2.56}$$

which shows that $\dot{\boldsymbol{Q}}\boldsymbol{Q}^T$ is antisymetric. Let us set

$$\dot{\boldsymbol{Q}}\boldsymbol{Q}^T = \boldsymbol{\Omega}\,, \tag{2.57}$$

where $\boldsymbol{\Omega}$ is an antisymmetric rotation tensor. Then, using the dual vector of this tensor (see eqn. (1.104))

$$\omega_k = -\frac{1}{2}\varepsilon_{kij}\Omega_{ij}\,, \tag{2.58}$$

we obtain, successively, with the help of (1.107)

$$\Omega_{ij} = -\varepsilon_{ijk}\omega_k = \varepsilon_{ikj}\omega_k \tag{2.59}$$

$$\boldsymbol{\Omega x} = \Omega_{ij}\, x_j \boldsymbol{e}_i = \varepsilon_{ikj}\omega_k x_j \boldsymbol{e}_i = \boldsymbol{\omega} \times \boldsymbol{x}\,. \tag{2.60}$$

Using (2.60), we can rewrite equation (2.55) in the form

$$\boldsymbol{v} = \boldsymbol{\omega} \times (\boldsymbol{x} - \boldsymbol{c}) + \dot{\boldsymbol{c}}\,. \tag{2.61}$$

Setting $\boldsymbol{r} = \boldsymbol{x} - \boldsymbol{c}$, where $\boldsymbol{r}$ is the position vector of the material point offset by a translation $\boldsymbol{c}$, we obtain

$$\boldsymbol{v} = \boldsymbol{\omega} \times \boldsymbol{r} + \dot{\boldsymbol{c}}\,. \tag{2.62}$$

This relation shows that the spatial velocity $\boldsymbol{v}$ of any material point of a rigid body is the sum of the angular velocity of rigid body rotation, $\boldsymbol{\omega} \times \boldsymbol{r}$, and a translation velocity $\dot{\boldsymbol{c}}$ of an arbitrarily chosen material point.

## 2.7 Gradient and Deformation Tensors

### 2.7.1 Definition

Consider a given particle whose position in the configuration $\mathcal{R}_0$ is $\boldsymbol{X}^0$, and whose neighborhood is $\mathcal{V}$. Its position in the configuration $\mathcal{R}$ is given by (2.1). If $\mathcal{V}$ is sufficiently small, the relation (2.1) for the other particles of $\mathcal{V}$ can be approximated using a Taylor expansion. Let $\boldsymbol{X}$ and $\boldsymbol{x}$ be the initial and current positions of a particle. Then, if the function $\boldsymbol{\chi}$ is sufficiently regular, we have

$$\begin{aligned} x_i &= \chi_i(X_k, t) \\ &= \chi_i(X_k^0, t) + \left.\frac{\partial \chi_i}{\partial X_j}\right|_{X_k^0} (X_j - X_j^0) + O(\|\boldsymbol{X} - \boldsymbol{X}^0\|^2)\,, \end{aligned} \tag{2.63}$$

where the last term signifies that

$$O\left(\|\boldsymbol{X} - \boldsymbol{X}^0\|^2\right) \sim C\|\boldsymbol{X} - \boldsymbol{X}^0\|^2 + \cdots\,, \tag{2.64}$$

with $C$ being a bounded constant. The tensor $\boldsymbol{F}$ whose components are given by

$$F_{ij} = \frac{\partial \chi_i}{\partial X_j} \tag{2.65}$$

is called the ***deformation gradient tensor***.

In the following, we will omit the variables of the functions, vectors, and tensors in order to simplify the notation.

If the distance $\|\boldsymbol{X} - \boldsymbol{X}^0\|$ between $\boldsymbol{X}$ and $\boldsymbol{X}^0$ is much less than unity, relation (2.63) can be closely approximated by

$$x_i \cong x_i^0 + F_{ij}(X_j - X_j^0) \qquad \text{with} \qquad x_i^0 = \chi_i(X_k^0, t)\,. \tag{2.66}$$

In this case, the tensor $\boldsymbol{F}$ is represented (abusively) by

$$F_{ij} = \frac{\partial x_i}{\partial X_j}\,. \tag{2.67}$$

Let $J$ be the ***Jacobian*** of $\boldsymbol{F}$:

$$J = \det\left(\frac{\partial \chi_i}{\partial X_j}\right) = \det \boldsymbol{F}\,. \tag{2.68}$$

The assumption of continuity of the material and thus the existence of a continuous derivative for the deformation of the medium implies that

$$0 < J < \infty\,. \tag{2.69}$$

This ensures the existence of the inverse $\boldsymbol{F}^{-1}$ of $\boldsymbol{F}$ with $\det \boldsymbol{F}^{-1} = 1/J$ . Using (2.8), (2.10), and (2.67), we can express $\boldsymbol{F}$ and $\boldsymbol{F}^{-1}$ in terms of displacement:

$$F_{ij} = \delta_{ij} + \frac{\partial U_i}{\partial X_j} \qquad F_{ij}^{-1} = \frac{\partial X_i}{\partial x_j} = \delta_{ij} - \frac{\partial u_i}{\partial x_j} \tag{2.70}$$

or

$$\boldsymbol{F} = \mathbf{grad}\,\chi(\boldsymbol{X}, t) = \boldsymbol{I} + \boldsymbol{\nabla U} \qquad \boldsymbol{F}^{-1} = \mathbf{grad}\,\chi^{-1}(\boldsymbol{x}, t) = \boldsymbol{I} - \boldsymbol{\nabla u}\,. \tag{2.71}$$

Another way to write (2.66) is

$$d\boldsymbol{x} = \boldsymbol{F}\,d\boldsymbol{X}\,, \tag{2.72}$$

where we see that $\boldsymbol{F}$ transforms a vector $d\boldsymbol{X}$ associated with $\boldsymbol{X}^0$ into a vector $d\boldsymbol{x}$ associated with $\boldsymbol{x}$ (fig. 2.1). To guarantee the existence of $\boldsymbol{F}$, derived from the deformation (2.1) or (2.63), the condition

$$\frac{\partial^2 x_i}{\partial X_l \partial X_k} = \frac{\partial^2 x_i}{\partial X_k \partial X_l}$$

is necessary. This compatibility condition is also sufficient in a simply connected region to ensure that (2.1) exists such that $\boldsymbol{F}$ is given by (2.67).

According to the ***polar decomposition theorem*** (1.132), there exists a unique rotation tensor $\boldsymbol{R}$ and two unique symmetric positive definite tensors $\boldsymbol{U}$ and $\boldsymbol{V}$, such that

$$\boldsymbol{F} = \boldsymbol{R}\boldsymbol{U} = \boldsymbol{V}\boldsymbol{R}\,. \tag{2.73}$$

For $\boldsymbol{F} = \boldsymbol{R}\boldsymbol{U}$, the ***right*** polar decomposition of $\boldsymbol{F}$, and for $\boldsymbol{F} = \boldsymbol{V}\boldsymbol{R}$, the ***left*** polar decomposition of $\boldsymbol{F}$, the tensors $\boldsymbol{U}$ and $\boldsymbol{V}$ are called the right and left stretch tensors. When $\boldsymbol{R} = \boldsymbol{I}$, (2.73) reduces to $\boldsymbol{F} = \boldsymbol{U} = \boldsymbol{V}$ and the deformation is called ***pure deformation***.

Inserting (2.73) in (2.72), we obtain

$$d\boldsymbol{x} = \boldsymbol{R}\boldsymbol{U}\, d\boldsymbol{X}\,. \tag{2.74}$$

We will see later that this relation allows us to conclude that the configuration change in the neighborhood of the material particle is obtained by the transformation of vector $d\boldsymbol{X}$ to a vector $\boldsymbol{U}\, d\boldsymbol{X}$ by a pure deformation $\boldsymbol{U}$ followed by a local rotation $\boldsymbol{R}$.

### 2.7.2 Deformation Tensors

Let us write (2.72) in the index notation

$$dx_i = F_{ij}\, dX_j\,. \tag{2.75}$$

Now, the square of the norm $ds$ of the vector $d\boldsymbol{x}$ is given by

$$ds^2 = \|d\boldsymbol{x}\|^2 = dx_m\, dx_m = F_{mi}F_{mj}\, dX_i\, dX_j\,. \tag{2.76}$$

The tensor $\boldsymbol{C}$, defined by

$$\boldsymbol{C} = \boldsymbol{F}^T\boldsymbol{F} = \left(\boldsymbol{F}^T\boldsymbol{F}\right)^T \qquad C_{ij} = F_{mi}F_{mj}\,, \tag{2.77}$$

is the ***symmetric right Cauchy-Green deformation tensor***.

Augustin Louis Cauchy (1789–1857) was a French mathematician born in Paris. He was a professor at the École Polytechnique in Paris. His extensive body of work treated all the major mathematical problems of his time. He is notably responsible for holomorphic functions and convergence criteria for series. His name is one of the 72 names of distinguished engineers and scientists inscribed on the Eiffel tower.

**Fig. 2.12** Augustin Louis Cauchy

This symmetric tensor is a metric tensor. As indicated by relation (2.76), it can be used to calculate the length of $d\boldsymbol{x}$ as a function of the components of $d\boldsymbol{X}$. Inversely, the length $dS$ of $d\boldsymbol{X}$ can be calculated in terms of the components of $d\boldsymbol{x}$:

$$dS^2 = \|d\boldsymbol{X}\|^2 = dX_m\, dX_m = F^{-1}_{mi}F^{-1}_{mj}\, dx_i\, dx_j\,. \tag{2.78}$$

With the notation $\boldsymbol{F}^{-T} = \left(\boldsymbol{F}^{-1}\right)^T$, the tensor $\boldsymbol{c}^{-1}$, defined by

$$\boldsymbol{c}^{-1} = \boldsymbol{F}^{-T}\boldsymbol{F}^{-1} = \left(\boldsymbol{F}^{-T}\boldsymbol{F}^{-1}\right)^T \qquad c_{ij}^{-1} = F_{mi}^{-1}F_{mj}^{-1}, \tag{2.79}$$

is the inverse of the ***symmetric left Cauchy-Green deformation tensor*** $\boldsymbol{c}$. The difference between $\|d\boldsymbol{x}\|^2$ and $\|d\boldsymbol{X}\|^2$ can be expressed in one of the two following forms:

$$\|d\boldsymbol{x}\|^2 - \|d\boldsymbol{X}\|^2 = C_{ij}\,dX_i\,dX_j - dX_m\,dX_m = 2E_{ij}\,dX_i\,dX_j \tag{2.80}$$

$$\|d\boldsymbol{x}\|^2 - \|d\boldsymbol{X}\|^2 = dx_m\,dx_m - c_{ij}^{-1}\,dx_i\,dx_j = 2e_{ij}\,dx_i\,dx_j\,. \tag{2.81}$$

The tensor $\boldsymbol{E}$ introduced in (2.80), whose components are

$$E_{ij} = \frac{1}{2}\left(C_{ij} - \delta_{ij}\right), \tag{2.82}$$

is called the ***Green-Lagrange strain tensor***.

George Green (1793–1841) was an English mathematician born in Sneinton, near Nottingham. Practically self-taught, he obtained a degree at the age of 44. He contributed to potential theory by introducing the functions which are now known by his name. He worked in optics, acoustics, and hydrodynamics. His work, not well known during his lifetime, was exposed to a larger public in 1846 by Lord Kelvin.

**Fig. 2.13** George Green

The tensor $\boldsymbol{e}$ introduced in (2.81), whose components are

$$e_{ij} = \frac{1}{2}\left(\delta_{ij} - c_{ij}^{-1}\right), \tag{2.83}$$

is the ***Euler-Almansi strain tensor*** .

In terms of material and spatial displacements $\boldsymbol{U}$ and $\boldsymbol{u}$, the deformation tensors above are written as

$$\begin{aligned} C_{ij} &= F_{mi}F_{mj} = \left(\delta_{mi} + \frac{\partial U_m}{\partial X_i}\right)\left(\delta_{mj} + \frac{\partial U_m}{\partial X_j}\right) \\ &= \delta_{ij} + \frac{\partial U_i}{\partial X_j} + \frac{\partial U_j}{\partial X_i} + \frac{\partial U_m}{\partial X_i}\frac{\partial U_m}{\partial X_j} \end{aligned} \tag{2.84}$$

$$\begin{aligned} c_{ij}^{-1} &= F_{mi}^{-1}F_{mj}^{-1} = \left(\delta_{mi} - \frac{\partial u_m}{\partial x_i}\right)\left(\delta_{mj} - \frac{\partial u_m}{\partial x_j}\right) \\ &= \delta_{ij} - \frac{\partial u_i}{\partial x_j} - \frac{\partial u_j}{\partial x_i} + \frac{\partial u_m}{\partial x_i}\frac{\partial u_m}{\partial x_j} \end{aligned} \tag{2.85}$$

$$E_{ij} = \frac{1}{2}\left(\frac{\partial U_i}{\partial X_j} + \frac{\partial U_j}{\partial X_i} + \frac{\partial U_m}{\partial X_i}\frac{\partial U_m}{\partial X_j}\right) \tag{2.86}$$

$$e_{ij} = \frac{1}{2}\left(\frac{\partial u_i}{\partial x_j} + \frac{\partial u_j}{\partial x_i} - \frac{\partial u_m}{\partial x_i}\frac{\partial u_m}{\partial x_j}\right). \tag{2.87}$$

The deformation tensors can also be written in terms of $\boldsymbol{U}$ and $\boldsymbol{V}$. Directly applying the polar decomposition (2.73), they are:

- right Cauchy-Green deformation tensor

$$\boldsymbol{C} = \boldsymbol{F}^T\boldsymbol{F} = \boldsymbol{U}\boldsymbol{R}^T\boldsymbol{R}\boldsymbol{U} = \boldsymbol{U}^2 \; ; \tag{2.88}$$

- left Cauchy-Green deformation tensor and its inverse

$$\boldsymbol{c} = \boldsymbol{F}\boldsymbol{F}^T = \boldsymbol{V}\boldsymbol{R}\boldsymbol{R}^T\boldsymbol{V}^T = \boldsymbol{V}^2 \tag{2.89}$$

$$\boldsymbol{c}^{-1} = \boldsymbol{F}^{-T}\boldsymbol{F}^{-1} = \mathbf{V}^{-2} \; ; \tag{2.90}$$

- Green-Lagrange strain tensor

$$\boldsymbol{E} = \frac{1}{2}(\boldsymbol{C} - \boldsymbol{I}) = \frac{1}{2}\left(\boldsymbol{U}^2 - \boldsymbol{I}\right) ; \tag{2.91}$$

- Euler-Almansi strain tensor

$$\boldsymbol{e} = \frac{1}{2}\left(\boldsymbol{I} - \boldsymbol{c}^{-1}\right) = \frac{1}{2}\left(\boldsymbol{I} - \boldsymbol{V}^{-2}\right). \tag{2.92}$$

It is important to observe that the rotation tensor $\boldsymbol{R}$ does not affect any of the deformation tensors considered. It is necessarily so for the measurement of deformation to have meaning; otherwise, a rigid body would be considered deformable. In addition, we can easily verify that for rigid body motion (2.51), $\boldsymbol{F} = \boldsymbol{Q}$ and $\boldsymbol{C} = \boldsymbol{c} = \boldsymbol{I}$ and thus $\boldsymbol{E} = \boldsymbol{e} = 0$.

Based on the deformation gradient tensor $\boldsymbol{F}$ and the associated deformation tensors, we can express the change in length of a linear element, a surface element, and a volume element during a motion of the body (fig. 2.14). A linear element $d\boldsymbol{X}$ in the reference configuration has norm $\|d\boldsymbol{X}\| = (d\boldsymbol{X} \cdot d\boldsymbol{X})^{1/2}$. After the body moves (relation (2.1)), it becomes the element $d\boldsymbol{x}$ with norm $\|d\boldsymbol{x}\| = (d\boldsymbol{x} \cdot d\boldsymbol{x})^{1/2}$. Taking into account (2.72), the relation between the squares of the norms in these two configurations is given by

$$\frac{\|d\boldsymbol{x}\|^2}{\|d\boldsymbol{X}\|^2} = \frac{\boldsymbol{F}\,d\boldsymbol{X} \cdot \boldsymbol{F}\,d\boldsymbol{X}}{\|d\boldsymbol{X}\|^2} = \frac{d\boldsymbol{X} \cdot \boldsymbol{F}^T\boldsymbol{F}\,d\boldsymbol{X}}{\|d\boldsymbol{X}\|^2} = \frac{d\boldsymbol{X} \cdot \boldsymbol{C}\,d\boldsymbol{X}}{\|d\boldsymbol{X}\|^2} \,. \tag{2.93}$$

As $d\boldsymbol{X} = \boldsymbol{N}\|d\boldsymbol{X}\|$ where $\boldsymbol{N}$ is the unit vector in the direction $d\boldsymbol{X}$, (2.93) is written as

$$\frac{\|d\boldsymbol{x}\|^2}{\|d\boldsymbol{X}\|^2} = \frac{d\boldsymbol{X} \cdot \boldsymbol{C}\,d\boldsymbol{X}}{\|d\boldsymbol{X}\|\,\|d\boldsymbol{X}\|} = \boldsymbol{N} \cdot \boldsymbol{C}\boldsymbol{N} = \lambda_N^2 \,. \tag{2.94}$$

The parameter $\lambda_N$ is called ***stretch*** or ***stretch ratio*** at $\boldsymbol{X}$ in the direction of $\boldsymbol{N}$. Stretch parameters are used in non-linear elastic response of solids subjected to large deformation. This subject is treated in chapter 6.

By the relation (2.88) linking $\boldsymbol{C}$ and $\boldsymbol{U}$, the stretch ratio can be expressed with the equations

$$\frac{\|d\boldsymbol{x}\|}{\|d\boldsymbol{X}\|} = \left(\boldsymbol{N} \cdot \boldsymbol{U}^2\boldsymbol{N}\right)^{1/2} = (\boldsymbol{U}\boldsymbol{N} \cdot \boldsymbol{U}\boldsymbol{N})^{1/2} = \|\boldsymbol{U}\boldsymbol{N}\| = \lambda_N \,. \tag{2.95}$$

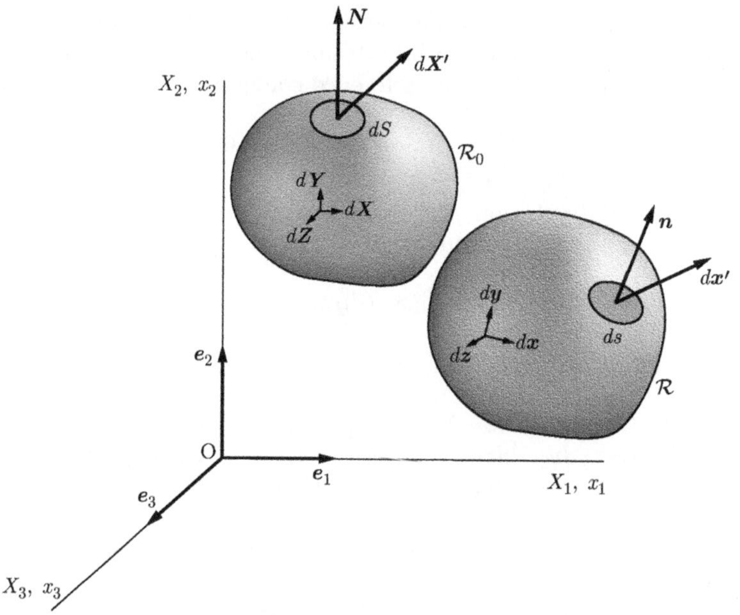


**Fig. 2.14** Linear and surface elements in the configurations $\mathcal{R}_0$ and $\mathcal{R}_t$ of the body $\mathcal{B}$

We can also express the angle between two linear elements similarly with the following procedures. Suppose that two linear elements $d\boldsymbol{X}$ and $d\boldsymbol{Y}$ intersect with an angle between them of $\Theta$ in the reference configuration. Then,

$$\cos\Theta = \frac{d\boldsymbol{X}\cdot d\boldsymbol{Y}}{\|d\boldsymbol{X}\|\,\|d\boldsymbol{Y}\|}\,. \tag{2.96}$$

After the motion, these two elements become $d\boldsymbol{x}$ and $d\boldsymbol{y}$ and the angle $\theta$ between them is given by the expression

$$\cos\theta = \frac{d\boldsymbol{x}\cdot d\boldsymbol{y}}{\|d\boldsymbol{x}\|\,\|d\boldsymbol{y}\|}\,. \tag{2.97}$$

Using successively (2.72) and (2.88), this last relation becomes

$$\begin{aligned}\cos\theta &= \frac{\boldsymbol{F}\,d\boldsymbol{X}\cdot\boldsymbol{F}\,d\boldsymbol{Y}}{\|\boldsymbol{F}\,d\boldsymbol{X}\|\,\|\boldsymbol{F}\,d\boldsymbol{Y}\|} = \frac{d\boldsymbol{X}\cdot\boldsymbol{F}^T\boldsymbol{F}\,d\boldsymbol{Y}}{\|\boldsymbol{F}d\boldsymbol{X}\|\,\|\boldsymbol{F}\,d\boldsymbol{Y}\|}\\ &= \frac{d\boldsymbol{X}\cdot\boldsymbol{C}\,d\boldsymbol{Y}}{\|\boldsymbol{F}\,d\boldsymbol{X}\|\,\|\boldsymbol{F}\,d\boldsymbol{Y}\|}\,.\end{aligned} \tag{2.98}$$

Since $d\boldsymbol{X} = \boldsymbol{N}_x\|d\boldsymbol{X}\|$ and $d\boldsymbol{Y} = \boldsymbol{N}_y\|d\boldsymbol{Y}\|$ where $\boldsymbol{N}_x$ and $\boldsymbol{N}_y$ are the unit vectors aligned with the linear elements, and as $\|\boldsymbol{F}\,d\boldsymbol{X}\| = (\boldsymbol{F}\,d\boldsymbol{X}\cdot\boldsymbol{F}\,d\boldsymbol{X})^{1/2} = (d\boldsymbol{X}\cdot\boldsymbol{C}\,d\boldsymbol{X})^{1/2}$, (2.98) yields

$$\cos\theta = \frac{\boldsymbol{N}_x\cdot\boldsymbol{C}\boldsymbol{N}_y}{(\boldsymbol{N}_x\cdot\boldsymbol{C}\boldsymbol{N}_x)^{1/2}(\boldsymbol{N}_y\cdot\boldsymbol{C}\boldsymbol{N}_y)^{1/2}}\,. \tag{2.99}$$

The angular difference $\Theta - \theta$ is normally attributed to shear.

In order to express the relation between the volume elements of the two configurations, we consider three infinitesimal, non-coplanar, linear elements $d\boldsymbol{X}$, $d\boldsymbol{Y}$, and $d\boldsymbol{Z}$ (fig. 2.14) in the reference configuration such that

$$dV = d\boldsymbol{X} \cdot (d\boldsymbol{Y} \times d\boldsymbol{Z}) > 0 \,. \tag{2.100}$$

In the deformed configuration, the three linear elements become $d\boldsymbol{x}$, $d\boldsymbol{y}$, and $d\boldsymbol{z}$, and the corresponding volume is given by

$$dv = d\boldsymbol{x} \cdot (d\boldsymbol{y} \times d\boldsymbol{z}) \,. \tag{2.101}$$

Since the volume is expressed as a determinant, from (2.75) we have

$$dv = \det \begin{pmatrix} dx_1 & dy_1 & dz_1 \\ dx_2 & dy_2 & dz_2 \\ dx_3 & dy_3 & dz_3 \end{pmatrix} = \det \begin{pmatrix} F_{1j}\, dX_j & F_{1j}\, dY_j & F_{1j}\, dZ_j \\ F_{2j}\, dX_j & F_{2j}\, dY_j & F_{2j}\, dZj \\ F_{3j}\, dX_j & F_{3j}\, dY_j & F_{3j}\, dZ_j \end{pmatrix} . \tag{2.102}$$

With this determinant and (2.68), the volume element is transformed according to the relation

$$dv = \det \boldsymbol{F}\, dV = J\, dV \,. \tag{2.103}$$

To evaluate the change in a surface element (fig. 2.14), we start with the expression of the volume element in the reference and deformed configurations

$$dV = d\boldsymbol{X'} \cdot \boldsymbol{N}\, dS \qquad dv = d\boldsymbol{x'} \cdot \boldsymbol{n}\, ds \,, \tag{2.104}$$

where the surface elements are indicated by $\boldsymbol{N}\, dS$ and $\boldsymbol{n}\, ds$ with $\boldsymbol{N}$ and $\boldsymbol{n}$ the unit normal vectors corresponding to the surface elements. Taking into account (2.72) and (2.104), (2.103) becomes

$$dv = \boldsymbol{F}\, d\boldsymbol{X'} \cdot \boldsymbol{n}\, ds = J\, d\boldsymbol{X'} \cdot \boldsymbol{N}\, dS \tag{2.105}$$

or

$$\left(\boldsymbol{F}^T \boldsymbol{n}\, ds - J\, \boldsymbol{N}\, dS\right) \cdot d\boldsymbol{X'} = 0 \,. \tag{2.106}$$

Since this relation is valid for an arbitrary choice of $d\boldsymbol{X'}$, we conclude that

$$\boldsymbol{n}\, ds = J\boldsymbol{F}^{-T} \boldsymbol{N}\, dS \qquad \text{or} \qquad d\boldsymbol{s} = J\boldsymbol{F}^{-T} \boldsymbol{N}\, dS \,. \tag{2.107}$$

This expression, called ***Nanson's formula***, relates the corresponding surface elements between the reference and deformed configurations.

### 2.7.3 Geometric Interpretation

We can interpret (2.73) geometrically. In order to do so, we first need to examine a few properties of the eigenvalues of $\boldsymbol{U}$ and $\boldsymbol{V}$. Let $\lambda_i$ $(i = 1, 2, 3)$ be the eigenvalues of $\boldsymbol{U}$ corresponding to the unit eigenvectors $\boldsymbol{A}_i$, then

$$\boldsymbol{U}\boldsymbol{A}_i = \lambda_i \boldsymbol{A}_i \qquad (\text{no sum over } i) \,. \tag{2.108}$$

As $\boldsymbol{U}$ is symmetric and positive definite, the $\lambda_i$ are real and $\lambda_i > 0$. In addition, by spectral decomposition (1.125), we have

$$\boldsymbol{U} = \lambda_1 \boldsymbol{A}_1 \otimes \boldsymbol{A}_1 + \lambda_2 \boldsymbol{A}_2 \otimes \boldsymbol{A}_2 + \lambda_3 \boldsymbol{A}_3 \otimes \boldsymbol{A}_3 \quad \text{with} \quad \boldsymbol{A}_i \cdot \boldsymbol{A}_j = \delta_{ij} \,. \tag{2.109}$$

Using (2.88) and (2.108), we can write

$$\boldsymbol{C}\boldsymbol{A}_i = \lambda_i^2 \boldsymbol{A}_i \qquad \text{(no sum over } i\text{)} \,. \tag{2.110}$$

Thus, the tensor $\boldsymbol{C}$ has $\lambda_i^2$ for eigenvalues and $\boldsymbol{A}_i$ $(i = 1, 2, 3)$ for eigenvectors. As for the eigenvalues and eigenvectors of $\boldsymbol{V}$, the use of (2.73) and (2.108) leads to

$$\boldsymbol{V}(\boldsymbol{R}\boldsymbol{A}_i) = \boldsymbol{R}\boldsymbol{U}\boldsymbol{A}_i = \lambda_i(\boldsymbol{R}\boldsymbol{A}_i) \qquad \text{(no sum over } i\text{)} \,. \tag{2.111}$$

This shows that the $\lambda_i$ ***are also the eigenvalues of*** $\boldsymbol{V}$, corresponding to the unit eigenvectors

$$\boldsymbol{b}_i = \boldsymbol{R}\boldsymbol{A}_i \,. \tag{2.112}$$

Thus the deformation transforms, with a rotation, the eigenvectors of $\boldsymbol{U}$ into those of $\boldsymbol{V}$. Similarly to (2.109), we can write

$$\boldsymbol{V} = \lambda_1 \boldsymbol{b}_1 \otimes \boldsymbol{b}_1 + \lambda_2 \boldsymbol{b}_2 \otimes \boldsymbol{b}_2 + \lambda_3 \boldsymbol{b}_3 \otimes \boldsymbol{b}_3 \quad \text{with} \quad \boldsymbol{b}_i \cdot \boldsymbol{b}_j = \delta_{ij} \,. \tag{2.113}$$

To obtain the eigenvalues and eigenvectors of $\boldsymbol{c}$, we use (2.89) and (2.110) to see that

$$\boldsymbol{V}^2(\boldsymbol{R}\boldsymbol{A}_i) = \boldsymbol{c}(\boldsymbol{R}\boldsymbol{A}_i) = \lambda_i^2(\boldsymbol{R}\boldsymbol{A}_i) \qquad \text{(no sum over } i\text{)} \,, \tag{2.114}$$

which proves that the tensors $\boldsymbol{V}$ and $\boldsymbol{c}$ have $\lambda_i$ and $\lambda_i^2$ for eigenvalues, respectively, and the same eigenvectors $\boldsymbol{b}_i$ $(i = 1, 2, 3)$. In the literature, the $\lambda_i$ are also called the ***principal stretches*** of the tensor $\boldsymbol{U}$, $\boldsymbol{A}_i$ the ***principal material directions***, and $\boldsymbol{b}_i$ the ***principal spatial directions***. In the case of a pure deformation, the difference between $\boldsymbol{A}_i$ and $\boldsymbol{b}_i$ disappears.

Now it is possible to give a geometric interpretation to (2.73). Consider a body whose initial, reference configuration is a unit sphere centered at the origin (fig. 2.15), and that undergoes the following homogeneous transformation (cf. sect. 2.8):

$$\boldsymbol{x} = \boldsymbol{F}\boldsymbol{A} \qquad \text{with} \qquad \|\boldsymbol{A}\| \leq 1 \,. \tag{2.115}$$

Substituting (2.73) in (2.115), we obtain

$$\boldsymbol{x} = \boldsymbol{R}\boldsymbol{U}\boldsymbol{A} = \boldsymbol{V}\boldsymbol{R}\boldsymbol{A} \qquad \text{with} \qquad \|\boldsymbol{A}\| \leq 1 \,. \tag{2.116}$$

From this expression, and in the light of the previous discussion on the properties of the eigenvalues of $\boldsymbol{U}$ and $\boldsymbol{V}$, the right and left polar decompositions can be geometrically interpreted as follows (fig. 2.15), where we show only the plane $(Ox_1x_2)$:

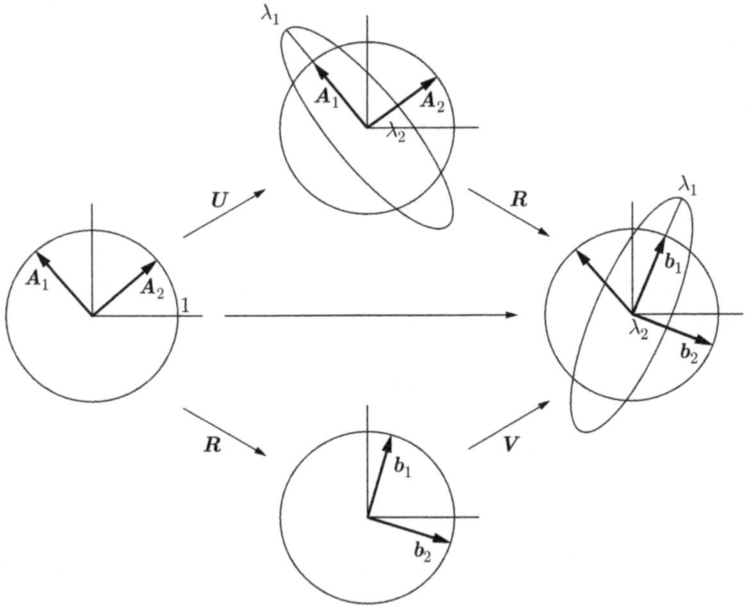


**Fig. 2.15** Geometric interpretation of $\boldsymbol{F} = \boldsymbol{RU} = \boldsymbol{VR}$

*Right Polar Decomposition:*

- First, applying the right stretch tensor $\boldsymbol{U}$ to the unit sphere, the radii in the directions of the principal material directions $\boldsymbol{A}_i$ are lengthened (or shortened) to their final lengths $\lambda_i$, remaining aligned with the principal material directions $\boldsymbol{A}_i$. Thus the sphere becomes an ellipsoid.
- Then, applying the rotation tensor $\boldsymbol{R}$ to the ellipsoid, the ellipsoid undergoes rotation and takes its final orientation with the axes, originally aligned with the principal material stretch directions, moving to alignment with $\boldsymbol{b}_i$.

*Left Polar Decomposition:*

- First, applying the rotation tensor $\boldsymbol{R}$ to the unit sphere, the axes aligned with the principal material directions $\boldsymbol{A}_i$ undergo a rotation and move to their final spatial orientations $\boldsymbol{b}_i$. The unit sphere turns around itself.
- Then, applying the left stretch tensor $\boldsymbol{V}$ to the rotated unit sphere, the radii aligned with the principal spatial directions $\boldsymbol{b}_i$ are lengthened (or shortened) to their final lengths $\lambda_i$, and the sphere becomes an ellipsoid.

As we have seen, mutually orthogonal lines which are along the principal material directions $\boldsymbol{A}_i$ before the transformation, remain mutually orthogonal after, and are aligned with the principal spatial directions $\boldsymbol{b}_i$. The rotation tensor $\boldsymbol{R}$ can thus be expressed as a function of $\boldsymbol{A}_i$ and $\boldsymbol{b}_i$:

$$\boldsymbol{R} = \boldsymbol{b}_1 \otimes \boldsymbol{A}_1 + \boldsymbol{b}_2 \otimes \boldsymbol{A}_2 + \boldsymbol{b}_3 \otimes \boldsymbol{A}_3 \,. \tag{2.117}$$

This relation can be easily established as follows. The rotation tensor is expressed as $\boldsymbol{R} = \boldsymbol{R}\boldsymbol{I}$ and the identity tensor as $\boldsymbol{I} = \boldsymbol{A}_i \otimes \boldsymbol{A}_i$. Taking into account (2.112) and the property (1.65), we obtain

$$\boldsymbol{R} = \boldsymbol{R}\boldsymbol{I} = \boldsymbol{R}(\boldsymbol{A}_i \otimes \boldsymbol{A}_i) = (\boldsymbol{R}\boldsymbol{A}_i) \otimes \boldsymbol{A}_i = \sum_{i=1}^{3} \boldsymbol{b}_i \otimes \boldsymbol{A}_i \,. \tag{2.118}$$

The deformation gradient tensor $\boldsymbol{F}$ can be expressed in terms of the vectors $\boldsymbol{A}_i$, $\boldsymbol{b}_i$, and the principal stretches $\lambda_i$ as

$$\boldsymbol{F} = \sum_{i=1}^{3} \lambda_i \boldsymbol{b}_i \otimes \boldsymbol{A}_i \,. \tag{2.119}$$

Next we present a method to determine the various kinematic tensors $\boldsymbol{C}$, $\boldsymbol{U}$, $\boldsymbol{c}$, $\boldsymbol{V}$, and $\boldsymbol{R}$. We can perform the concrete calculation of $\boldsymbol{U}$, $\boldsymbol{V}$, and $\boldsymbol{R}$ for a given deformation gradient $\boldsymbol{F}$. The tensors $\boldsymbol{U}$, $\boldsymbol{V}$, and $\boldsymbol{R}$ are determined from the relations

$$\boldsymbol{U} = \sqrt{\boldsymbol{F}^T\boldsymbol{F}} \qquad \boldsymbol{V} = \sqrt{\boldsymbol{F}\boldsymbol{F}^T} \qquad \boldsymbol{R} = \boldsymbol{F}\boldsymbol{U}^{-1} = \boldsymbol{V}^{-1}\boldsymbol{F} \,. \tag{2.120}$$

The principal difficulty is found in the calculation of $\boldsymbol{U}$ or the square root of $\boldsymbol{C} = \boldsymbol{F}^T\boldsymbol{F}$. Let $[P]$ be the orthogonal matrix that diagonalizes the matrix $[C]$ of the tensor $\boldsymbol{C}$, such that

$$[\Lambda^2] = [P][C][P]^T \qquad \text{or} \qquad \Lambda^2_{ij} = P_{im}C_{mn}P_{jn} \,, \tag{2.121}$$

with $[\Lambda^2] = \operatorname{diag}(\lambda_1^2, \lambda_2^2, \lambda_3^2)$, where the notation diag denotes the diagonal matrix, as in

$$\operatorname{diag}(a, b, c) = \begin{pmatrix} a & & \\ & b & \\ & & c \end{pmatrix} . \tag{2.122}$$

Then, the matrix $[U]$ of the tensor $\boldsymbol{U}$ is given by

$$[U] = [P]^T[\Lambda][P] \qquad \text{or} \qquad U_{ij} = P_{mi}\Lambda_{mn}P_{nj} \,, \tag{2.123}$$

with $[\Lambda] = \operatorname{diag}(\lambda_1, \lambda_2, \lambda_3)$. Having found $\boldsymbol{U}$, we can calculate $\boldsymbol{R}$ by (1.135) and $\boldsymbol{V}$ by (1.137) or with $\boldsymbol{V} = \boldsymbol{F}\boldsymbol{R}^T$.

## 2.8 Homogeneous Deformations

The deformation or transformation $\boldsymbol{x}$ of a body $\mathcal{B}$ is called ***homogeneous*** if the corresponding deformation gradient $\boldsymbol{F}$ is independent of the particle's position

$\boldsymbol{X}$. Geometrically, a homogeneous deformation transforms a straight line $P^0P$ of $\mathcal{R}_0$ to a straight line $p^0p$ of $\mathcal{R}$ (fig. 2.16). Such a deformation $\boldsymbol{x}$ is an ***affine transformation*** and has the following general form, with the notation $x_i^0 = \chi_i(X_j^0, t)$,

$$x_i = x_i^0(t) + M_{ij}(t)\big(X_j - X_j^0\big)\,. \tag{2.124}$$

In vector form, we have

$$\boldsymbol{x} = \boldsymbol{x}^0(t) + \boldsymbol{M}(t)\big(\boldsymbol{X} - \boldsymbol{X}^0\big)\,, \tag{2.125}$$

with $0 < \det \boldsymbol{M} < \infty$. The inverse relation is written as

$$\boldsymbol{X} = \boldsymbol{X}^0 + \boldsymbol{M}^{-1}(t)\big(\boldsymbol{x} - \boldsymbol{x}^0\big)\,. \tag{2.126}$$

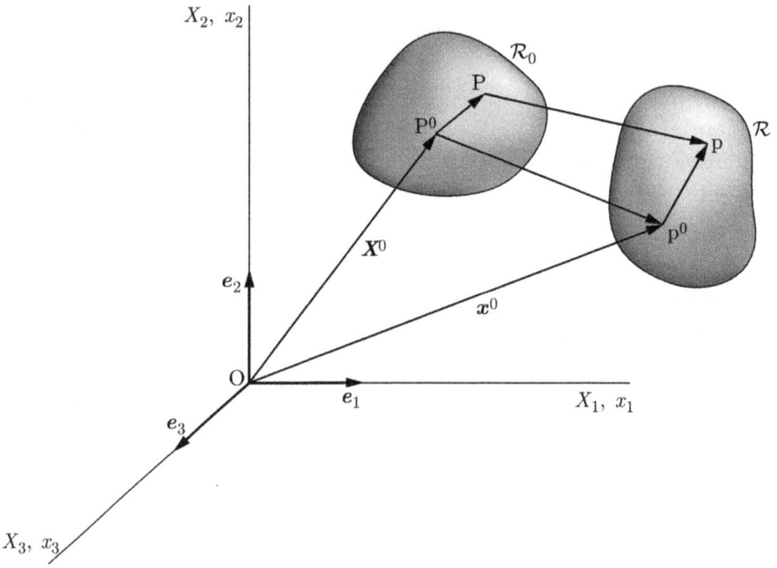


**Fig. 2.16** Transformation of a vector in homogeneous deformation

Below we present some important examples of homogeneous deformations obtained from relations (2.125) and (2.126).

*Translation:*

$\boldsymbol{M}$ is the unit tensor $\boldsymbol{I}$; without loss of generality, we set $\boldsymbol{X}^0 = \boldsymbol{0}$. We obtain

$$\boldsymbol{x} = \boldsymbol{x}^0(t) + \boldsymbol{X}\,. \tag{2.127}$$

*Rotation About the Origin:*

$\boldsymbol{X}^0 = \boldsymbol{x}^0 = \boldsymbol{0}$, and $\boldsymbol{M}$ is the ***rotation tensor*** $\boldsymbol{R}$ with the property

$$\boldsymbol{R}^T\boldsymbol{R} = \boldsymbol{R}\boldsymbol{R}^T = \boldsymbol{I} \qquad \text{and} \qquad \det(\boldsymbol{R}) = 1\,. \tag{2.128}$$

In this case, (2.125) and (2.126) become

$$\boldsymbol{x} = \boldsymbol{R}\boldsymbol{X} \qquad \boldsymbol{X} = \boldsymbol{R}^T\boldsymbol{x}\,. \tag{2.129}$$

Thus ***rigid body motion*** can be decomposed into a rotation followed by a translation.

*Uniform Expansion or Compression:*

$\boldsymbol{M} = m\boldsymbol{I}$ and (2.124) takes the form (fig. 2.17)

$$x_i = x_i^0 + m\big(X_i - X_i^0\big)\,. \tag{2.130}$$

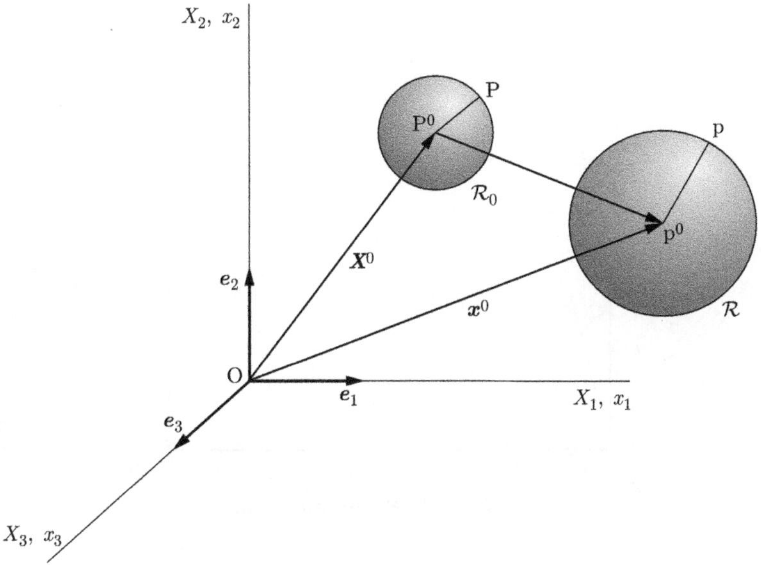


**Fig. 2.17** Uniform expansion

*Simple Shear:*

in a Cartesian coordinate system (fig. 2.18), the matrix $[M]$ is given by

$$[M] = \begin{pmatrix} 1 & k & 0 \\ 0 & 1 & 0 \\ 0 & 0 & 1 \end{pmatrix}, \tag{2.131}$$

or in vector notation, we have

$$\boldsymbol{x} = \boldsymbol{M}\boldsymbol{X} = (\boldsymbol{I} + k\,\boldsymbol{e}_1 \otimes \boldsymbol{e}_2)\boldsymbol{X}\ , \tag{2.132}$$

taking the origin as fixed, $\boldsymbol{X}^0 = \boldsymbol{x}^0 = \boldsymbol{0}$. Explicitly, (2.124) and (2.131) give

$$x_1 = X_1 + kX_2 \qquad x_2 = X_2 \qquad x_3 = X_3\,. \tag{2.133}$$

EXAMPLE 2.3

As an example, let us calculate $\boldsymbol{F}$, $\boldsymbol{C}$, $\boldsymbol{U}$, $\boldsymbol{V}$, and $\boldsymbol{R}$ for simple shear.

The matrices of $\boldsymbol{F}$ and $\boldsymbol{C}$ can be calculated directly, while obtaining those for $\boldsymbol{U}$, $\boldsymbol{V}$, and $\boldsymbol{R}$ requires solving an eigenvalue problem. The application of the definitions of $\boldsymbol{F}$ and $\boldsymbol{C}$ yields

$$[F] = \begin{pmatrix} 1 & k & 0 \\ 0 & 1 & 0 \\ 0 & 0 & 1 \end{pmatrix}$$

$$[C] = \begin{pmatrix} 1 & k & 0 \\ k & 1+k^2 & 0 \\ 0 & 0 & 1 \end{pmatrix}.$$

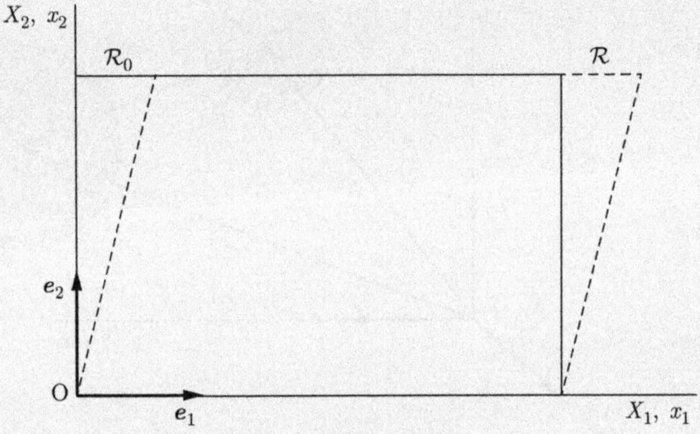


**Fig. 2.18** Simple shear

To calculate $\boldsymbol{U} = \sqrt{\boldsymbol{C}}$, we seek the diagonal matrix $[\Lambda^2]$ of $\boldsymbol{C}$. For this, we solve the eigenvalue problem (2.110). The corresponding characteristic equation is

$$\det\big([C] - \lambda_i^2[I]\big) = \big(\lambda_i^4 - (2+k^2)\lambda_i^2 + 1\big)\,\big(1 - \lambda_i^2\big) = 0\,. \tag{2.134}$$

The three solutions to this equation are

$$\begin{aligned} \lambda_1^2 &= 1 + \frac{1}{2}\,k^2 + k\sqrt{1 + \frac{1}{4}\,k^2} \\ \lambda_2^2 &= 1 + \frac{1}{2}k^2 - k\sqrt{1 + \frac{1}{4}k^2} \\ \lambda_3^2 &= 1\,. \end{aligned} \tag{2.135}$$

From which, we find the matrix $[\Lambda]$ defined by (2.123) such that

$$[\Lambda] = \operatorname{diag}(\lambda_1, \lambda_2, \lambda_3)\,. \tag{2.136}$$

Now we have to find the matrix $[P]$ in (2.123) to obtain the matrix of $\boldsymbol{U}$ with respect to the basis $\{\boldsymbol{e}_1, \boldsymbol{e}_2, \boldsymbol{e}_3\}$ for which the axes are the normalized eigenvectors $\boldsymbol{A}_i$ of $\boldsymbol{C}$. The eigenvectors can be determined from equation (2.110). After the calculations, we have the matrix rows

$$
\begin{aligned}
(A_1) &= \left(2+\frac{1}{2}k^2+k\sqrt{1+\frac{1}{4}k^2}\right)^{-1/2}\left(1,\ \sqrt{1+\frac{1}{4}k^2}+\frac{1}{2}k\,,\,0\right)\\
(A_2) &= \left(2+\frac{1}{2}k^2-k\sqrt{1+\frac{1}{4}k^2}\right)^{-1/2}\left(-1,\ \sqrt{1+\frac{1}{4}k^2}-\frac{1}{2}k\,,\,0\right)\\
(A_3) &= (0,0,1)\,,
\end{aligned}
$$

and the matrix $[P]$ is written as

$$
[P] = \begin{pmatrix} (A_1) \\ (A_2) \\ (A_3) \end{pmatrix}. \tag{2.137}
$$

Then we obtain

$$
\begin{aligned}
[U] = [P]^T[\Lambda][P] &= \frac{1}{\sqrt{1+\dfrac{k^2}{4}}}\begin{pmatrix} 1 & k/2 & 0 \\ k/2 & 1+k^2/2 & 0 \\ 0 & 0 & \sqrt{1+k^2/4} \end{pmatrix}\\
[R] = [FU^{-1}] &= \frac{1}{\sqrt{1+\dfrac{k^2}{4}}}\begin{pmatrix} 1 & k/2 & 0 \\ -k/2 & 1 & 0 \\ 0 & 0 & \sqrt{1+k^2/4} \end{pmatrix}\\
[V] = [FR^T] &= \frac{1}{\sqrt{1+\dfrac{k^2}{4}}}\begin{pmatrix} 1+k^2/2 & k/2 & 0 \\ k/2 & 1 & 0 \\ 0 & 0 & \sqrt{1+k^2/4} \end{pmatrix}.
\end{aligned} \tag{2.138}
$$

## 2.9 Small Displacements and Infinitesimal Strain Tensors

### 2.9.1 Small Displacements

Consider a displacement field dependent on a small real number $\varepsilon$ ($\varepsilon \ll 1$) such that

$$
\boldsymbol{U}(\boldsymbol{X}) = \varepsilon \boldsymbol{W}(\boldsymbol{X})\,, \tag{2.139}
$$

where $\boldsymbol{W}(\boldsymbol{X})$ is a ***given*** material displacement field, considered to be fixed for simplicity, to which the spatial field $\boldsymbol{w}(\boldsymbol{x})$ corresponds. From (2.86) and (2.87), the Green-Lagrange and Euler-Almansi strain tensors are given by

$$E_{ij} = \varepsilon \frac{1}{2} \left( \frac{\partial W_i}{\partial X_j} + \frac{\partial W_j}{\partial X_i} \right) + \varepsilon^2 \frac{1}{2} \frac{\partial W_m}{\partial X_i} \frac{\partial W_m}{\partial X_j} \tag{2.140}$$

$$e_{ij} = \varepsilon \frac{1}{2} \left( \frac{\partial w_i}{\partial x_j} + \frac{\partial w_j}{\partial x_i} \right) - \varepsilon^2 \frac{1}{2} \frac{\partial w_m}{\partial x_i} \frac{\partial w_m}{\partial x_j} \,. \tag{2.141}$$

When $\varepsilon$ approaches zero, the terms of order 2 become negligible compared to those of order 1. Thus we have

$$E_{ij} \simeq \varepsilon \frac{1}{2} \left( \frac{\partial W_i}{\partial X_j} + \frac{\partial W_j}{\partial X_i} \right) = \frac{1}{2} \left( \frac{\partial U_i}{\partial X_j} + \frac{\partial U_j}{\partial X_i} \right) \tag{2.142}$$

$$e_{ij} \simeq \varepsilon \frac{1}{2} \left( \frac{\partial w_i}{\partial x_j} + \frac{\partial w_j}{\partial x_i} \right) = \frac{1}{2} \left( \frac{\partial u_i}{\partial x_j} + \frac{\partial u_j}{\partial x_i} \right) . \tag{2.143}$$

In addition, starting from $x_i = X_i + U_i = X_i + \varepsilon W_i$ and $W_i(X_k) = w_i(x_k)$, we can write

$$\begin{aligned} \frac{\partial U_i}{\partial X_j} &= \varepsilon \frac{\partial W_i}{\partial X_j} = \varepsilon \frac{\partial w_i}{\partial x_k} \frac{\partial x_k}{\partial X_j} \\ &= \varepsilon \frac{\partial w_i}{\partial x_k} \left( \delta_{kj} + \varepsilon \frac{\partial W_k}{\partial X_j} \right) = \frac{\partial u_i}{\partial x_j} + \varepsilon^2 \frac{\partial w_i}{\partial x_k} \frac{\partial W_k}{\partial X_j} \end{aligned} \tag{2.144}$$

or

$$\frac{\partial U_i}{\partial X_j} = \frac{\partial u_i}{\partial x_j} + O(\varepsilon^2) \simeq \frac{\partial u_i}{\partial x_j} \,. \tag{2.145}$$

Consequently, in the case of small displacements, $O(\varepsilon^2) \to 0$ and the distinction between the Green-Lagrange and Euler-Almansi description of strains is negligible. Thus we can express the displacement derivatives with respect to the position before or after the deformation.

Experimental results show that in most engineering applications, the displacement gradient is also very small, such that

$$\left\| \frac{\partial U_i}{\partial X_j} \right\| = O(\varepsilon) \ll 1 \,. \tag{2.146}$$

In these conditions, (2.70) and (2.68) can be written as

$$F_{ij} = \delta_{ij} + O(\varepsilon), \quad F_{ij}^{-1} = \delta_{ij} - O(\varepsilon), \quad J \approx 1 + O(\varepsilon) \,. \tag{2.147}$$

For a tensor $\boldsymbol{L}$ of order 2, we can write

$$\begin{aligned} \frac{\partial \boldsymbol{L}}{\partial X_j} &= \frac{\partial x_i}{\partial X_j} \frac{\partial \boldsymbol{L}}{\partial x_i} = F_{ij} \frac{\partial \boldsymbol{L}}{\partial x_i} \\ &= \left( \delta_{ij} + \frac{\partial U_i}{\partial X_j} \right) \frac{\partial \boldsymbol{L}}{\partial x_i} = \frac{\partial \boldsymbol{L}}{\partial x_j} + \frac{\partial U_i}{\partial X_j} \frac{\partial \boldsymbol{L}}{\partial x_i} \,. \end{aligned} \tag{2.148}$$

Since $\partial U_i / \partial X_j$ is very small, the last term of (2.148) is negligible. Thus the material and spatial derivatives of the tensor $\boldsymbol{L}$ are approximately equal

$$\frac{\partial \boldsymbol{L}}{\partial X_j} \approx \frac{\partial \boldsymbol{L}}{\partial x_j} \,. \tag{2.149}$$

Relations (2.145)–(2.147) and (2.149) are the results of ***kinematic linearization***.

### 2.9.2 Infinitesimal Strain Tensor

We can now derive an important result from the preceding kinematic linearization. Relation (2.145) shows that if the terms of order $\varepsilon^2$ are negligible, there is no difference between the Green-Lagrange and Euler-Almansi strain tensors. It is thus natural to introduce the ***infinitesimal strain tensor*** $\boldsymbol{\varepsilon}$:

$$\varepsilon_{ij} = \frac{1}{2}\left(\frac{\partial U_i}{\partial X_j} + \frac{\partial U_j}{\partial X_i}\right) = \frac{1}{2}\left(\frac{\partial u_i}{\partial x_j} + \frac{\partial u_j}{\partial x_i}\right) \tag{2.150}$$

$$\boldsymbol{\varepsilon} = \frac{1}{2}\left(\boldsymbol{\nabla U} + (\boldsymbol{\nabla U})^T\right) = \frac{1}{2}\left(\boldsymbol{\nabla u} + (\boldsymbol{\nabla u})^T\right).$$

Note that $\boldsymbol{\varepsilon}$ is a tensor since the gradient of the displacement vector is a tensor (see sec. 1.4.3). Thus the transformation law for its components is given by (1.52) and the eigenvalues, which correspond to the principal infinitesimal strains, are from the solutions of equation (1.120) with $\boldsymbol{L} = \boldsymbol{\varepsilon}$. It is also interesting to note that the tensor $\boldsymbol{\varepsilon}$ is linear in $\nabla \boldsymbol{u}$. Therefore for the strains $\boldsymbol{\varepsilon}^{(1)}, \boldsymbol{\varepsilon}^{(2)}, \ldots$ resulting from the displacements $\boldsymbol{u}^{(1)}, \boldsymbol{u}^{(2)}, \ldots$, the principle of superposition applies, i.e., the total strain $\boldsymbol{\varepsilon} = \boldsymbol{\varepsilon}^{(1)} + \boldsymbol{\varepsilon}^{(2)} + \ldots$ corresponds to $\boldsymbol{u} = \boldsymbol{u}^{(1)} + \boldsymbol{u}^{(2)} + \ldots$.

We can interpret the components $\varepsilon_{ij}$ geometrically. Consider the small displacements between $\mathcal{R}_0$ and $\mathcal{R}$ (fig. 2.19). Let $d\boldsymbol{X}$ be an infinitesimal vector attached to the point $\boldsymbol{X}$ with components $(dX_1, 0, 0)$. The length $\|d\boldsymbol{x}\|$ of the corresponding vector $d\boldsymbol{x}$ in $\mathcal{R}$ is given by (2.80):

$$\|d\boldsymbol{x}\|^2 = \|d\boldsymbol{X}\|^2 + 2E_{ij}\, dX_i\, dX_j = (1 + 2E_{11})\, dX_1^2\,. \tag{2.151}$$

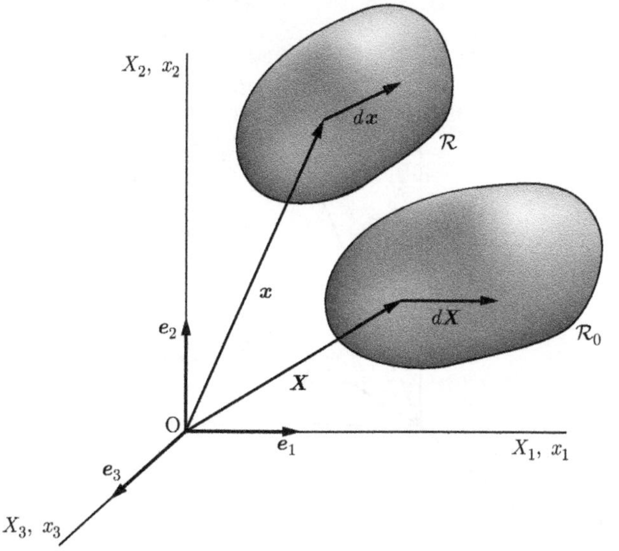


**Fig. 2.19** Deformation of an infinitesimal vector aligned with $\boldsymbol{e}_1$

Assuming small displacement gradients as discussed earlier, we can write

$$\begin{aligned} \|d\boldsymbol{x}\|^2 &\cong (1+2\varepsilon_{11})\|d\boldsymbol{X}\|^2 \\ \|d\boldsymbol{x}\| &\cong (1+2\varepsilon_{11})^{1/2}\, dX_1 = (1+\varepsilon_{11})\|d\boldsymbol{X}\|\ , \end{aligned} \tag{2.152}$$

from which

$$\varepsilon_{11} \cong \frac{\|d\boldsymbol{x}\| - \|d\boldsymbol{X}\|}{\|d\boldsymbol{X}\|}\,. \tag{2.153}$$

Thus, $\varepsilon_{11}$ measures the ***relative extension*** of a material line element aligned with direction 1. The other diagonal components $\varepsilon_{22}$ and $\varepsilon_{33}$ of $\boldsymbol{\varepsilon}$ have similar interpretations.

As for the geometric meaning of $\varepsilon_{12}$, consider two orthogonal vectors in $\mathcal{R}_0$ (fig. 2.20):

$$dX = (dX_1, 0, 0) \qquad \text{and} \qquad dY = (0, dY_2, 0)\,. \tag{2.154}$$

In $\mathcal{R}$, they are deformed and become the vectors $d\boldsymbol{x}$ and $d\boldsymbol{y}$ with components

$$dx_i = F_{i1}\, dX_1 \qquad\qquad dy_i = F_{i2}\, dY_2\,. \tag{2.155}$$

The lengths of $d\boldsymbol{x}$ and $d\boldsymbol{y}$ are

$$\|d\boldsymbol{x}\| \cong (1+\varepsilon_{11})\, dX_1 \qquad\qquad \|d\boldsymbol{y}\| \cong (1+\varepsilon_{22})\, dY_2\,. \tag{2.156}$$

Denoting by $\gamma_{12}$ the angle between $d\boldsymbol{x}$ and $d\boldsymbol{y}$, we have

$$\cos\gamma_{12} = \frac{d\boldsymbol{x}\cdot d\boldsymbol{y}}{\|d\boldsymbol{x}\|\,\|d\boldsymbol{y}\|} \cong \frac{2\varepsilon_{12}}{(1+\varepsilon_{11})(1+\varepsilon_{22})} \cong 2\varepsilon_{12}\,. \tag{2.157}$$

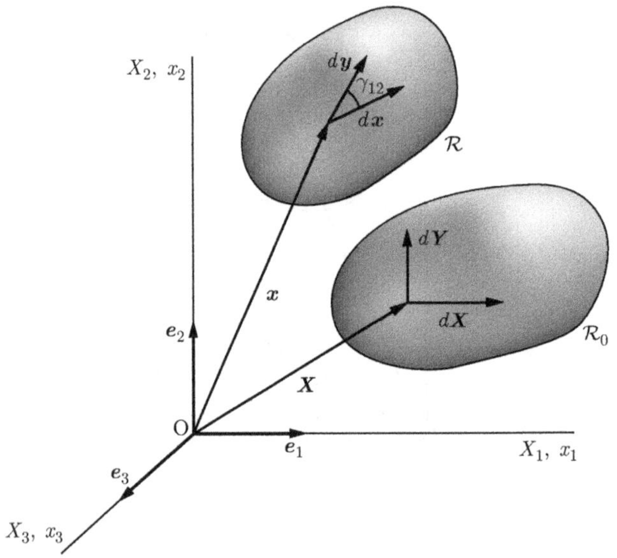


**Fig. 2.20** Modification of the angles between two vectors

Introducing the ***slip angle*** $\phi_{12}$ between two orthogonal directions $d\boldsymbol{X}$ and $d\boldsymbol{Y}$ (fig. 2.20), defined by

$$\phi_{12} = \frac{\pi}{2} - \gamma_{12}\,, \tag{2.158}$$

relation (2.157) can then be written as follows:

$$\cos\gamma_{12} = \sin\phi_{12} \cong \phi_{12} \cong 2\varepsilon_{12}\,. \tag{2.159}$$

That is, $\varepsilon_{12}$ is half of the cosine of the angle between the directions of the deformed infinitesimal vectors aligned with the directions 1 and 2 in $\mathcal{R}_0$. Similar interpretations can be given to $\varepsilon_{23}$ and $\varepsilon_{31}$.

The relative variation of volume is expressed in terms of the relative extensions. Consider three orthogonal vectors $d\boldsymbol{X} = dX\,\boldsymbol{e}_1, d\boldsymbol{Y} = dY\,\boldsymbol{e}_2, d\boldsymbol{Z} = dZ\,\boldsymbol{e}_3$ in the reference configuration. The volume of this cube is $dV = dXdYdZ$. After deformation, each element is modified as follows:

$$dx = (1+\varepsilon_{11})dX, \quad dy = (1+\varepsilon_{22})dY, \quad dz = (1+\varepsilon_{33})dZ\;, \tag{2.160}$$

and the volume after deformation is expressed as

$$\begin{aligned} dv &= dxdydz = (1+\varepsilon_{11})(1+\varepsilon_{22})(1+\varepsilon_{33})dXdYdZ \\ &= (1+\varepsilon_{11})(1+\varepsilon_{22})(1+\varepsilon_{33})dV\;. \end{aligned} \tag{2.161}$$

Neglecting the higher order terms of the deformation, we obtain the relative variation of the volume

$$\frac{dv - dV}{dV} = \varepsilon_{11} + \varepsilon_{22} + \varepsilon_{33} = \varepsilon_{ii}\;. \tag{2.162}$$

Note that in this case of infinitesimal strain, expression (2.162) is the trace of the gradient of the displacement vector, and using (2.150) we have

$$\varepsilon_{ii} = \operatorname{div}\boldsymbol{u} = \operatorname{tr}(\boldsymbol{\nabla u})\;. \tag{2.163}$$

As we have seen with (2.150), the infinitesimal strain tensor $\varepsilon_{ij}$ corresponds to the symmetric part of the displacement gradient $\partial u_i/\partial x_j$. Thus we have

$$du_i = \frac{\partial u_i}{\partial x_j}dx_j = \frac{1}{2}\left(\frac{\partial u_i}{\partial x_j} + \frac{\partial u_j}{\partial x_i}\right)dx_j + \frac{1}{2}\left(\frac{\partial u_i}{\partial x_j} - \frac{\partial u_j}{\partial X_i}\right)dx_j \tag{2.164}$$

$$d\boldsymbol{u} = \boldsymbol{\nabla u}\,d\boldsymbol{x} = \frac{1}{2}\left(\boldsymbol{\nabla u} + (\boldsymbol{\nabla u})^T\right)d\boldsymbol{x} + \frac{1}{2}\left(\boldsymbol{\nabla u} - (\boldsymbol{\nabla u})^T\right)d\boldsymbol{x}\;, \tag{2.165}$$

and we can thus define the antisymmetric part

$$\omega_{ij} = \frac{1}{2}\left(\frac{\partial u_i}{\partial x_j} - \frac{\partial u_j}{\partial x_i}\right) \tag{2.166}$$

$$\boldsymbol{\omega} = \frac{1}{2}\left(\boldsymbol{\nabla u} - (\boldsymbol{\nabla u})^T\right) \tag{2.167}$$

as the ***infinitesimal rotation tensor*** $\boldsymbol{\omega}$ and write

$$du_i = \varepsilon_{ij}\, dx_j + \omega_{ij}\, dx_j\,.$$

Note that the curl of the displacement vector $\boldsymbol{u}$ is related to a rotation vector of the body, whose components are those of the tensor $\boldsymbol{\omega}$ multiplied by the factor 2, that is,

$$\frac{1}{2}\boldsymbol{\nabla}\times\boldsymbol{u} = \omega_{32}\boldsymbol{e}_1 + \omega_{13}\boldsymbol{e}_2 + \omega_{21}\boldsymbol{e}_3\,. \tag{2.168}$$

Consequently, the infinitesimal displacement can be decomposed into a sum of a pure strain tensor and a pure rotation. However, an additive decomposition of the displacement gradient is not possible for large strains ($\boldsymbol{E} \neq \boldsymbol{\varepsilon}$). In this case we can use (2.73). A relation between the rotation tensor $\boldsymbol{R}$ (eqn. (2.73)) and the infinitesimal rotation tensor $\boldsymbol{\omega}$ can be easily established (exercise 2.13).

### 2.9.3 Compatibility Equations for the Infinitesimal Strain Tensor

For a given displacement field $\boldsymbol{u}$, the components of the infinitesimal strain tensor are easily calculated:

$$\varepsilon_{ij} = \frac{1}{2}\left(\frac{\partial u_i}{\partial x_j} + \frac{\partial u_j}{\partial x_i}\right) = \frac{1}{2}\left(u_{i,j} + u_{j,i}\right). \tag{2.169}$$

However, for a given $\varepsilon_{ij}$, a corresponding displacement field does not necessarily exist. Now we will determine the conditions that the components $\varepsilon_{ij}$ must meet to ensure the existence of a displacement $\boldsymbol{u}$ that satisfies (2.169). Differentiating (2.169), we obtain

$$\varepsilon_{ij,kl} = \frac{1}{2}\left(u_{i,jkl} + u_{j,ikl}\right), \tag{2.170}$$

where the indices $k$ and $l$ that follow a comma indicate, for example, the partial derivatives with respect to $x_k$ and $x_l$, respectively. Interchanging the indices, we have

$$\varepsilon_{kl,ij} = \frac{1}{2}\left(u_{k,lij} + u_{l,kij}\right) \tag{2.171}$$

$$\varepsilon_{jl,ik} = \frac{1}{2}\left(u_{j,lik} + u_{l,jik}\right) \tag{2.172}$$

$$\varepsilon_{ik,jl} = \frac{1}{2}\left(u_{i,kjl} + u_{k,ijl}\right). \tag{2.173}$$

Using the index symmetries of the partial derivatives of $\boldsymbol{u}$, it is not difficult to verify that

$$\varepsilon_{ij,kl} + \varepsilon_{kl,ij} - \varepsilon_{jl,ik} - \varepsilon_{ik,jl} = 0\,. \tag{2.174}$$

These are the Saint-Venant ***compatibility equations***. Among the 81 equations represented by (2.174), only six are independent due to the symmetry of

$\varepsilon_{ij}$ and their derivatives. Explicitly, these six compatibility equations are

$$\begin{aligned}
\frac{\partial^2 \varepsilon_{11}}{\partial x_2 \partial x_3} &= \frac{\partial}{\partial x_1}\left(-\frac{\partial \varepsilon_{23}}{\partial x_1} + \frac{\partial \varepsilon_{31}}{\partial x_2} + \frac{\partial \varepsilon_{12}}{\partial x_3}\right) \\
\frac{\partial^2 \varepsilon_{22}}{\partial x_3 \partial x_1} &= \frac{\partial}{\partial x_2}\left(-\frac{\partial \varepsilon_{31}}{\partial x_2} + \frac{\partial \varepsilon_{12}}{\partial x_3} + \frac{\partial \varepsilon_{23}}{\partial x_1}\right) \\
\frac{\partial^2 \varepsilon_{33}}{\partial x_1 \partial x_2} &= \frac{\partial}{\partial x_3}\left(-\frac{\partial \varepsilon_{12}}{\partial x_3} + \frac{\partial \varepsilon_{23}}{\partial x_1} + \frac{\partial \varepsilon_{31}}{\partial x_2}\right) \\
\frac{\partial^2 \varepsilon_{12}}{\partial x_1 \partial x_2} &= \frac{1}{2}\left(\frac{\partial^2 \varepsilon_{11}}{\partial {x_2}^2} + \frac{\partial^2 \varepsilon_{22}}{\partial {x_1}^2}\right) \\
\frac{\partial^2 \varepsilon_{23}}{\partial x_2 \partial x_3} &= \frac{1}{2}\left(\frac{\partial^2 \varepsilon_{22}}{\partial {x_3}^2} + \frac{\partial^2 \varepsilon_{33}}{\partial {x_2}^2}\right) \\
\frac{\partial^2 \varepsilon_{31}}{\partial x_3 \partial x_1} &= \frac{1}{2}\left(\frac{\partial^2 \varepsilon_{33}}{\partial {x_1}^2} + \frac{\partial^2 \varepsilon_{11}}{\partial {x_3}^2}\right) .
\end{aligned} \tag{2.175}$$

It is clear from the procedure we have followed to obtain (2.175) that when the displacement field is known, relations (2.175) are automatically satisfied. When $\varepsilon_{ij}$ are given, these six equations are necessary and sufficient to ensure the existence of a unique displacement field $\boldsymbol{u}$ satisfying (2.169), with possibly an additional rigid body motion, only if the body $\mathcal{B}$ is ***simply connected.*** For a ***multiply connected*** elastic solid, they are not sufficient and additional conditions are necessary.

An interpretation of the compatibility conditions (2.175) is given by the following thought experiment. Consider a two-dimensional body, for example, a plate of uniform thickness, cut into small square pieces. When there is no deformation of the plate, the pieces adjust perfectly to form the plate. Then we impose on each piece an arbitrary strain field and we attempt to assemble them again to reform the plate. While reconstructing the plate we see that, in general, they do not all fit, as they are separated by spaces between some or all of them. A perfect fit is not obtained unless the imposed strain on each square satisfies (2.175).

## 2.10 Velocity Gradient and Associated Tensors

In numerous problems in mechanics of continuous media, an interesting kinematic quantity is not only the change in the shape of the body, but ***the rate at which this change is produced.*** This is especially the case in fluid mechanics.

Let $\mathcal{V}$ be the neighborhood of the point P with coordinates $x_i$, and $Q$ an arbitrary point belonging to $\mathcal{V}$ with coordinates $x_i + dx_i$. The spatial velocity of $Q$ is given by

$$v_i(x_j + dx_j, t) = v_i(x_j, t) + \frac{\partial v_i(x_j, t)}{\partial x_j}\, dx_j + \cdots \ . \tag{2.176}$$

The tensor $\boldsymbol{L}$ whose components are

$$L_{ij} = \frac{\partial v_i}{\partial x_j} = (\boldsymbol{\nabla v})_{ij} \tag{2.177}$$

is called the ***velocity gradient***. We establish a relation between $\boldsymbol{L}$ and $\boldsymbol{F}$ as follows:

$$\begin{aligned} \dot{F}_{ij} &= \frac{D}{Dt}\left(\frac{\partial x_i}{\partial X_j}\right) = \frac{\partial \dot{x}_i}{\partial X_j} = \frac{\partial \dot{x}_i}{\partial x_m}\frac{\partial x_m}{\partial X_j} \\ &= \frac{\partial v_i}{\partial x_m}\frac{\partial x_m}{\partial X_j} = L_{im}F_{mj}\,. \end{aligned} \tag{2.178}$$

Then we have

$$\dot{\boldsymbol{F}} = \boldsymbol{LF}\,. \tag{2.179}$$

The symmetric part of $\boldsymbol{L}$, that is,

$$d_{ij} = \frac{1}{2}\left(\frac{\partial v_i}{\partial x_j} + \frac{\partial v_j}{\partial x_i}\right) \tag{2.180}$$

$$\boldsymbol{d} = \frac{1}{2}\left(\boldsymbol{\nabla v} + (\boldsymbol{\nabla v})^T\right) \tag{2.181}$$

is called the ***strain rate tensor*** or the ***rate of deformation tensor***, and the antisymmetric part of $\boldsymbol{L}$, that is,

$$\dot{\omega}_{ij} = \frac{1}{2}\left(\frac{\partial v_i}{\partial x_j} - \frac{\partial v_j}{\partial x_i}\right) \tag{2.182}$$

$$\dot{\boldsymbol{\omega}} = \frac{1}{2}\left(\boldsymbol{\nabla v} - (\boldsymbol{\nabla v})^T\right) \tag{2.183}$$

is called the ***rotation rate tensor*** or the ***rate of rotation tensor***. Thus we can write

$$\boldsymbol{L} = \boldsymbol{d} + \dot{\boldsymbol{\omega}}\,. \tag{2.184}$$

With definitions (2.180) and (2.182), it follows from (2.176) that

$$v_i(x_j + dx_j, t) - v_i(x_j, t) \cong d_{ij}\,dx_j + \dot{\omega}_{ij}\,dx_j\,. \tag{2.185}$$

Comparison of (2.180) with the definition of $\varepsilon_{ij}$ gives

$$d_{ij} = \frac{d\varepsilon_{ij}}{dt}\,. \tag{2.186}$$

And it is for this reason that $d_{ij}$ is called the rate of deformation tensor. However, we need to notice that the linearity of $d_{ij}$ with respect to $v_i$ in (2.180) is exact, no assumption of small displacements having been made during its calculation. The dual vector $\dot{\Omega}_i$ (1.104) associated with the rotation rate tensor, that is,

$$\dot{\Omega}_i = -\frac{1}{2}\varepsilon_{ijk}\dot{\omega}_{jk} = \frac{1}{2}\left(\mathbf{curl}\,\boldsymbol{v}\right)_i, \tag{2.187}$$

is called the ***rotation rate vector***. Note that in fluid mechanics, one typically introduces the vorticity vector $\boldsymbol{\omega}$ with the definition as the curl of the velocity. Then

$$\boldsymbol{\omega} = \mathbf{curl}\, \boldsymbol{v} = \boldsymbol{\nabla} \times \boldsymbol{v} \tag{2.188}$$

or

$$\omega_i = \varepsilon_{ijk} \frac{\partial v_k}{\partial x_j} \; . \tag{2.189}$$

And we easily deduce that

$$\boldsymbol{\omega} = 2\dot{\boldsymbol{\Omega}}. \tag{2.190}$$

To better understand the vorticity vector, consider the decomposition of a local motion of a fluid. Let P be a point at position $\boldsymbol{x}$ and P′ a neighboring point as shown in figure 2.21.

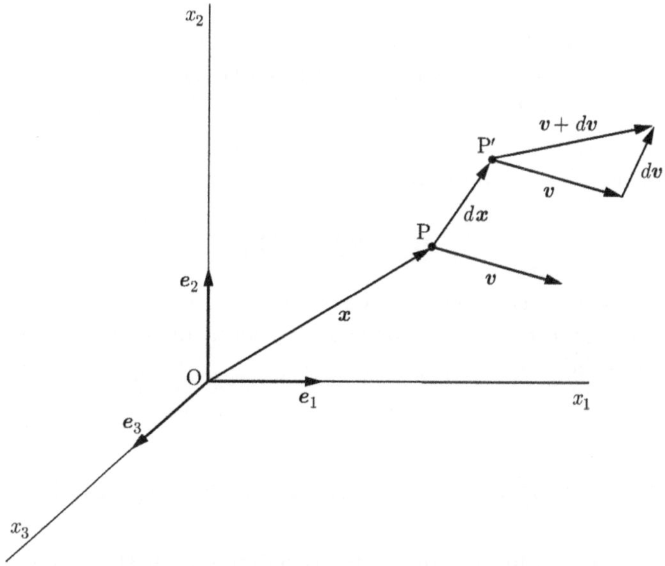


**Fig. 2.21** Relative motion of two fluid particles

The vector position of P′ relative to P is $d\boldsymbol{x}$. After an infinitesimal lapse of time, P and P′ occupy new positions. P moves with the local velocity $\boldsymbol{v}$ and P′ with the velocity $\boldsymbol{v} + d\boldsymbol{v}$. We consider P to be the principal fluid particle and, subtracting its translational velocity, we describe the motion of P′ as observed from this principal particle. This reasoning is thus valid only when the distance $d\boldsymbol{x}$ is very small. We can decompose the motion of P and P′ into three distinct parts: a translation, a rigid body rotation, and a strain. The translational motion is given by the velocity $\boldsymbol{v}$ of P. All the other motions, taken together, are given by $d\boldsymbol{v}$, the velocity of P′ with respect to P. We then have

$$d\boldsymbol{v} = \frac{\partial \boldsymbol{v}}{\partial \boldsymbol{x}} d\boldsymbol{x} = \boldsymbol{L}\, d\boldsymbol{x} \; . \tag{2.191}$$

By (2.184), the strain motions (stretching, shortening, ...) of P′ with respect to P are described by $\boldsymbol{d}$. Consequently the rotational motion of P′ with respect to P is taken into account by $\dot{\boldsymbol{\omega}}$. We can write

$$d\boldsymbol{v}^{(r)} = \dot{\boldsymbol{\omega}}\, d\boldsymbol{x} \ , \tag{2.192}$$

where the superscript $r$ refers to rotation.

The rigid body rotational motion of P′ with respect to P must have the form of the equation $\boldsymbol{v} = \boldsymbol{\Omega} \times \boldsymbol{x}$, where $\boldsymbol{\Omega}$ is the rate of angular rotation (a vector). By (2.187) and (2.190), we have $-\dot{\omega}_{ij} = \frac{1}{2}\varepsilon_{ijk}\omega_k = \dot{\omega}_{ji}$. Thus the rotational component of motion is given by

$$\begin{aligned} dv_j^{(r)} &= \dot{\omega}_{ji}\, dx_i = \frac{1}{2}\varepsilon_{ijk}\omega_k\, dx_i \\ &= \frac{1}{2}\varepsilon_{jki}\,(\omega_k)\, dx_i \ . \end{aligned} \tag{2.193}$$

This last equation is of the form $d\boldsymbol{v} = \boldsymbol{\Omega} \times d\boldsymbol{x}$. The vorticity vector $\boldsymbol{\omega}$ corresponds to an angular velocity such that the vorticity $\boldsymbol{\omega}$ is equal to $2\boldsymbol{\Omega}$, that is two times the vector rate of rigid body rotation of P′ with respect to P.

Note that in the case of rotation of a rigid body, the tensor $\boldsymbol{L}$ is obtained from equation (2.55). With definition (2.57), we have

$$\boldsymbol{L} = \dot{\boldsymbol{Q}}\boldsymbol{Q}^T = \boldsymbol{\Omega} \ , \tag{2.194}$$

which is an antisymmetric tensor. This shows in this case that $\boldsymbol{d} = 0$ and $\boldsymbol{L} = \dot{\boldsymbol{\omega}}$. The rotation rate tensor is thus entirely determined by the instantaneous rotation of the solid.

## 2.11 Objectivity of the Kinematic Quantities

The description of a physical quantity associated with the motion of a body generally depends on the choice of observer or reference frame.

In physics, we frequently use the inertial reference frame for which space is homogeneous and time is uniform. In this reference frame the Newtonian laws of motion are valid. A body in uniform straight motion during an interval of time experiences no force. Stated otherwise, in this reference frame we observe that the center of mass of a body $\mathcal{B}$ moves along a straight line at uniform velocity if no force is applied to the body.

In chapter 1, we developed the consequences induced by a change of the coordinate system for the same vector (event) in the case of a single observer. This development is the basis of tensor analysis and finds its origin in the requirement that all laws of continuum physics must be independent of the choice of coordinate system by the observer. However, when the same event in space is seen by two observers or reference frames, the relations between the coordinates and the various kinematic parameters are different in the two reference frames.

In the following, the ***observer*** or the ***reference frame*** will be denoted by $\mathrm{R} = (\mathrm{O}, \boldsymbol{x}, t)$. We want to distinguish the kinematic parameters, scalars, vectors, or tensors, which depend intrinsically on the observer from those that are essentially independent. This will be a preparation for the future discussion of the objectivity of the general or constitutive laws of physics. In mechanics of continuous media, an event, that is, a physical process, is defined by its coordinates in space $\boldsymbol{x}$ and the observation time $t$.

Consider an event viewed by two observers R and R$^*$, and noted respectively by $(\boldsymbol{x}, t)$ and $(\boldsymbol{x}^*, t^*)$. The motion between two observers is a function of space and time. If the effects due to relativity are negligible and we assume that the observers measure the same distance between two events as well as the same time intervals between events, we can show that the most general transformation between the observations $(\boldsymbol{x}, t)$ and $(\boldsymbol{x}^*, t^*)$ of the same event is given by

$$\boldsymbol{x}^* = \boldsymbol{Q}(t)\boldsymbol{x} + \boldsymbol{c}(t) \qquad t^* = t - \alpha\,, \tag{2.195}$$

where $\boldsymbol{Q}(t)$ is an orthogonal rotation tensor with the time $t$ as parameter, $\boldsymbol{c}(t)$ a vector, and $\alpha$ a scalar constant.

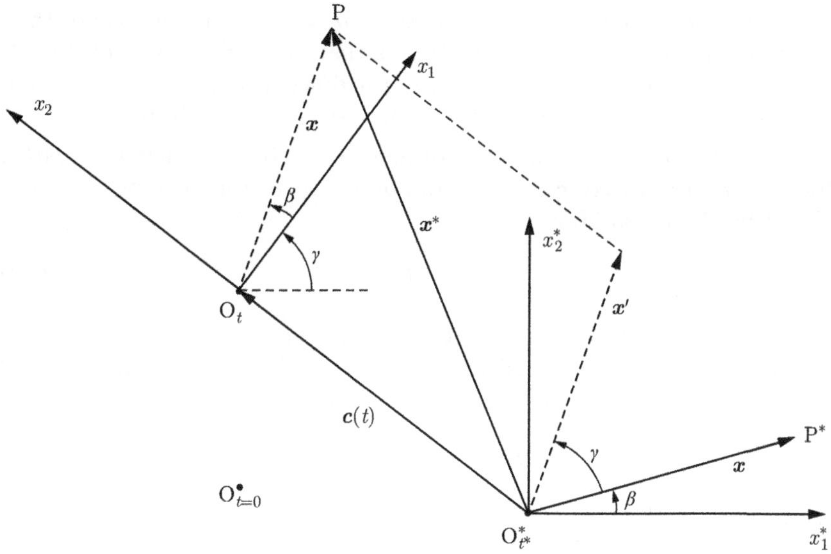


**Fig. 2.22** Interpretation of (2.195). Two observers initially positioned at $O^\bullet_{t=0}$ move in two reference frames

The interpretation of (2.195) is shown in figure 2.22, where we have two observers R and R$^*$ at time $t$. Suppose that an event (i.e., an experiment) takes place in P. The vector position of P with respect to the observer R is $\boldsymbol{x}$. The same event viewed by the observer R$^*$ is not simply given by the vector addition $\boldsymbol{x}^* = \boldsymbol{c}(t) + \boldsymbol{x}$, but rather by the general expression (2.195). We must take into account the rigid body rotation of observer R with respect to observer R*, for observer R* to see the same event. This rotation is accomplished by the tensor $\boldsymbol{Q}(t)$, for which the components are functions of two angles, $\beta$ and $\gamma$, and time. We note that the change of reference frame represents more than a simple coordinate transformation.

Therefore, the motion of a body $\mathcal{B}$, described by $\chi(\boldsymbol{X}, t)$ according to the first observer, is described by the second observer as $\chi^*(\boldsymbol{X}, t^*)$, and these two descriptions are related as follows:

$$\chi^*(\boldsymbol{X}, t^*) = \boldsymbol{Q}(t)\chi(\boldsymbol{X}, t) + \boldsymbol{c}(t) \qquad t^* = t - \alpha\,. \tag{2.196}$$

In order to examine the ramifications of the transformation (2.196), consider two events simultaneously recorded by R as $(\boldsymbol{x}_1, t)$ and $(\boldsymbol{x}_2, t)$, and as $(\boldsymbol{x}_1^*, t)$ and $(\boldsymbol{x}_2^*, t)$ by R*. For these two events, the relative positions viewed by the two observers are $\boldsymbol{u} = \boldsymbol{x}_2 - \boldsymbol{x}_1$ and $\boldsymbol{u}^* = \boldsymbol{x}_2^* - \boldsymbol{x}_1^*$, respectively. From relation (2.196), we obtain

$$\boldsymbol{u}^* = \boldsymbol{Q}\boldsymbol{u}\,, \quad t^* = t - \alpha\,. \tag{2.197}$$

Since the ***tensor Q is orthogonal***, we can write

$$\boldsymbol{u}^* \cdot \boldsymbol{u}^* = (\boldsymbol{Q}\boldsymbol{u}) \cdot (\boldsymbol{Q}\boldsymbol{u}) = \boldsymbol{u} \cdot (\boldsymbol{Q}^T\boldsymbol{Q})\boldsymbol{u} = \boldsymbol{u} \cdot \boldsymbol{u}\,. \tag{2.198}$$

This last equation shows that the norms of $\boldsymbol{u}$ and $\boldsymbol{u}^*$ are the same, that is, $\|\boldsymbol{u}^*\| = \|\boldsymbol{u}\|$ and that the transformation is that of a rigid body rotation (sec. 2.6.2). The vector fields which are transformed according to (2.197) are called spatially objective or indifferent with respect to the reference frame.

By the definition of a ***spatially objective vector***, we define a ***spatially objective tensor of order*** 2. Let $\boldsymbol{v}$ and $\boldsymbol{w}$ be two vectors seen by the observer R, related by the tensor $\boldsymbol{L}$ of order 2

$$\boldsymbol{w} = \boldsymbol{L}\boldsymbol{v}\,. \tag{2.199}$$

Since $\boldsymbol{v}$ and $\boldsymbol{w}$ are objective, the observer R* sees them as $\boldsymbol{w}^* = \boldsymbol{Q}\boldsymbol{w}$ and $\boldsymbol{v}^* = \boldsymbol{Q}\boldsymbol{v}$. This observer considers the tensor of order 2 as $\boldsymbol{L}^*$, such that $\boldsymbol{w}^* = \boldsymbol{L}^*\boldsymbol{v}^*$. To relate $\boldsymbol{L}$ and $\boldsymbol{L}^*$, we notice that

$$\boldsymbol{w}^* = \boldsymbol{Q}\boldsymbol{w} = \boldsymbol{Q}\boldsymbol{L}\boldsymbol{v} = \boldsymbol{Q}\boldsymbol{L}\boldsymbol{Q}^T\boldsymbol{v}^*\,. \tag{2.200}$$

From this last relation we deduce the equation

$$\boldsymbol{L}^* = \boldsymbol{Q}\boldsymbol{L}\boldsymbol{Q}^T\,. \tag{2.201}$$

Tensor fields of order 2, which are transformed according to (2.201) when the observer is changed, are called spatially objective, or independent of the reference frame. As for a scalar field, it is called objective or independent of the reference frame when

$$f^*(\boldsymbol{x}^*, t) = f(\boldsymbol{x}, t)\,. \tag{2.202}$$

In the following, we say that:

- a scalar quantity $\phi$ is ***objective*** if and only if (iff) $\phi^* = \phi$;
- a vector quantity $\boldsymbol{f}$ is ***materially objective*** iff $\boldsymbol{f}^* = \boldsymbol{f}$;
- a vector quantity $\boldsymbol{f}$ is ***spatially objective*** iff $\boldsymbol{f}^* = \boldsymbol{Q}\boldsymbol{f}$;
- a tensor quantity $\boldsymbol{T}$ is ***materially objective*** iff $\boldsymbol{T}^* = \boldsymbol{T}$;
- a tensor quantity $\boldsymbol{T}$ is ***spatially objective*** iff $\boldsymbol{T}^* = \boldsymbol{Q}\boldsymbol{T}\boldsymbol{Q}^T$.

In order to examine the objectivity of the velocity and acceleration of a motion, we note that when differentiating (2.196) with respect to time, the velocity $\boldsymbol{V}(\boldsymbol{X},t) = \dot{\boldsymbol{\chi}}(\boldsymbol{X},t)$ and acceleration $\boldsymbol{A}(\boldsymbol{X},t) = \ddot{\boldsymbol{\chi}}(\boldsymbol{X},t)$ of a particle $\boldsymbol{X}$ are transformed as follows:

$$\boldsymbol{V}^*(\boldsymbol{X},t^*) = \boldsymbol{Q}(t)\boldsymbol{V}(\boldsymbol{X},t) + \dot{\boldsymbol{c}}(t) + \dot{\boldsymbol{Q}}(t)\boldsymbol{\chi}(\boldsymbol{X},t) \tag{2.203}$$

$$\begin{aligned}\boldsymbol{A}^*(\boldsymbol{X},t^*) &= \ddot{\boldsymbol{\chi}}^*(\boldsymbol{X},t^*) \\ &= \boldsymbol{Q}(t)\ddot{\boldsymbol{\chi}}(\boldsymbol{X},t) + \ddot{\boldsymbol{c}}(t) + \ddot{\boldsymbol{Q}}(t)\boldsymbol{\chi}(\boldsymbol{X},t) \\ &\quad + 2\dot{\boldsymbol{Q}}(t)\boldsymbol{V}(X,t).\end{aligned} \tag{2.204}$$

Consequently, the definitions of the velocity and acceleration are ***relative*** and inextricably linked to the observer. Applying the definition of the ***deformation gradient*** to (2.196), we obtain the transformation law

$$\begin{aligned}\boldsymbol{F}^*(\boldsymbol{X},t^*) &= \frac{\partial\boldsymbol{\chi}^*(\boldsymbol{X},t)}{\partial\boldsymbol{X}} = \frac{\partial\boldsymbol{\chi}^*(\boldsymbol{X},t)}{\partial\boldsymbol{\chi}(\boldsymbol{X},t)}\frac{\partial\boldsymbol{\chi}(\boldsymbol{X},t)}{\partial\boldsymbol{X}} \\ &= \boldsymbol{Q}(t)\boldsymbol{F}(\boldsymbol{X},t).\end{aligned} \tag{2.205}$$

This relation will play an important role in later discussions. Note that

$$J^* = \det\boldsymbol{F}^*(\boldsymbol{X},t^*) = \det\boldsymbol{F}(\boldsymbol{X},t) = J. \tag{2.206}$$

Thus the scalar quantity $J$ is not affected by a change of observer. Starting from the definitions (2.77), (2.79), (2.82), and (2.83) and using (2.205), we have the following transformation laws for deformation tensors:

$$\boldsymbol{C}^* = \boldsymbol{C} \qquad \boldsymbol{E}^* = \boldsymbol{E} \tag{2.207}$$

$$\boldsymbol{c}^* = \boldsymbol{Q}\boldsymbol{c}\boldsymbol{Q}^T \qquad \boldsymbol{e}^* = \boldsymbol{Q}\boldsymbol{e}\boldsymbol{Q}^T. \tag{2.208}$$

Thus, from these definitions, the Jacobian $J$ is objective; the right Cauchy-Green deformation tensor $\boldsymbol{C}$ and the Green-Lagrange strain tensor $\boldsymbol{E}$ are materially objective; and the Cauchy deformation tensor $\boldsymbol{c}$ and the Euler-Almansi strain tensor $\boldsymbol{e}$ are spatially objective. On the other hand, the velocity, the acceleration, and the deformation gradient ***are not objective***. It is also important to notice that the material derivative of an objective material quantity remains materially objective, whereas the material derivative of a spatially objective quantity is generally not spatially objective. For example,

$$\dot{\boldsymbol{E}}^* = \dot{\boldsymbol{E}} \tag{2.209}$$

$$\dot{\boldsymbol{c}}^* = \boldsymbol{Q}\dot{\boldsymbol{c}}\boldsymbol{Q}^T + \dot{\boldsymbol{Q}}\boldsymbol{c}\boldsymbol{Q}^T + \boldsymbol{Q}\boldsymbol{c}\dot{\boldsymbol{Q}}^T \neq \boldsymbol{Q}\dot{\boldsymbol{c}}\boldsymbol{Q}^T. \tag{2.210}$$

In the context of a change of reference frame (2.195), we examine how the velocity gradient tensor is transformed. Rewrite (2.203) and (2.204) in the spatial representation

$$\boldsymbol{v}^* = \dot{\boldsymbol{c}} + \dot{\boldsymbol{Q}}\boldsymbol{x} + \boldsymbol{Q}\boldsymbol{v} \tag{2.211}$$

$$\boldsymbol{a}^* = \ddot{\boldsymbol{c}} + \boldsymbol{Q}\boldsymbol{a} + 2\dot{\boldsymbol{Q}}\boldsymbol{v} + \ddot{\boldsymbol{Q}}\boldsymbol{x}\,. \tag{2.212}$$

From equation (2.211), we evaluate the velocity gradient tensor $\boldsymbol{L}^*$. We obtain

$$L^*_{ij} = \frac{\partial v^*_i}{\partial x^*_j} = \frac{\partial v^*_i}{\partial x_k}\frac{\partial x_k}{\partial x^*_j}\,.$$

With (2.195), we have

$$\frac{\partial x^*_j}{\partial x_k} = Q_{jk}\,.$$

Its inverse, $\partial x_k/\partial x^*_j$, is $Q^{-1}_{kj} = Q^T_{kj}$. The evaluation of $\partial v^*_i/\partial x_k$ is performed via equation (2.211). We have

$$\frac{\partial v^*_i}{\partial x_k} = \dot{Q}_{ik} + Q_{il}\frac{\partial v_l}{\partial x_k}\,.$$

Assembling these various relations, we write

$$\boldsymbol{L}^* = (\boldsymbol{Q}\boldsymbol{L} + \dot{\boldsymbol{Q}})\boldsymbol{Q}^T = \boldsymbol{Q}\boldsymbol{L}\boldsymbol{Q}^T + \boldsymbol{\Omega}\,, \tag{2.213}$$

where we used definition (2.57).

Now we establish the relation between the notion of objectivity and rigid body motion. For this, we compare the equation that describes rigid body motion (2.52) and the transformation (2.195). Setting $\boldsymbol{b} = \boldsymbol{0}$ to simplify (sec. 2.6.3), the motion described by (2.195) is rigid body motion when

$$\boldsymbol{x} = \boldsymbol{Q}(t)\boldsymbol{X} + \boldsymbol{c}(t)\;. \tag{2.214}$$

This last equation can be generalized for two different motions $\boldsymbol{x}$ and $\boldsymbol{y}$ of a continuous media. These two motions differ by a rigid body motion if

$$\boldsymbol{y} = \boldsymbol{Q}(t)\boldsymbol{x} + \boldsymbol{c}(t)\,. \tag{2.215}$$

The form of (2.215) resembles that of (2.195) with the following difference: $\boldsymbol{x}$ and $\boldsymbol{y}$ are two distinct events for a single observer, whereas $\boldsymbol{x}$ and $\boldsymbol{x}^*$ are the positions of a single event recorded by two different observers. Thus, we can consider the change of the observer, as defined in (2.195), as a rigid body motion superimposed on the actual configuration of the medium.

The importance of the objectivity or the non-objectivity of a quantity will appear when we discuss the constitutive equations of materials subject to large transformations or displacements even when accompanied by small strains.

To conclude, we write relations (2.211) and (2.212) in another form. Equation (2.195) gives

$$\boldsymbol{Q}^T(\boldsymbol{x}^* - \boldsymbol{c}) = \boldsymbol{x}. \tag{2.216}$$

Inserting (2.216) in (2.211), we have

$$\boldsymbol{v}^* = \boldsymbol{Q}\boldsymbol{v} + \dot{\boldsymbol{c}} + \dot{\boldsymbol{Q}}\boldsymbol{Q}^T(\boldsymbol{x}^* - \boldsymbol{c}). \tag{2.217}$$

Combining (2.216), (2.217), and (2.212), the acceleration becomes

$$\boldsymbol{a}^* = \boldsymbol{Q}\boldsymbol{a} + \ddot{\boldsymbol{c}} + 2\dot{\boldsymbol{Q}}\boldsymbol{Q}^T(\boldsymbol{v}^* - \dot{\boldsymbol{c}}) - 2\left(\dot{\boldsymbol{Q}}\boldsymbol{Q}^T\right)^2(\boldsymbol{x}^* - \boldsymbol{c}) + \ddot{\boldsymbol{Q}}\boldsymbol{Q}^T(\boldsymbol{x}^* - \boldsymbol{c}) \ . \tag{2.218}$$

Using the rotation tensor (2.57), the velocity is written as

$$\boldsymbol{v}^* - \boldsymbol{Q}\boldsymbol{v} = \dot{\boldsymbol{c}} + \boldsymbol{\Omega}(\boldsymbol{x}^* - \boldsymbol{c}) \ , \tag{2.219}$$

where the two terms on the right-hand side correspond to the translation and rotation velocities of the two reference frames, while the acceleration takes the form

$$\boldsymbol{a}^* - \boldsymbol{Q}\boldsymbol{a} = \ddot{\boldsymbol{c}} + 2\boldsymbol{\Omega}(\boldsymbol{v}^* - \dot{\boldsymbol{c}}) + \left(\dot{\boldsymbol{\Omega}} - \boldsymbol{\Omega}^2\right)(\boldsymbol{x}^* - \boldsymbol{c}) \ . \tag{2.220}$$

On the right-hand side of (2.220), the first term is the translation acceleration, the second the Coriolis acceleration, and the following terms the rotational and centripetal acceleration of the reference frame, respectively. To obtain (2.220), we use the relation

$$\dot{\boldsymbol{\Omega}} = \ddot{\boldsymbol{Q}}\boldsymbol{Q}^T + \dot{\boldsymbol{Q}}\dot{\boldsymbol{Q}}^T = \ddot{\boldsymbol{Q}}\boldsymbol{Q}^T + \dot{\boldsymbol{Q}}\boldsymbol{Q}^T\boldsymbol{Q}\dot{\boldsymbol{Q}}^T = \ddot{\boldsymbol{Q}}\boldsymbol{Q}^T - \boldsymbol{\Omega}^2 . \tag{2.221}$$

If the reference frame is Galilean, a special case of the inertial reference frame such that $\dot{\boldsymbol{Q}} = \ddot{\boldsymbol{Q}} = \boldsymbol{0}$ and $\ddot{\boldsymbol{c}} = \boldsymbol{0}$, then the acceleration is objective. In the Galilean reference frame, $\boldsymbol{Q} = \boldsymbol{Q}_0 = cnst$ and $\boldsymbol{c} = \boldsymbol{c}_0 + \boldsymbol{c}_1 t$ where $\boldsymbol{c}_1$ is a constant velocity. In this case relation (2.195) simplifies to

$$\boldsymbol{x}^* = \boldsymbol{Q}_0\boldsymbol{x} + \boldsymbol{c}_0 + \boldsymbol{c}_1 t \ , \tag{2.222}$$

which is the Galilean transformation.

## 2.12 Exercises

**2.1** A continuous medium in its deformed configuration is given by the relations

$$x_1 = \frac{1}{2}X_1 \qquad x_2 = X_2 \qquad x_3 = X_3 \, . \tag{2.223}$$

Calculate the displacement field in material and spatial coordinates.

**2.2** Consider simple shear as given by the matrix $\boldsymbol{M}$ in equation (2.132). Calculate $\boldsymbol{M}^{-1}$ and the deformation tensors $\boldsymbol{C}$, $\boldsymbol{c}$, $\boldsymbol{E}$, $\boldsymbol{e}$.

**2.3** Repeat the same exercise for the case of pure dilation given by $\boldsymbol{M} = m\boldsymbol{I}$, where $\boldsymbol{I}$ is the unit tensor.

**2.4** A cube vibrates around its equilibrium position. The Lagrangian description of motion is given by the equations

$$\begin{aligned} x_1 &= X_1 + a\cos 2\pi(\frac{t}{T} - \frac{X_1}{L}) \\ x_2 &= X_2 + b\sin 2\pi(\frac{t}{T} - \frac{X_1}{L}) \\ x_3 &= X_3 \,. \end{aligned}$$

The symbols $a$, $b$, and $L$ are constants with a dimension of length, and $T$ is a constant with a dimension of time.

1) Determine the trajectories of the material points.
2) Calculate the components of the velocity and acceleration.
3) Calculate the deformation gradient tensor $\boldsymbol{F}$ and its material derivative $D\boldsymbol{F}/Dt$.
4) In the Lagragian representation, calculate the velocity gradient tensor $\boldsymbol{L}$.
5) Calculate the rate of deformation tensor $\boldsymbol{d}$, the rate of rotation $\dot{\boldsymbol{\omega}}$, and the vorticity $\dot{\boldsymbol{\Omega}}$.

**2.5** For the deformation given in example 2.1:

1) Calculate the matrix of the deformation gradient $\boldsymbol{F}$. Is the deformation homogeneous? Does the deformation take place at constant volume? For which values of $a$ is the transformation invertible?
2) Calculate the matrices of $\boldsymbol{C}, \boldsymbol{E}$, and $\boldsymbol{\varepsilon}$. Compare $\boldsymbol{E}$ and $\boldsymbol{\varepsilon}$ for the case $0 < a \ll 1$.
3) Verify that the vectors aligned with the axis $x_3$ and the diagonals $AH$ and $DE$ are the eigenvectors of $\boldsymbol{C}$. Using these results, calculate $\boldsymbol{U} = \sqrt{\boldsymbol{C}}$.
4) Calculate the matrix of the rotation tensor $\boldsymbol{R}$ in the polar decomposition $\boldsymbol{F} = \boldsymbol{R}\boldsymbol{U}$.

**2.6** Use equation (2.205) to prove relations (2.207) and (2.208).

**2.7** Use (2.205) and the third relation of (2.120) to show that $\boldsymbol{R}^* = \boldsymbol{Q}\boldsymbol{R}$, $\boldsymbol{U}^* = \boldsymbol{U}$, and $\boldsymbol{V}^* = \boldsymbol{Q}\boldsymbol{V}\boldsymbol{Q}^T$.

**2.8** Show that $\dot{\boldsymbol{C}} = 2\dot{\boldsymbol{E}} = 2\boldsymbol{F}^T\boldsymbol{d}\,\boldsymbol{F}$ where $\boldsymbol{d}$ is the strain rate tensor (2.180).

**2.9** For the following motion: $x_1 = \lambda_1 X_1$, $x_2 = \lambda_2 X_2$, $x_3 = \lambda_3 X_3$, determine the matrices of the tensors $\boldsymbol{C}$, $\boldsymbol{R}$, $\boldsymbol{U}$, and $\boldsymbol{E}$, the principal vectors of $\boldsymbol{C}$ and $\boldsymbol{U}$ and their invariants.

**2.10** Prove that $\boldsymbol{e} = \boldsymbol{F}^{-T}\boldsymbol{E}\boldsymbol{F}^{-1}$ and $\boldsymbol{c} = \boldsymbol{R}\boldsymbol{C}\boldsymbol{R}^T$.

**2.11** Demonstrate that $\boldsymbol{F}\boldsymbol{A}_i = \lambda_i \boldsymbol{b}_i$ with $\lambda_i$ the principal values of the tensor $\boldsymbol{U}$ and with $\boldsymbol{A}_i$ $(i = 1, 2, 3)$, the associated principal vectors, and then prove relation (2.119).

**2.12** Using kinematic linearization (sec.2.9.1) and Nanson's formula (2.107) show that the corresponding area elements before and after deformation are approximately the same.

**2.13** Show that kinematic linearization (sec.2.9.1) results in the following relations between the deformation tensors $\boldsymbol{U}, \boldsymbol{\varepsilon}$ and rotation tensors $\boldsymbol{R}, \boldsymbol{\omega}$:

$$\boldsymbol{U} \approx \boldsymbol{I} + \boldsymbol{\varepsilon}$$
$$\boldsymbol{R} \approx \boldsymbol{I} + \boldsymbol{\omega}\,.$$

**2.14** Derive relation (2.157).

**2.15** The displacement of a body is described by the equations

$$\begin{aligned} u_1 &= u_1(x_1, x_2) \\ u_2 &= u_2(x_1, x_2) \\ u_3 &= 0\,. \end{aligned}$$

Let ABCD be an infinitesimal element with sides $dx_1\, dx_2$ as shown in figure 2.23. From the deformed configuration A′B′C′D′, deduce the deformation-displacement relations in the case of infinitesimal strain.

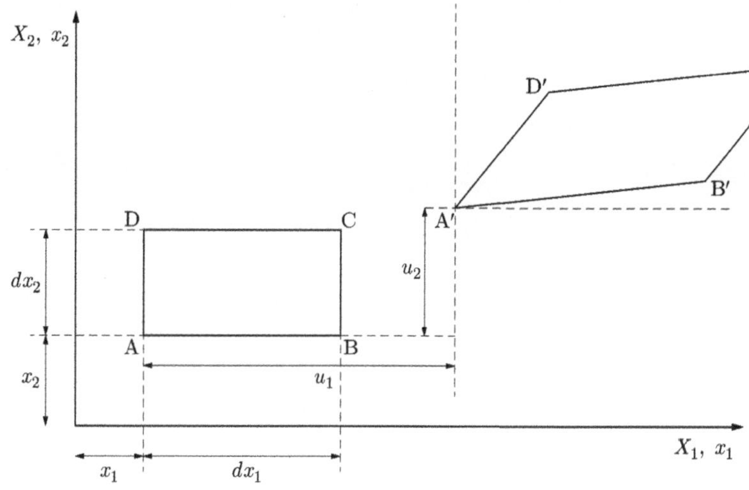


**Fig. 2.23** Deformation of an infinitesimal element

**2.16** A plate with unit thickness in the plane $Ox_1x_2$ is subject to a uniform deformation field given by

$$\varepsilon_{ij} = 10^{-3} \begin{pmatrix} 2 & 1 & 0 \\ 1 & 2 & 0 \\ 0 & 0 & -1 \end{pmatrix}. \tag{2.224}$$

Find the change in length of a linear element with initial length equal to 1:

1) parallel to the axis $x_1$;
2) parallel to the axis $x_2$;
3) that is at an angle of $45^0$ degrees with the axis $x_1$.

**2.17** Prove that the strain $\varepsilon_N$ for a linear element $dS$ in the plane $Ox_1x_2$, in direction $\boldsymbol{N}$, which makes an angle $\theta$ with the horizontal axis is given by

$$\varepsilon_N = \varepsilon_{11}\cos^2\theta + \varepsilon_{22}\sin^2\theta + 2\varepsilon_{12}\cos\theta\sin\theta \ , \tag{2.225}$$

where $\varepsilon_{11}, \varepsilon_{22}, \varepsilon_{12}$ are the components of the infinitesimal strain tensor.

CHAPTER 3

# Dynamics of Continuous Media

## 3.1 Introduction

After the study of the motion of a body and introduction to the various parameters that describes it, the next step is the study of the conservation of certain quantities, such as mass, momentum, and angular momentum during the motion. In this context we will introduce the concept of stress and its measurement; its properties will be studied in detail. In the mechanics of continuous media, the stress is the parameter that characterizes the mechanical interaction with the environment and is the cause of motion. It is therefore essential to develop the mechanics of continuous media as the generalization of point mechanics and Newton's laws. As was the case for kinematics in chapter 2, the type of material and the specification of the motion will not figure in the development of this chapter.

The dynamics of continuous media is also developed in the following references: [15, 20, 33, 34, 36].

## 3.2 Reynolds Transport Theorem

### 3.2.1 Background

The time derivative of a volume integral plays a very important role in the formulation of the laws for mechanics of continuous media. For example, consider the function $I(t)$ given by

$$I(t) = \int_{\Omega} \Phi(\boldsymbol{X}, t)\, dV\,, \tag{3.1}$$

where $\Phi(\boldsymbol{X}, t)$ is a scalar, vector, or tensor quantity, and $\Omega \subseteq \mathcal{R}_0$ represents the volume of the body in the initial configuration at time $t = 0$, or a part $\Pi$

of it. We define the ***material time derivative of a volume integral*** by the expression

$$\frac{DI(t)}{Dt} = \frac{d}{dt}\int_{\Omega} \Phi(\boldsymbol{X},t)\, dV\,. \tag{3.2}$$

In the case where the boundary does not change with time, we can bring the time derivative inside the integral and (3.2) becomes

$$\frac{DI(t)}{Dt} = \frac{d}{dt}\int_{\Omega} \Phi(\boldsymbol{X},t)\, dV = \int_{\Omega} \frac{\partial \Phi(\boldsymbol{X},t)}{\partial t}\, dV\,. \tag{3.3}$$

When the boundary of the body does change in time, the derivative applied earlier is no longer valid and we need another method. That case will be presented next.

Suppose that the scalar, vector, or tensor quantity $\varphi(\boldsymbol{x},t)$ expresses a characteristic of the body and that the body, or a part $\Pi \subseteq \mathcal{B}$ of it, occupies a volume $\omega(t) \subseteq \mathcal{R}$ with boundary $\partial\omega(t) \subseteq \partial\mathcal{R}$ at time $t$. The balance equation can be formulated in the following way: the rate of variation of the integral of $\varphi(\boldsymbol{x},t)$ in the volume $\omega(t)$ as a function of time is equal to the rate of variation of $\varphi(\boldsymbol{x},t)$ in $\omega(t)$ plus the total flux of $\varphi(\boldsymbol{x},t)$ passing through the surface $\partial\omega(t)$.

**Reynolds Transport Theorem**

*Mathematically, the balance equation is written as*

$$\frac{d}{dt}\int_{\omega(t)} \varphi(\boldsymbol{x},t)\, dv = \int_{\omega(t)} \frac{\partial \varphi(\boldsymbol{x},t)}{\partial t}\, dv + \int_{\partial\omega(t)} \varphi(\boldsymbol{x},t)\, \boldsymbol{v}\cdot\boldsymbol{n}\, ds\,. \tag{3.4}$$

*This expression defines the Reynolds transport theorem. Using the divergence theorem (sec. 1.4.13), we can also write it in the form*

$$\frac{d}{dt}\int_{\omega(t)} \varphi(\boldsymbol{x},t)\, dv = \int_{\omega(t)} \left(\frac{\partial \varphi(\boldsymbol{x},t)}{\partial t} + \frac{\partial}{\partial x_i}\left(\varphi(\boldsymbol{x},t)v_i\right)\right) dv \tag{3.5}$$

*or with (2.27),*

$$\frac{d}{dt}\int_{\omega(t)} \varphi(\boldsymbol{x},t)\, dv = \int_{\omega(t)} \left(\frac{D\varphi(\boldsymbol{x},t)}{Dt} + \varphi(\boldsymbol{x},t)\,\operatorname{div}\boldsymbol{v}\right) dv\,. \tag{3.6}$$

In the following equations, we denote $\omega(t)$ by $\omega$ and $\partial\omega(t)$ by $\partial\omega$. To further simplify the notation, we will on some occasions omit the arguments of tensor, vector, and scalar functions.

PROOF.
1) To demonstrate the Reynolds transport theorem, the time variation of the Jacobian is required. We know from (2.103) that

$$dv = J(\boldsymbol{X},t)\, dV\,, \tag{3.7}$$

where the Jacobian $J$ (2.68) can be expressed as

$$J(\boldsymbol{X},t) = \det \boldsymbol{F} = \varepsilon_{ijk}\, F_{i1}F_{j2}F_{k3} = \varepsilon_{ijk}\, F_{1i}F_{2j}F_{3k}\,, \tag{3.8}$$

since by writing $\varepsilon_{ijk}F_{i1}F_{j2}F_{k3}$ and using the properties of $\varepsilon_{ijk}$, we obtain

$$\begin{aligned}\varepsilon_{ijk}F_{i1}F_{j2}F_{k3} &= F_{11}F_{22}F_{33} + F_{21}F_{32}F_{13} + F_{31}F_{12}F_{23}\\ &\quad - F_{11}F_{32}F_{23} - F_{31}F_{22}F_{13} - F_{21}F_{12}F_{33}\,.\end{aligned} \tag{3.9}$$

The calculation of the material derivative of $J(\boldsymbol{X},t)$ is easily carried out as follows. We have

$$\dot{J} = \varepsilon_{ijk}\frac{\partial\dot{\chi}_1}{\partial X_i}\frac{\partial\chi_2}{\partial X_j}\frac{\partial\chi_3}{\partial X_k} + \varepsilon_{ijk}\frac{\partial\chi_1}{\partial X_i}\frac{\partial\dot{\chi}_2}{\partial X_j}\frac{\partial\chi_3}{\partial X_k} + \varepsilon_{ijk}\frac{\partial\chi_1}{\partial X_i}\frac{\partial\chi_2}{\partial X_j}\frac{\partial\dot{\chi}_3}{\partial X_k}\,. \tag{3.10}$$

Introducing the velocity (2.20), and using the implicit function theorem, we obtain

$$\begin{aligned}\dot{J} &= \varepsilon_{ijk}\left(\frac{\partial v_1}{\partial x_p}\frac{\partial x_p}{\partial X_i}\frac{\partial\chi_2}{\partial X_j}\frac{\partial\chi_3}{\partial X_k} + \frac{\partial\chi_1}{\partial X_i}\frac{\partial v_2}{\partial x_p}\frac{\partial x_p}{\partial X_j}\frac{\partial\chi_3}{\partial X_k}\right.\\ &\quad\left. + \frac{\partial\chi_1}{\partial X_i}\frac{\partial\chi_2}{\partial X_j}\frac{\partial v_3}{\partial x_p}\frac{\partial x_p}{\partial X_k}\right)\\ &= \frac{\partial v_1}{\partial x_p}\,\delta_{1p}J + \frac{\partial v_2}{\partial x_p}\,\delta_{2p}J + \frac{\partial v_3}{\partial x_p}\,\delta_{3p}J\,.\end{aligned} \tag{3.11}$$

Finally, we have

$$\dot{J} = \frac{\partial v_i}{\partial x_i}\,J(\boldsymbol{X},t) = \nabla\cdot\boldsymbol{v}\Big|_{\boldsymbol{x}=\boldsymbol{\chi}(\boldsymbol{X},t)} J(\boldsymbol{X},t)\,. \tag{3.12}$$

This expression, sometimes written as $\dot{J}/J = \operatorname{div}\boldsymbol{v}$, is the rate of volume dilation. As a final result of the previous developments, we can obtain the material derivative of a volume element by combining (3.7) with (3.12). Then, successively,

$$\frac{D\,dv}{Dt} = \dot{J}\,dV = \frac{\dot{J}}{J}\,dv = \operatorname{div}\boldsymbol{v}dv\,. \tag{3.13}$$

2) We can derive the Reynolds transport theorem defined by (3.5) or (3.6). Let the integral $I(t)$ be defined by the relation

$$I(t) = \int_\omega \varphi(\boldsymbol{x},t)\,dx_1\,dx_2\,dx_3\,, \tag{3.14}$$

and evaluate the integral

$$\frac{DI(t)}{Dt} = \frac{d}{dt}\int_\omega \varphi(\boldsymbol{x},t)\,dx_1\,dx_2\,dx_3\,. \tag{3.15}$$

To do so, we cannot permute the derivative with respect to time with the integral, since the latter is over a volume which depends on time. Thus we change to the material representation

$$\frac{DI(t)}{Dt} = \frac{d}{dt}\int_\Omega \varphi(\boldsymbol{\chi}(\boldsymbol{X},t),t)\,J(\boldsymbol{X},t)\,dX_1\,dX_2\,dX_3\,. \tag{3.16}$$

Osborne Reynolds (1842–1912) was an English mathematician born in Belfast. He was appointed professor of engineering at Owens College, now the University of Manchester, in 1868. He made contributions in hydrodynamics and fluid mechanics. He introduced the Reynolds number, as well as the Reynolds decomposition for the modeling of turbulence.

**Fig. 3.1** Osborne Reynolds

Using (2.103), we have

$$dx_1\,dx_2\,dx_3 = J(\boldsymbol{X},t)\,dX_1\,dX_2\,dX_3 \quad \text{or} \quad dv = J(\boldsymbol{X},t)\,dV\,. \tag{3.17}$$

Then, (3.16) can be expressed as

$$\frac{DI(t)}{Dt} = \frac{d}{dt}\int_\Omega \Phi(\boldsymbol{X},t)J(\boldsymbol{X},t)\,dX_1\,dX_2\,dX_3\,, \tag{3.18}$$

where

$$\Phi(\boldsymbol{X},t) = \varphi\big(\boldsymbol{\chi}(\boldsymbol{X},t),t\big)\,. \tag{3.19}$$

Thus it is easier with relation (3.18) to express the derivative of the integral

$$\begin{aligned}\frac{DI(t)}{Dt} = \int_\Omega \Bigg( &\frac{\partial\Phi(\boldsymbol{X},t)}{\partial t}\bigg|_{\boldsymbol{X}} J(\boldsymbol{X},t) \\ &+ \Phi(\boldsymbol{X},t)\frac{\partial J(\boldsymbol{X},t)}{\partial t}\bigg|_{\boldsymbol{X}}\Bigg)\,dX_1\,dX_2 dX_3\,.\end{aligned} \tag{3.20}$$

Using relation (3.12), integral (3.20) can be written as follows:

$$\begin{aligned}\frac{DI(t)}{Dt} = \int_\Omega \Bigg( &\frac{\partial\Phi(\boldsymbol{X},t)}{\partial t}\bigg|_{\boldsymbol{X}} \\ &+ \Phi(\boldsymbol{X},t)(\nabla\cdot\boldsymbol{v})\Big|_{\boldsymbol{x}=\boldsymbol{\chi}(\boldsymbol{X},t)}\Bigg) J(\boldsymbol{X},t)\,dX_1\,dX_2\,dX_3\,.\end{aligned} \tag{3.21}$$

Making the appropriate changes in the last integral and using (2.3), (3.17), (3.19), and

$$\frac{\partial\Phi(\boldsymbol{X},t)}{\partial t}\bigg|_{\boldsymbol{X}=\boldsymbol{\chi}^{-1}(\boldsymbol{x},t)} = \frac{D\varphi(\boldsymbol{x},t)}{Dt}\,, \tag{3.22}$$

we have the final result

$$\frac{DI(t)}{Dt} = \int_\omega \left(\frac{D\varphi(\boldsymbol{x},t)}{Dt} + \varphi(\boldsymbol{x},t)\nabla\cdot\boldsymbol{v}(\boldsymbol{x},t)\right)dx_1\,dx_2\,dx_3\,. \tag{3.23}$$

Expression (3.23) is the ***Reynolds transport theorem*** applied to a scalar function $\varphi(\boldsymbol{x},t)$.

We can consider the Reynolds theorem to be a generalization of Leibnitz' theorem:

$$\frac{d}{dt}\int_{a(t)}^{b(t)} \varphi(x,t)\,dx = \int_a^b \frac{\partial \varphi}{\partial t}\,dx + \frac{db}{dt}\,\varphi(x=b,t) - \frac{da}{dt}\,\varphi(x=a,t)\,. \tag{3.24}$$

Since

$$\frac{D\varphi(\boldsymbol{x},t)}{Dt} = \frac{\partial \varphi(\boldsymbol{x},t)}{\partial t} + \boldsymbol{v}(\boldsymbol{x},t)\cdot\nabla\varphi(\boldsymbol{x},t)\,, \tag{3.25}$$

integral (3.23) can be written as

$$\begin{aligned}\frac{DI(t)}{Dt} &= \int_\omega \left(\frac{\partial \varphi(\boldsymbol{x},t)}{\partial t} + \boldsymbol{v}(\boldsymbol{x},t)\cdot\nabla\varphi(\boldsymbol{x},t) + \nabla\cdot\boldsymbol{v}(\boldsymbol{x},t)\varphi(\boldsymbol{x},t)\right)dv \\ &= \int_\omega \left(\frac{\partial \varphi(\boldsymbol{x},t)}{\partial t} + \left(\boldsymbol{v}(\boldsymbol{x},t)\cdot\nabla\varphi(\boldsymbol{x},t) + \nabla\cdot\boldsymbol{v}(\boldsymbol{x},t)\varphi(\boldsymbol{x},t)\right)\right)dv \\ &= \int_\omega \frac{\partial \varphi(\boldsymbol{x},t)}{\partial t}\,dv + \int_\omega \operatorname{div}\big(\boldsymbol{v}(\boldsymbol{x},t)\varphi(\boldsymbol{x},t)\big)\,dv\,. \end{aligned} \tag{3.26}$$

Using the divergence theorem (1.228) for the second integral, we obtain

$$\frac{DI(t)}{Dt} = \int_\omega \frac{\partial \varphi(\boldsymbol{x},t)}{\partial t}\,dv + \int_{\partial\omega} \varphi(\boldsymbol{x},t)\,\boldsymbol{v}\cdot\boldsymbol{n}\,ds\,, \tag{3.27}$$

which is none other than (3.4).

## 3.3 Conservation of Mass

In the mechanics of continuous media, the ***mass*** is treated as a fundamental concept and considered as inherent to the body. The mass of the body $\mathcal{B}$ signifies, first of all, the quantity of material that $\mathcal{B}$ contains. Experience shows that this quantity does not depend on time, nor the transformation of the material (for example, by a phase change), nor the deformation of $\mathcal{B}$. In addition, the quantity of material in two arbitrary parts of $\mathcal{B}$ is the quantity of material in the first part plus that in the second. These aspects can be formulated and thus rendered more precise in two different ways, depending on whether we adopt the material or spatial descriptions to describe the motion of $\mathcal{B}$.

### 3.3.1 Material Form

Let $\mathcal{R}_0$ and $\mathcal{R}$ be, respectively, the initial (or reference) and the current configurations of $\mathcal{B}$ (fig. 3.2). The ***initial mass density*** of the body $\mathcal{B}$ in the ***material*** description is a ***positive*** and ***integrable scalar function*** $P_0(\boldsymbol{X})$,

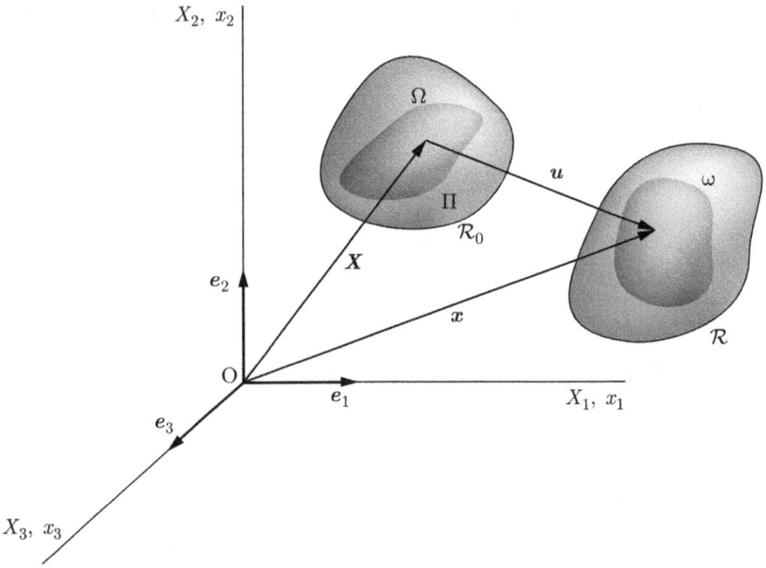


**Fig. 3.2** Motion of an arbitrary part $\Pi$ of $\mathcal{B}$

defined on $\mathcal{R}_0$, such that the mass $m(\Omega)$ of an arbitrary part $\Pi$ of $\mathcal{B}$ at time $t = 0$ is given by

$$m(\Omega) = \int_{\Omega} P_0(\boldsymbol{X})\, dV = \int_{\Omega} P_0(\boldsymbol{X})\, dX_1\, dX_2\, dX_3\,, \tag{3.28}$$

where $\Omega \subseteq \mathcal{R}_0$ is the initial configuration of $\Pi \subseteq \mathcal{B}$. Similarly, ***at later times*** the ***current density*** of $\mathcal{B}$ in the ***spatial*** description is a ***scalar***, ***positive***, and ***integrable function*** $\rho(\boldsymbol{x}, t)$ defined on $\mathcal{R}$, such that the mass $m_t(\omega)$ of any part $\Pi$ of $\mathcal{B}$ at a given instant $t \geq 0$ is given by

$$m_t(\omega) = \int_{\omega} \rho(\boldsymbol{x}, t)\, dv = \int_{\omega} \rho(\boldsymbol{x}, t)\, dx_1\, dx_2\, dx_3\,, \tag{3.29}$$

where $\omega \subseteq \mathcal{R}$ is the configuration at time $t$ of $\Pi$.

Principle of conservation of mass *The mass contained in an arbitrary part* $\Pi$ *of the body* $\mathcal{B}$ *remains constant in time, that is,*

$$m_t(\omega) = m(\Omega)\,. \tag{3.30}$$

By introducing (3.28) and (3.29) in (3.30), we obtain

$$\int_{\omega} \rho(\boldsymbol{x}, t)\, dv = \int_{\Omega} P_0(\boldsymbol{X})\, dV\,. \tag{3.31}$$

In this expression the vector position at time $t$, $\boldsymbol{x}$, of a particle in $\Pi$ is related to the initial vector position, $\boldsymbol{X}$, by the equation of motion of $\mathcal{B}$, given by (2.1).

Taking (2.1) into account, (3.31) is mathematically equivalent to changing variables. Changing variables from $\boldsymbol{x}$ to $\boldsymbol{X}$, the corresponding volume elements $dv$ and $dV$ are related by (2.103). Then, using (2.1) in (3.31), we obtain

$$\int_{\Omega} \big(J(\boldsymbol{X},t)\rho\big(\chi(\boldsymbol{X},t),t\big)\big) - P_0(\boldsymbol{X})\big)\, dV = 0. \tag{3.32}$$

If the ***density at time*** $t$, $P(\boldsymbol{X},t)$, in the ***material*** description is defined by

$$P(\boldsymbol{X},t) = \rho\big(\chi(\boldsymbol{X},t),t\big)\,, \tag{3.33}$$

integral (3.32) can be written as

$$\int_{\Omega} \big(J(\boldsymbol{X},t)P(\boldsymbol{X},t) - P_0(\boldsymbol{X})\big)\, dV = 0\,. \tag{3.34}$$

Expression (3.34) is the ***global material form*** of the principle of conservation of mass. In order to obtain the ***local*** form of the principle of conservation of mass, we exploit the fact that (3.34) is valid for an ***arbitrary*** initial part $\Omega$ of $\mathcal{R}_0$, as long as the integrand of (3.34) is ***continuous*** with respect to $\boldsymbol{X}$, and we invoke the localization theorem.(1)

**Localization Theorem**

*Let $f$ be a continuous scalar, vector, or tensor function defined on an open domain $\mathcal{D}$ of a three-dimensional Euclidean space. If*

$$\int_{\Omega} f\, dV = 0 \tag{3.35}$$

*for any closed sub-domain $\Omega$ of $\mathcal{D}$, then*

$$f = 0 \tag{3.36}$$

*for every point of $\mathcal{D}$.*

Consequently, the integrand of (3.34) must be zero. We obtain

$$J(\boldsymbol{X},t)P(\boldsymbol{X},t) = P_0(\boldsymbol{X})\,. \tag{3.37}$$

This equality represents the local ***material form*** of the principle of conservation of mass.

The body $\mathcal{B}$ or the material constituting $\mathcal{B}$ is said to be ***incompressible*** if the density is invariable, that is, it does not depend on space or time. In this case, $P(\boldsymbol{X},t) = P_0(\boldsymbol{X})$ and consequently

$$J(\boldsymbol{X},t) = 1 \tag{3.38}$$

(1) The localization theorem is based on the Dubois-Reymond lemma used in the calculus of variations.

for every point $\boldsymbol{X}$ of $\mathcal{R}_0$ and at every instant $t$. This condition is frequently met in fluid mechanics and in the study of rubber-like solid materials. It follows from (3.37) and (3.33) that the incompressibility condition (3.38) is equivalent to

$$P(\boldsymbol{X},t) = \rho\big(\chi(\boldsymbol{X},t),t\big) = P_0(\boldsymbol{X}) . \tag{3.39}$$

A motion for which relation (3.38) is seen to be the case is a motion for which the volume remains constant (see (3.7)) and is called ***isochoric motion***.

### 3.3.2 Spatial Form

Consider equation (3.31) again. The derivative with respect to time, using the Reynolds transport theorem, is

$$\frac{d}{dt}\int_{\omega} \rho(\boldsymbol{x},t)\,dv = \int_{\omega}\left(\frac{D\rho(\boldsymbol{x},t)}{Dt} + \rho(\boldsymbol{x},t)\,\mathrm{div}\,\boldsymbol{v}(\boldsymbol{x},t)\right)dv = 0 . \tag{3.40}$$

This is the ***global spatial form*** of the principle of conservation of mass. Assuming that the integrand of (3.40) is continuous with respect to $\boldsymbol{x}$, and by applying the localization theorem to it, we obtain the ***local spatial form*** of the principle of conservation of mass

$$\frac{D\rho(\boldsymbol{x},t)}{Dt} + \rho(\boldsymbol{x},t)\,\mathrm{div}\,\boldsymbol{v}(\boldsymbol{x},t) = 0 . \tag{3.41}$$

This equation is also called the ***continuity equation***. Since

$$\frac{D\rho(\boldsymbol{x},t)}{Dt} = \frac{\partial\rho(\boldsymbol{x},t)}{\partial t} + \nabla\rho(\boldsymbol{x},t)\cdot\boldsymbol{v}(\boldsymbol{x},t) \tag{3.42}$$

$$\mathrm{div}\big(\rho(\boldsymbol{x},t)\boldsymbol{v}(\boldsymbol{x},t)\big) = \rho(\boldsymbol{x},t)\,\mathrm{div}\,\boldsymbol{v}(\boldsymbol{x},t) + \nabla\rho(\boldsymbol{x},t)\cdot\boldsymbol{v}(\boldsymbol{x},t) , \tag{3.43}$$

expression (3.41) can be written in the equivalent form

$$\frac{\partial\rho(\boldsymbol{x},t)}{\partial t} + \mathrm{div}\big(\rho(\boldsymbol{x},t)\boldsymbol{v}(\boldsymbol{x},t)\big) = 0 . \tag{3.44}$$

Equation (3.41), or (3.44), is a basic equation for fluid mechanics while its material equivalent (3.37) is used in solid mechanics. If the material is incompressible, then from (3.39), the density $\rho(\boldsymbol{x},t)$ is constant and $D\rho(\boldsymbol{x},t)/Dt = 0$. It also follows from (3.41) that the ***incompressibility condition*** can be expressed by one of the following equivalent forms:

$$\mathrm{div}\,\boldsymbol{v} = \frac{\partial v_i}{\partial x_i} = d_{ii} = \mathrm{tr}\,\boldsymbol{d} = I_1(\boldsymbol{d}) = 0 , \tag{3.45}$$

where we have used the definition (2.180) of the strain rate tensor $\boldsymbol{d}$. Note that the velocity field that satisfies (3.45) is solenoidal. As $\mathrm{div}\,\boldsymbol{v} = 0$, it follows from (3.12) that $\dot{J} = 0$ so $J$ remains constant over time. Since $J(\boldsymbol{X},0) = 1$, the motion of an incompressible material takes place with constant volume and is often called isochoric.

Another way to deduce the continuity equation consists of the following steps: consider in figure 3.3 a fluid that flows through a volume element $\Delta x_1 \Delta x_2 \Delta x_3$ with velocity $v_i(x_1, x_2, x_3)$.

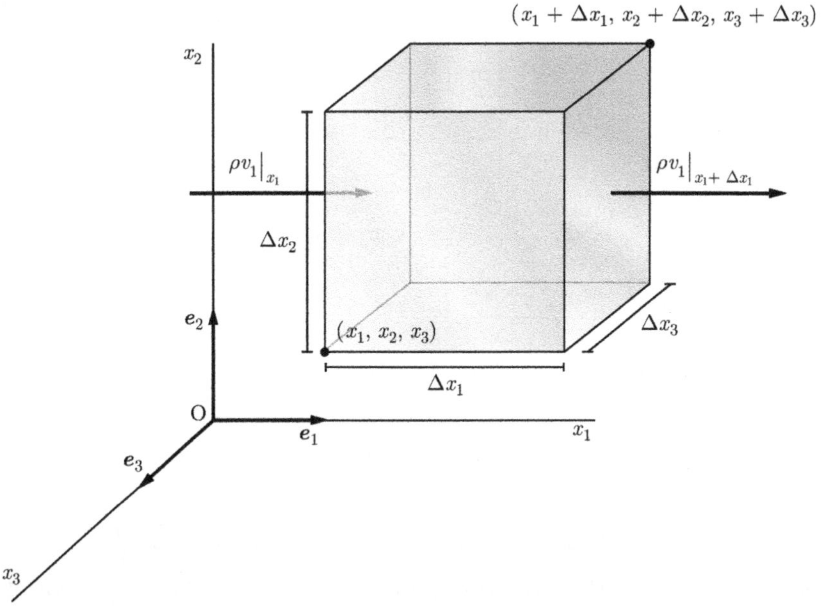


**Fig. 3.3** Mass balance in an elementary volume

If we suppose that the flow is oriented in the positive direction of the $x_1$ axis, the quantity of mass entering the volume through the surface at $x_1$ is given by

$$\rho v_1\big|_{x_1} \Delta x_2 \, \Delta x_3 \, . \tag{3.46}$$

The quantity of mass leaving the volume through the surface at $x_1 + \Delta x_1$ is given by

$$\rho v_1\big|_{x_1+\Delta x_1} \Delta x_2 \, \Delta x_3 \, . \tag{3.47}$$

In the same way, the quantities of mass entering and leaving in the directions $x_2$ and $x_3$ are given by

$$\begin{array}{lll} \text{in direction } x_2 & \rho v_2\big|_{x_2} \Delta x_3 \, \Delta x_1 & \rho v_2\big|_{x_2+\Delta x_2} \Delta x_3 \, \Delta x_1 \\ \text{in direction } x_3 & \rho v_3\big|_{x_3} \Delta x_1 \, \Delta x_2 & \rho v_3\big|_{x_3+\Delta x_3} \Delta x_1 \, \Delta x_2 \, . \end{array} \tag{3.48}$$

The rate of change of the mass in the element of volume $\Delta x_1 \, \Delta x_2 \, \Delta x_3$ is

$$(\Delta x_1 \, \Delta x_2 \, \Delta x_3) \frac{\partial \rho}{\partial t} \, . \tag{3.49}$$

The result for the flow in the volume under consideration is given by

$$\left[ \begin{array}{c} \text{rate of change} \\ \text{of mass} \end{array} \right] = \left[ \begin{array}{c} \text{rate of} \\ \text{mass entering} \end{array} \right] - \left[ \begin{array}{c} \text{rate of} \\ \text{mass leaving} \end{array} \right] .$$

The balance equation for this result is written using relations (3.46)–(3.49)

$$\begin{aligned}(\Delta x_1 \Delta x_2 \Delta x_3)\frac{\partial \rho}{\partial t} &= \left(\rho v_1\big|_{x_1} - \rho v_1\big|_{x_1+\Delta x_1}\right)\Delta x_2\,\Delta x_3 \\ &+ \left(\rho v_2\big|_{x_2} - \rho v_2\big|_{x_2+\Delta x_2}\right)\Delta x_3\,\Delta x_1 \\ &+ \left(\rho v_3\big|_{x_3} - \rho v_3\big|_{x_3+\Delta x_3}\right)\Delta x_1\,\Delta x_2\,.\end{aligned} \tag{3.50}$$

Dividing (3.50) by the volume $\Delta x_1 \Delta x_2 \Delta x_3$ and taking the limits $\Delta x_1 \to 0$, $\Delta x_2 \to 0$, $\Delta x_3 \to 0$, we have

$$\begin{aligned}\frac{\partial \rho}{\partial t} &= \lim_{\Delta x_1 \to 0} \frac{\rho v_1\big|_{x_1} - \rho v_1\big|_{x_1+\Delta x_1}}{\Delta x_1} + \lim_{\Delta x_2 \to 0} \frac{\rho v_2\big|_{x_2} - \rho v_2\big|_{x_2+\Delta x_2}}{\Delta x_2} \\ &+ \lim_{\Delta x_3 \to 0} \frac{\rho v_3\big|_{x_3} - \rho v_3\big|_{x_3+\Delta x_3}}{\Delta x_3} \\ &= -\left(\frac{\partial(\rho v_1)}{\partial x_1} + \frac{\partial(\rho v_2)}{\partial x_2} + \frac{\partial(\rho v_3)}{\partial x_3}\right).\end{aligned} \tag{3.51}$$

Equation (3.51) can also be put in the form

$$\frac{\partial \rho}{\partial t} + v_1 \frac{\partial \rho}{\partial x_1} + v_2 \frac{\partial \rho}{\partial x_2} + v_3 \frac{\partial \rho}{\partial x_3} = -\rho\left(\frac{\partial v_1}{\partial x_1} + \frac{\partial v_2}{\partial x_2} + \frac{\partial v_3}{\partial x_3}\right). \tag{3.52}$$

The left-hand side of this equation represents the material derivative of the density and the right-hand side the divergence of the velocity field

$$\frac{D\rho}{Dt} = -\rho \operatorname{div} \boldsymbol{v}\,. \tag{3.53}$$

We thus find the continuity equation as established for an elementary control volume.

## 3.4 Volume Forces, Contact Forces and Cauchy's Postulate

Like mass, ***force*** is a fundamental concept in the mechanics of continuous media. The forces determine the interactions between the different internal parts of a body or between the body and its external environment. A force can only be detected by its effects; that is, it cannot be ***directly*** measured. For this reason, force is one of the most abstract concepts in mechanics.

To elaborate this point, we cite an extract from letter *LXXIV* from Euler to a German princess [14]: "for as a body, in virtue of it's nature, preserves the same state of motion, or of rest, and cannot be drawn out of it but by external causes, it follows that, in order to a body's changing it's state, it must be forced out of it by some external cause: without which it would always continue in the same state. Hence it is, that we give to this external cause the name of power or *force*. It is a term in common use, though many by whom it is employed have but a very imperfect idea of it."(2)

(2)Translation by Henry Hunter published in London in 1795.

We will treat two types of forces: ***volume or body forces*** (acting at a distance) such as gravity or electromagnetic forces, including the Lorentz force, and ***contact forces***. Let $\mathcal{B}$ be a body with the initial configuration $\mathcal{R}_0$ and currently in configuration $\mathcal{R}$. The volume force acting on $\mathcal{B}$ at time $t$ represents the action of its external environment on the ***interior*** points of $\mathcal{B}$ at that instant. More precisely, if $\Pi$ is an arbitrary, small part of $\mathcal{B}$ whose initial and current configurations are $\Omega$ and $\omega$ (fig. 3.2), the volume force acting on $\Pi$ at time $t$ is given by

$$\boldsymbol{f}^b(\omega,t) = \int_\omega \rho(\boldsymbol{x},t)\boldsymbol{b}(\boldsymbol{x},t)\,dv\,, \tag{3.54}$$

where $\boldsymbol{b}(\boldsymbol{x},t)$ is a vector function defined on $\mathcal{R}$ called the ***spatial volume force density*** (per unit mass) at time $t$. The material version of (3.54) takes the form

$$\boldsymbol{F}^b(\Omega,t) = \int_\Omega P_0(\boldsymbol{X})\boldsymbol{B}(\boldsymbol{X},t)\,dV, \tag{3.55}$$

where $\boldsymbol{B}(\boldsymbol{X},t)$ is the ***material volume force density*** at time $t$. As $\boldsymbol{f}^b(\omega,t)$ and $\boldsymbol{F}^b(\Omega,t)$ represent the same quantity, we can have $\boldsymbol{f}^b(\omega,t) = \boldsymbol{F}^b(\Omega,t)$, that is

$$\int_\omega \rho(\boldsymbol{x},t)\boldsymbol{b}(\boldsymbol{x},t)\,dv - \int_\Omega P_0(\boldsymbol{X})\boldsymbol{B}(\boldsymbol{X},t)\,dV = 0\,. \tag{3.56}$$

By changing the variables from $\boldsymbol{x}$ to $\boldsymbol{X}$, then using (3.33) and (3.37), and finally, by applying the localization theorem, we obtain the relation

$$\boldsymbol{B}(\boldsymbol{X},t) = \boldsymbol{b}\big(\chi(\boldsymbol{X},t),t\big)\,. \tag{3.57}$$

Contact forces can describe the interaction between two interior parts of a body $\mathcal{B}$ separated by a surface (that is, internal cohesive forces) or the action of external bodies directly in contact with the boundary of $\mathcal{B}$. Let $\Pi^-$ and $\Pi^+$ be two arbitrary parts of a body $\mathcal{B}$ such that their initial configurations $\Omega^- \subset \mathcal{R}_0$ and $\Omega^+ \subset \mathcal{R}_0$ are separated by a surface $\Gamma_0$ and their configurations at a later time, $\omega^- \subset \mathcal{R}$ and $\omega^+ \subset \mathcal{R}$, by a surface $\Gamma$ (fig. 3.4). In the spatial description, the action of $\Pi^+$ on $\Pi^-$ at the instant $t$ across a surface element $\delta s(\boldsymbol{x})$ of $\Gamma$ around $\boldsymbol{x}$ is represented by a contact force element $\delta\boldsymbol{f}^c(\boldsymbol{x},t,\Gamma)$. By writing this, we tacitly assume that the action of $\Pi^+$ on $\Pi^-$ is influenced only by the form of $\omega^+$ through its boundary $\Gamma$ with $\omega^-$. In addition we assume that the limit

$$\boldsymbol{t}(\boldsymbol{x},t,\Gamma) = \lim_{\delta s\to 0}\frac{\delta\boldsymbol{f}^c(\boldsymbol{x},t,\Gamma)}{\delta s(\boldsymbol{x})} \tag{3.58}$$

exists. The vector $\boldsymbol{t}(\boldsymbol{x},t,\Gamma)$ is the force, per unit surface (spatial), exerted by $\Pi^+$ through $\Gamma$ on $\Pi^-$. This contact force density is called the ***spatial stress vector*** (***contact***) or the ***surface stress vector***.

According to (3.58), the stress vector $\boldsymbol{t}$ at $\boldsymbol{x}$ depends on the whole of the surface $\Gamma$. However, in classical mechanics of continuous media, the following hypothesis is applied.

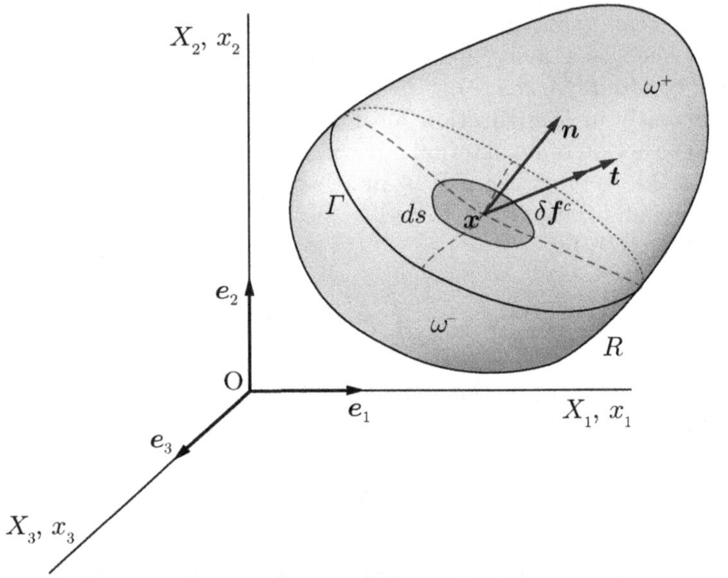


**Fig. 3.4** Contact force and the contact stress vector

CAUCHY'S POSTULATE *The stress vector* $\boldsymbol{t}$ *at* $\boldsymbol{x}$ *on the surface* $\Gamma$ *depends only on the outward unit normal* $\boldsymbol{n}$ *of* $\Gamma$ *at* $\boldsymbol{x}$*, that is,*

$$\boldsymbol{t}(\boldsymbol{x}, t, \Gamma) = \boldsymbol{t}(\boldsymbol{x}, t, \boldsymbol{n}) . \tag{3.59}$$

This postulate stipulates that if three different surfaces $\Gamma_1$, $\Gamma_2$, and $\Gamma_3$ share the same tangent plane at $\boldsymbol{x}$, the stress vectors acting on these surfaces at $\boldsymbol{x}$ are identical (fig. 3.5).

Finally, the action of $\Pi^+$ on $\Pi^-$ through the surface $\Gamma$ is described by the contact force vector

$$\boldsymbol{f}^c(\Gamma, t) = \int_\Gamma \boldsymbol{t}(\boldsymbol{x}, t, \boldsymbol{n}) \, ds . \tag{3.60}$$

By analogy, the action of the external environment on a body $\mathcal{B}$ through the boundary $\partial\mathcal{R}$ is given by

$$\boldsymbol{f}^c(\partial\mathcal{R}, t) = \int_{\partial\mathcal{R}} \boldsymbol{t}(\boldsymbol{x}, t, \boldsymbol{n}) \, ds . \tag{3.61}$$

In the following, the contact force is expressed with respect to the current, deformed configuration $\mathcal{R}$. However, in many important problems, in particular in solid mechanics, the deformed configuration is not known in advance. It is thus more convenient to express the contact force with respect to the initial, reference configuration $\mathcal{R}_0$. The notions of the ***nominal contact stress*** vector and ***material contact stress*** vector can be introduced. But, as the physical and geometrical interpretation of these vectors is not intuitive, we will present them in more detail in section 3.9.

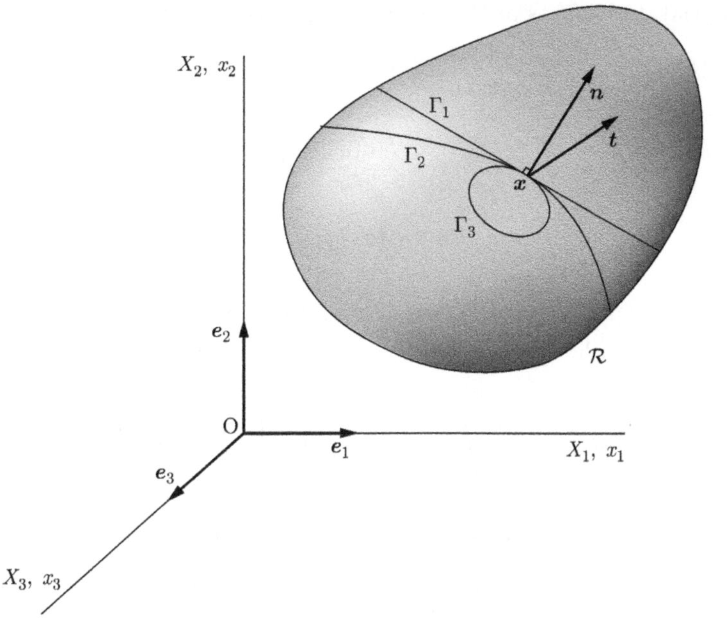


**Fig. 3.5** Interpretation of Cauchy's postulate

## 3.5 Conservation of Momentum and Angular Momentum

In physics, the ***momentum*** of a particle of mass $m$ and velocity $\boldsymbol{v}$ is defined by

$$\overline{\boldsymbol{m}} = m\boldsymbol{v} \qquad \overline{m}_i = mv_i\,, \tag{3.62}$$

and the ***angular momentum*** of the particle with respect to the origin 0 by

$$\widehat{\boldsymbol{m}} = m\boldsymbol{x} \times \boldsymbol{v} \qquad \widehat{m_i} = m\varepsilon_{ijk}x_jv_k\,. \tag{3.63}$$

For a part $\Pi$ of a body $\mathcal{B}$ in the initial configuration, $\mathcal{R}$, and the current configuration, $\mathcal{R}_t$, (fig. 3.2), we have the following definitions of the momentum and angular momentum with respect to the origin 0

$$\overline{\boldsymbol{m}}(\omega, t) = \int_\omega \rho(\boldsymbol{x}, t)\boldsymbol{v}(\boldsymbol{x}, t)\, dv$$
$$\overline{m}_i(\omega, t) = \int_\omega \rho(\boldsymbol{x}, t)v_i(\boldsymbol{x}, t)\, dv \tag{3.64}$$

$$\widehat{\boldsymbol{m}}(\omega, t) = \int_\omega \rho(\boldsymbol{x}, t)\, \boldsymbol{x} \times \boldsymbol{v}(\boldsymbol{x}, t)\, dv$$
$$\widehat{m_i}(\omega, t) = \int_\omega \rho(\boldsymbol{x}, t)\varepsilon_{ijk}x_jv_k(\boldsymbol{x}, t)\, dv. \tag{3.65}$$

The material derivatives of the preceding quantities are expressed as

$$\frac{D\overline{\boldsymbol{m}}(\omega,t)}{Dt} = \int_\omega \rho(\boldsymbol{x},t)\frac{D\boldsymbol{v}(\boldsymbol{x},t)}{Dt}\,dv = \int_\omega \rho(\boldsymbol{x},t)\boldsymbol{a}(\boldsymbol{x},t)\,dv \tag{3.66}$$

$$\begin{aligned}\frac{D\widehat{\boldsymbol{m}}(\omega,t)}{Dt} &= \int_\omega \rho(\boldsymbol{x},t)\boldsymbol{x}\times\frac{D\boldsymbol{v}(\boldsymbol{x},t)}{Dt}\,dv \\ &= \int_\omega \rho(\boldsymbol{x},t)\boldsymbol{x}\times\boldsymbol{a}(\boldsymbol{x},t)\,dv\ ,\end{aligned} \tag{3.67}$$

where $\boldsymbol{a}$ is the spatial acceleration defined by (2.33). The Reynolds transport theorem and the continuity equation (3.41) permit us to write equation (3.64) in the form

$$\begin{aligned}\frac{D\overline{m}_i(\omega,t)}{Dt} &= \frac{d}{dt}\int_\omega \rho v_i\,dv = \int_\omega\left(\frac{D(\rho v_i)}{Dt} + \rho v_i\frac{\partial v_m}{\partial x_m}\right)dv \\ &= \int_\omega\left(\frac{D\rho}{Dt}v_i + \rho\frac{Dv_i}{Dt} + \rho v_i\frac{\partial v_m}{\partial x_m}\right)dv \\ &= \int_\omega\left(\rho\frac{Dv_i}{Dt} + v_i\left(\frac{D\rho}{Dt} + \rho\frac{\partial v_m}{\partial x_m}\right)\right)dv \\ &= \int_\omega \rho a_i\,dv\,.\end{aligned}$$

The equality (3.66) is thus proved. Similarly, relation (3.67) is demonstrated by writing

$$\begin{aligned}&\frac{D\widehat{m}_i(\omega,t)}{Dt} \\ &= \frac{d}{dt}\int_\omega \rho\varepsilon_{ijk}x_jv_k\,dv \\ &= \int_\omega\left(\frac{D(\rho\varepsilon_{ijk}x_jv_k)}{Dt} + \rho\varepsilon_{ijk}x_jv_k\frac{\partial v_m}{\partial x_m}\right)dv \\ &= \int_\omega\left(\frac{D\rho}{Dt}\varepsilon_{ijk}x_jv_k + \rho\varepsilon_{ijk}\left(\frac{Dx_j}{Dt}v_k + x_j\frac{Dv_k}{Dt} + x_jv_k\frac{\partial v_m}{\partial x_m}\right)\right)dv \\ &= \int_\omega\left(\rho\varepsilon_{ijk}x_j\frac{Dv_k}{Dt} + \rho\varepsilon_{ijk}v_jv_k + \varepsilon_{ijk}x_jv_k\left(\frac{D\rho}{Dt} + \rho\frac{\partial v_m}{\partial x_m}\right)\right)dv \\ &= \int_\omega \rho\varepsilon_{ijk}x_ja_k\,dv\,,\end{aligned}$$

where we have used the fact that $\varepsilon_{ijk}v_jv_k = 0$. Now we will derive and state the two fundamental principles of mechanics of continuous media, known as ***Euler's laws of motion***.

PRINCIPLE OF CONSERVATION OF MOMENTUM *The rate of change of the momentum of an arbitrary part* $\Pi$ *of a body* $\mathcal{B}$ *at time* $t$ *is equal to the sum of the forces applied to* $\Pi$ *at that instant.*

The sum of the forces is composed of the volume forces acting on the particles Π and the contact forces acting on the boundary of Π. In the spatial description, that is equivalent to the sum $\boldsymbol{f}^b(\omega,t)+\boldsymbol{f}^c(\partial\omega,t)$. With (3.54), (3.60), and (3.64), the principle of conservation of momentum of Π has the following ***spatial*** formulation:

$$\frac{d}{dt}\int_\omega \rho(\boldsymbol{x},t)\boldsymbol{v}(\boldsymbol{x},t)\,dv=\int_\omega \rho(\boldsymbol{x},t)\boldsymbol{b}(\boldsymbol{x},t)\,dv+\int_{\partial\omega}\boldsymbol{t}(\boldsymbol{x},t,\boldsymbol{n})\,ds\,. \tag{3.68}$$

With (3.66), we can write (3.68) as

$$\int_\omega \rho(\boldsymbol{x},t)\boldsymbol{a}(\boldsymbol{x},t)\,dv=\int_\omega \rho(\boldsymbol{x},t)\boldsymbol{b}(\boldsymbol{x},t)\,dv+\int_{\partial\omega}\boldsymbol{t}(\boldsymbol{x},t,\boldsymbol{n})\,ds\,. \tag{3.69}$$

PRINCIPLE OF CONSERVATION OF ANGULAR MOMENTUM *The rate of change of angular momentum (with respect to the origin) of an arbitrary part Π of a body $\mathcal{B}$ at time t is equal to the moment (with respect to the origin) of the forces applied to Π at that instant.*

In the spatial description, that is equivalent to stating

$$\begin{aligned}&\frac{d}{dt}\int_\omega \rho(\boldsymbol{x},t)\boldsymbol{x}\times\boldsymbol{v}(\boldsymbol{x},t)\,dv\\&=\int_\omega \rho(\boldsymbol{x},t)\boldsymbol{x}\times\boldsymbol{b}(\boldsymbol{x},t)\,dv+\int_{\partial\omega}\boldsymbol{x}\times\boldsymbol{t}(\boldsymbol{x},t,\boldsymbol{n})\,ds\,.\end{aligned} \tag{3.70}$$

Applying (3.67), (3.70) becomes

$$\begin{aligned}&\int_\omega \rho(\boldsymbol{x},t)\boldsymbol{x}\times\boldsymbol{a}(\boldsymbol{x},t)\,dv\\&=\int_\omega \rho(\boldsymbol{x},t)\boldsymbol{x}\times\boldsymbol{b}(\boldsymbol{x},t)\,dv+\int_{\partial\omega}\boldsymbol{x}\times\boldsymbol{t}(\boldsymbol{x},t,\boldsymbol{n})\,ds\,.\end{aligned} \tag{3.71}$$

We must point out that the existence of an observer is implicitly assumed in these statements of Euler's laws of motion and in expressions (3.68) and (3.70). This then implies that the principles of conservation of momentum and of angular momentum are not objective, that is, they are not invariant when changing from one observer to another. This is because the velocity and acceleration are not objective quantities, as has been shown in section 2.11. Often, the observers for which (3.68) and (3.70) are invariants are qualified as ***inertial*** or ***Galilean*** (see sec. 2.11).

## 3.6 Cauchy's Theorem and Equation of Motion

We are now going to deduce important consequences of the principles of conservation of momentum and angular momentum. The first is an equivalent of Newton's third law.

CAUCHY'S LEMMA *If the stress vector $\mathbf{t}(\boldsymbol{x}, t, \boldsymbol{n})$ is continuous with respect to $\boldsymbol{x}$, then the principle of conservation of momentum (3.69) implies that*

$$\boldsymbol{t}(\boldsymbol{x}, t, -\boldsymbol{n}) = -\boldsymbol{t}(\boldsymbol{x}, t, \boldsymbol{n}) . \tag{3.72}$$

This is none other than the principle of action and reaction.

PROOF.
Let us cut the configuration $\omega$ of an arbitrary part $\Pi$ of a body $\mathcal{B}$ into two parts, $\omega^-$ and $\omega^+$, with a surface $\Gamma$ between them for which the unit normal exiting $\omega^-$ is $\boldsymbol{n}$ (fig. 3.6). The boundary of $\omega^-$ is composed of two surfaces $\Gamma$ and $\Gamma^-$, that is, $\partial\omega^- = \Gamma \cup \Gamma^-$. In the same way, the boundary of $\omega^+$ is $\partial\omega^+ = \Gamma \cup \Gamma^+$.

The principle of conservation of momentum is valid for $\omega^-$ and for $\omega^+$

$$\begin{aligned} &\int_{\omega^-} \rho(\boldsymbol{x}, t)\boldsymbol{a}(\boldsymbol{x}, t)\, dv \\ &\quad = \int_{\omega^-} \rho(\boldsymbol{x}, t)\boldsymbol{b}(\boldsymbol{x}, t)\, dv + \int_{\partial\omega^-} \boldsymbol{t}(\boldsymbol{x}, t, \boldsymbol{n})\, ds \\ &\quad = \int_{\omega^-} \rho(\boldsymbol{x}, t)\boldsymbol{b}(\boldsymbol{x}, t)\, dv + \int_{\Gamma^-} \boldsymbol{t}(\boldsymbol{x}, t, \boldsymbol{n}) ds + \int_{\Gamma} \boldsymbol{t}(\boldsymbol{x}, t, \boldsymbol{n})\, ds \end{aligned} \tag{3.73}$$

$$\begin{aligned} &\int_{\omega^+} \rho(\boldsymbol{x}, t)\boldsymbol{a}(\boldsymbol{x}, t)\, dv \\ &\quad = \int_{\omega^+} \rho(\boldsymbol{x}, t)\boldsymbol{b}(\boldsymbol{x}, t)\, dv + \int_{\partial\omega^+} \boldsymbol{t}(\boldsymbol{x}, t, \boldsymbol{n})\, ds \\ &\quad = \int_{\omega^+} \rho(\boldsymbol{x}, t)\boldsymbol{b}(\boldsymbol{x}, t)\, dv + \int_{\Gamma^+} \boldsymbol{t}(\boldsymbol{x}, t, \boldsymbol{n})\, ds + \int_{\Gamma} \boldsymbol{t}(\boldsymbol{x}, t, -\boldsymbol{n})\, ds . \end{aligned} \tag{3.74}$$

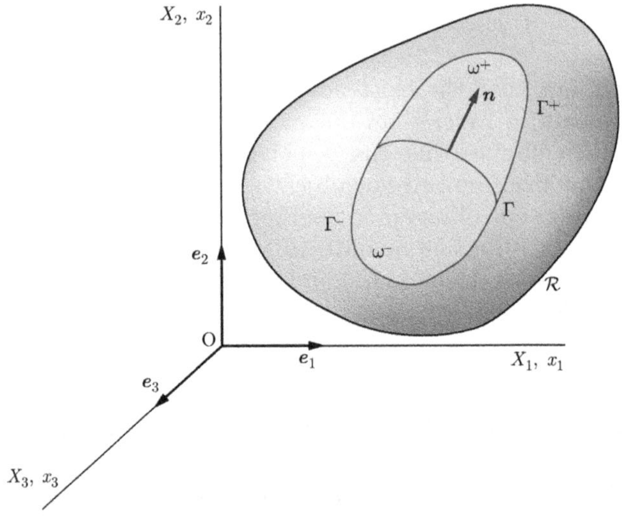


**Fig. 3.6** Partition of $\omega$ into $\omega^+$ and $\omega^-$ with a surface $\Gamma$

Now, equation (3.69) is valid with $\omega = \omega^- \cup \omega^+$ and $\partial\omega = \Gamma^- \cup \Gamma^+$. Combining (3.73) and (3.74) with (3.69), we obtain

$$\int_\Gamma \big(\boldsymbol{t}(\boldsymbol{x}, t, -\boldsymbol{n}) + \boldsymbol{t}(\boldsymbol{x}, t, \boldsymbol{n})\big)\, ds = 0\,. \tag{3.75}$$

Since $\boldsymbol{t}(\boldsymbol{x}, t, \boldsymbol{n})$ is continuous at $\boldsymbol{x}$ and the choice of $\Gamma$ is arbitrary, the application of the localization theorem to (3.75) yields (3.72).

With Cauchy's lemma, we are ready to state and prove one of the principal results of mechanics of continuous media.

**Cauchy's Theorem** (existence of the stress tensor)

*If the stress vector $\boldsymbol{t}(\boldsymbol{x}, t, \boldsymbol{n})$ is continuous with respect to $\boldsymbol{x}$ and if $\rho(\boldsymbol{x}, t)\boldsymbol{b}(\boldsymbol{x}, t)$ and $\rho(\boldsymbol{x}, t)\boldsymbol{a}(\boldsymbol{x}, t)$ are bounded, then the principle of conservation of momentum implies that there exists a stress tensor $\boldsymbol{\sigma}(\boldsymbol{x}, t)$ such that*

$$\boldsymbol{t}(\boldsymbol{x}, t, \boldsymbol{n}) = \boldsymbol{\sigma}(\boldsymbol{x}, t)\, \boldsymbol{n} \qquad \text{or} \qquad t_i(\boldsymbol{x}, t, \boldsymbol{n}) = \sigma_{ij}(\boldsymbol{x}, t) n_j\,. \tag{3.76}$$

PROOF.
Consider a tetrahedron $\omega_0$ for which the three faces $S_i$ are perpendicular to the unit vectors $\boldsymbol{e}_i$ $(i = 1, 2, 3)$ and join at the position $\boldsymbol{x}$ of a particle of a body $\mathcal{B}$ (fig. 3.7). Let the fourth face be $S_4$, with area $A$ and ***arbitrary*** unit normal $\boldsymbol{n} = (\cos\alpha_1, \cos\alpha_2, \cos\alpha_3)^T$. Then, a simple calculation shows that the area $A_i$ of the face $S_i$ with unit (outgoing) normal $\boldsymbol{e}_i$ is given by

$$A_i = A\cos\alpha_i \qquad\qquad \cos\alpha_i = \boldsymbol{n}\cdot\boldsymbol{e}_i\,. \tag{3.77}$$

Denoting the distance from $\boldsymbol{x}$ to $S_4$ by $h$, the volume of the tetrahedron is

$$V = \frac{1}{3}\, hA\,. \tag{3.78}$$

Consider another tetrahedron $\omega$ similar to the first, $\omega_0$ (fig. 3.7). Each linear dimension in $\omega$ is proportional to that in $\omega_0$ with the ratio $\lambda > 0$. Then the volume of $\omega$ is

$$v = \frac{1}{3}\, \lambda^3 hA\,. \tag{3.79}$$

The area $a$ of the face $s_4$ of $\omega$ with $\boldsymbol{n}$ as outgoing unit vector is

$$a = \lambda^2 A\,, \tag{3.80}$$

and the area $a_i$ of the face $s_i$ of $\omega$ with outgoing unit normal $\boldsymbol{e}_i$ is expressed as follows:

$$a_i = \lambda^2 A\boldsymbol{n}\cdot\boldsymbol{e}_i = \lambda^2 A\cos\alpha_i\,. \tag{3.81}$$

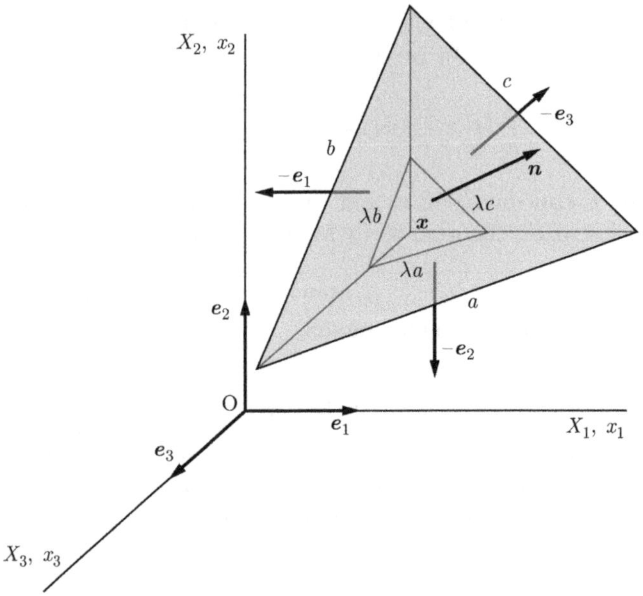


**Fig. 3.7** Cauchy tetrahedron

For the tetrahedron $\omega$, the principle of conservation of momentum (3.69) yields

$$\begin{aligned}&\int_\omega \big(\rho(\boldsymbol{y},t)\boldsymbol{a}(\boldsymbol{y},t) - \rho(\boldsymbol{y},t)\boldsymbol{b}(\boldsymbol{y},t)\big)\, dv \\ &\quad = \int_{s_4} \boldsymbol{t}(\boldsymbol{y},t,\boldsymbol{n})\, ds + \sum_{i=1}^{3} \int_{s_i} \boldsymbol{t}(\boldsymbol{y},t,-\boldsymbol{e}_i)\, ds\,.\end{aligned} \tag{3.82}$$

Since the stress vector $\boldsymbol{t}$ is, by hypothesis, continuous at $\boldsymbol{y}$, then the mean value theorem of integral calculus implies that there exists $\overline{\boldsymbol{y}}_i \in s_i$ such that

$$\int_{s_4} \boldsymbol{t}(\boldsymbol{y},t,\boldsymbol{n})\, ds = a\boldsymbol{t}(\overline{\boldsymbol{y}}_4,t,\boldsymbol{n}) = \lambda^2 A\, \boldsymbol{t}(\overline{\boldsymbol{y}}_4,t,\boldsymbol{n}) \tag{3.83}$$

$$\int_{s_i} \boldsymbol{t}(\boldsymbol{y},t,-\boldsymbol{e}_i)\, ds = a_i\boldsymbol{t}(\overline{\boldsymbol{y}}_i,t,-\boldsymbol{e}_i) = \lambda^2 A \cos\alpha_i \boldsymbol{t}(\overline{\boldsymbol{y}}_i,t,-\boldsymbol{e}_i)\,. \tag{3.84}$$

In this last relation there is no sum over $i$.

In addition, as $\rho(\boldsymbol{y},t)\boldsymbol{b}(\boldsymbol{y},t)$ and $\rho(\boldsymbol{y},t)\boldsymbol{a}(\boldsymbol{y},t)$ are assumed to be continuous and bounded, there exists a finite constant $M > 0$ such that

$$\left\| \int_\omega \big(\rho(\boldsymbol{y},t)\boldsymbol{a}(\boldsymbol{y},t) - \rho(\boldsymbol{y},t)\boldsymbol{b}(\boldsymbol{y},t)\big)\, dv \right\| \le Mv = \frac{1}{3} M\lambda^3 hA\,. \tag{3.85}$$

Taking into account (3.83), (3.84), and (3.85) in (3.82), we can write

$$0 \le \left\| \lambda^2 A\boldsymbol{t}(\overline{\boldsymbol{y}}_4,t,\boldsymbol{n}) + \sum_{i=1}^{3} \lambda^2 A\cos\alpha_i \boldsymbol{t}(\overline{\boldsymbol{y}}_i,t,-\boldsymbol{e}_i) \right\| \le \frac{1}{3} M\lambda^3 hA\,, \tag{3.86}$$

that is,

$$0 \leq \left\| \boldsymbol{t}(\overline{\boldsymbol{y}}_4, t, \boldsymbol{n}) + \sum_{i=1}^{3} \cos\alpha_i \boldsymbol{t}(\overline{\boldsymbol{y}}_i, t, -\boldsymbol{e}_i) \right\| \leq \frac{1}{3} M\lambda h \,. \tag{3.87}$$

As $\lambda \to 0$, then $\overline{\boldsymbol{y}}_i \to \boldsymbol{x}$ for every $i = 1, 2, 3, 4$ and (3.87) becomes

$$\left\| \boldsymbol{t}(\boldsymbol{x}, t, \boldsymbol{n}) + \sum_{i=1}^{3} \cos\alpha_i \boldsymbol{t}(\boldsymbol{x}, t, -\boldsymbol{e}_i) \right\| = 0 \,. \tag{3.88}$$

Using Cauchy's lemma (3.72) and the second relation of (3.77), it follows from (3.88) that

$$\boldsymbol{t}(\boldsymbol{x}, t, \boldsymbol{n}) - \sum_{i=1}^{3} (\boldsymbol{n} \cdot \boldsymbol{e}_i) \boldsymbol{t}(\boldsymbol{x}, t, \boldsymbol{e}_i) = 0 \,, \tag{3.89}$$

that is,

$$\boldsymbol{t}(\boldsymbol{x}, t, \boldsymbol{n}) = (\boldsymbol{n} \cdot \boldsymbol{e}_1)\boldsymbol{t}(\boldsymbol{x}, t, \boldsymbol{e}_1) + (\boldsymbol{n} \cdot \boldsymbol{e}_2)\boldsymbol{t}(\boldsymbol{x}, t, \boldsymbol{e}_2) + (\boldsymbol{n} \cdot \boldsymbol{e}_3)\boldsymbol{t}(\boldsymbol{x}, t, \boldsymbol{e}_3) \,. \tag{3.90}$$

The definition of the tensor product of two vectors (1.48) allows us to write (3.90) in the form

$$\boldsymbol{t}(\boldsymbol{x}, t, \boldsymbol{n}) = \big(\boldsymbol{t}(\boldsymbol{x}, t, \boldsymbol{e}_1) \otimes \boldsymbol{e}_1 + \boldsymbol{t}(\boldsymbol{x}, t, \boldsymbol{e}_2) \otimes \boldsymbol{e}_2 + \boldsymbol{t}(\boldsymbol{x}, t, \boldsymbol{e}_3) \otimes \boldsymbol{e}_3\big)\boldsymbol{n} \,. \tag{3.91}$$

Consequently, the existence of the ***Cauchy stress tensor***

$$\boldsymbol{\sigma}(\boldsymbol{x}, t) = \boldsymbol{t}(\boldsymbol{x}, t, \boldsymbol{e}_1) \otimes \boldsymbol{e}_1 + \boldsymbol{t}(\boldsymbol{x}, t, \boldsymbol{e}_2) \otimes \boldsymbol{e}_2 + \boldsymbol{t}(\boldsymbol{x}, t, \boldsymbol{e}_3) \otimes \boldsymbol{e}_3 \,, \tag{3.92}$$

such as (3.76), is valid and proved. A stress, being a force per unit surface, is expressed in Pascals (Pa) in SI units.

Cauchy's theorem expresses the ***linear*** dependence of $\boldsymbol{t}(\boldsymbol{x}, t, \boldsymbol{n})$ with respect to the unit normal. Thus, when the stress tensor $\boldsymbol{\sigma}(\boldsymbol{x}, t)$ is known, the stress vector acting at $\boldsymbol{x}$ on any surface with outgoing unit normal $\boldsymbol{n}$ is completely determined. Consequently, the ***state of stress*** at $\boldsymbol{x}$ (at time $t$) is characterized by the stress tensor $\boldsymbol{\sigma}(\boldsymbol{x}, t)$. Even if the main properties of the stress tensor will be studied further on, it is useful to give a geometric interpretation of its components $\sigma_{ij}$ in order to better understand its importance in continuum mechanics.

The components $\sigma_{ij}$ of the matrix of $\boldsymbol{\sigma}$ with respect to the basis $\{\boldsymbol{e}_1, \boldsymbol{e}_2, \boldsymbol{e}_3\}$ are obtained by

$$\sigma_{ij} = \boldsymbol{e}_i \cdot \boldsymbol{\sigma}\boldsymbol{e}_j = \boldsymbol{e}_i \cdot \boldsymbol{t}_{\boldsymbol{e}_j} \qquad \boldsymbol{t}_{\boldsymbol{e}_j} = \boldsymbol{\sigma}\boldsymbol{e}_j \,. \tag{3.93}$$

This relation shows that $\sigma_{ij}$ is the component of the stress vector $\boldsymbol{t}_{\boldsymbol{e}_j}$ in the direction $i$ acting on a spatial surface element whose unit normal is aligned in the direction of $\boldsymbol{e}_j$ (fig. 3.8).

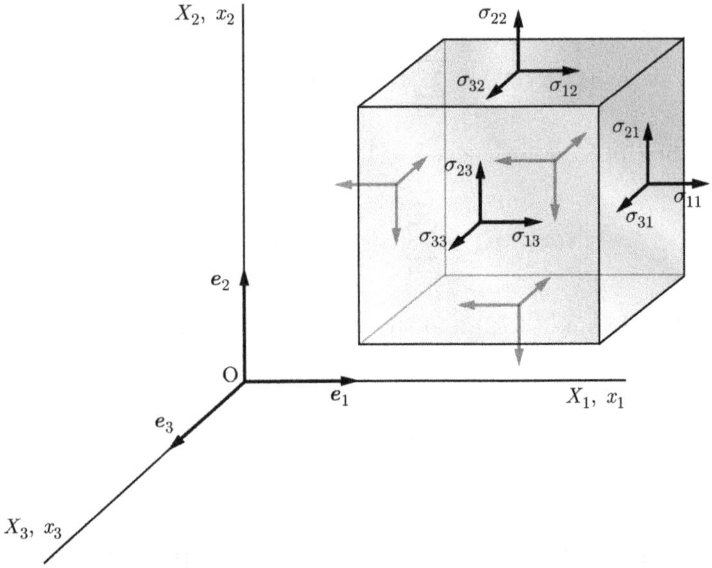


**Fig. 3.8** Stress components of the Cauchy stress tensor $\boldsymbol{\sigma}$

For example, $\sigma_{11}$ is the component in direction 1 of the stress vector acting on a surface element with unit normal $\boldsymbol{e}_1$, and $\sigma_{12}$ is the component in direction 1 of the stress vector acting on a surface element with unit normal $\boldsymbol{e}_2$. The normal component of $\boldsymbol{t}_{\boldsymbol{e}_j}$, that is,

$$\sigma_{jj} = \boldsymbol{e}_j \cdot \boldsymbol{t}_{\boldsymbol{e}_j} = \boldsymbol{e}_j \cdot \boldsymbol{\sigma} \boldsymbol{e}_j \qquad \text{(no sum over } j\text{)}\,, \tag{3.94}$$

is called the ***normal stress***. It corresponds to ***tension*** if it is positive or to ***compression*** if it is negative. The tangent components of $\boldsymbol{t}_{\boldsymbol{e}_j}$, that is,

$$\sigma_{ij} = \boldsymbol{e}_i \cdot \boldsymbol{t}_{\boldsymbol{e}_j} = \boldsymbol{e}_i \cdot \boldsymbol{\sigma} \boldsymbol{e}_j \qquad \text{with} \qquad \boldsymbol{e}_i \cdot \boldsymbol{e}_j = 0\,, \tag{3.95}$$

are called ***shear stresses***. For example, $\sigma_{11}$ is a normal stress and $\sigma_{12}$ is a shear stress.

We will now use Cauchy's theorem and the divergence theorem to derive the equations of motion for a continuous medium starting from the principle of conservation of momentum.

PRINCIPLE OF CONSERVATION OF MOMENTUM

**Theorem**

*Suppose that the stress tensor $\boldsymbol{\sigma}(\boldsymbol{x}, t)$ is continuously differentiable with respect to $\boldsymbol{x}$, and that $\rho(\boldsymbol{x}, t)\boldsymbol{b}(\boldsymbol{x}, t)$ and $\rho(\boldsymbol{x}, t)\boldsymbol{a}(\boldsymbol{x}, t)$ are continuous at $\boldsymbol{x}$. Then, the principle of conservation of momentum, that is (3.69), is satisfied if and only if, for an arbitrary point $\boldsymbol{x}$ of $\mathcal{R}$,*

$$\mathbf{div}\,\boldsymbol{\sigma}(\boldsymbol{x}, t) + \rho(\boldsymbol{x}, t)\boldsymbol{b}(\boldsymbol{x}, t) = \rho(\boldsymbol{x}, t)\boldsymbol{a}(\boldsymbol{x}, t) \quad \textit{or} \quad \sigma_{ij,j} + \rho b_i = \rho a_i\,. \tag{3.96}$$

PROOF.
*Necessity.* Introducing (3.76) in (3.69), we obtain

$$\int_\omega \rho(\boldsymbol{x},t)a_i(\boldsymbol{x},t)\,dv = \int_\omega \rho(\boldsymbol{x},t)b_i(\boldsymbol{x},t)\,dv + \int_{\partial\omega} \sigma_{ij}(\boldsymbol{x},t)n_j\,ds\,. \tag{3.97}$$

Applying the divergence theorem to the last term, we obtain

$$\int_\omega \big(\rho(\boldsymbol{x},t)a_i(\boldsymbol{x},t) - \rho(\boldsymbol{x},t)b_i(\boldsymbol{x},t) - \sigma_{ij,j}(\boldsymbol{x},t)\big)\,dv = 0\,. \tag{3.98}$$

Since the integrand is continuous at $\boldsymbol{x}$, (3.96) follows from the localization theorem.

*Sufficiency.* Suppose that (3.96) is valid for every interior point of $\mathcal{R}$. Then, for an arbitrary domain $\omega$ de $\mathcal{R}$,

$$\int_\omega \big(\rho(\boldsymbol{x},t)a_i(\boldsymbol{x},t) - \rho(\boldsymbol{x},t)b_i(\boldsymbol{x},t) - \sigma_{ij,j}(\boldsymbol{x},t)\big)\,dv = 0\,. \tag{3.99}$$

Applying Cauchy's theorem and the divergence theorem to this equation, we can conclude that (3.68) is verified.

Equation (3.96), derived by Cauchy, is called ***Cauchy's equation of motion***. When there is no acceleration, it is also called the ***equilibrium equation***. As we will see, equation (3.96) is one of the most often used equations in the mechanics of continuous media.

## PRINCIPLE OF CONSERVATION OF ANGULAR MOMENTUM

**Theorem**

*Suppose that the stress tensor $\boldsymbol{\sigma}(\boldsymbol{x},t)$ is continuously differentiable with respect to $\boldsymbol{x}$, and that $\rho(\boldsymbol{x},t)\boldsymbol{b}(\boldsymbol{x},t)$ and $\rho(\boldsymbol{x},t)\boldsymbol{a}(\boldsymbol{x},t)$ are continuous at $\boldsymbol{x}$. Then the principle of conservation of angular momentum (3.71) implies the symmetry of the Cauchy stress tensor,*

$$\boldsymbol{\sigma}^T = \boldsymbol{\sigma} \qquad or \qquad \sigma_{ij} = \sigma_{ji}\,. \tag{3.100}$$

PROOF.
Taking into account (3.76) in (3.71), we have

$$\begin{aligned}&\int_\omega \rho(\boldsymbol{x},t)\varepsilon_{ijk}x_j a_k(\boldsymbol{x},t)\,dv\\ &= \int_\omega \rho(\boldsymbol{x},t)\varepsilon_{ijk}x_j b_k(\boldsymbol{x},t)\,dv + \int_{\partial\omega} \varepsilon_{ijk}x_j\sigma_{km}(\boldsymbol{x},t)n_m\,ds\,.\end{aligned} \tag{3.101}$$

Applying the divergence theorem to the last term and since $x_{j,m} = \delta_{jm}$ (eqn. (1.27)), we obtain

$$\begin{aligned}\int_{\partial\omega} &\varepsilon_{ijk} x_j \sigma_{km}(\boldsymbol{x},t) n_m \, ds \\ &= \int_\omega \varepsilon_{ijk}\big(x_{j,m}\sigma_{km}(\boldsymbol{x},t) + x_j \sigma_{km,m}(\boldsymbol{x},t)\big)\, dv \\ &= \int_\omega \varepsilon_{ijk}\big(\sigma_{kj}(\boldsymbol{x},t) + x_j \sigma_{km,m}(\boldsymbol{x},t)\big)\, dv \,. \end{aligned} \tag{3.102}$$

The substitution of (3.102) in (3.101) yields

$$\begin{aligned}\int_\omega &\varepsilon_{ijk} x_j \big(\rho(\boldsymbol{x},t) a_k(\boldsymbol{x},t) - \rho(\boldsymbol{x},t) b_k(\boldsymbol{x},t) - \sigma_{km,m}(\boldsymbol{x},t)\big)\, dv \\ &= \int_\omega \varepsilon_{ijk}\sigma_{kj}(\boldsymbol{x},t)\, dv \,. \end{aligned} \tag{3.103}$$

From (3.96), the left-hand side of this expression is equal to zero. Thus (3.103) is reduced to

$$\int_\omega \varepsilon_{ijk}\sigma_{kj}(\boldsymbol{x},t)\, dv = 0 \,. \tag{3.104}$$

The localization theorem and expressing $\sigma_{ij}$ in symmetric and antisymmetric parts lead to

$$\varepsilon_{ijk}\sigma_{kj} = \frac{1}{2}\varepsilon_{ijk}(\sigma_{kj}+\sigma_{jk}) - \frac{1}{2}\varepsilon_{ijk}(\sigma_{jk}-\sigma_{kj}) = -\frac{1}{2}\,\varepsilon_{ijk}(\sigma_{jk}-\sigma_{kj}) = 0 \,. \tag{3.105}$$

This implies that $\sigma_{jk} = \sigma_{kj}$, that is, (3.100). Equation (3.100) tells us that

$$\sigma_{12} = \sigma_{21} \qquad \sigma_{23} = \sigma_{32} \qquad \sigma_{31} = \sigma_{13} \,. \tag{3.106}$$

Consequently, among the nine components of the Cauchy stress tensor $\boldsymbol{\sigma}$ (fig. 3.8), six are independent. In addition, because of its symmetry, $\boldsymbol{\sigma}$ possesses several properties which can be obtained by directly applying the results of linear algebra for symmetric tensors. In particular, the spectral decomposition theorem for a symmetric tensor can lead to a better understanding of $\boldsymbol{\sigma}$. Finally, we can point out that the symmetry of the tensor $\boldsymbol{\sigma}$ guarantees, by itself, the conservation of angular momentum. The proof of this affirmation is performed simply by inverting the order of the steps in the previous theorem. Thus, Euler's two laws of motion are satisfied if the stress tensor $\boldsymbol{\sigma}$ is symmetric and satisfies equation (3.96).

## 3.7 Properties of the Cauchy Stress Tensor

Now we will study the main properties of $\boldsymbol{\sigma}(\boldsymbol{x},t)$, starting with (3.76) and (3.100). In order to simplify the notation in this section, we will not consider the dependence of $\boldsymbol{\sigma}$ on $\boldsymbol{x}$ and $t$ as the properties of $\boldsymbol{\sigma}$ remain valid independent of the values of $\boldsymbol{x}$ and $t$. The stress vector is given by Cauchy's theorem expressed in relation (3.76)

$$\boldsymbol{t} = \boldsymbol{\sigma n} \qquad \text{or} \qquad t_i = \sigma_{ij} n_j \,. \tag{3.107}$$

In general, $\boldsymbol{t}$ does not act in the direction of the unit normal $\boldsymbol{n}$ with which it is associated. Thus, $\boldsymbol{t}$ not only has a ***normal component***

$$t_N = \boldsymbol{n} \cdot \boldsymbol{t} = \boldsymbol{n} \cdot \boldsymbol{\sigma n} \qquad \text{or} \qquad t_N = n_i t_i = \sigma_{ij} n_i n_j \,, \tag{3.108}$$

but also a ***tangential component*** associated with shear

$$t_T = \left\|\boldsymbol{t} - (\boldsymbol{n} \cdot \boldsymbol{t})\boldsymbol{n}\right\| = \left\|(\boldsymbol{I} - \boldsymbol{n} \otimes \boldsymbol{n})\boldsymbol{t}\right\| = (t_i t_i - t_N^2)^{1/2} \,. \tag{3.109}$$

Nonetheless, it is possible for $\boldsymbol{t}$ to act uniquely in the direction of $\boldsymbol{n}$. This possibility leads to the following eigenvalue problem:

$$\boldsymbol{\sigma n} = \lambda \boldsymbol{n} \qquad \text{or} \qquad \sigma_{ij} n_j = \lambda n_i \,. \tag{3.110}$$

Linear algebra (sec. 1.3.8) allows us to state that the characteristic equation associated with (3.110)

$$\det(\boldsymbol{\sigma} - \lambda \boldsymbol{I}) = 0 \qquad \text{or} \qquad \det(\sigma_{ij} - \lambda \delta_{ij}) = 0 \tag{3.111}$$

has three real roots $\sigma_i$ $(i = 1, 2, 3)$ because the tensor $\boldsymbol{\sigma}$ is symmetric. These roots are the eigenvalues of $\boldsymbol{\sigma}$ which, in mechanics, are called the ***principal stresses***. In general, $\boldsymbol{\sigma}$ has three distinct principal stresses $\sigma_1$, $\sigma_2$, and $\sigma_3$, which are usually ordered so that $\sigma_1 \geq \sigma_2 \geq \sigma_3$. The axes aligned with the principal vectors $\boldsymbol{n_i}$ associated with $\sigma_i$ are called the ***principal stress axes*** and the planes normal to these axes are called the ***principal planes***. In summary, a principal stress is the normal stress that acts on the principal plane, where no shear stress exists.

In linear algebra it is proven that the eigenvectors corresponding to the distinct eigenvalues of a symmetric tensor $\boldsymbol{\sigma}$ are mutually orthogonal. This means that the two axes or principal planes associated with any two distinct stresses are perpendicular. This property is used for the spectral decomposition of $\boldsymbol{\sigma}$

$$\boldsymbol{\sigma} = \sigma_1 \boldsymbol{n}_1 \otimes \boldsymbol{n}_1 + \sigma_2 \boldsymbol{n}_2 \otimes \boldsymbol{n}_2 + \sigma_3 \boldsymbol{n}_3 \otimes \boldsymbol{n}_3 \,, \tag{3.112}$$

where $\boldsymbol{n}_i \cdot \boldsymbol{n}_j = \delta_{ij}$ for $i, j = 1, 2, 3$. In other words, with respect to the basis $\{\boldsymbol{n}_1, \boldsymbol{n}_2, \boldsymbol{n}_3\}$ composed of the orthonormal principal vectors $\boldsymbol{n}_i$, the matrix $[\sigma]$ of the Cauchy stress tensor is diagonal:

$$[\sigma] = \begin{pmatrix} \sigma_1 & 0 & 0 \\ 0 & \sigma_2 & 0 \\ 0 & 0 & \sigma_3 \end{pmatrix} . \tag{3.113}$$

The state of stress of a particle of a body $\mathcal{B}$ is said to be ***three-dimensional*** if $\sigma_1$, $\sigma_2$, and $\sigma_3$ are all three non zero, ***two-dimensional***, or plane, if two of the stresses $\sigma_1$, $\sigma_2$ and $\sigma_3$ are non zero, and ***one-dimensional*** if only one of the stresses $\sigma_1$, $\sigma_2$, and $\sigma_3$ is non zero.

If equation (3.111) is developed, it becomes (eqn. (1.120))

$$\lambda^3 - \mathrm{I}_1(\boldsymbol{\sigma})\lambda^2 + \mathrm{I}_2(\boldsymbol{\sigma})\lambda - \mathrm{I}_3(\boldsymbol{\sigma}) = 0\,, \tag{3.114}$$

where

$$\mathrm{I}_1(\boldsymbol{\sigma}) = \mathrm{tr}\,\boldsymbol{\sigma} = \sigma_{ii} \tag{3.115}$$

$$\mathrm{I}_2(\boldsymbol{\sigma}) = \frac{1}{2}\left((\mathrm{tr}\,\boldsymbol{\sigma})^2 - \mathrm{tr}\,\boldsymbol{\sigma}^2\right) = \frac{1}{2}\left((\sigma_{ii})^2 - \sigma_{mn}\sigma_{nm}\right) \tag{3.116}$$

$$\mathrm{I}_3(\boldsymbol{\sigma}) = \det\boldsymbol{\sigma} = \varepsilon_{ijk}\sigma_{i1}\sigma_{j2}\sigma_{k3} \tag{3.117}$$

are the ***principal invariants*** of $\boldsymbol{\sigma}$ with respect to different orthonormal bases. As we will see, these invariants play a major role in the formulation of the constitutive equations of isotropic materials. With (3.113), they can be expressed in terms of the principal stresses

$$\begin{aligned} \mathrm{I}_1(\boldsymbol{\sigma}) &= \sigma_1 + \sigma_2 + \sigma_3 \\ \mathrm{I}_2(\boldsymbol{\sigma}) &= \sigma_1\sigma_2 + \sigma_2\sigma_3 + \sigma_3\sigma_1 \\ \mathrm{I}_3(\boldsymbol{\sigma}) &= \sigma_1\sigma_2\sigma_3\,. \end{aligned} \tag{3.118}$$

Note that the permutation of the indices 1, 2, and 3 does not change $\mathrm{I}_1(\boldsymbol{\sigma})$, $\mathrm{I}_2(\boldsymbol{\sigma})$, or $\mathrm{I}_3(\boldsymbol{\sigma})$. Note that since (3.118) are invariants, so are the principal stresses.

### Transformation of the Stress Tensor

In (3.93), the components $\sigma_{ij}$ of the matrix of $\boldsymbol{\sigma}$ are defined relative to the orthonormal basis $\{\boldsymbol{e}_1, \boldsymbol{e}_2, \boldsymbol{e}_3\}$ by

$$\sigma_{ij} = \boldsymbol{e}_i \cdot \boldsymbol{\sigma}\boldsymbol{e}_j\,. \tag{3.119}$$

Consider another orthonormal basis $\{\boldsymbol{e}'_1, \boldsymbol{e}'_2, \boldsymbol{e}'_3\}$ obtained by rotation of $\{\boldsymbol{e}_1, \boldsymbol{e}_2, \boldsymbol{e}_3\}$

$$\boldsymbol{e}'_i = c_{ij}\boldsymbol{e}_j \qquad (i = 1, 2, 3)\,, \tag{3.120}$$

where $c_{ij}$ is given by relation (1.6). Then the components of $\boldsymbol{\sigma}'$, with respect to the orthonormal basis $\{\boldsymbol{e}'_1, \boldsymbol{e}'_2, \boldsymbol{e}'_3\}$, are related to $\sigma_{ij}$ (see (1.52)) by

$$\sigma'_{ij} = \boldsymbol{e}'_i \cdot \boldsymbol{\sigma}\boldsymbol{e}'_j = c_{im}c_{jn}\boldsymbol{e}_m \cdot \boldsymbol{\sigma}\boldsymbol{e}_n = c_{im}c_{jn}\sigma_{mn}\,. \tag{3.121}$$

Expression (3.121) is the ***Cauchy stress matrix transformation rule***. It is seen that the scalars given by (3.118) are said to be invariant in the sense that

$$\begin{aligned} &\sigma_{ii} = \sigma'_{ii} \\ &(\sigma_{ii})^2 - \sigma_{mn}\sigma_{mn} = (\sigma'_{ii})^2 - \sigma'_{mn}\sigma'_{mn} \\ &\varepsilon_{ijk}\sigma_{i1}\sigma_{j2}\sigma_{k3} = \varepsilon_{ijk}\sigma'_{i1}\sigma'_{j2}\sigma'_{k3}\,. \end{aligned}$$

It is often useful to decompose $\boldsymbol{\sigma}$ in the following way (see (1.90) and (1.91)):

$$\boldsymbol{\sigma} = \boldsymbol{s} + \sigma_0 \boldsymbol{I} \qquad \text{or} \qquad \sigma_{ij} = s_{ij} + \sigma_0 \delta_{ij} \,, \tag{3.122}$$

where

$$\boldsymbol{s} = \boldsymbol{\sigma} - \sigma_0 \boldsymbol{I} \,, \qquad \sigma_0 = \frac{1}{3} I_1(\boldsymbol{\sigma}) = \frac{1}{3} \sigma_{kk} \,. \tag{3.123}$$

The tensor $\boldsymbol{s}$, so defined, is called the ***deviatoric stress tensor*** associated with $\boldsymbol{\sigma}$. By its construction, $\operatorname{tr} \boldsymbol{s} = s_{ii} = 0$. In other words, if the deviatoric part of $\boldsymbol{\sigma}$ is zero, then $\boldsymbol{\sigma}$ takes the form $\boldsymbol{\sigma} = -p\boldsymbol{I}$ with $\sigma_0 = -p$. Note that $p = -(1/3) \operatorname{tr} \boldsymbol{\sigma}$. In this case, we have a ***state of pure hydrostatic stress***, and $p$ is the ***hydrostatic pressure***. The negative sign comes from the conventional use of pressure in fluid mechanics, which is regarded as positive when it causes compression.

EXAMPLE 3.1

Let $\boldsymbol{\sigma}$ be the tensor whose elements are given by (1.122), which describes the state of stress of a continuous medium (arbitrary units). Then, the eigenvalue problem to solve is none other than that of finding the principal stresses and directions. Using this same state of stress, find the stress vector on the plane defined by the unit normal vector $\boldsymbol{n} = 2\boldsymbol{e}_1/3 + 2\boldsymbol{e}_2/3 - \boldsymbol{e}_3/3$:

$$[\sigma] = \begin{pmatrix} 2 & 1 & -2 \\ 1 & 4 & -3 \\ -2 & -3 & -2 \end{pmatrix} . \tag{3.124}$$

To find the vector components on the given plane, we use Cauchy's theorem (3.76)

$$\begin{pmatrix} t_1 \\ t_2 \\ t_3 \end{pmatrix} = \begin{pmatrix} 2 & 1 & -2 \\ 1 & 4 & -3 \\ -2 & -3 & -2 \end{pmatrix} \begin{pmatrix} 2/3 \\ 2/3 \\ -1/3 \end{pmatrix} = \frac{1}{3} \begin{pmatrix} 8 \\ 13 \\ -8 \end{pmatrix} .$$

The normal and shear stress components on this plane are given by (3.108) and (3.109); the values are, respectively,

$$t_N = n_i t_i = 5.55 \quad \text{and} \quad t_T = \left(t_i t_i - t_N^2\right)^{1/2} = 1.48 .$$

Now consider a coordinate system defined by the principal directions. With respect to this system, we define a plane with a normal vector given by $\boldsymbol{m} = (1/\sqrt{3})\boldsymbol{n}_1 + (1/\sqrt{3})\boldsymbol{n}_2 + (1/\sqrt{3})\boldsymbol{n}_3$. Using (3.76) the stress vector on this plane is

$$\begin{pmatrix} t_1 \\ t_2 \\ t_3 \end{pmatrix} = \begin{pmatrix} 6 & 0 & 0 \\ 0 & 1.65 & 0 \\ 0 & 0 & -3.65 \end{pmatrix} \begin{pmatrix} 1/\sqrt{3} \\ 1/\sqrt{3} \\ 1/\sqrt{3} \end{pmatrix} = \frac{1}{\sqrt{3}} \begin{pmatrix} 6 \\ 1.65 \\ -3.65 \end{pmatrix} .$$

The normal and shear components are

$$t_N = n_i t_i = 1.33 \quad \text{and} \quad t_T = \left(t_i t_i - t_N^2\right)^{1/2} = 3.94.$$

The plane defined by the unit vector above is called the octahedral plane and the associated stresses are called the normal and shear octahedral stresses or the comparison stresses. On this particular plane, it can be shown that

$$\begin{aligned} t_N &= I_1(\boldsymbol{\sigma})/3 \\ t_T &= \frac{1}{3}\sqrt{2I_1^2(\boldsymbol{\sigma}) - 6I_2(\boldsymbol{\sigma})} \,. \end{aligned}$$

The proof is left as an exercise for the reader.

The shear component can also be expressed in terms of the principal stresses or of the second invariant $I_2(\boldsymbol{s})$ of the stress deviatoric tensor $\boldsymbol{s}$ (3.123)

$$t_T = \frac{1}{3}\left[(\sigma_1 - \sigma_2)^2 + (\sigma_2 - \sigma_3)^2 + (\sigma_3 - \sigma_1)^2\right]^{1/2} = \sqrt{\frac{2}{3} I_2(\boldsymbol{s})} \,.$$

Note that an equivalent stress $\sigma_e$, the von Mises stress, proportional to $t_T$,

$$\sigma_e = \left[\frac{1}{2}\left((\sigma_1 - \sigma_2)^2 + (\sigma_2 - \sigma_3)^2 + (\sigma_3 - \sigma_1)^2\right)\right]^{1/2} = \sqrt{\frac{1}{3} I_2(\boldsymbol{s})} \,.$$

is frequently used in solid mechanics to characterize the onset of plastic deformation or rupture of materials.

## 3.8 Simplified Stress States

As has been mentioned, the equilibrium equations for a continuous medium correspond to equation (3.96) without acceleration

$$\begin{aligned} \frac{\partial \sigma_{11}}{\partial x_1} + \frac{\partial \sigma_{12}}{\partial x_2} + \frac{\partial \sigma_{13}}{\partial x_3} + \rho b_1 = 0 \\ \frac{\partial \sigma_{21}}{\partial x_1} + \frac{\partial \sigma_{22}}{\partial x_2} + \frac{\partial \sigma_{23}}{\partial x_3} + \rho b_2 = 0 \\ \frac{\partial \sigma_{31}}{\partial x_1} + \frac{\partial \sigma_{32}}{\partial x_2} + \frac{\partial \sigma_{33}}{\partial x_3} + \rho b_3 = 0 \end{aligned} \tag{3.125}$$

or

$$\frac{\partial \sigma_{ij}}{\partial x_j} + \rho b_i = 0 \,. \tag{3.126}$$

The volume force is often denoted as $f_i = \rho b_i$.

EXAMPLE 3.2
Let the state of stress of a body be given by the following expressions:

$$\begin{aligned}\sigma_{11} &= 10x_1^3 + x_2^2, \quad \sigma_{12} = 2x_3 \\ \sigma_{22} &= 20x_1^3 + 100, \quad \sigma_{13} = x_2^3 \\ \sigma_{33} &= 30x_2^2 + 10x_3^2, \quad \sigma_{23} = 52x_1^3 \,.\end{aligned}$$

We want to find the volume forces such that static equilibrium is satisfied. The static equilibrium equations are given by (3.125). With the given stress components, these equations yield

$$\begin{aligned}30x_1^2 + \rho b_1 &= 0 \\ 0 + \rho b_2 &= 0 \\ 20x_3 + \rho b_3 &= 0 \,.\end{aligned}$$

Thus the volume force which maintains the equilibrium is given by the vector $(-30x_1^2, 0, -20x_3)$. Equations (3.125) can be simplified if we assume negligible volume forces. These three equations are insufficient to determine the six components $\sigma_{ij}$ of $\boldsymbol{\sigma}$, but they must be met for any body in the absence of acceleration. Simple inspection reveals that if each component of $\boldsymbol{\sigma}$ is independent of $\boldsymbol{x}$, the three equations of (3.125), without volume forces, are trivially satisfied. A state of stress is said to be ***homogeneous*** if $\boldsymbol{\sigma}$ is independent of $\boldsymbol{x}$. Such states of stress are important, not only because a large number of static or quasi-static tests of continuous media are based on them, but also because a good understanding of these states is necessary before treating more complicated stress states.

**Uniform Tension or Compression**

Suppose that tension or compression is applied in direction 1. The tensor $\boldsymbol{\sigma}$ is given by

$$\boldsymbol{\sigma} = \sigma \boldsymbol{n}_1 \otimes \boldsymbol{n}_1 \qquad \text{or} \qquad [\sigma] = \begin{pmatrix} \sigma & 0 & 0 \\ 0 & 0 & 0 \\ 0 & 0 & 0 \end{pmatrix}, \tag{3.127}$$

where $\sigma$ is constant. This tensor characterizes the state of stress in a prismatic cylindrical bar parallel to $\boldsymbol{e}_1$, with no force on the lateral surfaces and with normal stress uniformly applied at the two ends. The bar is said to be in ***tension*** if $\sigma > 0$ and in ***compression*** if $\sigma < 0$. The principal axes of stress include the one parallel to $\boldsymbol{e}_1$ and all those which are normal to $\boldsymbol{e}_1$. More general than (3.127), uniform tension or compression in the direction defined by a unit vector $\boldsymbol{m}$ is expressed as

$$\boldsymbol{\sigma} = \sigma(\boldsymbol{m} \otimes \boldsymbol{m}) \qquad \text{or} \qquad \sigma_{ij} = \sigma m_i m_j \,, \tag{3.128}$$

with $\sigma$ as a constant.

### Uniform Shear Stress

Uniform shear stress is applied in direction 1 on the planes perpendicular to $\boldsymbol{e}_2$. The tensor $\boldsymbol{\sigma}$ is given by

$$\boldsymbol{\sigma} = \tau(\boldsymbol{e}_1 \otimes \boldsymbol{e}_2 + \boldsymbol{e}_2 \otimes \boldsymbol{e}_1) \qquad \text{or} \qquad [\sigma] = \begin{pmatrix} 0 & \tau & 0 \\ \tau & 0 & 0 \\ 0 & 0 & 0 \end{pmatrix}, \tag{3.129}$$

where $\tau \geq 0$ is constant. This state of stress can be found in the laminar flow of a viscous fluid parallel to $\boldsymbol{e}_1$ over a surface perpendicular to $\boldsymbol{e}_2$. The characteristic equation (3.114) for this state of stress takes the form

$$\lambda(\lambda^2 - \tau^2) = 0\,. \tag{3.130}$$

Consequently, the principal stresses are $\sigma_1 = \tau$, $\sigma_2 = 0$, and $\sigma_3 = -\tau$, and the corresponding principal directions are $\boldsymbol{n}_1 = (\boldsymbol{e}_1 + \boldsymbol{e}_2)/\sqrt{2}$, $\boldsymbol{n}_2 = \boldsymbol{e}_3$, and $\boldsymbol{n}_3 = (\boldsymbol{e}_1 - \boldsymbol{e}_2)/\sqrt{2}$ . In summary, the spectral decomposition (3.112) for a shear stress is

$$\boldsymbol{\sigma} = \tau(\boldsymbol{e}_1 + \boldsymbol{e}_2) \otimes \frac{\boldsymbol{e}_1 + \boldsymbol{e}_2}{2} - \tau(\boldsymbol{e}_1 - \boldsymbol{e}_2) \otimes \frac{\boldsymbol{e}_1 - \boldsymbol{e}_2}{2}\,. \tag{3.131}$$

### Hydrostatic Pressure

We have already seen that the stress tensor corresponds to the form

$$\boldsymbol{\sigma} = -p(\boldsymbol{x})\boldsymbol{I} \qquad \text{or} \qquad \sigma_{ij} = -p(\boldsymbol{x})\,\delta_{ij}\,, \tag{3.132}$$

and the equilibrium equation (3.125) reduces to

$$-\nabla p + \rho \boldsymbol{b} = 0 \qquad \text{or} \qquad -p_{,i} + \rho b_i = 0\,. \tag{3.133}$$

### Pure Bending

We suppose no volume forces and that $\boldsymbol{\sigma}$ is given by

$$\boldsymbol{\sigma} = \alpha(x_2 - h_0)\boldsymbol{e}_1 \otimes \boldsymbol{e}_1 \quad \text{or} \quad [\sigma] = \begin{pmatrix} \alpha(x_2 - h_0) & 0 & 0 \\ 0 & 0 & 0 \\ 0 & 0 & 0 \end{pmatrix}, \tag{3.134}$$

where $\alpha$ and $h_0$ are constant. The three equations (3.125) are directly satisfied by (3.134). The tensor $\boldsymbol{\sigma}$ gives an approximation of the stress field such as that which appears in a prismatic beam parallel to $\boldsymbol{e}_1$ with moments applied at the ends acting about the axis $\boldsymbol{e}_3$.

**Plane Stress**

In this case,

$$\boldsymbol{\sigma} = \sigma_{11}\boldsymbol{e_1} \otimes \boldsymbol{e_1} + \sigma_{22}\boldsymbol{e_2} \otimes \boldsymbol{e_2} + \sigma_{12}(\boldsymbol{e_1} \otimes \boldsymbol{e_2} + \boldsymbol{e_2} \otimes \boldsymbol{e_1}) \tag{3.135}$$

or

$$[\sigma] = \begin{pmatrix} \sigma_{11} & \sigma_{12} & 0 \\ \sigma_{12} & \sigma_{22} & 0 \\ 0 & 0 & 0 \end{pmatrix}, \tag{3.136}$$

where $\sigma_{11}$, $\sigma_{22}$, and $\sigma_{12}$ are functions only of $x_1$ and $x_2$. Then, with no volume forces, equations (3.125) simplify to

$$\frac{\partial \sigma_{11}}{\partial x_1} + \frac{\partial \sigma_{12}}{\partial x_2} = 0 \qquad \frac{\partial \sigma_{21}}{\partial x_1} + \frac{\partial \sigma_{22}}{\partial x_2} = 0\,. \tag{3.137}$$

The stress field (3.136) can be used to approximate the stresses in a thin plate, parallel to the plane, perpendicular to $\boldsymbol{e_3}$, on which forces parallel to the plane act.

## 3.9 Piola-Kirchhoff Stress Tensors

### 3.9.1 General Considerations

Until now, the contact force has been expressed per unit area in the deformed configuration $\mathcal{R}$. The Cauchy stress tensor, $\boldsymbol{\sigma}$, is expressed with respect to the current, deformed configuration. This is the reason for which we call it the real stress. Consequently, the principles of conservation of momentum and angular momentum have been formulated only with respect to the deformed configuration. As mentioned at the end of section 3.4, the solution of problems in solid mechanics requires a formulation with respect to the initial, or reference, configuration $\mathcal{R}_0$. This is not only because it is difficult to know the deformed condition of a solid beforehand, but also because it is more convenient to analyze the experimental response of a solid with respect to its undeformed configuration. However, there is not simply a change of variables in the equations of motion and the Cauchy stress components using (2.1), and so we need to express the contact force in the current configuration per unit surface element in the undeformed surface element. Consequently, measurements of stress defined with respect to the undeformed configuration have been proposed. Two of them, well known in the study of solids, are the Piola-Kirchhoff stress tensors. The starting point for their definition is the expression of the contact force ***actually*** acting on a surface in the deformed configuration by a stress vector ***hypothetically*** applied to the corresponding surface in the initial configuration. Such a definition of the stress permits us to reformulate the principles stated in section 3.5 and reach similar conclusions to those of section 3.6.

Gustav Kirchhoff (1824–1887) was born in Königsberg. He taught at the University of Breslau, then at Heidelberg and finally at Humboldt University in Berlin. He made important contributions to spectroscopy, black body radiation and in elasticity. He reworked the Lagrangian description of the stress tensor previously introduced by Gabrio Piola (1794–1850).

**Fig. 3.9** Gustav Kirchhoff

### 3.9.2 First and Second Piola-Kirchhoff Tensors

Let $\boldsymbol{t}(\boldsymbol{x},t,\boldsymbol{n})$ be the Cauchy stress vector acting on the actual surface element $\boldsymbol{n}\,ds$ at $\boldsymbol{x}$ (fig. 3.10). To this vector we associate the vector $\boldsymbol{T}(\boldsymbol{X},t,\boldsymbol{N})$, called the ***first Piola-Kirchhoff stress vector***, to the corresponding reference surface element $\boldsymbol{N}dS$, and related to $\boldsymbol{t}(\boldsymbol{x},t,\boldsymbol{n})$ as follows:

$$\boldsymbol{T}\big(\boldsymbol{X},t,\boldsymbol{N}(\boldsymbol{X})\big)\,dS = \boldsymbol{t}\big(\boldsymbol{x},t,\boldsymbol{n}(\boldsymbol{x},t)\big)\,ds\,. \tag{3.138}$$

Since $dS$ and $ds$ are both positive scalars, (3.138) implies that $\boldsymbol{T}$ and $\boldsymbol{t}$ have the ***same direction***. However, the stress vector $\boldsymbol{T}$ does not represent the actual intensity at the current time $t$; it is often called the pseudo-stress vector and is a function of $\boldsymbol{X}$ and the normal $\boldsymbol{N}$ on $dS$ in the initial configuration. Relation (3.138) yields the elementary contact force applied in both configurations. In addition, $dS$ and $ds$ being different in general, $\|\boldsymbol{T}\|$ and $\|\boldsymbol{t}\|$ generally are also different.

Introducing Cauchy's relation (3.76) in (3.138) and then using Nanson's formula (2.107), we obtain

$$\begin{aligned}\boldsymbol{T}(\boldsymbol{X},t,\boldsymbol{N})\,dS &= \boldsymbol{t}\big(\boldsymbol{x},t,\boldsymbol{n}\big)\,ds = \boldsymbol{\sigma}\big(\boldsymbol{x},t\big)\boldsymbol{n}\,ds\\ &= J(\boldsymbol{X},t)\boldsymbol{\sigma}\big(\chi(\boldsymbol{X},t),t\big)\boldsymbol{F}^{-T}\boldsymbol{N}\,dS\,.\end{aligned} \tag{3.139}$$

As $dS > 0$, we conclude that

$$\boldsymbol{T}(\boldsymbol{X},t,\boldsymbol{N}) = \boldsymbol{P}(\boldsymbol{X},t)\boldsymbol{N}\,, \tag{3.140}$$

where

$$\boldsymbol{P}(\boldsymbol{X},t) = J(\boldsymbol{X},t)\boldsymbol{\sigma}\big(\chi(\boldsymbol{X},t),t\big)\boldsymbol{F}^{-T} \tag{3.141}$$

is the ***first Piola-Kirchhoff stress tensor***. Equation (3.140) is none other than an equivalent statement of Cauchy's theorem (3.76). This result can be deduced directly from the principle of conservation of momentum and is written in the following form:

$$\begin{aligned}&\int_\Omega P_0(\boldsymbol{X})\boldsymbol{a}\big(\chi(\boldsymbol{X},t),t\big)\,dV\\ &\quad= \int_\Omega P_0(\boldsymbol{X})\boldsymbol{B}(\boldsymbol{X},t)\,dV + \int_{\partial\Omega}\boldsymbol{T}(\boldsymbol{X},t,\boldsymbol{N})\,dS\,,\end{aligned} \tag{3.142}$$

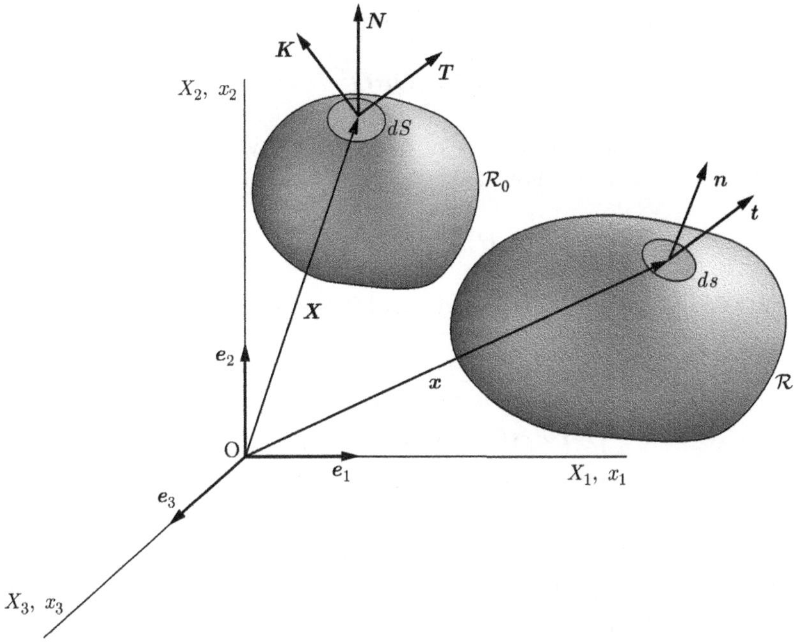


**Fig. 3.10** Relations between the Cauchy stress vector $\boldsymbol{t}$ and the first and second Piola-Kirchhoff stress vectors $\boldsymbol{T}$ and $\boldsymbol{K}$

where $P_0(\boldsymbol{X})$ is the initial density defined by (3.28) and $\boldsymbol{B}(\boldsymbol{X},t)$ the volume force density defined by (3.55).

Substituting (3.140) in (3.142) and employing the same arguments used in the proof of (3.96), we can deduce the ***motion equation*** for a continuous medium

$$\mathbf{div}\,\boldsymbol{P}(\boldsymbol{X},t) + P_0(\boldsymbol{X})\boldsymbol{B}(\boldsymbol{X},t) = P_0(\boldsymbol{X})\boldsymbol{A}(\boldsymbol{X},t)\,. \tag{3.143}$$

Note that we performed the derivation here with respect to the material variable $\boldsymbol{X}$. Let us examine the properties of the tensor $\boldsymbol{P}$. Using (3.100) and (3.141), we can easily show that

$$\boldsymbol{P}\boldsymbol{F}^T = \boldsymbol{F}\boldsymbol{P}^T\,. \tag{3.144}$$

Thus, not being symmetric, $\boldsymbol{P}$ does not possess the properties of the Cauchy stress tensor $\boldsymbol{\sigma}$ presented in section 3.7, and the principle of conservation of angular momentum is only satisfied if $\boldsymbol{P}$ meets condition (3.144).

Now consider the relation from Cauchy's theorem (3.76) as seen by two observers R and R$^*$. Assuming that the vectors $\boldsymbol{t}$ and $\boldsymbol{n}$ are objective and that they are transformed according to (2.195), we can reason as follows. Starting with $\boldsymbol{t}^* = \boldsymbol{\sigma}^*\boldsymbol{n}^*$ and the objectivity of $\boldsymbol{t}^*$ and $\boldsymbol{n}^*$, we can write

$$\boldsymbol{Q}\boldsymbol{t} = \boldsymbol{\sigma}^*\boldsymbol{Q}\boldsymbol{n}\,. \tag{3.145}$$

Moreover, from (3.76), we have

$$\boldsymbol{Q}\boldsymbol{t} = \boldsymbol{Q}\boldsymbol{\sigma}\boldsymbol{n}\,. \tag{3.146}$$

Comparing these two equations, we obtain

$$\boldsymbol{\sigma}^* = \boldsymbol{Q}\boldsymbol{\sigma}\boldsymbol{Q}^T . \tag{3.147}$$

Thus, the Cauchy stress tensor is objective. Now examine the first Piola-Kirchhoff stress tensor $\boldsymbol{P}$. For that, we write (3.141) for the observer $\mathrm{R}^*$ as

$$\boldsymbol{P}^*\boldsymbol{F}^{*T} = J^*\boldsymbol{\sigma}^* . \tag{3.148}$$

Using (2.205), (2.206), (3.141), and (3.147) in (3.148), we can write, successively,

$$\begin{aligned} &\boldsymbol{P}^*(\boldsymbol{Q}\boldsymbol{F})^T = J\boldsymbol{Q}\boldsymbol{\sigma}\boldsymbol{Q}^T \\ &\boldsymbol{P}^*\boldsymbol{F}^T\boldsymbol{Q}^T = \boldsymbol{Q}J\boldsymbol{\sigma}\boldsymbol{Q}^T = \boldsymbol{Q}\boldsymbol{P}\boldsymbol{F}^T\boldsymbol{Q}^T \\ &\boldsymbol{P}^* = \boldsymbol{Q}\boldsymbol{P} . \end{aligned} \tag{3.149}$$

Thus, the tensor $\boldsymbol{P}$ is not objective when changing observers.

Although the first Piola-Kirchhoff stress tensor $\boldsymbol{P}$ can be used for problems in solid mechanics, it is not symmetric and is not objective for a change of observers. To avoid these inconveniences in the formulation of the constitutive laws of solid materials undergoing large deformations, we often use the ***second Piola-Kirchhoff stress tensor*** $\boldsymbol{S}$, which is objective. This tensor can be introduced as follows.

The ***second Piola-Kirchhoff stress vector*** $\boldsymbol{K}$ applied at $\boldsymbol{X}$ and acting on the reference surface element $\boldsymbol{N}dS$ is defined by

$$\boldsymbol{K}(\boldsymbol{X},t,\boldsymbol{N})\,dS = \boldsymbol{F}^{-1}(\boldsymbol{X},t)\,\boldsymbol{t}\big(\boldsymbol{\chi}(\boldsymbol{X},t),\,t,\,\boldsymbol{n}(\boldsymbol{X},t)\big)\,ds . \tag{3.150}$$

This is the case as, by definition, $\boldsymbol{K}$ expresses the contact force per unit reference surface "transformed" by $\boldsymbol{F}^{-1}$ (physically, such a vector is not natural). Then, with the same arguments used to derive (3.140) and (3.141), this vector is written as

$$\boldsymbol{K}(\boldsymbol{X},t,\boldsymbol{N}) = \boldsymbol{S}(\boldsymbol{X},t)\boldsymbol{N} \tag{3.151}$$

with

$$\begin{aligned} \boldsymbol{S}(\boldsymbol{X},t) &= J(\boldsymbol{X},t)\boldsymbol{F}^{-1}(\boldsymbol{X},t)\boldsymbol{\sigma}\big(\boldsymbol{\chi}(\boldsymbol{X},t),t\big)\boldsymbol{F}^{-T}(\boldsymbol{X},t) \\ &= \boldsymbol{F}^{-1}(\boldsymbol{X},t)\boldsymbol{P}(\boldsymbol{X},t) . \end{aligned} \tag{3.152}$$

As $\boldsymbol{\sigma}$ is symmetric, it is easy to show that $\boldsymbol{S}$ is too. However, unlike $\boldsymbol{\sigma}$, $\boldsymbol{S}$ has no physically significant interpretation. The equation of motion for a continuous medium can also be expressed as a function of $\boldsymbol{S}$; we see that $\boldsymbol{P}(\boldsymbol{X},t) = \boldsymbol{F}(\boldsymbol{X},t)\boldsymbol{S}(\boldsymbol{X},t)$ and we only have to substitute this expression in (3.143). Note that, when $\boldsymbol{S}$ is employed, conservation of angular momentum is automatically satisfied due to the symmetry of $\boldsymbol{S}$.

### 3.9.3 Linearization of the Stress Tensors

Now we examine the effects on the stress tensors of kinematic linearization introduced in section 2.9. The Piola-Kirchhoff tensor expressed by (3.144) is written in index notation as

$$P_{mk} = F_{mi}(P_{ij})^T(F_{jk})^{-T} = F_{mi}P_{ji}F_{kj}^{-1}\,. \tag{3.153}$$

Using (2.70) and (2.145) in (3.153), we obtain

$$P_{mk} = P_{km} - P_{jm}\frac{\partial U_k}{\partial X_j} + P_{ki}\frac{\partial U_m}{\partial X_i} - P_{ji}\frac{\partial U_m}{\partial X_i}\frac{\partial U_k}{\partial X_j}\,. \tag{3.154}$$

Similarly, with (3.152), (2.145), and the second equality from (2.70), the second Piola-Kirchhoff tensor is expressed as

$$S_{ij} = F_{ik}^{-1}P_{kj} = \left(\delta_{ik} - \frac{\partial U_i}{\partial X_k}\right)P_{kj} = P_{ij} - P_{kj}\frac{\partial U_i}{\partial X_k}\,. \tag{3.155}$$

Finally, for the Cauchy stress tensor, we write (3.141) as

$$\sigma_{ij} = J^{-1}P_{ik}(F_{kj})^T = J^{-1}P_{ik}F_{jk}\,. \tag{3.156}$$

From (2.70) and (2.147), we have

$$\begin{aligned}\sigma_{ij} &= J^{-1}P_{ik}\left(\delta_{jk} + \frac{\partial U_j}{\partial X_k}\right)\\ &= J^{-1}(P_{ij} + P_{ik}\frac{\partial U_j}{\partial X_k}) \approx P_{ij} + P_{ik}\frac{\partial U_j}{\partial X_k}\,.\end{aligned} \tag{3.157}$$

Neglecting the terms with the displacement gradient in (3.154), (3.155), and (3.157), we end up with

$$P_{mk} \approx P_{km} \qquad S_{ij} \approx P_{ij} \qquad \sigma_{ij} \approx P_{ij}\,. \tag{3.158}$$

Thus, the result of kinematic linearization, for small displacements and infinitesimal gradients, is expressed by the symmetry of the first Piola-Kirchhoff tensor and by the approximate equality of the three stress tensors.

EXAMPLE 3.3

To understand the differences between the three stress tensors, consider the case of a circular prismatic bar, fixed at end A and subjected to a force $P_t$ at B, as shown in figure 3.11.

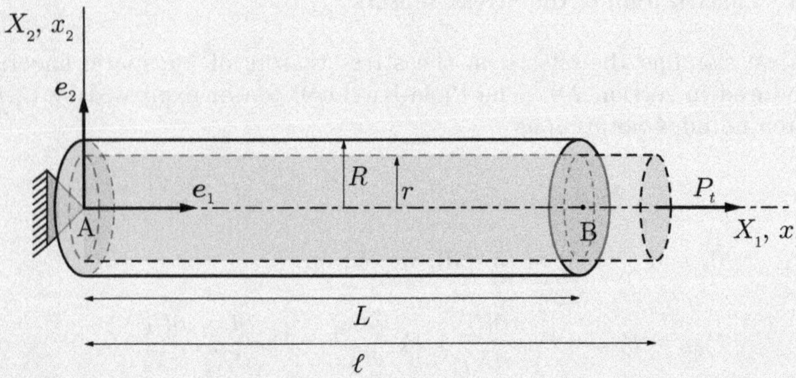


**Fig. 3.11** Cylindrical bar subjected to an axial force

The motion is given by the following expressions:

$$
\begin{aligned}
x_1 &= aX_1 \\
x_2 &= bX_2 \\
x_3 &= bX_3 \ .
\end{aligned}
\tag{3.159}
$$

The length and radius before deformation are $L$ and $R$, respectively. At time $t$, these parameters become $\ell$ and $r$. For end B of the bar and according to (3.159), we can write

$$
\begin{aligned}
\ell &= aL \quad \Rightarrow \quad a = \ell/L \\
r &= bR \quad \Rightarrow \quad b = r/R \ .
\end{aligned}
\tag{3.160}
$$

From (3.160), relations (3.159) are expressed as

$$
\begin{aligned}
x_1 &= \frac{\ell}{L} X_1 \\
x_2 &= \frac{r}{R} X_2 \\
x_3 &= \frac{r}{R} X_3 \ .
\end{aligned}
\tag{3.161}
$$

Thus, the matrix of the deformation gradient tensor (2.65) is

$$
[F] = \begin{pmatrix} \ell/L & 0 & 0 \\ 0 & r/R & 0 \\ 0 & 0 & r/R \end{pmatrix}
\tag{3.162}
$$

and its Jacobian $J$

$$
J = \det[F] = \frac{\ell}{L}\left(\frac{r}{R}\right)^2 = \frac{\ell}{L}\frac{A_t}{A_0} \ ,
\tag{3.163}
$$

where $A_0$ and $A_t$ are the areas of the section at times $t = 0$ and $t$, respectively.

A force $P_t$, parallel to axis 1, acts at the center of gravity of the section at end B. Thus the matrix of the Cauchy stress tensor is defined by the force at time $t$, $P_t$, and the area of the section at time $t$:

$$[\sigma] = \frac{P_t}{A_t}\begin{pmatrix} 1 & 0 & 0 \\ 0 & 0 & 0 \\ 0 & 0 & 0 \end{pmatrix} \quad \text{or} \quad \sigma_{11} = \frac{P_t}{A_t}\,. \tag{3.164}$$

With (3.141) and (3.152), the matrices of the first and second Piola-Kirchhoff stress tensors are

$$\begin{aligned}
[P] &= J[\sigma][F]^{-T} = \frac{\ell}{L}\frac{A_t}{A_0}\begin{pmatrix} \frac{P_t}{A_t} & 0 & 0 \\ 0 & 0 & 0 \\ 0 & 0 & 0 \end{pmatrix}\begin{pmatrix} \frac{L}{\ell} & 0 & 0 \\ 0 & \frac{R}{r} & 0 \\ 0 & 0 & \frac{R}{r} \end{pmatrix} \\
&= \frac{P_t}{A_0}\begin{pmatrix} 1 & 0 & 0 \\ 0 & 0 & 0 \\ 0 & 0 & 0 \end{pmatrix} \quad \text{or} \quad P_{11} = \frac{P_t}{A_0}\,. \\
[S] &= J[F]^{-1}[\sigma][F]^{-T} \\
&= \frac{\ell}{L}\frac{A_t}{A_0}\begin{pmatrix} \frac{L}{\ell} & 0 & 0 \\ 0 & \frac{R}{r} & 0 \\ 0 & 0 & \frac{R}{r} \end{pmatrix}\begin{pmatrix} \frac{P_t}{A_t} & 0 & 0 \\ 0 & 0 & 0 \\ 0 & 0 & 0 \end{pmatrix}\begin{pmatrix} \frac{L}{\ell} & 0 & 0 \\ 0 & \frac{R}{r} & 0 \\ 0 & 0 & \frac{R}{r} \end{pmatrix} \\
&= \frac{P_t}{A_0}\frac{L}{\ell}\begin{pmatrix} 1 & 0 & 0 \\ 0 & 0 & 0 \\ 0 & 0 & 0 \end{pmatrix} \quad \text{or} \quad S_{11} = \frac{P_t}{A_0}\frac{L}{\ell}\,.
\end{aligned}$$

These expressions show that the component of the first Piola-Kirchhoff tensor is given by the force at time $t$ and the area of the section at time $t = 0$ (or the initial section) and is often called engineering stress. The component of the second tensor has no direct interpretation, unlike the other two tensors. Nonetheless, it is useful for the modeling of solids undergoing large deformations. This subject will be treated in chapter 6.

Since the deformation is homogeneous, the components of infinitesimal strain are expressed as

$$\begin{aligned}
\frac{\ell}{L} &= \frac{L + \Delta\ell}{L} = 1 + \varepsilon_{11} \\
\frac{r}{R} &= \frac{R + \Delta R}{R} = 1 + \varepsilon_{22}\,.
\end{aligned}$$

Thus,

$$[F] = \begin{pmatrix} 1+\varepsilon_{11} & 0 & 0 \\ 0 & 1+\varepsilon_{22} & 0 \\ 0 & 0 & 1+\varepsilon_{22} \end{pmatrix}.$$

When $\varepsilon_{11}, \varepsilon_{22} \ll 1$, the three measures of stress are approximately the same since $A = A_0(1+\varepsilon_{22})^2 \approx A_0$, $S_{11} = \frac{P_t}{A_0}(1+\varepsilon_{11}) \approx \frac{P_t}{A_0} = P_{11}$, and $\sigma_{11} = \frac{P_t}{A_t} \approx \frac{P_t}{A_0}$, which are the results of kinematic linearization (see (2.146) and (2.147)).

## 3.10 Exercises

**3.1** Show that the velocity field $v_i = Ax_i/r^3$, where $x_i x_i = r^2$ and $A$ is an arbitrary constant, satisfies the conservation of mass equation for an incompressible fluid.

**3.2** For a velocity field $v_i = x_i/(1+t)$, show that

$$\rho x_1 x_2 x_3 = \rho_0 X_1 X_2 X_3 \,.$$

**3.3** Show that the flow given by the velocity field

$$\begin{aligned} v_r &= \frac{(1-r^2)\cos\theta}{r^2} \\ v_\theta &= \frac{(1+r^2)\sin\theta}{r^2} \\ v_z &= 0 \end{aligned}$$

satisfies the incompressibility equation when the density is constant.

**3.4** The state of stress in a body is given by the following stress matrix:

$$[\sigma] = \begin{pmatrix} 0 & Cx_1 & 0 \\ Cx_1 & 0 & -Cx_2 \\ 0 & -Cx_2 & 0 \end{pmatrix} , \tag{3.165}$$

where $C$ is an arbitrary constant.

1) Determine the volume force in order to satisfy static equilibrium.
2) Calculate at point P, with coordinates $(4, -4, 7)$, the stress vector on the plane defined by the equation $2x_1 + 2x_2 - x_3 = -7$, and on the sphere $x_1^2 + x_2^2 + x_3^2 = 81$ passing through P.
3) Determine the principal stresses, the maximum shear stress, and the principal deviatoric stress at P.

**3.5** In the absence of volume forces, determine if the following stress field satisfies equilibrium:

$$\begin{aligned} \sigma_{11} &= 4x_1^2 + 8x_1x_2 - 5x_2^2 & \sigma_{22} &= 5x_1^2 + \frac{1}{2}x_1x_2 + 4x_2^2 \\ \sigma_{12} &= -\frac{1}{4}x_1^2 - 8x_1x_2 - 4x_2^2 & \sigma_{33} &= \sigma_{32} = \sigma_{31} = 0 \,. \end{aligned} \tag{3.166}$$

**3.6** Let $\mathcal{B}$ be a weightless three-dimensional body, subject to a uniform pressure (normal) on its entire external surface. Show that $\mathcal{B}$ is in equilibrium.

**3.7** For each state of stress at a point given by the following matrices:

$$[\sigma] = \begin{pmatrix} p & p & p \\ p & p & p \\ p & p & p \end{pmatrix} \tag{3.167}$$

$$[\sigma] = \begin{pmatrix} p & p & p \\ p & p & p \\ p & p & -2p \end{pmatrix} \tag{3.168}$$

$$[\sigma] = \begin{pmatrix} 0 & p & p \\ p & 0 & p \\ p & p & 0 \end{pmatrix}, \tag{3.169}$$

with $p$ a constant, determine the principal stresses. To which state of stress do each of these cases correspond?

**3.8** Show that the invariants of the stress deviatoric tensor $\boldsymbol{s}$ are related to those of the stress tensor $\boldsymbol{\sigma}$ by the following expressions:

$$I_1(\boldsymbol{s}) = 0 \tag{3.170}$$

$$I_2(\boldsymbol{s}) = \frac{1}{3}I_1^2(\boldsymbol{\sigma}) - I_2(\boldsymbol{\sigma}) \tag{3.171}$$

$$I_3(\boldsymbol{s}) = \frac{2}{27}I_1^3(\boldsymbol{\sigma}) - \frac{1}{3}I_1(\boldsymbol{\sigma})I_2(\boldsymbol{\sigma}) + I_3(\boldsymbol{\sigma}) \,. \tag{3.172}$$

Note that in the characteristic equation (1.120), the second invariant $I_2(\boldsymbol{s})$ is given by the negative of (1.121). This change results in a positive definite form of $I_2(\boldsymbol{s})$.

**3.9** If $P_{ij\ldots}(\boldsymbol{x},t)$ is an arbitrary scalar, vector, or tensor function, prove that

$$\int_{\partial\omega} P_{ij\ldots}\sigma_{pq}n_q \, ds = \int_{\omega} \left(\sigma_{pq}P_{ij\ldots,q} + \rho\, P_{ij\ldots}(\dot{v}_p - b_p)\right) dv \,. \tag{3.173}$$

**3.10** Using (3.141), prove equality (3.144).

**3.11** Show that the second Piola-Kirchhoff stress tensor $\boldsymbol{S}$ satisfies the relations

$$\boldsymbol{S} = \boldsymbol{S}^T \qquad \text{and} \qquad \boldsymbol{S}^* = \boldsymbol{S} \,. \tag{3.174}$$

Chapter 4

# Energy

## 4.1 Introduction

Having described the principles of conservation of mass, momentum, and angular momentum, we will now introduce the principles related to the thermodynamics of continuous media in motion and the conservation of energy.

We can recall that all deformations in a material produce a thermal effect in the same way that a thermal effect produces a deformation. This is easily observed by heating a metal bar which lengthens under the action of the heat.

In this chapter, we will generally work in the spatial or Eulerian representation. The principle of conservation of total energy is first established. It leads to the principle of conservation of internal energy. Then, we will consider the conservation of mechanical energy in the Lagragian representation. Later, we will show that from the principle of conservation of total energy, for which objectivity is imposed, we can infer the other conservation laws. Finally, the chapter ends with the introduction of entropy and the second law of thermodynamics, which is based on the Clausius–Duhem inequality, a measure of the irreversibility of the phenomena associated with the physics of continuous media.

Continuous media thermodynamics is covered in detail by the following authors: [15, 17, 18, 22, 58, 68].

## 4.2 Conservation of Energy

Let $\omega(t)$ be the material volume of a continuous medium at the instant $t$, such that $\omega(t) \subseteq \mathcal{R}$, the deformed configuration of the body $\mathcal{B}$. We generalize the concept of kinetic energy by defining it as the integral over the deformed volume $\omega(t)$ of half the density, $\rho(\boldsymbol{x}, t)$, multiplied by the square of the local spatial velocity, $\boldsymbol{v}(\boldsymbol{x}, t)$. The kinetic energy of $\omega(t)$, which we denote $E_k(t)$, is a scalar given by the relation

$$E_k(t) = \int_{\omega(t)} \rho(\boldsymbol{x}, t) \frac{\boldsymbol{v}(\boldsymbol{x}, t) \cdot \boldsymbol{v}(\boldsymbol{x}, t)}{2} \, dv \, . \tag{4.1}$$

To simplify, the dependence of $\omega$ with respect to time will no longer be explicitly shown in the following. Also for the sake of simplicity in the expressions, we will omit the arguments of the functions on several occasions. Besides the kinetic energy, the energy in a material region contains contributions from motion at the microscopic scale such as random translational motion, molecular vibrations and rotations, and other microscopic energy modes. All these energies contribute to the ***internal energy*** $E_{\text{int}}(t)$. For example, we know that for two bodies $\mathcal{B}_1$ and $\mathcal{B}_2$ at rest (zero kinetic energy), if the temperature of the first is higher that that of the second, then $\mathcal{B}_1$ contains more energy than $\mathcal{B}_2$. The internal energy $E_{\text{int}}(t)$ of $\mathcal{B}$ is expressed as the volume integral of the internal energy density $u(\boldsymbol{x},t)$ per unit mass. We have

$$E_{\text{int}}(t) = \int_{\omega} \rho(\boldsymbol{x},t)\, u(\boldsymbol{x},t)\, dv\,. \tag{4.2}$$

The sum of the kinetic and internal energies (the latter of which, for materials, is the analog of the potential energy in classical mechanics) is the ***total energy*** of $\mathcal{B}$. The total energy can vary over time under the action of work done by forces that act on $\mathcal{B}$ and by external contributions of heat energy. Before discussing precisely the concept of work for a continuous medium, we will recall its formulation in classical mechanics. Newton's law for a particle of mass $m$ moving at velocity $\boldsymbol{v}$ is written as

$$m\,\frac{d\boldsymbol{v}}{dt} = \boldsymbol{F}\,. \tag{4.3}$$

Taking the scalar product of the two sides of this relation with $\boldsymbol{v}$, we obtain

$$m\boldsymbol{v}\cdot\frac{d\boldsymbol{v}}{dt} = m\frac{d}{dt}\left(\frac{1}{2}\,\|\boldsymbol{v}\|^2\right) = m\frac{d}{dt}\left(\frac{1}{2}\,v^2\right) = \boldsymbol{F}\cdot\boldsymbol{v}\ ,$$

which is a form of the theorem for kinetic energy.

The ***power***, that is, the variation of work with respect to time, results from the scalar product of the force $\boldsymbol{F}$ with the velocity. Only the force component in the direction of the velocity increases the kinetic energy $\frac{1}{2}mv^2$ of the particle. The components of force orthogonal to the velocity induce a curvature in the trajectory, but do not increase the kinetic energy.

In continuous media, power is thus given by the scalar product of the force and the velocity of the material.

Consider again the body $\mathcal{B}$. For the volume forces, this power is written as

$$\int_{\omega} \rho(\boldsymbol{x},t)\, \boldsymbol{b}(\boldsymbol{x},t)\cdot\boldsymbol{v}(\boldsymbol{x},t)\, dv\,. \tag{4.4}$$

The power provided by the surface forces is given by the relation

$$\int_{\partial\omega} \boldsymbol{t}\cdot\boldsymbol{v}\, ds = \int_{\partial\omega} \boldsymbol{\sigma}\boldsymbol{n}\cdot\boldsymbol{v}\, ds\,, \tag{4.5}$$

where we have used Cauchy's relation (3.107). The integral on the right-hand side of (4.5) can be transformed by the divergence theorem. Using (1.228) and by taking into account (1.69) and the symmetry of $\boldsymbol{\sigma}$, we have

$$\int_{\partial\omega} \boldsymbol{\sigma n} \cdot \boldsymbol{v}\, ds = \int_{\partial\omega} \boldsymbol{n} \cdot \boldsymbol{\sigma}^T \boldsymbol{v}\, ds = \int_\omega \operatorname{div}(\boldsymbol{\sigma}^T \boldsymbol{v})\, dv = \int_\omega \operatorname{div}(\boldsymbol{\sigma v})\, dv\,. \tag{4.6}$$

In index notation, we have

$$\int_{\partial\omega} \sigma_{ij}\, v_j\, n_i\, ds = \int_\omega \frac{\partial}{\partial x_i} (\sigma_{ij} v_j)\, dv = \int_\omega \left( \frac{\partial \sigma_{ij}}{\partial x_i} v_j + \sigma_{ij} \frac{\partial v_j}{\partial x_i} \right) dv\,, \tag{4.7}$$

and by using the definition of the scalar product of two tensors of order 2 (1.94), the last integral of (4.7) becomes

$$\int_\omega \left( \frac{\partial \sigma_{ij}}{\partial x_i} v_j + \sigma_{ij} \frac{\partial v_j}{\partial x_i} \right) dv = \int_\omega \left( (\mathbf{div}\, \boldsymbol{\sigma}) \cdot \boldsymbol{v} + \boldsymbol{\sigma} : \boldsymbol{\nabla v} \right) dv\,. \tag{4.8}$$

In vector form, we write

$$\int_{\partial\omega} \boldsymbol{\sigma n} \cdot \boldsymbol{v}\, ds = \int_\omega \operatorname{div}(\boldsymbol{\sigma v})\, dv = \int_\omega \left( (\mathbf{div}\, \boldsymbol{\sigma}) \cdot \boldsymbol{v} + \boldsymbol{\sigma} : \boldsymbol{\nabla v} \right) dv\,. \tag{4.9}$$

Heat transfer is the second way in which energy is passed to the material. We will first take into account a production/consumption term, in the form

$$\int_\omega r(\boldsymbol{x}, t)\, dv\,, \tag{4.10}$$

where $r(\boldsymbol{x}, t)$ represents the ***heat produced*** or ***received*** per unit time and volume. This could be the ***heat produced*** or consumed by a chemical reaction in the material or heating by the Joule effect (a carbon electrode in the material). It can also take into account the heat received by radiation from possible external sources. Its dimensions are $\mathrm{M\,L^{-1}T^{-3}}$ with the symbols M, L, T designating mass, length, and time, respectively; the corresponding SI unit is $\mathrm{W/m^3}$. The external contribution of heat is most often by conduction through the surface $\partial\omega$. Of course, other forms of heat transfer can be found, radiation for example. We will ignore them from here on.

Let $q$ be the scalar quantity that represents the heat that enters into $\mathcal{B}$ per unit time and unit surface $ds$. Let $\boldsymbol{n}$ be the normal to $ds$. By analogy with Cauchy's postulate, we assume that $q$ at the point $\boldsymbol{x}$ depends only on the outgoing unit normal at point $\boldsymbol{x}$, that is,

$$q = q(\boldsymbol{x}, t, \boldsymbol{n})\,. \tag{4.11}$$

Denote by $q_1$, $q_2$, $q_3$ the ***heat flux*** obtained at a material point P when the normal $\boldsymbol{n}$ is directed along the basis vectors $\boldsymbol{e}_1$, $\boldsymbol{e}_2$, $\boldsymbol{e}_3$ respectively. Then,

$$q_i = q(\boldsymbol{x}, t, \boldsymbol{e}_i)\,. \tag{4.12}$$

With our reasoning based on the conservation of energy in a tetrahedral material element, we can show in a way analogous to that for Cauchy's theorem for the stress tensor (sec. 3.6) that the heat $q$ is a linear combination of the fluxes $q_i$ multiplied by the components $n_i$

$$q = q_i n_i = \boldsymbol{q} \cdot \boldsymbol{n} . \tag{4.13}$$

By definition, $\boldsymbol{q}(\boldsymbol{x}, t)$ is the heat flux vector. The rate of heat received by conduction for the whole body is equal to

$$-\int_{\partial\omega} \boldsymbol{q} \cdot \boldsymbol{n} \, ds . \tag{4.14}$$

The negative sign introduced in (4.14) signifies that a positive heat rate is obtained when $\boldsymbol{q}$ points into the interior of the material volume. The quantity $-\boldsymbol{q} \cdot \boldsymbol{n}$ is thus the surface density of the rate of heat received by conduction through $\partial\omega$.

If for any evolution of the material we have $\boldsymbol{q} = 0$ and $r = 0$, we say that the evolution of the medium is adiabatic; there is no heat exchange with the exterior. If we apply the divergence theorem to the integral (4.14), we have

$$-\int_{\partial\omega} \boldsymbol{q} \cdot \boldsymbol{n} \, ds = -\int_{\omega} \operatorname{div} \boldsymbol{q} \, dv . \tag{4.15}$$

We are now ready to state the law of conservation of energy, which is the first principle of thermodynamics.

First principle of thermodynamics *The time derivative of the total energy in $\mathcal{B}$ is equal to the sum of the power of the volume and contact forces and the rate of heat received by the material.*

Combining equations (4.1), (4.2), (4.4)–(4.6), (4.10), and (4.15), we write

$$\frac{d}{dt} \int_{\omega} \rho \left( \frac{\boldsymbol{v} \cdot \boldsymbol{v}}{2} + u \right) dv = \int_{\omega} \left( \rho \boldsymbol{b} \cdot \boldsymbol{v} + \operatorname{div}(\boldsymbol{\sigma} \boldsymbol{v}) - \operatorname{div} \boldsymbol{q} + r \right) dv \tag{4.16}$$

or, with definitions (4.1) and (4.2), and relation (4.8),

$$\begin{aligned} &\frac{D}{Dt} \left( E_k(t) + E_{\text{int}}(t) \right) \\ &\qquad = \int_{\omega} \left( \rho \boldsymbol{b} \cdot \boldsymbol{v} + (\mathbf{div}\, \boldsymbol{\sigma}) \cdot \boldsymbol{v} + \boldsymbol{\sigma} : \boldsymbol{\nabla} \boldsymbol{v} - \operatorname{div} \boldsymbol{q} + r \right) dv . \end{aligned} \tag{4.17}$$

Using the Reynolds transport theorem and the conservation of mass (3.41), the first term of (4.16) becomes

$$\begin{aligned} \frac{D}{Dt} \left( E_k(t) + E_{\text{int}}(t) \right) &= \int_{\omega} \left( \boldsymbol{v} \cdot \boldsymbol{a} + \frac{Du}{Dt} \right) \rho \, dv \\ &= \int_{\omega} \rho \frac{D}{Dt} \left( \frac{\boldsymbol{v} \cdot \boldsymbol{v}}{2} + u \right) dv . \end{aligned} \tag{4.18}$$

The body forces $\boldsymbol{b}$ can always be derived from a time independent potential $W(\boldsymbol{x})$. By definition,

$$\boldsymbol{b} = -\boldsymbol{\nabla} W\,, \qquad b_i = -\frac{\partial W}{\partial x_i}\,. \tag{4.19}$$

Thus for $\rho\,\boldsymbol{b}\cdot\boldsymbol{v}$,

$$\rho b_i v_i = -\rho v_i \frac{\partial W}{\partial x_i} = -\rho\frac{DW}{Dt}\,. \tag{4.20}$$

Combining relations (4.16), (4.18), (4.19) and using the localization theorem, we obtain the local form of conservation of total energy

$$\rho\frac{D}{Dt}\left(u + \frac{\boldsymbol{v}\cdot\boldsymbol{v}}{2} + W\right) = \operatorname{div}(\boldsymbol{\sigma}\boldsymbol{v}) - \operatorname{div}\boldsymbol{q} + r\,. \tag{4.21}$$

If we do not take into account the potential $W$, in relations (4.16)–(4.18), we can obtain

$$\int_\omega (\rho\boldsymbol{a} - \rho\boldsymbol{b} - \mathbf{div}\,\boldsymbol{\sigma})\cdot\boldsymbol{v}\,dv + \int_\omega\left(\rho\frac{Du}{Dt} - \boldsymbol{\sigma} : \boldsymbol{\nabla}\boldsymbol{v} + \operatorname{div}\,\boldsymbol{q} - r\right)dv = 0\,. \tag{4.22}$$

The first integral of (4.22) is zero due to the principle of conservation of momentum (3.96). Invoking the localization theorem for the second volume integral of (4.22), the law of conservation of internal energy becomes

$$\rho\frac{Du}{Dt} = \boldsymbol{\sigma} : \boldsymbol{\nabla}\boldsymbol{v} - \operatorname{div}\boldsymbol{q} + r\,. \tag{4.23}$$

The first term on the right-hand side of (4.23), which we denote as $\mathfrak{D}$, can be rewritten taking into account the symmetry of $\sigma_{ij}$ and relation (2.180)

$$\mathfrak{D} = \sigma_{ij}\frac{\partial v_i}{\partial x_j} = \sigma_{ji}\frac{\partial v_i}{\partial x_j} = \sigma_{ij}\frac{\partial v_j}{\partial x_i} = \frac{1}{2}\sigma_{ij}\left(\frac{\partial v_i}{\partial x_j} + \frac{\partial v_j}{\partial x_i}\right) = \sigma_{ij}d_{ij} \tag{4.24}$$

or

$$\mathfrak{D} = \boldsymbol{\sigma} : \boldsymbol{\nabla}\boldsymbol{v} = \boldsymbol{\sigma} : \boldsymbol{d}\,. \tag{4.25}$$

This term can be interpreted as the ***contact force power*** acting on the material. We can also write $\mathfrak{D}$ as $\operatorname{tr}(\boldsymbol{\sigma}\boldsymbol{L})$ from (1.95), or $\boldsymbol{\sigma} : \boldsymbol{L}$, where the notation $\boldsymbol{L}$ designates the velocity gradient tensor $\partial\boldsymbol{v}/\partial\boldsymbol{x}$ defined by (2.177). Equation (4.23) shows that the increase in internal energy is equal to the sum of the power developed by the contact forces, the conductive heat transfer, and the volume production of heat inside $\mathcal{B}$.

We can obtain the ***kinetic energy theorem*** by subtracting relation (4.23) from (4.21) and taking into account (4.20)

$$\rho\frac{D}{Dt}\left(\frac{\boldsymbol{v}\cdot\boldsymbol{v}}{2}\right) = \rho\boldsymbol{b}\cdot\boldsymbol{v} + (\mathbf{div}\,\boldsymbol{\sigma})\cdot\boldsymbol{v}\,. \tag{4.26}$$

This theorem states that the time variation of kinetic energy is equal to the power of the volume forces (first term on the right-hand side) and the contact forces (second term). We note that this relation is none other than the conservation of momentum (3.96) as a scalar product with $\boldsymbol{v}$.

EXAMPLE 4.1
For a linear elastic body in equilibrium, subjected to volume forces $\boldsymbol{b}$ and surface forces $\boldsymbol{t}$, prove the following equality:

$$\int_\omega \boldsymbol{\sigma} : \boldsymbol{\varepsilon}\, dv = \int_\omega \rho\, \boldsymbol{b} \cdot \boldsymbol{u}\, dv + \int_{\partial\omega} \boldsymbol{t} \cdot \boldsymbol{u}\, ds\,, \tag{4.27}$$

where $\boldsymbol{\sigma}, \boldsymbol{\varepsilon}$ are the stress and strain tensors, $\rho$ the density, and $\boldsymbol{u}$ the displacement vector.
In index notation,

$$\int_\omega \sigma_{ij}\varepsilon_{ij}\, dv = \int_\omega \rho\, b_i u_i dv + \int_{\partial\omega} t_i u_i ds\ .$$

Using (3.76), the surface integral is converted to a volume integral as follows

$$\int_{\partial\omega} t_i u_i ds = \int_{\partial\omega} \sigma_{ij} n_j u_i ds = \int_{\partial\omega} \sigma_{ij} u_i n_j ds = \int_\omega \frac{\partial(\sigma_{ij}u_i)}{\partial x_j} dv$$

or

$$\int_\omega \frac{\partial(\sigma_{ij}u_i)}{\partial x_j} dv = \int_\omega \left( \frac{\partial(\sigma_{ij})}{\partial x_j} u_i + \sigma_{ij} \frac{\partial u_i}{\partial x_j} \right) dv = \int_\omega \left(\sigma_{ij,j} u_i + \sigma_{ij} u_{i,j}\right) dv\ .$$

Taking into account the equilibrium equations (3.126), the right-hand side of (4.27) is written as

$$\begin{aligned}\int_\omega \left(\rho\, b_i u_i + \sigma_{ij,j} u_i + \sigma_{ij} u_{i,j}\right) dv &= \int_\omega \left(u_i(\rho\, b_i + \sigma_{ij,j}) + \sigma_{ij} u_{i,j}\right) dv \\ &= \int_\omega \sigma_{ij} u_{i,j} dv\ .\end{aligned}$$

Because of the symmetry of the stress tensor, the integrand on the right side can be modified,

$$\begin{aligned}\sigma_{ij} u_{i,j} &= \frac{1}{2}(\sigma_{ij} u_{i,j} + \sigma_{ij} u_{i,j}) = \frac{1}{2}(\sigma_{ij} u_{i,j} + \sigma_{ji} u_{j,i}) = \frac{1}{2}(\sigma_{ij} u_{i,j} + \sigma_{ij} u_{j,i}) \\ &= \frac{1}{2}\sigma_{ij}(u_{i,j} + u_{j,i}) = \sigma_{ij}\varepsilon_{ij}\ .\end{aligned}$$

Finally we have

$$\int_\omega \sigma_{ij} u_{i,j} dv = \int_\omega \sigma_{ij}\varepsilon_{ij} dv\ .$$

## 4.3 Conservation of Mechanical Energy in the Material Representation

The analysis of the conservation of energy shown earlier was done in the Eulerian representation. A similar development to describe the different energy components can be carried out in the material representation; we will do so in this section. To simplify, we can ignore the heat flux $\boldsymbol{q}$ and the volume term $r$. Using (4.5) and (4.6), the equation of conservation of energy (4.16) becomes

$$\begin{aligned}\frac{D}{Dt}\left(E_k(t)+E_{\text{int}}(t)\right) &= \int_\omega \rho\,\frac{D}{Dt}\left(\frac{\boldsymbol{v}\cdot\boldsymbol{v}}{2}\right)dv+\int_\omega \rho\,\frac{Du}{Dt}\,dv\\ &= \int_\omega \rho\,\boldsymbol{b}\cdot\boldsymbol{v}\,dv+\int_{\partial\omega}\boldsymbol{t}\cdot\boldsymbol{v}\,ds\,.\end{aligned} \tag{4.28}$$

Taking into account (4.23) with $\boldsymbol{q}=r=0$ and (4.25), we have

$$\int_\omega \rho\,\frac{Du}{Dt}\,dv=\int_\omega \boldsymbol{\sigma}:\boldsymbol{\nabla v}\,dv=\int_\omega \boldsymbol{\sigma}:\boldsymbol{d}\,dv\,. \tag{4.29}$$

This last equation shows that by ignoring all thermal effects, the rate of change of the internal energy is equal to the power of the internal forces. Since the volume is arbitrary, we deduce the local form which is written as

$$\rho\dot{u}=\boldsymbol{\sigma}:\boldsymbol{d}\,. \tag{4.30}$$

Finally, the conservation of mechanical energy becomes

$$\int_\omega \rho\,\frac{D}{Dt}\left(\frac{\boldsymbol{v}\cdot\boldsymbol{v}}{2}\right)dv+\int_\omega \boldsymbol{\sigma}:\boldsymbol{d}\,dv=\int_\omega \rho\,\boldsymbol{b}\cdot\boldsymbol{v}\,dv+\int_{\partial\omega}\boldsymbol{t}\cdot\boldsymbol{v}\,ds\,. \tag{4.31}$$

In order to write this last relation in material coordinates, first consider the kinetic energy. From (2.103), (3.33), and the equation of conservation of mass (3.37), we deduce that $\rho\,dv=P_0\,dV$. In addition, considering equation (2.20), we have

$$\int_\omega \rho\,\frac{\boldsymbol{v}\cdot\boldsymbol{v}}{2}\,dv=\int_\Omega P_0\,\frac{\boldsymbol{V}\cdot\boldsymbol{V}}{2}\,dV\,. \tag{4.32}$$

For the second term of the left-hand side of (4.31), according to (2.179), $\boldsymbol{L}=\dot{\boldsymbol{F}}\boldsymbol{F}^{-1}$. Since $\boldsymbol{\sigma}:\boldsymbol{d}=\boldsymbol{\sigma}:\left(\dot{\boldsymbol{F}}\boldsymbol{F}^{-1}\right)$, we obtain

$$\int_\omega \boldsymbol{\sigma}:\boldsymbol{d}\,dv=\int_\omega \boldsymbol{\sigma}:\left(\dot{\boldsymbol{F}}\boldsymbol{F}^{-1}\right)dv\,. \tag{4.33}$$

Using tensor identity (1.97), the right-hand side of (4.33) is modified as follows:

$$\int_\omega \boldsymbol{\sigma}:\left(\dot{\boldsymbol{F}}\boldsymbol{F}^{-1}\right)dv=\int_\omega \left(\boldsymbol{\sigma}\boldsymbol{F}^{-T}\right):\dot{\boldsymbol{F}}\,dv\,. \tag{4.34}$$

With (3.141) and (2.103), we can write (4.34) in the material configuration with the Piola-Kirchhoff tensor $\boldsymbol{P}$

$$\begin{aligned}\int_\omega \boldsymbol{\sigma}:\left(\dot{\boldsymbol{F}}\boldsymbol{F}^{-1}\right)dv &= \int_\omega \left(\boldsymbol{\sigma}\boldsymbol{F}^{-T}\right):\dot{\boldsymbol{F}}\,dv\\ &= \int_\Omega \left(J\,\boldsymbol{\sigma}\boldsymbol{F}^{-T}\right):\dot{\boldsymbol{F}}\,dV=\int_\Omega \boldsymbol{P}:\dot{\boldsymbol{F}}\,dV\,.\end{aligned} \tag{4.35}$$

The contribution of the volume forces is easily expressed in material coordinates using relations (2.20), (3.37), (3.57), and (2.103)

$$\int_\omega \rho\, \boldsymbol{b} \cdot \boldsymbol{v}\, dv = \int_\Omega P_0\, \boldsymbol{B} \cdot \boldsymbol{V}\, dV\,. \tag{4.36}$$

For the contact forces, we use (2.20) and (3.138) so that

$$\int_{\partial\omega} \boldsymbol{t} \cdot \boldsymbol{v}\, ds = \int_{\partial\Omega} \boldsymbol{T} \cdot \boldsymbol{V}\, dS\,. \tag{4.37}$$

Finally, the principle of conservation of mechanical energy is expressed in the reference configuration as

$$\begin{aligned} &\int_\Omega P_0 \frac{D}{Dt}\left(\frac{\boldsymbol{V}\cdot\boldsymbol{V}}{2}\right) dV + \int_\Omega \boldsymbol{P} : \dot{\boldsymbol{F}}\, dV \\ &\quad = \int_\Omega P_0\, \boldsymbol{B}\cdot\boldsymbol{V}\, dV + \int_{\partial\Omega} \boldsymbol{T}\cdot\boldsymbol{V}\, dS\,. \end{aligned} \tag{4.38}$$

According to (4.28), the material derivative of the internal energy is expressed as

$$\begin{aligned} \frac{D}{Dt} E_{\text{int}}(t) &= \int_\omega \rho \frac{Du}{Dt}\, dv \\ &= \int_\Omega \frac{D}{Dt} u\big(\chi(\boldsymbol{X},t),t\big)\, J(\boldsymbol{X},t) P(\boldsymbol{X},t)\, dV\,. \end{aligned} \tag{4.39}$$

Setting $u\big(\chi(\boldsymbol{X},t),t\big) = U(\boldsymbol{X},t)$, we have

$$\begin{aligned} &\int_\Omega \frac{D}{Dt} u\big(\chi(\boldsymbol{X},t),t\big)\,) J(\boldsymbol{X},t) P(\boldsymbol{X},t)\, dV \\ &\quad = \int_\Omega \frac{D}{Dt} U(\boldsymbol{X},t)\, P_0(\boldsymbol{X})\, dV\,. \end{aligned} \tag{4.40}$$

Consequently, the second term on the left-hand side of (4.38) can be expressed as

$$\int_\Omega \boldsymbol{P} : \dot{\boldsymbol{F}}\, dV = \int_\Omega P_0 \frac{DU}{Dt}\, dV \tag{4.41}$$

and locally as

$$P_0 \dot{U} = \boldsymbol{P} : \dot{\boldsymbol{F}}\,. \tag{4.42}$$

The expression of the internal energy can be modified using $\boldsymbol{d} = \boldsymbol{F}^{-T} \dot{\boldsymbol{E}} \boldsymbol{F}^{-1}$ (see problem 2.8), equations (3.152) and (1.97). Then

$$\begin{aligned} \int_\omega \boldsymbol{\sigma} : \boldsymbol{d}\, dv &= \int_\Omega J \boldsymbol{\sigma} : \big(\boldsymbol{F}^{-T} \dot{\boldsymbol{E}} \boldsymbol{F}^{-1}\big)\, dV \\ &= \int_\Omega J\big(\boldsymbol{\sigma} \boldsymbol{F}^{-T}\big) : \big(\boldsymbol{F}^{-T} \dot{\boldsymbol{E}}\big)\, dV \\ &= \int_\Omega (\boldsymbol{F}\boldsymbol{S}) : \big(\boldsymbol{F}^{-T} \dot{\boldsymbol{E}}\big)\, dV = \int_\Omega \boldsymbol{S} : \dot{\boldsymbol{E}}\, dV\,. \end{aligned} \tag{4.43}$$

It is interesting to point out that mechanical power can be equally expressed as the doubly contracted product of the Cauchy stress tensor and the strain rate tensor ($\boldsymbol{\sigma} : \boldsymbol{d}$), or as the first Piola-Kirchhoff stress tensor and the deformation gradient rate tensor $\left(\boldsymbol{P} : \dot{\boldsymbol{F}}\right)$, or as the second Piola-Kirchhoff stress tensor and the Green-Lagrange strain rate tensor $\left(\boldsymbol{S} : \dot{\boldsymbol{E}}\right)$. Consequently, we can write the following equality for the power produced by the internal stresses and the strain per unit volume

$$J\rho\dot{u} = P_0\dot{U} = J\boldsymbol{\sigma} : \boldsymbol{d} = \boldsymbol{P} : \dot{\boldsymbol{F}} = \boldsymbol{S} : \dot{\boldsymbol{E}} \,. \tag{4.44}$$

Such pairs of parameters are called ***conjugate parameters*** because their inner (scalar) products yield an energy. This is a very important result of mechanics of continuous media which will allow us to deduce the constitutive equations from a potential function. This subject will be treated in chapter 6.

## 4.4 Interpretation of the Conservation Laws by the First Principle

In this section, we will revisit the laws of conservation of mass, momentum, and angular momentum starting from the principles of conservation of energy and objectivity.

The first principle of thermodynamics can be written according to relations (4.16) and (4.17):

$$\frac{D}{Dt}\left(E_k(t) + E_{\text{int}}(t)\right) = \int_\omega \left(\rho\boldsymbol{b}\cdot\boldsymbol{v} + \operatorname{div}(\boldsymbol{\sigma}\boldsymbol{v}) - \operatorname{div}\,\boldsymbol{q} + r\right) dv \,. \tag{4.45}$$

With the transport theorem in the form (3.5), the left-hand side of (4.45) becomes

$$\begin{aligned}&\frac{D}{Dt}\left(E_k(t) + E_{\text{int}}(t)\right)\\ &= \int_\omega \left(\frac{\partial}{\partial t}\rho\left(\frac{1}{2}\boldsymbol{v}\cdot\boldsymbol{v} + u\right) + \operatorname{div}\left(\rho\left(\frac{1}{2}\boldsymbol{v}\cdot\boldsymbol{v} + u\right)\boldsymbol{v}\right)\right) dv \,.\end{aligned} \tag{4.46}$$

By applying the localization theorem, relation (4.45) can be written as

$$\begin{aligned}&\frac{\partial}{\partial t}\rho\left(u + \frac{1}{2}\boldsymbol{v}\cdot\boldsymbol{v}\right) + \operatorname{div}\left(\rho\left(\frac{1}{2}\boldsymbol{v}\cdot\boldsymbol{v} + u\right)\boldsymbol{v}\right)\\ &\quad = \rho\boldsymbol{b}\cdot\boldsymbol{v} + r - \operatorname{div}\boldsymbol{q} + \operatorname{div}(\boldsymbol{\sigma}\boldsymbol{v}) \,.\end{aligned} \tag{4.47}$$

This last equation can easily be put in the form

$$\begin{aligned}&\rho\dot{u} - \boldsymbol{\sigma} : \boldsymbol{\nabla}\boldsymbol{v} + \operatorname{div}\boldsymbol{q} - r + \boldsymbol{v}\cdot\left(\rho\boldsymbol{a} - \mathbf{div}\,\boldsymbol{\sigma} - \rho\boldsymbol{b}\right)\\ &\quad + \left(\frac{1}{2}\boldsymbol{v}\cdot\boldsymbol{v} + u\right)\left(\dot{\rho} + \rho\operatorname{div}\boldsymbol{v}\right) = 0 \,,\end{aligned} \tag{4.48}$$

where $\dot{\rho}$ denotes the material derivative of $\rho$.

Consider two reference frames, $\mathrm{R} = (0, \boldsymbol{x}, t)$ and $\mathrm{R}^* = (0^*, \boldsymbol{x}^*, t^*)$, moving with respect to each other such that the relative motion is described by (2.195).

If, in addition to (2.195), (2.211), (2.205), and (2.213), the following transformation rules are valid:

$$\rho^* = \rho \tag{4.49}$$
$$u^* = u \tag{4.50}$$
$$\boldsymbol{q}^* = \boldsymbol{Q}\boldsymbol{q} \tag{4.51}$$
$$\boldsymbol{\sigma}^* = \boldsymbol{Q}\boldsymbol{\sigma}\boldsymbol{Q}^T \tag{4.52}$$
$$r^* = r\ , \tag{4.53}$$

the quantities $\rho$, $u$, $\boldsymbol{q}$, $\boldsymbol{\sigma}$, and $r$ are said to be objective (see sec. 2.11 and 3.9).

We show that the objectivity of the conservation of energy (4.48) remains valid if we write it with the starred quantities, implying the laws of conservation of mass, momentum and angular momentum.

### 4.4.1 First Case: Uniform Translation

Choose the reference frame $\mathrm{R}^*$ in translation with respect to R at constant translation velocity $\dot{\boldsymbol{c}}(t)$. Let

$$\dot{\boldsymbol{c}}(t) = \boldsymbol{u} \tag{4.54}$$
$$\boldsymbol{Q} = \boldsymbol{I}\,. \tag{4.55}$$

Then relation (2.211) becomes

$$\boldsymbol{v}^* = \boldsymbol{u} + \boldsymbol{v}\,. \tag{4.56}$$

Next we rewrite (4.48) with the starred quantities, replace $\boldsymbol{v}^*$ with its value (4.56), and from the resulting equation, subtract (4.48). Using relations (4.49)–(4.53), we have

$$\frac{\boldsymbol{u}\cdot\boldsymbol{u}}{2}(\dot{\rho} + \rho \operatorname{div} \boldsymbol{v}) + \boldsymbol{u}\cdot\boldsymbol{v}(\dot{\rho} + \rho \operatorname{div} \boldsymbol{v}) + \boldsymbol{u}\cdot(\rho\boldsymbol{a} - \mathbf{div}\,\boldsymbol{\sigma} - \rho\boldsymbol{b}) = 0\,. \tag{4.57}$$

If we change the scale of $\boldsymbol{u}$ to $\alpha\boldsymbol{u}$, we can impose that (4.57) is valid for any $\alpha$. We obtain

$$\dot{\rho} + \rho \operatorname{div} \boldsymbol{v} = 0 \tag{4.58}$$
$$\rho\frac{D\boldsymbol{v}}{Dt} - \mathbf{div}\,\boldsymbol{\sigma} - \rho\boldsymbol{b} + \boldsymbol{v}(\dot{\rho} + \rho \operatorname{div} \boldsymbol{v}) = 0\,. \tag{4.59}$$

This last relation can be put in the form

$$\frac{\partial}{\partial t}(\rho\boldsymbol{v}) + \mathbf{div}(\rho\boldsymbol{v}\otimes\boldsymbol{v}) = \mathbf{div}\,\boldsymbol{\sigma} + \rho\boldsymbol{b}\,. \tag{4.60}$$

Equation (4.60) is the equation of the conservation of momentum where the acceleration term is modified by the conservation of mass (4.58).

### 4.4.2 Second Case: Rigid Body Rotation

If we now consider the reference frame R* in rotational motion with respect to R such that we superimpose a rigid body rotation upon the existing velocity field

$$\boldsymbol{c}(t) = \boldsymbol{0} \tag{4.61}$$

$$\boldsymbol{Q}(t) = \boldsymbol{I} \tag{4.62}$$

$$\dot{\boldsymbol{Q}}(t) = \boldsymbol{\Omega}\,, \tag{4.63}$$

the velocity $\boldsymbol{v}^*$ is written, taking into account (2.57) and (2.60), as

$$\boldsymbol{v}^* = \boldsymbol{v} + \boldsymbol{\omega} \times \boldsymbol{x}\,. \tag{4.64}$$

The vector $\boldsymbol{\omega}$ is the ***dual vector*** of $\boldsymbol{\Omega}$ (sec. 2.6.3). We apply the same reasoning as before. The principle of conservation of energy yields the conservation of angular momentum, such that

$$\boldsymbol{\sigma} = \boldsymbol{\sigma}^T\,. \tag{4.65}$$

Note that, in general, the volume force term where $\boldsymbol{b}$ has the dimensions of an acceleration, is expressed by (2.212) in the form

$$\boldsymbol{b}^* = \boldsymbol{Q}\boldsymbol{b} + \ddot{\boldsymbol{c}} + 2\dot{\boldsymbol{Q}}\boldsymbol{v} + \ddot{\boldsymbol{Q}}\boldsymbol{x}\,. \tag{4.66}$$

In the case where the reference frame R* rotates at constant angular velocity ($\ddot{\boldsymbol{c}} = \ddot{\boldsymbol{Q}} = \boldsymbol{0}$), this last relation becomes, taking into account (2.60),

$$\boldsymbol{b}^* = \boldsymbol{b} + 2\dot{\boldsymbol{Q}}\boldsymbol{v} = \boldsymbol{b} + 2\boldsymbol{\Omega}\boldsymbol{v} = \boldsymbol{b} + 2\boldsymbol{\omega} \times \boldsymbol{v}\,, \tag{4.67}$$

where the ***Coriolis force*** term appears as $\boldsymbol{\omega} \times \boldsymbol{v}$.

## 4.5 The Notion of Entropy

The entropy of a system can be considered to be a measure of its disorder. Due to the unending incoherent agitation of the molecules in a continuous medium, two observation levels need to be distinguished. At the microscopic (molecular) level, the physical system ($\mathcal{X}$) passes through (or can pass through) a very large number of different states $\mathcal{X}_i$, whereas at the macroscopic level, where we habitually observe the system, these states are indistinguishable. We could say that the disorder of the system comes from the number of states $\mathcal{X}_i$, equivalent from the macroscopic viewpoint, and that its entropy is related to this number. Note that the usual kinematic and dynamic values of the material particle (for example mass, velocity, contact forces) are measured at the macroscopic level, but in reality correspond to averages of measurements made at the microscopic level.

More precisely, associating with each state $\mathcal{X}_i$ a probability $p_i$, an exact measurement of the disorder of $\mathcal{X}$ is given by its entropy which we denote $s$, defined, within a multiplicative constant, by the relation

$$s(\mathcal{X}) = -\sum_{i=1}^{N} p_i \log p_i \,. \tag{4.68}$$

Note that this definition is close to that used in communication theory to measure the entropy of information. It is evident that in the case of $N$ different states $\mathcal{X}_i$, the maximum entropy is attained when these results are equally probable. Thus as $p_i = 1/N$, the maximum is

$$s(\mathcal{X}) = \log N \,. \tag{4.69}$$

Inversely, entropy is minimal (zero) if one state is certain and the others are impossible. Another feature to notice in (4.68) is that for two independent systems $\mathcal{X}$ and $\mathcal{X}'$, the probability of the microscopic state ($\mathcal{X}_i$ and $\mathcal{X}'_j$) is $p_i p'_j$, so that the entropy of the union of $\mathcal{X}$ and $\mathcal{X}'$ is given by

$$s(\mathcal{X} \cup \mathcal{X}') = -\sum_{i,j=1}^{N} p_i p'_j \log(p_i p'_j) = s(\mathcal{X}) + s(\mathcal{X}') \,, \tag{4.70}$$

as we always have

$$\sum_{i}^{N} p_i = \sum_{j}^{N} p'_j = 1 \,. \tag{4.71}$$

We thus see that entropy is an extensive parameter.

We also see that if a certain physical quantity takes the value $A_i$ in the state $\mathcal{X}_i$, its macroscopic value is given by the formula

$$A = \sum_{i}^{N} p_i A_i \,, \tag{4.72}$$

which shows the mathematical relation between the two observation levels.

The concept of temperature appears first in common observation, but its relation with statistical mechanics can be approached as follows. The internal energy of a system is the total quantity of disordered energy that it contains, that is, the energy differently distributed, from one state $\mathcal{X}_i$ to another, among its molecules.

It is important to point out that in this definition, certain components of the internal energy present in chemical reactions (and also in gas dynamics) are excluded. Statistical analysis shows that the kinetic energy is distributed, on average, in an equal way to each molecule, and over each degree of freedom of disordered motion.

The absolute temperature $T$ is thus, within a multiplicative constant, the disordered energy per molecule per degree of freedom. We can interpret the fact that two different bodies placed in contact tend toward uniform temperature by the statistical principle: that their disordered energy per molecule per degree of freedom must become equal as they must be equal for each degree of freedom.

A relation obviously exists between absolute temperature and entropy which can be developed with statistical mechanics. It is expressed by the relation

$$\delta u(\mathcal{X}) = T\,\delta s(\mathcal{X}) , \tag{4.73}$$

which links the increases of $\delta u(\mathcal{X})$, the internal energy, and $\delta s(\mathcal{X})$, the system entropy, when all the characteristics (density, deformation, etc.) of the system remain constant.

We do not attempt to interpret this relation, which should be taken as fundamental when we are at the macroscopic level.

At the same time, the irreversibility of physical phenomena is expressed by the fact that the increase of entropy of a system is always greater than a minimum equal to the heat transferred to the system $\delta q(\mathcal{X})$ divided by the absolute temperature $T$, that is, we always have the inequality

$$\delta s(\mathcal{X}) \geq \frac{\delta q(\mathcal{X})}{T} , \tag{4.74}$$

where equality only holds for reversible transformations. This inequality is the basis for the formulation of the second principle of thermodynamics in mechanics of continuous media.

An interesting viewpoint for the irreversibility of physical phenomena is given by Boltzmann's theory for hydrodynamics. Starting from a description at the atomic scale of macroscopic systems by Newtonian mechanics, one is led to resolve a set of $N$ non-linear ordinary differential equations

$$m_i \frac{d^2 \boldsymbol{x}_i}{dt^2} = \boldsymbol{F}_i \qquad i = 1, \ldots, N , \tag{4.75}$$

where $N$ is of the order of Avogadro's number, $N_A \sim 6 \times 10^{23}$. The symbol $m_i$ denotes the molecular mass, $\boldsymbol{v}_i = d\boldsymbol{x}_i/dt$ is the molecular velocity, and $\boldsymbol{F}_i$ the force acting on the $i^{\text{th}}$ molecule due to molecular interactions. This problem is obviously unsolvable because of its enormous size ($O(N_A)$ equations), and we move from the atomic level to kinetic theory of $N$ bodies, which is established from the Newton-Hamilton equations. This theory employs distribution functions $f_N(\boldsymbol{x}_1, \boldsymbol{v}_1, \ldots, \boldsymbol{x}_N, \boldsymbol{v}_N)$ which give the joint probability of finding molecule 1 at position $\boldsymbol{x}_1$ with velocity $\boldsymbol{v}_1$ and molecule 2 at position $\boldsymbol{x}_2$ with velocity $\boldsymbol{v}_2$ and so on until molecule $N$. The individual trajectories in the Newtonian approach are replaced by a notion of phase space where the dynamics is described by a partial differential equation known as the ***Liouville equation*** with dimension $6N$. We see that the mass of information has not been reduced from that of the Newtonian approach. Nonetheless, the Liouville equation is a base for the application of a powerful and elegant procedure which eliminates the redundant information. This leads to the definition of distribution functions $f_M \equiv f_{12\ldots M}, M < N$, which become a chain of equations known by the name of the BBGKY (Bogoliubov-Born-Green-Kirkwood-Yvon) hierarchy. For the interesting macroscopic quantities, such as density, pressure, temperature, etc., distributions with one or two bodies are sufficient and we thus choose $M = 1, 2$ in the BBGKY hierarchy.

The most important single body equation is Boltzmann's equation:

$$\frac{\partial f}{\partial t} + \boldsymbol{v} \cdot \frac{\partial f}{\partial \boldsymbol{x}} + \boldsymbol{a} \cdot \frac{\partial f}{\partial \boldsymbol{v}} = C[f, f]\,. \tag{4.76}$$

Here $\boldsymbol{a}$ is the molecular acceleration. The function $f(\boldsymbol{x}, \boldsymbol{v}, t)$ is the probability density of finding a classical point particle at position $\boldsymbol{x}$ at time $t$ with velocity $\boldsymbol{v}$. The left-hand side of (4.76) represents the free motion of the particles in the phase space, while $C[f, f]$ is a binary collision operator in the form of an integral that takes into account the molecular interactions; its definition is beyond the scope of this discussion. Boltzmann's equation is constructed on the hypothesis of molecular chaos

$$f_{12}(\boldsymbol{x}_1, \boldsymbol{v}_1, \boldsymbol{x}_2, \boldsymbol{v}_2, t) = f(\boldsymbol{x}_1, \boldsymbol{v}_1, t) f(\boldsymbol{x}_2, \boldsymbol{v}_2, t)\ , \tag{4.77}$$

which breaks the time symmetry and reversibility that applies to Newtonian mechanics at the atomic level, thus opening the door to irreversible behavior. The irreversibility is measured by a quantity called $H$, (see [52]), related to entropy by the relation $s = -k_B H$, where $k_B$ is Boltzmann's constant, which is defined by

$$H = \int f(\boldsymbol{x}, \boldsymbol{v}, t) \ln f(\boldsymbol{x}, \boldsymbol{v}, t)\, d\boldsymbol{v}\, d\boldsymbol{x}\,. \tag{4.78}$$

The $H$ theorem shows that $dH/dt \leq 0$. Note that the definition of entropy that we gave in (4.68) is a numerical approximation of the value $-H$.

We obtain the macroscopic variables such as density and velocity by integration on velocity space

$$\rho(\boldsymbol{x}, t) = m \int f(\boldsymbol{x}, \boldsymbol{v}, t)\, d\boldsymbol{v} \tag{4.79}$$

$$\rho(\boldsymbol{x}, t)\boldsymbol{u}(\boldsymbol{x}, t) = m \int f(\boldsymbol{x}, \boldsymbol{v}, t)\boldsymbol{v}\, d\boldsymbol{v}\,, \tag{4.80}$$

where $\boldsymbol{u}$ denotes the velocity in physical space and $m$ is the atomic or molecular mass. Starting from these equations and (4.76), one can obtain the equations of hydrodynamics, and in particular, the Navier-Stokes equations.

## 4.6 Second Principle of Thermodynamics

The second principle of thermodynamics in continuous media is also known by the name of the Clausius-Duhem inequality.

Rudolf Clausius (1822–1888) was born in Koslin (Prussia), known today as Koszalin (Poland). He was professor at the Artillary and Engineering School in Berlin, then at the Swiss Institute of Technology in Zürich, then at the Universities of Würzburg and Bonn. He made very important contributions to thermodynamics, where he introduced the notion of entropy, and to the kinetic theory of gases.

**Fig. 4.1** Rudolf Clausius

SECOND PRINCIPLE OF THERMODYNAMICS *For a material volume, the material derivative of entropy is always greater than or equal to the sum of the volume distribution of entropy sources in the body and the entropy flux across the surface.*

We can generalize (4.74) for a non-homogeneous medium, which, for a material volume $\omega$, takes the form

$$\frac{d}{dt}\int_\omega \rho s\,dv \geq \int_\omega \frac{r}{T}\,dv - \int_{\partial\omega} \frac{\boldsymbol{q}\cdot\boldsymbol{n}}{T}\,ds\,, \tag{4.81}$$

where $s$ is the entropy per unit mass. It is necessary to take into account the temperatures at which the components of $(r\,dv)$ and $(-q_i n_i\,ds)$ are conveyed to $\omega$.

The ***local form*** of the second principle is obtained by the application of the transport theorem to the material derivative, taking into account the conservation of mass, the divergence theorem for the surface integral (the last term), and finally, with the application of the localization theorem,

$$\rho\frac{Ds}{Dt} \geq \frac{r}{T} - \operatorname{div}\left(\frac{\boldsymbol{q}}{T}\right). \tag{4.82}$$

This inequality must be satisfied at every point, at all times, by every process. The equality sign holds only for reversible processes.

To study the consequences of (4.82), we must eliminate the term in $r$, the heat produced per unit mass and unit time, using the local form of the equation of conservation of internal energy (4.23). This elimination is necessary, because $r$ is arbitrary since it refers to an action at a distance. Thus we find the Clausius-Duhem inequality:

$$\rho\frac{Ds}{Dt} \geq \frac{1}{T}\left(\rho\frac{Du}{Dt} - \boldsymbol{\sigma} : \boldsymbol{d}\right) + \frac{1}{T^2}\,\boldsymbol{q}\cdot\boldsymbol{\nabla}T\,, \tag{4.83}$$

which must be satisfied by every thermodynamic process.

Pierre Duhem (1861–1916) was born in Paris. He was named professor at the University of Bordeaux. His work in hydrodynamics and thermodynamics shows that he was a pioneer in the study of irreversible phenomena in thermodynamics. His principal publication was *Traité de l'énergétique* published in 1911.

**Fig. 4.2** Pierre Duhem

If we introduce the ***Helmholtz specific free energy***,

$$f = u - Ts\,, \tag{4.84}$$

the Clausius-Duhem inequality (4.83) takes the form

$$\rho \frac{Df}{Dt} \leq \boldsymbol{\sigma} : \boldsymbol{d} - \rho s \frac{DT}{Dt} - \frac{\boldsymbol{q} \cdot \boldsymbol{\nabla} T}{T}\,. \tag{4.85}$$

The Clausius-Duhem inequality (4.83) can also be easily expressed in the material description. In that case, the contact force power $\boldsymbol{\sigma} : \boldsymbol{d}$ (4.25) can be expressed as $\boldsymbol{P} : \dot{\boldsymbol{F}}$ (see (4.44)).

In finishing this chapter we note that the second principle of thermodynamics is not a conservation principle but an inequality indicating the irreversible nature (or direction) of the physical process that must always be satisfied. The consequences (4.83) and (4.85) will be studied in sections 6.8 to 6.11 for some simple cases.

Hermann von Helmholtz (1821–1894) was born in Potsdam. He was named professor at the University of Berlin. His work in electrophysiology led him to write a book called *Physiological Basis for the Theory of Music.* He made major contributions in the domains of physics and chemistry: potential energy, laws of vorticity, and the Helmholtz decomposition (Helmholtz-Hodge theorem) for a vector field.

**Fig. 4.3** Hermann von Helmholtz

## 4.7 Exercises

**4.1** Let $Q$ be a scalar field defined in the deformed configuration $\omega$ of a body in motion. Applying the Reynolds transport theorem and the continuity equation, show that

$$\frac{d}{dt}\int_\omega \rho(\boldsymbol{x},t)\,Q(\boldsymbol{x},t)\,dv = \int_\omega \rho(\boldsymbol{x},t)\,\frac{DQ(\boldsymbol{x},t)}{Dt}\,dv\,. \tag{4.86}$$

Using this result, and knowing that the kinetic energy of the body is defined by (4.1), derive the expression $DE_k/Dt$.

**4.2** Using the Reynolds theorem and the principle of conservation of mass, show that the temporal derivative of the total energy is written as

$$\frac{D}{Dt}\left(E_k + E_{\text{int}}\right) = \int_\omega \rho\left(\boldsymbol{v}\cdot\boldsymbol{a} + \frac{Du}{Dt}\right)dv\,. \tag{4.87}$$

Express the term $\boldsymbol{v}\cdot\boldsymbol{a}$ as a function only of $\boldsymbol{v}$ and explain this result.

**4.3** The second principle of thermodynamics applied to a homogeneous medium occupying a volume $\omega$ is expressed by equation (4.81).

1) State the local form of the second principle of thermodynamics.
2) Eliminate the term $r$ for heat per unit mass, using the local form of the law of conservation of energy, to establish the Clausius-Duhem inquality (4.83).
3) What happens to this inequality if we introduce the Helmholtz specific free energy (4.84)?

**4.4** For a perfect fluid:

1) express the principle of conservation of internal energy for a perfect fluid whose stress tensor is given by $\boldsymbol{\sigma} = -p\boldsymbol{I}$ ;
2) rewrite the equation using the enthalpy per unit mass, defined as $h = u + p/\rho$ ;
3) show that for an adiabatic flow, the conservation of energy takes the form

$$\rho\,\frac{Dh}{Dt} = \frac{Dp}{Dt}\,. \tag{4.88}$$

**4.5** Prove the symmetry of the stress tensor by performing the derivation in detail for the case of rigid body rotation in paragraph 4.4.2.

*Suggestion:* in order to do this, first calculate $\boldsymbol{v}^*$, then write the conservation of energy equation in the starred reference frame and substitute the expression $\boldsymbol{v}^*$. Lastly, subtract the original equation from the expression obtained and discuss the result.

Chapter 5

# Constitutive Equations: Basic Principles

## 5.1 Introduction

We have used tensor formalism to present the description of the properties of a continuous medium. This permits us to reason in general terms, regardless of the coordinate system to which we refer.

In chapter 2, we examined the local description of the motion of a medium, which can be characterized by various tensors. Those that feature displacements as variables will be more appropriate for the description of solids, whereas tensors whose variables are velocities will be better applied to fluids.

The mechanics of continuous media is an axiomatic approach which leads to a phenomenological model. With this tool, the objective is to predict the motion of a material, taking into account the initial and boundary conditions. We associate a thermodynamic variable with this motion; most often, this variable will be the temperature.

Whichever model is chosen to describe the medium (perfect or viscous fluid, elastic solid, viscoelastic solid, etc.), the conservation laws and the first principle of thermodynamics must always be respected. The laws for mass, momentum, angular momentum, and energy lead to a set of partial differential equations in the local form. With respect to a Cartesian coordinate system, these equations are as follows.

*Conservation of mass:*

$$\frac{D\rho}{Dt} + \rho \operatorname{div} \boldsymbol{v} = 0 \tag{5.1}$$

or

$$\frac{\partial \rho}{\partial t} + v_j \frac{\partial \rho}{\partial x_j} + \rho \frac{\partial v_i}{\partial x_i} = 0\,. \tag{5.2}$$

*Conservation of momentum:*

$$\rho\frac{D\boldsymbol{v}}{Dt} = \mathbf{div}\,\boldsymbol{\sigma} + \rho\boldsymbol{b} \tag{5.3}$$

or

$$\rho\left(\frac{\partial v_i}{\partial t} + v_j\frac{\partial v_i}{\partial x_j}\right) = \frac{\partial \sigma_{ji}}{\partial x_j} + \rho b_i\,. \tag{5.4}$$

*Conservation of angular momentum:*

$$\boldsymbol{\sigma}^T = \boldsymbol{\sigma} \tag{5.5}$$

or

$$\sigma_{ij} = \sigma_{ji}\,. \tag{5.6}$$

*Conservation of energy:*

$$\rho\frac{Du}{Dt} = \mathrm{tr}\,(\boldsymbol{\sigma}\boldsymbol{L}) - \mathrm{div}\,\boldsymbol{q} + r \tag{5.7}$$

or

$$\rho\left(\frac{\partial u}{\partial t} + v_j\frac{\partial u}{\partial x_j}\right) = \sigma_{ij}\frac{\partial v_j}{\partial x_i} - \frac{\partial q_i}{\partial x_i} + r\,. \tag{5.8}$$

This system contains eight independent equations $(1+3+3+1)$, with the body forces, $\boldsymbol{b}$, and the volume production of heat, $r$, as given for the problem.

The unknown functions are the motion vector $\boldsymbol{\chi}$ (2.1) and the temperature. In the field equations we find the unknown variables $\rho$, $\sigma_{ij}$, $u$, and $q_i$. Note that $\boldsymbol{v}$ is calculated from $\boldsymbol{\chi}$. Altogether, we have eighteen unknowns : $\boldsymbol{\chi}(3)$, $T(1)$, $\rho(1)$, $\boldsymbol{\sigma}(9)$, $u(1)$, $\boldsymbol{q}(3)$.

If we assume that the conservation of angular momentum is satisfied, then $\boldsymbol{\sigma}$ has only six unknowns. We thus have five partial differential equations for fifteen unknowns.

In any case, our problem is incompletely posed. Until now, we have stated the principles of conservation in general without reference to the model for the continuous medium. In order to characterize the behavior of a material, we need to take into account the constitutive equations. The nature and form of these equations are based on the results from experiments obtained in the laboratory. Figures 5.1 and 5.2 show typical stress-strain and stress-strain rate curves. The constitutive equations are also proposed as axioms based on mathematical or physical arguments. These equations give the dependence of $\sigma_{ij}$, $u$, and $q_i$ with respect to the history of the material deformation and temperature. For a given model, we choose a certain number of kinematic and thermal variables and express how $\sigma_{ij}$, $u$, and $q_i$ depend on these variables.

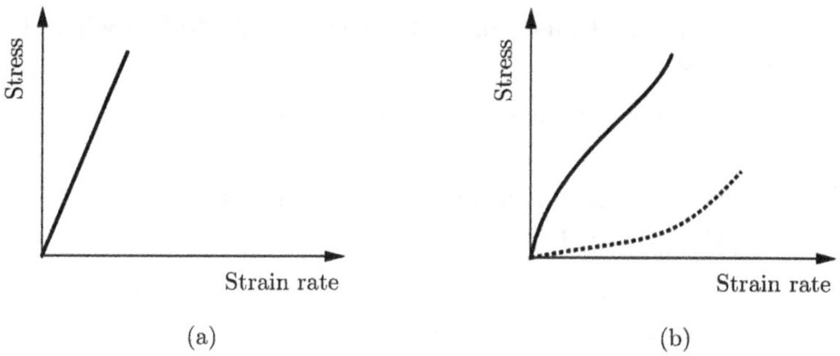


**Fig. 5.1** Stress-strain rate relations: (a) linear; (b) nonlinear

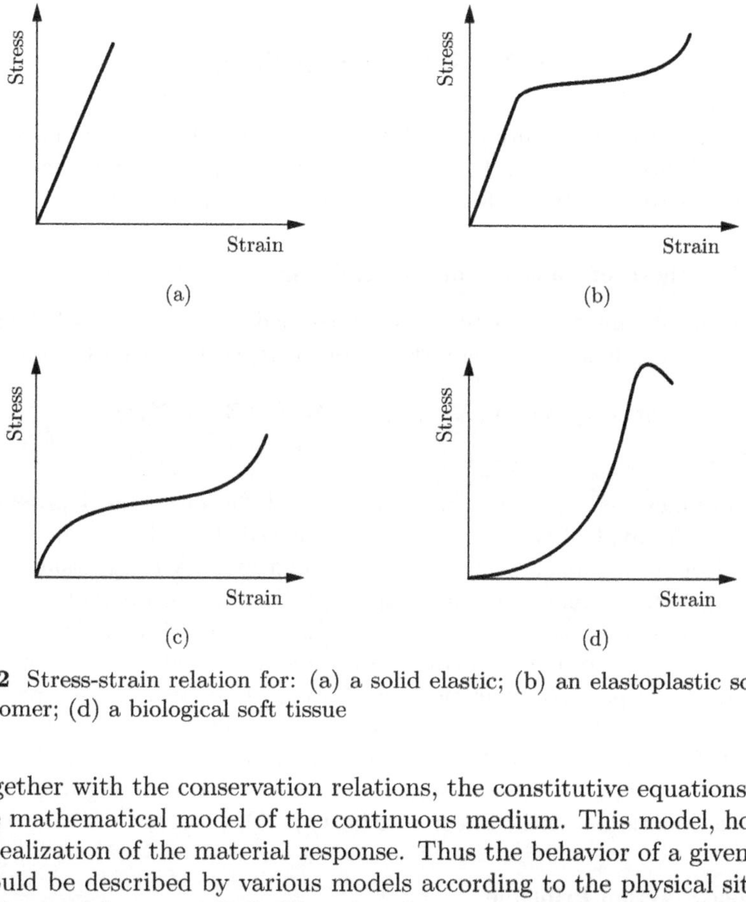


**Fig. 5.2** Stress-strain relation for: (a) a solid elastic; (b) an elastoplastic solid; (c) an elastomer; (d) a biological soft tissue

Together with the conservation relations, the constitutive equations establish the mathematical model of the continuous medium. This model, however, is an idealization of the material response. Thus the behavior of a given material should be described by various models according to the physical situation where the models are applied. For example, at room temperature, glass is an elastic material. In the transformation zone, around 600°C, it is viscoelastic. In a glass melting furnace, we can consider it to be like an incompressible Newtonian viscous fluid. Another example is that of polymers, which in slow-motion act like a viscous fluid, but act like an elastic solid when forced into rapid motion.

The constitutive equations must satisfy ***at least*** the following three fundamental principles:

1) objectivity and independence with respect to the observer or the reference frame;
2) the properties of material symmetry for the medium;
3) the second principle of thermodynamics.

The founding principles for the development of the constitutive equations are also largely developed in the following texts: [12, 17, 18, 19, 22, 49, 57]. The text by Truesdell and Toupin [60] constitutes an important and very elaborate contribution to the theory of continuous media.

## 5.2 General Principles

We will now introduce many general principles (including those above) which must be satisfied absolutely by the constitutive equations that we establish. Recall that $\chi$ represents the motion of a body $\mathcal{B}$ given by (2.1).

### 5.2.1 Hypothesis of Causality or Determinism

The stress at instant $t$ in a material is determined by the history of the motion of the material. Thus the Cauchy stress tensor $\boldsymbol{\sigma}(\boldsymbol{x}, t)$ is given by

$$\boldsymbol{\sigma}(\boldsymbol{x}, t) = \boldsymbol{\sigma}\big(\chi(\boldsymbol{X}, t), t\big) = \underset{\substack{\tau \leq t \\ \boldsymbol{Z} \in \mathcal{B}}}{\boldsymbol{\Sigma}} \big(\chi(\boldsymbol{Z}, \tau); \boldsymbol{X}, t\big) , \tag{5.9}$$

where the functional $\boldsymbol{\Sigma}$ characterizes in general the mechanical properties of the body $\mathcal{B}$. An explicit time dependence is allowed. In addition, the functional can vary from one particle to another (dependence on $\boldsymbol{X}$) expressing an inhomogeneous distribution of the material properties. In a simplified way, we can interpret a functional as a function of another function (here, the function of motion $\chi$). In the mechanics of continuous media, it is often represented by an integral over the past configurations $\chi(\boldsymbol{Z}, \tau)$ ($\boldsymbol{Z} \in \mathcal{B}$, $\tau < t$) thus permitting the description of the constitutive relations for different classes of materials. Note that the way (5.9) is written expresses non-locality since there is a dependence upon every point $\boldsymbol{Z}$ belonging to the body $\mathcal{B}$. To simplify, we sometimes denote the functional with respect to $\boldsymbol{\sigma}(\boldsymbol{x}, t)$ as $\boldsymbol{\Sigma}(\chi; \boldsymbol{X}, t)$.

### 5.2.2 Local Action Principle

For a given particle $\boldsymbol{X}$ of the material, the functional $\boldsymbol{\Sigma}(\chi; \boldsymbol{X}, t)$ depends only on the neighborhood of $\boldsymbol{X}$. For two arbitrary motions $\chi$ and $\overline{\chi}$ that coincide in a neighborhood $\mathcal{V}(\boldsymbol{X}) \subset \mathcal{B}$ at any time $\tau \leq t$, the value of $\boldsymbol{\Sigma}$ is the same. Formally, we write

$$\boldsymbol{\Sigma}(\chi; \boldsymbol{X}, t) = \boldsymbol{\Sigma}(\overline{\chi}; \boldsymbol{X}, t) , \tag{5.10}$$

as long as there exists a neighborhood $\mathcal{V}(\boldsymbol{X})$ such that

$$\chi(\boldsymbol{Z},\tau) = \overline{\chi}(\boldsymbol{Z},\tau) \qquad \forall \boldsymbol{Z} \in \mathcal{V}(\boldsymbol{X}) \quad \forall \tau \leq t\,. \tag{5.11}$$

We observe that $\boldsymbol{\Sigma}(\chi; \boldsymbol{X}, t)$ is a functional of the function $\chi$ of two variables: time $\tau$ and the particle $\boldsymbol{Z}$ in the neighborhood of the chosen particule $\boldsymbol{X}$. In this case, the tensor $\boldsymbol{\sigma}(\boldsymbol{x},t)$ is given by a relation analogous to (5.9) such that

$$\overline{\boldsymbol{\sigma}}(\boldsymbol{x},t) = \boldsymbol{\sigma}(\boldsymbol{x},t) = \underset{\substack{\tau \leq t \\ \boldsymbol{Z} \in \mathcal{V}(\boldsymbol{X})}}{\boldsymbol{\Sigma}} \left(\chi(\boldsymbol{Z},\tau); \boldsymbol{X}, t\right). \tag{5.12}$$

### 5.2.3 Principle of Objectivity

Not every constitutive equation that satisfies the local action principle is admissible. It must also satisfy the principle of objectivity, or frame indifference, which requires that the functional $\boldsymbol{\Sigma}$ be invariant in any change of the reference frame. We wish to write constitutive equations that are independent of the observer, and, in particular, independent of superimposed rigid body motion. More precisely, we have (sec. 2.11)

$$\boldsymbol{x}^* = \chi^*(\boldsymbol{X},t^*) = \boldsymbol{c}(t) + \boldsymbol{Q}(t)\chi(\boldsymbol{X},t) \tag{5.13}$$

$$\boldsymbol{\sigma}^*(\boldsymbol{X},t^*) = \boldsymbol{Q}(t)\boldsymbol{\sigma}(\boldsymbol{X},t)\boldsymbol{Q}^T(t) \tag{5.14}$$

$$t^* = t - \alpha\,. \tag{5.15}$$

Using (5.13)–(5.15), the principle of objectivity becomes

$$\boldsymbol{\sigma}^*(\boldsymbol{X},t^*) = \underset{\substack{\tau^* \leq t^* \\ \boldsymbol{Z} \in \mathcal{V}(\boldsymbol{X})}}{\boldsymbol{\Sigma}} \left(\chi^*(\boldsymbol{Z},\tau^*); \boldsymbol{X}, t^*\right) \tag{5.16}$$

such that

$$\underset{\substack{\tau^* \leq t^* \\ \boldsymbol{Z} \in \mathcal{V}(\boldsymbol{X})}}{\boldsymbol{\Sigma}} \left(\chi^*(\boldsymbol{Z},\tau^*); \boldsymbol{X}, t^*\right) = \boldsymbol{Q}(t) \underset{\substack{\tau \leq t \\ \boldsymbol{Z} \in \mathcal{V}(\boldsymbol{X})}}{\boldsymbol{\Sigma}} \left(\chi(\boldsymbol{Z},\tau); \boldsymbol{X}, t\right)\boldsymbol{Q}^T(t)\,. \tag{5.17}$$

In order to clarify the impact of this principle on the statement of the functional $\boldsymbol{\Sigma}$, we study three particular reference frame changes in succession.

In the following we take $\tau$ as the time variable; $t$ will indicate the instant at which the stress is evaluated. Then, $\boldsymbol{Q}(\tau)\boldsymbol{Q}^T(\tau) = \boldsymbol{Q}^T(\tau)\boldsymbol{Q}(\tau) = \boldsymbol{I}$, $\det \boldsymbol{Q}(\tau) = 1$, $\forall \tau$.

*Rigid translation of the moving observer without changing the time scale:*

Let us set $\boldsymbol{Q}(\tau) = \boldsymbol{I}$, $\alpha = 0$ and

$$\boldsymbol{c}(\tau) = -\chi(\boldsymbol{X},\tau)\,. \tag{5.18}$$

This means that the reference frame is in rigid translation such that, after the frame change, the material point $\boldsymbol{X}$ at time $\tau$ remains at the origin. In addition, $t^* = t$. From (5.13), we have for $\boldsymbol{Z} \in \mathcal{V}(\boldsymbol{X})$

$$\chi^*(\boldsymbol{Z},\tau) = \chi(\boldsymbol{Z},\tau) - \chi(\boldsymbol{X},\tau)$$

and from (5.12), (5.14), and (5.16)

$$\boldsymbol{\sigma}^*(\boldsymbol{X},t^*)=\boldsymbol{\sigma}(\boldsymbol{X},t)=\underset{\substack{\tau\le t\\ \boldsymbol{Z}\in\mathcal{V}(\boldsymbol{X})}}{\boldsymbol{\Sigma}}\big(\chi(\boldsymbol{Z},\tau)-\chi(\boldsymbol{X},\tau);\boldsymbol{X},t\big)\,. \tag{5.19}$$

*Change of the time scale in a fixed reference frame:*

This situation corresponds to the following choice:

$$\boldsymbol{Q}(\tau)=\boldsymbol{I}\qquad \boldsymbol{c}(\tau)=\boldsymbol{0}\qquad t=\alpha\,. \tag{5.20}$$

The instant $t$ is the reference time after the change. Thus we have from (5.15) and (5.20)

$$\tau^*=\tau-\alpha=\tau-t\,. \tag{5.21}$$

At the instant $t^*$, using (5.21) and (5.14), we have

$$t^*=t-t=0\qquad \boldsymbol{\sigma}^*(\boldsymbol{X},t^*)=\boldsymbol{\sigma}^*(\boldsymbol{X},0)=\boldsymbol{\sigma}(\boldsymbol{X},t)\,. \tag{5.22}$$

From (5.13), (5.20), and (5.21) we find

$$\chi^*(\boldsymbol{Z},\tau^*)=\chi(\boldsymbol{Z},\tau)=\chi(\boldsymbol{Z},t+\tau^*)\,. \tag{5.23}$$

And from (5.22), (5.16), and (5.23), we obtain

$$\begin{aligned}\boldsymbol{\sigma}(\boldsymbol{X},t)=\boldsymbol{\sigma}^*(\boldsymbol{X},0)&=\underset{\substack{\tau^*\le t^*\\ \boldsymbol{Z}\in\mathcal{V}(\boldsymbol{X})}}{\boldsymbol{\Sigma}}\big(\chi(\boldsymbol{Z},\tau^*+t);\boldsymbol{X},0\big)\\ &=\underset{\substack{\tau-t\le 0\\ \boldsymbol{Z}\in\mathcal{V}(\boldsymbol{X})}}{\boldsymbol{\Sigma}}\big(\chi(\boldsymbol{Z},\tau);\boldsymbol{X},0\big)\,.\end{aligned} \tag{5.24}$$

Consequently, the functional $\boldsymbol{\Sigma}$ does not depend explicitly on $t$. Now introduce

$$\tau=t-s\qquad 0\le s\le\infty\,. \tag{5.25}$$

Combining (5.19) and (5.24), we obtain

$$\boldsymbol{\sigma}(\boldsymbol{X},t)=\underset{\substack{s\ge 0\\ \boldsymbol{Z}\in\mathcal{V}(\boldsymbol{X})}}{\boldsymbol{\Sigma}}\big(\chi(\boldsymbol{Z},t-s)-\chi(\boldsymbol{X},t-s);\boldsymbol{X}\big)\,. \tag{5.26}$$

Thus the functional $\boldsymbol{\Sigma}$ only depends on the relative motion starting from the reference time of all the particles $\boldsymbol{Z}$ in $\mathcal{B}$, that is $\boldsymbol{Z}\in\mathcal{V}(\boldsymbol{X})$.

*Rigid rotation of the reference frame:*

We choose $\boldsymbol{c}(\tau)=\boldsymbol{0}$, $\alpha=0$, and arbitrary $\boldsymbol{Q}(\tau)$. This corresponds to unsteady rotation of the reference frame. In this rotation, the stress tensor is transformed according to the relation

$$\boldsymbol{\sigma}^*(\boldsymbol{X},t)=\boldsymbol{Q}(t)\boldsymbol{\sigma}(\boldsymbol{X},t)\boldsymbol{Q}^T(t)\,. \tag{5.27}$$

By combining (5.13), (5.14), (5.16), (5.26), and (5.27), we can write

$$\begin{aligned}&\boldsymbol{Q}(t)\boldsymbol{\Sigma}\big(\chi(\boldsymbol{Z},t-s)-\chi(\boldsymbol{X},t-s),\boldsymbol{X}\big)\boldsymbol{Q}^T(t)\\ &\quad=\boldsymbol{\Sigma}\big(\chi^*(\boldsymbol{Z},t-s)-\chi^*(\boldsymbol{X},t-s),\boldsymbol{X}\big)\\ &\quad=\boldsymbol{\Sigma}\left(\boldsymbol{Q}(t-s)\big(\chi(\boldsymbol{Z},t-s)-\chi(\boldsymbol{X},t-s)\big),\boldsymbol{X}\right).\end{aligned} \tag{5.28}$$

This last equation is the restriction that we must impose on the functional $\boldsymbol{\Sigma}$ to ensure that it is objective. We can easily see that, inversely, all constitutive equations of the form (5.26) that obey the condition (5.28) satisfy the principle of frame indifference. This is due to the fact that any general change of the reference frame can be obtained by a sequence of the three particular changes described above. Hence, equation (5.26), satisfying the condition (5.28), is the most general constitutive equation for the theory of mechanics of continuous media. More precisely, we observe that the stress is represented by an isotropic tensor function with a tensor value (sec. 1.3.11).

### 5.2.4 Principle of Material Invariance

Solid materials have symmetry properties because of their crystallographic characteristics: cubic, rhombohedral, etc. Certain fluids also possess this type of property, for example, liquid crystal fluids. In this case, the constitutive laws will not change form when the material coordinates $(X_1, X_2, X_3)$ become $(X_1, X_2, -X_3)$. This represents a reflection operation on the coordinates with respect to the plane $X_3 = 0$. However, this condition imposes restrictions on the equations.

Reflection through a symmetry plane including the origin 0 that is orthogonal to the unit vector $\boldsymbol{n}$ is defined by the tensor $\boldsymbol{R}$. We have

$$\overline{\boldsymbol{X}} = \boldsymbol{R}\boldsymbol{X} = (\boldsymbol{I} - 2\boldsymbol{n} \otimes \boldsymbol{n})\,\boldsymbol{X} \ .$$

In index notation we have

$$R_{ij} = \delta_{ij} - 2\,n_i n_j \qquad \text{with} \qquad \det[R] = -1\,. \tag{5.29}$$

We denote by $\{\boldsymbol{O}\}$ a sub-group of the complete group of orthogonal transformations for the material axes, and by $\{\boldsymbol{B}\}$ the group of translations of these axes. Then the principle of material invariance is stated as follows.

The constitutive equation must be formally invariant with respect to a group of orthogonal transformations $\{\boldsymbol{O}\}$ and a group of translations $\{\boldsymbol{B}\}$ of the material coordinates. These restrictions come from the symmetry conditions induced by $\{\boldsymbol{O}\}$ and $\{\boldsymbol{B}\}$ in the coordinate system $\boldsymbol{X}$.

We then have a transformation of the form

$$\overline{\boldsymbol{X}} = \boldsymbol{O}\boldsymbol{X} + \boldsymbol{B}\,, \tag{5.30}$$

with

$$\boldsymbol{O}\boldsymbol{O}^T = \boldsymbol{O}^T\boldsymbol{O} = \boldsymbol{I} \qquad \text{and} \qquad \det \boldsymbol{O} = \pm 1\,. \tag{5.31}$$

These conditions express the geometric symmetries represented by $\{\boldsymbol{O}\}$ and the inhomogeneities represented by $\{\boldsymbol{B}\}$, at $\boldsymbol{X}$, of the physical properties of the material body. When $\{\boldsymbol{O}\}$ is the proper orthogonal group characterized by the matrice $[O]$ such that $\det[O] = +1$, the material is ***hemitropic***; we can not perform the reflection of the axis $X_i$ with respect to the plane $X_i = 0$. When $\{O\}$ is the complete group ($\det[O] = \pm 1$), the material is said to be ***isotropic***. A material that is not hemitropic is called ***anisotropic***.

When the functions do not depend on the translations, $\{\boldsymbol{B}\}$, of the origin of the material coordinates, we say that the material is ***homogeneous***. If these functions change with certain translations $\{\boldsymbol{B}\}$ of the material axes, then the material is ***inhomogeneous***.

If we combine the principle of material invariance and the principle of objectivity with a transformation relative to the material coordinates $X_i$, we have the condition

$$\begin{aligned}&\boldsymbol{\Sigma}\big(\boldsymbol{\chi}(\boldsymbol{Z},t-s)-\boldsymbol{\chi}(\boldsymbol{X},t-s)\,,\boldsymbol{X}\big)\\&\quad=\boldsymbol{\Sigma}\big(\boldsymbol{\chi}(\boldsymbol{O}\boldsymbol{Z}+\boldsymbol{B},t-s)-\boldsymbol{\chi}(\boldsymbol{O}\boldsymbol{X}+\boldsymbol{B},t-s),\boldsymbol{O}\boldsymbol{X}+\boldsymbol{B}\big)\,.\end{aligned} \tag{5.32}$$

### 5.2.5 Principle of Memory

The values of the variables in the constitutive relations in the remote past do not have an appreciable impact on the current values of these variables.

We will return to this principle later and introduce the concept of fading memory.

### 5.2.6 Principle of Admissibility

All the constitutive equations must be coherent with the fundamental principles of mechanics of continuous media, that is, they must obey the conservation laws of mass, momentum, and energy, as well as the Clausius-Duhem inequality.

## 5.3 Consequence of the Principle of Local Action

Assume, for simplicity, that the vector function $\boldsymbol{\chi}(\boldsymbol{Z},t)$ can be expanded in a Taylor series about $\boldsymbol{Z}$ for every $\tau \leq t$ and $\boldsymbol{Z} \in \mathcal{V}(\boldsymbol{X})$:

$$\boldsymbol{\chi}(\boldsymbol{Z},\tau)=\boldsymbol{\chi}(\boldsymbol{X},\tau)+(\boldsymbol{Z}-\boldsymbol{X})\frac{\partial\boldsymbol{\chi}}{\partial\boldsymbol{X}}(\boldsymbol{X},\tau)+O\big(\|\boldsymbol{Z}-\boldsymbol{X}\|^2\big)\,. \tag{5.33}$$

The deformation gradient tensor $\boldsymbol{F}$, introduced by the relation (2.65), appears in this expression. We limit ourselves here to the case of materials that depend only on the first order of the gradient and thus only consider the first term of the expansion (5.33). We can then describe $\boldsymbol{\sigma}(\boldsymbol{X},t)$ as

$$\boldsymbol{\sigma}(\boldsymbol{X},t)=\boldsymbol{\Sigma}\Big(\frac{\partial\boldsymbol{\chi}}{\partial\boldsymbol{X}};\boldsymbol{X},t\Big)\,. \tag{5.34}$$

These are said to be ***simple*** materials.

Combining what we have written for the result of the principle of local action (5.34), as well as equation (5.26), we have

$$\boldsymbol{\sigma}(\boldsymbol{X},t) = \underset{s\geq 0}{\boldsymbol{\Sigma}}\left(\frac{\partial \boldsymbol{\chi}}{\partial \boldsymbol{X}}(t-s); \boldsymbol{Z}-\boldsymbol{X}, \boldsymbol{X}\right). \tag{5.35}$$

In (5.35) we have written the explicit dependence with respect to the directional vectors $\boldsymbol{Z}-\boldsymbol{X}$ which have $\boldsymbol{X}$ as the origin in three spatial dimensions. More simply, we can check that this dependence can be expressed as a function of the vector basis $\boldsymbol{e}_k$ of the material coordinates of $\boldsymbol{X}$. This statement underlines the dependence of the functional $\boldsymbol{\Sigma}$ on the choice of basis. The introduction of $\boldsymbol{Z}-\boldsymbol{X}$ in (5.35) represents the directional dependence of the material properties at point $\boldsymbol{X}$. When there is no ambiguity, we can drop $\boldsymbol{Z}-\boldsymbol{X}$ in the arguments of the functional. In addition, equation (5.35) expresses that the stress at time $t$ depends on the deformation gradient history. We have

$$\boldsymbol{\sigma}(\boldsymbol{X},t) = \underset{s\geq 0}{\boldsymbol{\Sigma}}\big(\boldsymbol{F}(\boldsymbol{X},t-s), \boldsymbol{X}\big). \tag{5.36}$$

If we impose the objectivity condition (5.28) on relation (5.36), we obtain

$$\boldsymbol{Q}(t)\boldsymbol{\Sigma}\big(\boldsymbol{F}(\boldsymbol{X},t-s),\boldsymbol{X}\big)\boldsymbol{Q}^T(t) = \boldsymbol{\Sigma}\big(\boldsymbol{Q}(t-s)\boldsymbol{F}(\boldsymbol{X},t-s),\boldsymbol{X}\big). \tag{5.37}$$

With the polar decomposition theorem (2.73), we have

$$\boldsymbol{F}(\boldsymbol{X},t-s) = \boldsymbol{R}(t-s)\boldsymbol{U}(\boldsymbol{X},t-s),$$

where $\boldsymbol{R}(t-s)$ and $\boldsymbol{U}(\boldsymbol{X},t-s)$ are the rotation tensor and the right symmetric stretch tensor histories, respectively. If we make the particular choice that $\boldsymbol{Q}(t-s) = \boldsymbol{R}^T(t-s)$, equation (5.37) becomes

$$\boldsymbol{R}^T\boldsymbol{\Sigma}\big(\boldsymbol{F}(\boldsymbol{X},t-s),\boldsymbol{X}\big)\boldsymbol{R} = \boldsymbol{\Sigma}\big(\boldsymbol{U}(\boldsymbol{X},t-s),\boldsymbol{X}\big). \tag{5.38}$$

We can leave out the explicit dependence of $\boldsymbol{\Sigma}$ with respect to $\boldsymbol{X}$ as this is already taken into account by $\boldsymbol{F}$. This eases the expression without, however, affecting the generality of what follows. Consequently, expression (5.36) for the stress tensor can be written in the form

$$\boldsymbol{\sigma}(\boldsymbol{X},t) = \boldsymbol{R}(t)\boldsymbol{\Sigma}\big(\boldsymbol{U}(\boldsymbol{X},t-s)\big)\boldsymbol{R}^T(t). \tag{5.39}$$

This last relation is the general form of the constitutive equation of a simple material. It yields the general solution of the functional relation (5.37). In addition, it shows that the stress in a simple material is affected by the rotation at the time under consideration, but past rotations have no influence.

Recalling the polar decomposition theorem and (2.88), we can put (5.39) in the form

$$\overline{\boldsymbol{\sigma}}(t) = \mathcal{F}\big(\boldsymbol{C}(t-s)\big), \tag{5.40}$$

with

$$\overline{\boldsymbol{\sigma}} \equiv \boldsymbol{F}^T\boldsymbol{\sigma}\boldsymbol{F} \qquad \mathcal{F}\left(\big(\boldsymbol{U}(t-s)\big)^2\right) \equiv \boldsymbol{U}(t)\boldsymbol{\Sigma}\big(\boldsymbol{U}(t-s)\big)\boldsymbol{U}(t).$$

The tensor $\overline{\boldsymbol{\sigma}}$ is the transported or convected stress tensor. Defining

$$\mathcal{L}\big(\boldsymbol{C}(t-s)\big) \equiv \boldsymbol{C}^{-1}(t)\boldsymbol{\mathcal{F}}\big(\boldsymbol{C}(t-s)\big)\boldsymbol{C}^{-1}(t)\,,$$

then equation (5.40) becomes

$$\boldsymbol{\sigma}(t) = \boldsymbol{F}\mathcal{L}\big(\boldsymbol{C}(t-s)\big)\boldsymbol{F}^T\,. \tag{5.41}$$

## 5.4 Thermomechanical Constitutive Equations

Until now we have considered materials in an isothermal situation. As soon as we want to take into account thermal effects, we need to introduce a new primary variable analogous to that of motion or deformation. For this purpose, we use temperature, which we take as

$$T = T(\boldsymbol{X}, t)\,. \tag{5.42}$$

This means that in a thermomechanical problem, the independent constitutive variables are motion $\chi$ and temperature $T$. The velocity will be obtained simply by the time derivative of the motion and then the velocity gradient tensor from $\boldsymbol{v}$. The density is related to the motion by the continuity equation (3.37) written in the form $\det \boldsymbol{F} = P_0/P = \rho_0/\rho$ in spatial coordinates.

### 5.4.1 Principle of Determinism

Besides the motion history of the material, the stress is also influenced by the temperature. Thus, equation (5.9) is generalized by the relation

$$\boldsymbol{\sigma}(\boldsymbol{x}, t) = \underset{\substack{\tau \leq t \\ \boldsymbol{Z} \in \mathcal{B}}}{\Sigma} \big(\chi(\boldsymbol{Z}, \tau), T(\boldsymbol{Z}, t); \boldsymbol{X}, t\big)\,. \tag{5.43}$$

Since we take into account thermal effects, we also need to define constitutive relations for the heat flux vector, $\boldsymbol{q}$, the internal energy, $u$, and the entropy density, $s$. To (5.43) we add constitutive equations, such as the principle of determinism, generalized as follows: the values of the thermomechanical functions ($\boldsymbol{\sigma}, \boldsymbol{q}, u$, and $s$), at a material point $\boldsymbol{X}$ at time $t$, are determined by the motion and temperature histories for all the points in the body $\mathcal{B}$. Thus we have

$$\boldsymbol{\sigma}(\boldsymbol{X}, t) = \boldsymbol{\Sigma}(\chi, T; \boldsymbol{X}, t) = \boldsymbol{\Sigma}\big(\boldsymbol{x}(\boldsymbol{Z}, \tau), T(\boldsymbol{Z}, \tau), \boldsymbol{X}, t\big) \tag{5.44}$$

$$\boldsymbol{q}(\boldsymbol{X}, t) = \boldsymbol{\mathcal{Q}}(\chi, T; \boldsymbol{X}, t) \tag{5.45}$$

$$u(\boldsymbol{X}, t) = \mathcal{U}(\chi, T; \boldsymbol{X}, t) \tag{5.46}$$

$$s(\boldsymbol{X}, t) = \mathcal{S}(\chi, T; \boldsymbol{X}, t)\,. \tag{5.47}$$

### 5.4.2 Principle of Equipresence

We assume that all the constitutive functionals are expressed as functions of the same set of independent constitutive parameters, until proven otherwise.

### 5.4.3 Principle of Local Action

The reasoning that we have applied in section 5.3 also applies to the temperature field. We perform a Taylor series expansion of $T(\boldsymbol{Z},\tau)$ around $T(\boldsymbol{X},\tau)$ to write

$$T(\boldsymbol{Z},\tau) = T(\boldsymbol{X},\tau) + (\boldsymbol{Z}-\boldsymbol{X})\frac{\partial T}{\partial \boldsymbol{X}} + O(\|\boldsymbol{Z}-\boldsymbol{X}\|^2)\,. \tag{5.48}$$

Limiting ourselves to simple materials, the functional $\boldsymbol{\Sigma}$ only depends on gradients of order 1. Then

$$T(\boldsymbol{Z},\tau) \simeq T(\boldsymbol{X},\tau) + (\boldsymbol{Z}-\boldsymbol{X})\frac{\partial T}{\partial \boldsymbol{X}}$$

and

$$\boldsymbol{\sigma}(\boldsymbol{X},t) = \underset{\substack{\tau\leq t\\ \boldsymbol{Z}\in\mathcal{V}(\boldsymbol{X})}}{\boldsymbol{\Sigma}} \left(\chi(\boldsymbol{Z},\tau), \frac{\partial\chi(\boldsymbol{Z},\tau)}{\partial \boldsymbol{X}}, T(\boldsymbol{Z},\tau), \frac{\partial T(\boldsymbol{Z},\tau)}{\partial \boldsymbol{X}}; \boldsymbol{X}, t\right). \tag{5.49}$$

### 5.4.4 Principle of Objectivity

We have seen in chapter 2 that a scalar field is objective if and only if

$$T^* = T\,. \tag{5.50}$$

For the rigid translation of the reference frame (5.18), the functional of $\boldsymbol{\sigma}$ is written as

$$\boldsymbol{\sigma}(\boldsymbol{X},t) = \underset{\substack{\tau\leq t\\ \boldsymbol{Z}\in\mathcal{V}(\boldsymbol{X})}}{\boldsymbol{\Sigma}} \left(\chi(\boldsymbol{Z},\tau) - \chi(\boldsymbol{X},\tau), T(\boldsymbol{Z},\tau), \boldsymbol{X}, t\right). \tag{5.51}$$

The considerations related to the change of the time scale lead to the relation

$$\boldsymbol{\sigma}(\boldsymbol{X},t) = \boldsymbol{\Sigma}\big(\chi(\boldsymbol{Z},t-s) - \chi(\boldsymbol{X},t-s), T(\boldsymbol{Z},t-s), \boldsymbol{X}\big)\,. \tag{5.52}$$

Finally, taking into account the rotation of the reference frame imposes the condition

$$\begin{aligned}&\boldsymbol{Q}(t)\boldsymbol{\Sigma}\big(\chi(\boldsymbol{Z},t-s) - \chi(\boldsymbol{X},t-s), T(\boldsymbol{Z},t-s), \boldsymbol{X}\big)\boldsymbol{Q}^T(t)\\ &\quad = \boldsymbol{\Sigma}\left(\boldsymbol{Q}(t-s)\big(\chi(\boldsymbol{Z},t-s) - \chi(\boldsymbol{X},t-s)\big), T(\boldsymbol{Z},t-s), \boldsymbol{X}\right).\end{aligned} \tag{5.53}$$

Combining the results of the principles of local action (5.49) and objectivity, we obtain the relation

$$\boldsymbol{\sigma}(\boldsymbol{X},t) = \boldsymbol{\Sigma}\left(\frac{\partial\chi}{\partial \boldsymbol{X}}(t-s), T(t-s), \frac{\partial T}{\partial \boldsymbol{X}}(t-s), \boldsymbol{Z}-\boldsymbol{X}, \boldsymbol{X}\right), \tag{5.54}$$

which generalizes (5.35).

When the theory takes into account thermal effects, we should expect density to vary. We can show that, in general, the use of the polar decomposition theorem in order to impose objectivity reveals the Cauchy-Green deformation tensor, but also the scalar invariant associated with the deformation gradient tensor $\boldsymbol{F}$, that is,

$$\det \boldsymbol{F}(t-s) = \big(\det \boldsymbol{C}(t-s)\big)^{1/2} = \frac{\rho_0}{\rho(t-s)} . \qquad (5.55)$$

(see (2.68), (2.77), and (3.37)).

Consequently, the most general equation for simple materials is of the form

$$\boldsymbol{\sigma}(\boldsymbol{X},t) = \boldsymbol{F}\mathcal{L}\left(\boldsymbol{C}(t-s), \rho^{-1}(t-s), T(t-s), \frac{\partial T}{\partial \boldsymbol{X}}(t-s), \boldsymbol{X}\right)\boldsymbol{F}^T , \qquad (5.56)$$

which is a generalization of (5.41).

## 5.5 Definition of a Fluid and a Solid

We define a fluid as a simple material for which we presume that the reference configuration is, most often, that present at the time it is being considered. A fluid is also a material incapable of "resisting" an applied shear: subject to this stress it responds by flowing. Classical Newtonian fluids have an infinitesimally short memory. This means that in the functional (5.56), the stress only depends on $\boldsymbol{C}(t-s)$, for example, for $0 \leq s \leq \varepsilon$ with $\varepsilon$ tending to zero. The stress tensor has a quasi instantaneous memory.

We define a solid as a simple material medium that possesses preferred configurations. One of these can be taken as the reference configuration and we call it the reference state. In most cases, the material is not under stress in this state ($\boldsymbol{\sigma} = 0$). We call this the natural state of the material. If however, in this reference state $\boldsymbol{\sigma} \neq 0$, we say that the material is prestressed.

## 5.6 Principle of Regular Memory

We assume that the thermomechanical histories $\chi(\boldsymbol{X},\tau)$ and $T(\boldsymbol{X},\tau)$ can be expanded in a Taylor series with respect to $\tau$ around $t$ and $\forall \boldsymbol{X} \in \mathcal{B}$. We have

$$\chi(\boldsymbol{X},\tau) = \chi(\boldsymbol{X},t) + (\tau - t)\dot{\chi}(\boldsymbol{X},t) + \cdots \qquad (5.57)$$

and

$$T(\boldsymbol{X},\tau) = T(\boldsymbol{X},t) + (\tau - t)\dot{T}(\boldsymbol{X},t) + \cdots \qquad (5.58)$$

with the notation

$$\dot{\chi} = \left.\frac{\partial \chi}{\partial t}\right|_{\boldsymbol{X}} , \qquad \dot{T} = \left.\frac{\partial T}{\partial t}\right|_{\boldsymbol{X}} . \qquad (5.59)$$

Recall that the first relation of (5.59) is identical to (2.17). To obtain the principle of regular memory, we suppose that the functionals are as regular as possible in order to smooth the discontinuities in these functions and their time derivatives. Consequently, the axiom of regular memory leads to the replacement of (5.56) by

$$\begin{aligned}\boldsymbol{\sigma}(\boldsymbol{X},t) = \boldsymbol{F}\mathcal{L}\Big(&\boldsymbol{C},\dot{\boldsymbol{C}},\ddot{\boldsymbol{C}},\ldots;\\ &\rho^{-1},\dot{\rho},\ddot{\rho},\ldots;T,\dot{T},\ddot{T},\ldots;\frac{\partial T}{\partial \boldsymbol{X}},\frac{\partial \dot{T}}{\partial \boldsymbol{X}},\ldots,\boldsymbol{X}\Big)\boldsymbol{F}^T,\end{aligned} \tag{5.60}$$

where the time derivatives of various variables appear.

The ***concept of fading memory*** makes use of functions that vanish over time, introduced in the constitutive equation in order to give more weight to recent events near the present time $t$ and less to the distant past. Typically these functions are written in a form like $e^{-\beta s}$ with $\beta$ constant. They are especially useful in viscoelasticity to take into account the phenomena of creep, that is, strain under constant stress; and stress relaxation, strain maintained constant under time variable stress. Although the subject is important for a large class of materials, it is beyond the introductory scope of this book.

## 5.7 Exercises

**5.1** Let $\boldsymbol{u}(\boldsymbol{x},t)$ be an objective vector field. Show that its spatial gradient is also objective, i.e., that it satisfies

$$(\nabla \boldsymbol{u})^* = \boldsymbol{Q}\,\nabla \boldsymbol{u}\,\boldsymbol{Q}^T, \tag{5.61}$$

where $(\nabla \boldsymbol{u})^* = \partial \boldsymbol{u}^*/\partial \boldsymbol{x}^*$ denotes the spatial gradient of the vector $\boldsymbol{u}^*$.

**5.2** Prove that the strain rate tensor $\boldsymbol{d}$ (and the rotation rate $\dot{\boldsymbol{\omega}}$) is (is not) objective.

**5.3** Let $\boldsymbol{T}$ be an arbitrary spatially objective tensor of order 2. Is the material derivative of $\boldsymbol{T}$ objective?

**5.4** Let $\boldsymbol{T}$ be an arbitrary spatially objective tensor of order 2. Prove that the expression

$$\dot{\boldsymbol{T}} + \boldsymbol{T}\dot{\boldsymbol{\omega}} - \dot{\boldsymbol{\omega}}\boldsymbol{T} \tag{5.62}$$

is objective, where $\dot{\boldsymbol{T}}$ denotes the material derivative of $\boldsymbol{T}$ and $\dot{\boldsymbol{\omega}}$ the rotation rate tensor.

**5.5** Prove that the tensor $\boldsymbol{T}$ of order 2, defined by the relation

$$\boldsymbol{T} = 2\dot{\boldsymbol{d}} + 2\boldsymbol{d}\boldsymbol{L} + 2\boldsymbol{L}^T\boldsymbol{d} \tag{5.63}$$

is spatially objective. The tensor $\boldsymbol{d}$ is the rate of deformation tensor and $\dot{\boldsymbol{d}}$ is its material derivative. The tensor $\boldsymbol{L}$ is that for the velocity gradient. For the proof, use the equation (also to be proven)

$$\dot{\boldsymbol{d}}^* = \dot{\boldsymbol{Q}}\boldsymbol{d}\boldsymbol{Q}^T + \boldsymbol{Q}\boldsymbol{d}\dot{\boldsymbol{Q}}^T + \boldsymbol{Q}\dot{\boldsymbol{d}}\boldsymbol{Q}^T\,. \tag{5.64}$$

CHAPTER 6

# Classical Constitutive Equations

## 6.1 Introduction

In this chapter we will examine the classical constitutive laws of Newtonian viscous fluids, elastic and hyperelastic solids, and heat conduction. We will evaluate some of these constitutive equations in the context of the second principle of thermodynamics to verify that their formulations satisfy that inequality. The section on the thermodynamics of an ideal fluid establishes the methodological link between thermodynamics of continuous media and classical thermodynamics in order to show that these two points of view are complementary. This chapter ends with some considerations on the subject of thermoelasticity.

For the behavior of fluids, additional reading is proposed in [2, 12, 24, 38, 42, 57, 59, 60]. For solids, the reader is referred to [4, 6, 7, 15, 20, 32, 36, 41, 42, 47, 55].

## 6.2 Simple Fluids

In general, we can say that a fluid is a continuous medium such that in any deformed configuration that leaves the density unchanged, the fluid retains no memory of its past states. We can then propose the following definition: a fluid is a material such that each configuration of the body that leaves the density at a prescribed value may be considered to be the reference configuration [12].

If we limit ourselves to first-order partial derivatives (with respect to time or space) in (5.60), we have the equation for simple thermomechanical materials

$$\boldsymbol{\sigma}(\boldsymbol{X},t) = \boldsymbol{F}\mathcal{L}\left(\boldsymbol{C},\, \dot{\boldsymbol{C}},\, \rho^{-1},\, \dot{\rho},\, T, \dot{T},\, \frac{\partial T}{\partial \boldsymbol{X}},\, \frac{\partial \dot{T}}{\partial \boldsymbol{X}},\, \boldsymbol{X}\right)\boldsymbol{F}^T . \tag{6.1}$$

Since every configuration can be a reference configuration, we choose the current deformed configuration as the reference configuration and we can write $\boldsymbol{x} = \boldsymbol{X} = \chi$ with $\rho$ given. From (2.179) and (2.181), it follows that

$$\boldsymbol{F} = \boldsymbol{I} \qquad \boldsymbol{C} = \boldsymbol{I} \qquad \dot{\boldsymbol{C}} = \dot{\boldsymbol{F}}^T\boldsymbol{F} + \boldsymbol{F}^T\dot{\boldsymbol{F}} = 2\boldsymbol{d} \tag{6.2}$$

$$\frac{\partial T}{\partial \boldsymbol{X}} = \frac{\partial T}{\partial \boldsymbol{x}} \qquad \text{and} \qquad \frac{\partial \dot{T}}{\partial \boldsymbol{X}} = \frac{\partial \dot{T}}{\partial \boldsymbol{x}} . \tag{6.3}$$

Using the equation of conservation of mass, $\dot{\rho} = -\rho\, d_{ii}$ (3.41), we eliminate the dependence with respect to $\dot{\rho}$. Relation (6.1) thus becomes

$$\boldsymbol{\sigma}(\boldsymbol{x},t) = \boldsymbol{K}\left(\boldsymbol{d},\, \rho^{-1},\, T,\, \dot{T},\, \frac{\partial T}{\partial \boldsymbol{x}},\, \frac{\partial \dot{T}}{\partial \boldsymbol{x}},\, \boldsymbol{x}\right). \tag{6.4}$$

In the following, we assume that the fluid behavior is independent of the temperature gradient $\partial T/\partial \boldsymbol{x}$ and of its temporal variation $\partial \dot{T}/\partial \boldsymbol{x}$. This simplification is consistent with experimental observations of fluid behavior. Equation (6.4) becomes

$$\boldsymbol{\sigma}(\boldsymbol{x},t) = \boldsymbol{K}\big(\boldsymbol{d},\, \rho^{-1},\, T,\, \dot{T},\, \boldsymbol{x}\big). \tag{6.5}$$

The steps from (6.1) to (6.5) caused the objective character of (6.1) to disappear. We can restore it by applying the principle of objectivity to (6.5) which shows that the functional $\boldsymbol{K}$ does not depend explicitly on $\boldsymbol{x}$. We consider that the stress depends on the instantaneous value of the temperature, and that the history term, $\dot{T}$, is obtained from the principle of conservation of energy. Consequently, a fluid is a medium where the constitutive law is of the form

$$\boldsymbol{\sigma}(\boldsymbol{x},t) = \boldsymbol{K}\big(\boldsymbol{d}, \rho^{-1}, T\big)\ , \tag{6.6}$$

with the condition imposed by objectivity

$$\boldsymbol{Q}\boldsymbol{K}\boldsymbol{Q}^T = \boldsymbol{K}\big(\boldsymbol{Q}\boldsymbol{d}\boldsymbol{Q}^T, \rho^{-1}, T\big). \tag{6.7}$$

Condition (6.7) imposes that the functional, now reduced to the function $\boldsymbol{K}$, be an isotropic function of the symmetric tensor $\boldsymbol{d}$. By application of the representation theorem (sec. 1.3.11) for isotropic functions of symmetric tensors, relation (6.7) becomes

$$\boldsymbol{\sigma}(\boldsymbol{x},t) = K_0\boldsymbol{I} + K_1\boldsymbol{d} + K_2\boldsymbol{d}^2. \tag{6.8}$$

The scalar functions $K_i$ $(i = 0, 1, 2)$ will be functions of the invariants of $\boldsymbol{d}$, $\rho^{-1}$ and $T$.

When the fluid is incompressible, its density is invariant. Thus, $\rho_0 = \rho$ and $\det \boldsymbol{F} = 1$. In addition, the first invariant of $\boldsymbol{d}$ is zero. Therefore, the constitutive equation of an *incompressible fluid* has the form

$$\boldsymbol{\sigma}(\boldsymbol{x},t) = -p\boldsymbol{I} + K_1\big(\mathrm{I}_2(\boldsymbol{d}), \mathrm{I}_3(\boldsymbol{d})\big)\boldsymbol{d} + K_2\big(\mathrm{I}_2(\boldsymbol{d}), \mathrm{I}_3(\boldsymbol{d})\big)\boldsymbol{d}^2\ , \tag{6.9}$$

with $p$, the undetermined scalar pressure. We note that when the fluid is at rest, $\boldsymbol{d} = 0$ and $\boldsymbol{\sigma} = -p\boldsymbol{I}$. The behavior of the fluid is obtained from hydrostatic equilibrium (eqn. (3.133)). Note also that the dependence of stress on velocity is a function only of the symmetric strain rate tensor. This is a direct consequence of the principle of objectivity which excludes the use of the rate of rotation or vorticity tensor, as the latter is not objective and is also antisymmetric.

We notice that equation (6.9) is non-linear in $\boldsymbol{d}$. It is in this case a non-Newtonian fluid as opposed to a classical Newtonian fluid. The relation describes a fluid known by the name of a Rivlin-Ericksen fluid of complexity one (see [59]).

## 6.3 Classical Fluids or Newtonian Viscous Fluids

A ***classical fluid*** is a medium for which the components of the stress tensor are *linear* functions of the strain rate tensor. This definition imposes $K_2 = 0$ in equation (6.9). We note the particular case of *perfect fluids*, or *inviscid fluids*, for which, by definition, the stress tensor is independent of $\boldsymbol{d}$. The stress tensor is thus spherical (non-diagonal components are zero). We write

$$\boldsymbol{\sigma} = -p\boldsymbol{I} \tag{6.10}$$

$$p = p(\rho, T)\,. \tag{6.11}$$

In (6.11), we have used the definition that $1/\rho = v$ with $v$ being the specific volume as encountered in the laws of the physics of gases. If the fluid is *incompressible*, the pressure is determined by the solution of the equations of motion. Conversely, if the fluid is compressible, the pressure is given by a state equation derived from thermodynamic considerations. If the fluid is not perfect, it is called viscous. We pose

$$\boldsymbol{\sigma} = -p\boldsymbol{I} + \boldsymbol{T}\,, \tag{6.12}$$

with $\boldsymbol{T}$, the deviatoric part of the stress tensor that is called the *extra-stress*, such that $\boldsymbol{T} = K_1\big(\mathrm{I}_1(\boldsymbol{d}), \mathrm{I}_2(\boldsymbol{d}), \mathrm{I}_3(\boldsymbol{d})\big)\boldsymbol{d}$.

For classical fluids, $\boldsymbol{T}$ is a linear function of $\boldsymbol{d}$ and it can be shown that $\boldsymbol{T}$ is necessarily of the form

$$\boldsymbol{T} = \lambda\,\mathrm{tr}\,\boldsymbol{d}\boldsymbol{I} + 2\mu\boldsymbol{d}\,, \tag{6.13}$$

where $\lambda$ and $\mu$ are scalar coefficients. The tensor $\boldsymbol{T}$ depends on the strain rate tensor (and not the strain tensor) in a way such that it vanishes when the fluid is in rigid body motion (that is, zero deformation rate). The pressure is a scalar field which does not depend explicitly on the strain rate. The coefficient $\lambda$ is the *volume viscosity*, while $\mu$ is the coefficient of *shear or dynamic viscosity.* We will see that $\lambda$ and $\mu$ are always positive. These coefficients have dimensions $\mathrm{M\,L^{-1}T^{-1}}$ and the corresponding SI units are $\mathrm{N\,s/m^2}$ or $\mathrm{Pa\cdot s}$.

**Table 6.1** Material constants for viscous fluids

| | $\mu$ (Pa·s) | $\rho$ (kg/m$^3$) | $\nu$(m$^2$ s$^{-1}$) |
|---|---|---|---|
| air | $1.776\,10^{-5}$ | 1.225 | $14.5\,10^{-6}$ |
| water | 0.0011 | 999.2 | $1.138\,10^{-6}$ |

For example, in table 6.1 we give the material constants at room temperature of two fluids widely used in industrial applications: air and water. The *kinematic viscosity* $\nu$ is defined by $\nu = \mu/\rho$. In index notation, the constitutive equation for a compressible Newtonian fluid is written as

$$\sigma_{ij} = -p\,\delta_{ij} + \lambda\,d_{kk}\,\delta_{ij} + 2\mu\,d_{ij}\ , \tag{6.14}$$

with

$$\begin{aligned} p &= p(\rho, T) \\ \lambda &= \lambda(\rho, T) \\ \mu &= \mu(\rho, T) \\ d_{ij} &= \frac{1}{2}\left(\frac{\partial v_i}{\partial x_j} + \frac{\partial v_j}{\partial x_i}\right). \end{aligned} \tag{6.15}$$

The constitutive equation of an incompressible viscous fluid reduces to

$$\sigma_{ij} = -p\,\delta_{ij} + 2\mu\,d_{ij}\ , \tag{6.16}$$

with $p$ being an undetermined scalar field and $\mu$ is a constant in most cases. Note that taking the trace of (6.16), we obtain $p = -(1/3)\,\mathrm{tr}\,\boldsymbol{\sigma}$ and thus that the pressure is the mean of the diagonal components of the stress tensor.

EXAMPLE 6.1 (Simple Shear Flow)
Consider a simple shear flow established between two parallel walls. The lower wall is fixed and the upper wall moves along its plane at a constant velocity $U$. The velocity field is such that

$$\begin{aligned} v_1 &= kx_2 \\ v_2 &= 0 \\ v_3 &= 0\,. \end{aligned}$$

Only the component $d_{12}$ of the tensor $\boldsymbol{d}$ is non-zero. Then, at an arbitrary point $M$ in the fluid, as shown in figure 6.1, we have $\sigma_{22} = -p$ and $\sigma_{12} = \mu k$. The fluid above point $M$ exerts on the fluid below a shear force proportional to $\mu$ and to the velocity gradient $k$ in the direction $x_2$. Due to the presence of viscosity, the fastest upper layers of fluid tend to speed up the lower layers that they are in contact with; inversely, the slower moving layers tend to slow down the faster flowing layers.

This physical interpretation confirms that it is reasonable to suppose that $\mu > 0$. This also shows that in a perfect fluid the different layers of fluid exert no accelerating or decelerating influence on each other. They only experience the forces due to pressure.

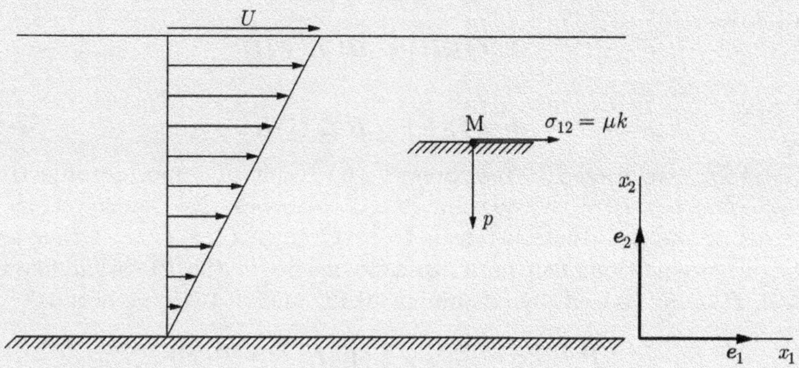


**Fig. 6.1** Shear flow

It is understood that we have presented a drastic idealization of reality as all real fluids are viscous and they may also be turbulent. When they are slightly viscous, we can model them using the concept of a perfect fluid.

## 6.4 Isothermal Isotropic Elastic Media

We have seen previously that a solid has a special memory: it remembers its reference configuration.

An ***isothermal elastic medium*** is a simple material such that, by (5.36),

$$\boldsymbol{\sigma}(\boldsymbol{X}, t) = \boldsymbol{\Sigma}\big(\boldsymbol{F}(\boldsymbol{X}, t-s), \boldsymbol{X}\big)\,. \tag{6.17}$$

Taking into account the fact that $\boldsymbol{F}$ is the deformation gradient tensor that relates the reference state to the deformed state, for an elastic material, we can write

$$\boldsymbol{\sigma}(\boldsymbol{X}, t) = \boldsymbol{\Sigma}\big(\boldsymbol{F}(t)\big)\,. \tag{6.18}$$

To simplify, in the following we consider that the dependence of $\boldsymbol{\sigma}$ on time $t$ is carried by $\boldsymbol{F}$ and we will no longer explicitly note $t$ in $\boldsymbol{\sigma}$. The principle of objectivity requires

$$\boldsymbol{\sigma}^* = \boldsymbol{\Sigma}\big(\boldsymbol{F}^*\big)\,. \tag{6.19}$$

Using (2.205) and (3.147), we obtain

$$\boldsymbol{\sigma}^* = \boldsymbol{\Sigma}\big(\boldsymbol{F}^*\big) = \boldsymbol{\Sigma}(\boldsymbol{QF}) \qquad \boldsymbol{\sigma}^* = \boldsymbol{Q}\boldsymbol{\sigma}\boldsymbol{Q}^T = \boldsymbol{Q}\boldsymbol{\Sigma}(\boldsymbol{F})\boldsymbol{Q}^T \tag{6.20}$$

or

$$\boldsymbol{\Sigma}(\boldsymbol{QF}) = \boldsymbol{Q}\boldsymbol{\Sigma}(\boldsymbol{F})\boldsymbol{Q}^T\,. \tag{6.21}$$

With the polar decomposition theorem and by setting $\boldsymbol{Q} = \boldsymbol{R}^T$, this last equation becomes

$$\boldsymbol{\Sigma}(\boldsymbol{QRU}) = \boldsymbol{R}^T \boldsymbol{\Sigma}(\boldsymbol{F})\boldsymbol{R} \tag{6.22}$$

and

$$\boldsymbol{\sigma} = \boldsymbol{\Sigma}(\boldsymbol{F}) = \boldsymbol{R}\boldsymbol{\Sigma}(\boldsymbol{U})\boldsymbol{R}^T \tag{6.23}$$

$\forall \boldsymbol{F}$ and $\boldsymbol{R}$. Relation (6.23) expresses the result of imposing objectivity on (6.18). The constitutive equation (6.23) expresses the Cauchy stress tensor in terms of the deformation tensor $\boldsymbol{U}$ or $\boldsymbol{C}$ since $\boldsymbol{C} = \boldsymbol{U}^2$ . Other forms of constitutive equations can be written by means of the Piola-Kirchhoff stress tensors $\boldsymbol{P}$ or $\boldsymbol{S}$. Indeed, by combining (6.18) and (3.141), we obtain

$$\boldsymbol{P} = J\boldsymbol{\sigma}\boldsymbol{F}^{-T} = J\boldsymbol{\Sigma}(\boldsymbol{F})\boldsymbol{F}^{-T} = \boldsymbol{\mathcal{P}}(\boldsymbol{F}), \tag{6.24}$$

where $\boldsymbol{\mathcal{P}}$ is a tensor function of $\boldsymbol{F}$. Following steps similar to those that led to (6.23) and taking into account (3.149), we obtain

$$\boldsymbol{\mathcal{P}}(\boldsymbol{F}) = \boldsymbol{R}\boldsymbol{\mathcal{P}}(\boldsymbol{U}). \tag{6.25}$$

To write the constitutive equation as a function of the tensor $\boldsymbol{S}$, we introduce (6.23) in (3.152) and we proceed in a similar way to arrive at

$$\boldsymbol{S} = J\boldsymbol{F}^{-1}\boldsymbol{\sigma}\boldsymbol{F}^{-T} = J\boldsymbol{F}^{-1}\boldsymbol{R}\boldsymbol{\Sigma}(\boldsymbol{U})\boldsymbol{R}^T\boldsymbol{F}^{-T}. \tag{6.26}$$

Employing the polar decomposition theorem, this last relation can be written as

$$\boldsymbol{S} = J\boldsymbol{U}^{-1}\boldsymbol{\Sigma}(\boldsymbol{U})\boldsymbol{U}^{-T} \tag{6.27}$$

or

$$\boldsymbol{S} = \boldsymbol{\mathcal{S}}(\boldsymbol{U}), \tag{6.28}$$

where $\boldsymbol{\mathcal{S}}$ is a tensor function of $\boldsymbol{U}$. Since $\boldsymbol{U}^* = \boldsymbol{U}$ and $\boldsymbol{S}^* = \boldsymbol{S}$, the conditions imposed by objectivity are satisfied. Equation (6.28) can be expressed as a function of the Cauchy-Green tensor $\boldsymbol{C}$ since $\boldsymbol{C} = \boldsymbol{U}^2$.

In the following we will construct a theory of isothermal, isotropic elastic media. If the material is isotropic with respect to its reference configuration, by the material invariance principle, we must have (sec. 5.2.4)

$$\boldsymbol{\Sigma}(\boldsymbol{F}) = \boldsymbol{\Sigma}(\overline{\boldsymbol{F}}), \tag{6.29}$$

where $\overline{\boldsymbol{F}}$ is the deformation gradient tensor calculated for a material coordinate system

$$\overline{\boldsymbol{X}} = \boldsymbol{O}\boldsymbol{X} + \boldsymbol{B}, \tag{5.30}$$

a relation in which $\{\boldsymbol{O}\}$ and $\{\boldsymbol{B}\}$ take into account, respectively, the symmetries and the translations of the material axes (sec. 5.2.4). The deformation gradient tensor $\overline{\boldsymbol{F}}$, according to (5.30), can be expressed as

$$\overline{\boldsymbol{F}} = \frac{\partial \boldsymbol{x}}{\partial \boldsymbol{X}}\frac{\partial \boldsymbol{X}}{\partial \overline{\boldsymbol{X}}} = \boldsymbol{F}\boldsymbol{O}^T. \tag{6.30}$$

Combining (6.29) and (6.30), leads to

$$\boldsymbol{\Sigma}(\boldsymbol{F}) = \boldsymbol{\Sigma}\big(\boldsymbol{F}\boldsymbol{O}^T\big)\,. \tag{6.31}$$

The polar decomposition theorem yields the following solution: $\boldsymbol{F} = \boldsymbol{V}\boldsymbol{R}$ $(= \boldsymbol{R}\boldsymbol{U})$. To make $\boldsymbol{V}$ appear in the functional $\boldsymbol{\Sigma}$, we must choose $\boldsymbol{O}$ such that $\boldsymbol{F}\boldsymbol{O}^T = \boldsymbol{V}$, thus $\boldsymbol{O} = \boldsymbol{R}$. We see from the polar decomposition that this is an appropriate choice and we obtain, according to (6.18),

$$\boldsymbol{\sigma} = \boldsymbol{\Sigma}(\boldsymbol{V})\,. \tag{6.32}$$

From objectivity (5.28) and isotropy (6.31), for every orthogonal matrix $\boldsymbol{O}$, we have

$$\boldsymbol{O}\boldsymbol{\Sigma}(\boldsymbol{U})\boldsymbol{O}^T = \boldsymbol{\Sigma}(\boldsymbol{O}\boldsymbol{U}) = \boldsymbol{\Sigma}\big(\boldsymbol{O}\boldsymbol{U}\boldsymbol{O}^T\big) = \boldsymbol{\Sigma}\big(\boldsymbol{F}\boldsymbol{O}^T\big) = \boldsymbol{\Sigma}(\boldsymbol{V})\,. \tag{6.33}$$

We conclude that the functional $\boldsymbol{\Sigma}$ is an isotropic function of $\boldsymbol{V}$. For an isotropic material, constitutive equations (6.18) and (6.32) can also be written in other forms.

Taking into account that $\boldsymbol{V}^2 = \boldsymbol{F}\boldsymbol{F}^T$ (eqn. (1.134)), the constitutive law of an isotropic elastic material can be expressed by one of the equalities

$$\boldsymbol{\sigma} = \boldsymbol{\Sigma}(\boldsymbol{V}) \qquad \boldsymbol{\sigma} = \boldsymbol{\mathcal{H}}(\boldsymbol{V}^2) \qquad \boldsymbol{\sigma} = \boldsymbol{\mathcal{K}}(\boldsymbol{e})\,, \tag{6.34}$$

where $\boldsymbol{e}$ is the Euler-Almansi tensor defined by (2.83). As the tensor functions $\boldsymbol{\Sigma}$ and $\boldsymbol{\mathcal{H}}$ are *isotropic* and depend on symmetric tensors, we can write, for example (see (1.140))

$$\begin{aligned} \boldsymbol{\sigma} = {} & k_0\big(\mathrm{I}_1(\boldsymbol{e}), \mathrm{I}_2(\boldsymbol{e}), \mathrm{I}_3(\boldsymbol{e})\big)\boldsymbol{I} + k_1\big(\mathrm{I}_1(\boldsymbol{e}), \mathrm{I}_2(\boldsymbol{e}), \mathrm{I}_3(\boldsymbol{e})\big)\boldsymbol{e} \\ & + k_2\big(\mathrm{I}_1(\boldsymbol{e}), \mathrm{I}_2(\boldsymbol{e}), \mathrm{I}_3(\boldsymbol{e})\big)\boldsymbol{e}^2 \end{aligned} \tag{6.35}$$

with

$$k_p = k_p\big(\mathrm{I}_1(\boldsymbol{e}), \mathrm{I}_2(\boldsymbol{e}), \mathrm{I}_3(\boldsymbol{e})\big) \qquad p = 0, 1, 2\ ,$$

which are scalar functions of the invariants of the tensor $\boldsymbol{e}$. We can also notice that the stress tensor $\boldsymbol{\sigma}$ and the strain tensor $\boldsymbol{e}$ have the same principal directions. In addition, relation (6.35) shows that for an isotropic material, three parameters are necessary to characterize the constitutive response.

## 6.5 Hyperelastic Materials

Relation (6.35), presented in the preceding section, is the most general form of the constitutive equation for an isotropic elastic material. We obtained this constitutive equation from purely theoretical considerations without any reference to thermodynamics. This theory of elasticity, developed around (6.35), is called *Cauchy elasticity* in the literature and the corresponding material is a ***Cauchy elastic material***. In this section, we develop a theory of constitutive equations based on the hypothesis of the existence of an energy function. The theory is adapted to non-linear elastic materials where the deformations can be large, that is, finite. It is normally called finite hyperelasticity or simply ***hyperelasticity***. The materials that it describes are called hyperelastic, or ***Green elastic materials***.

Assuming that the processes are isothermal and considering only mechanical effects, we introduce the free energy function per unit volume in the material description. Thus, we define the ***energy function*** $\mathcal{W}(\boldsymbol{X},t)$ such that

$$\mathcal{W}(\boldsymbol{X},t) = P_0(\boldsymbol{X})U(\boldsymbol{X},t)\,. \tag{6.36}$$

From (4.44), we obtain

$$\frac{D\mathcal{W}(\boldsymbol{X},t)}{Dt} = \dot{\mathcal{W}}(\boldsymbol{X},t) = P_0(\boldsymbol{X})\dot{U}(\boldsymbol{X},t) = \boldsymbol{P} : \dot{\boldsymbol{F}}\,. \tag{6.37}$$

A hyperelastic, or Green elastic, material is one for which the elastic energy is given by the free energy function such that

$$\mathcal{W}(\boldsymbol{X},t) = P_0(\boldsymbol{X})U\big(\boldsymbol{F}(\boldsymbol{X},t),\boldsymbol{X}\big)\,, \tag{6.38}$$

which, for a homogeneous material, becomes

$$\mathcal{W}(\boldsymbol{X},t) = \mathcal{W}\big(\boldsymbol{F}(\boldsymbol{X},t),\boldsymbol{X}\big) = \mathcal{W}(\boldsymbol{F})\,. \tag{6.39}$$

The free energy is zero for the reference configuration, that is $\mathcal{W}(\boldsymbol{I}) = 0$, and also satisfies the condition $\mathcal{W}(\boldsymbol{F}) \geq 0$.

In order to establish the relation between the energy and the stresses, we proceed as follows. First, we express the time derivative of $\mathcal{W}(\boldsymbol{F})$ defined in (1.167)

$$\dot{\mathcal{W}}(\boldsymbol{F}) = \frac{D\mathcal{W}(\boldsymbol{F})}{Dt} = \frac{D\mathcal{W}(\boldsymbol{F})}{D\boldsymbol{F}} : \frac{D\boldsymbol{F}}{Dt} = \frac{\partial \mathcal{W}(\boldsymbol{F})}{\partial \boldsymbol{F}} : \dot{\boldsymbol{F}}\,. \tag{6.40}$$

Then, we combine (6.40) with (6.37) to obtain

$$\frac{\partial \mathcal{W}(\boldsymbol{F})}{\partial \boldsymbol{F}} : \dot{\boldsymbol{F}} = \boldsymbol{P} : \dot{\boldsymbol{F}} \qquad \text{or} \qquad \left(\frac{\partial \mathcal{W}(\boldsymbol{F})}{\partial \boldsymbol{F}} - \boldsymbol{P}\right) : \dot{\boldsymbol{F}} = 0\,. \tag{6.41}$$

Relation (6.41) is valid for arbitrary values of $\dot{\boldsymbol{F}}$. Thus, for a hyperelastic material, the constitutive equation is written as

$$\boldsymbol{P} = \frac{\partial \mathcal{W}(\boldsymbol{F})}{\partial \boldsymbol{F}}\,. \tag{6.42}$$

Strictly speaking, both ***linear*** and ***non-linear*** elastic materials are hyperelastic. However, the name hyperelastic is conventionally used for non-linear elastic behavior.

Expressing $\boldsymbol{\sigma}$ as a function of $\boldsymbol{P}$ as in (3.141), the constitutive equation becomes

$$\boldsymbol{\sigma} = J^{-1}\frac{\partial \mathcal{W}(\boldsymbol{F})}{\partial \boldsymbol{F}}\,\boldsymbol{F}^T\,. \tag{6.43}$$

We assume that the strain energy function is independent of the reference frame. This implies that for two observers in relative motion as described by (5.13), we have

$$\mathcal{W}(\boldsymbol{F}) = \mathcal{W}(\boldsymbol{F}^\star) = \mathcal{W}(\boldsymbol{QF}) . \tag{6.44}$$

Replacing $\boldsymbol{F}$ by its right polar decomposition $\boldsymbol{F} = \boldsymbol{RU}$ and setting $\boldsymbol{Q} = \boldsymbol{R}^T$, we obtain

$$\mathcal{W}(\boldsymbol{F}) = \mathcal{W}(\boldsymbol{R}^T\boldsymbol{RU}) = \mathcal{W}(\boldsymbol{U}) , \tag{6.45}$$

which expresses the necessary and sufficient conditions for the objectivity of the energy function $\mathcal{W}(\boldsymbol{F})$. Recalling that $\boldsymbol{U} = \boldsymbol{C}^{1/2}$, we can write

$$\mathcal{W}(\boldsymbol{F}) = \mathcal{W}(\boldsymbol{U}) = \widehat{\mathcal{W}}(\boldsymbol{C}) . \tag{6.46}$$

In solid mechanics, we focus our attention on the formulation of the constitutive relations as a function of the metric tensor $\boldsymbol{C}$. It is thus necessary to express $\partial\mathcal{W}(\boldsymbol{F})/\partial\boldsymbol{F}$ in (6.42) or (6.43) as a function of $\boldsymbol{C}$. Differentiating (6.46) with respect to time, leads to

$$\frac{\partial\mathcal{W}(\boldsymbol{F})}{\partial\boldsymbol{F}} : \dot{\boldsymbol{F}} = \frac{\partial\widehat{\mathcal{W}}(\boldsymbol{C})}{\partial\boldsymbol{C}} : \dot{\boldsymbol{C}} . \tag{6.47}$$

Taking into account $\boldsymbol{C} = \boldsymbol{C}^T = \boldsymbol{F}^T\boldsymbol{F}$ and using (1.95) it can be shown that

$$\left(\frac{\partial\mathcal{W}(\boldsymbol{F})}{\partial\boldsymbol{F}}\right)^T = 2\frac{\partial\widehat{\mathcal{W}}(\boldsymbol{C})}{\partial\boldsymbol{C}}\boldsymbol{F}^T . \tag{6.48}$$

Due to the symmetry of $\boldsymbol{C}$, $\partial\widehat{\mathcal{W}}(\boldsymbol{C})/\partial\boldsymbol{C}$ is also symmetric. Consequently, we have

$$\frac{\partial\mathcal{W}(\boldsymbol{F})}{\partial\boldsymbol{F}} = 2\boldsymbol{F}\frac{\partial\widehat{\mathcal{W}}(\boldsymbol{C})}{\partial\boldsymbol{C}} , \tag{6.49}$$

and we write (6.42) in the form

$$\boldsymbol{P} = 2\boldsymbol{F}\frac{\partial\widehat{\mathcal{W}}(\boldsymbol{C})}{\partial\boldsymbol{C}} . \tag{6.50}$$

Inserting (6.50) in (3.152), the constitutive equation as a function of the second Piola-Kirchhoff tensor becomes

$$\boldsymbol{S} = 2\frac{\partial\widehat{\mathcal{W}}(\boldsymbol{C})}{\partial\boldsymbol{C}} . \tag{6.51}$$

Equations (6.50) and (6.51) represent the general forms of the constitutive equations for a hyperelastic material that satisfies objectivity.

### 6.5.1 Isotropic Hyperelastic Materials

Let us examine the constitutive equations of an isotropic elastic medium. Since the symmetry of the material is not taken into account in the preceding equations, we will now study its consequences on the constitutive equation. According to (6.30), the strain energy must satisfy the relation

$$\mathcal{W}(\boldsymbol{F}) = \mathcal{W}(\overline{\boldsymbol{F}}) = \mathcal{W}(\boldsymbol{FO}^T) . \tag{6.52}$$

As $\boldsymbol{O}$ can be identified with $\boldsymbol{Q}$ (sec. 5.2.4), we can also have $\overline{\boldsymbol{F}} = \boldsymbol{F}\boldsymbol{Q}^T$, so that

$$\overline{\boldsymbol{C}} = \overline{\boldsymbol{F}}^T\overline{\boldsymbol{F}} = \boldsymbol{Q}\boldsymbol{F}^T\boldsymbol{F}\boldsymbol{Q}^T = \boldsymbol{Q}\boldsymbol{C}\boldsymbol{Q}^T . \tag{6.53}$$

Taking into account (6.46), a material undergoing the transformation (5.30) satisfies the following equalities:

$$\mathcal{W}(\overline{\boldsymbol{F}}) = \widehat{\mathcal{W}}(\overline{\boldsymbol{C}}) = \widehat{\mathcal{W}}(\boldsymbol{Q}\boldsymbol{C}\boldsymbol{Q}^T) \tag{6.54}$$

$$\widehat{\mathcal{W}}(\boldsymbol{C}) = \widehat{\mathcal{W}}(\boldsymbol{Q}\boldsymbol{C}\boldsymbol{Q}^T) . \tag{6.55}$$

Relation (6.55) is the isotropic condition for the strain energy function (see eq. (1.141)). It also implies that $\widehat{\mathcal{W}}(\boldsymbol{C})$ is an isotropic scalar function of the symmetric tensor $\boldsymbol{C}$. Because of the representation theorem for invariants [17], the scalar function (6.55) can be written as a function of the principal invariants of its argument $\boldsymbol{C}$. We can thus replace (6.55) with the function

$$\widehat{\mathcal{W}}(\boldsymbol{C}) = \Phi\big(I_1(\boldsymbol{C}),\, I_2(\boldsymbol{C}),\, I_3(\boldsymbol{C})\big) . \tag{6.56}$$

Since the principal values of $\boldsymbol{C}$ are $\lambda_1^2$, $\lambda_2^2$, $\lambda_3^2$, the corresponding invariants are given by

$$\begin{aligned} I_1(\boldsymbol{C}) &= \lambda_1^2 + \lambda_2^2 + \lambda_3^2 \\ I_2(\boldsymbol{C}) &= \lambda_1^2\lambda_2^2 + \lambda_2^2\lambda_3^2 + \lambda_3^2\lambda_1^2 \\ I_3(\boldsymbol{C}) &= \lambda_1^2\lambda_2^2\lambda_3^2 . \end{aligned} \tag{6.57}$$

To simplify, we use the same symbol for the energy function $\Phi$, and the invariants (6.57) will be specified without reference to the tensor $\boldsymbol{C}$.

Inspecting function (6.56) and relations (6.50) and (6.51), the next step in the formulation of the isotropic material constitutive equations is to differentiate $\Phi(\boldsymbol{C})$ with respect to the invariants (6.57). Assuming that $\Phi(\boldsymbol{C})$ has continuous derivatives with respect to the invariants, we have

$$\frac{\partial\Phi(\boldsymbol{C})}{\partial\boldsymbol{C}} = \frac{\partial\Phi}{\partial I_1}\frac{\partial I_1}{\partial\boldsymbol{C}} + \frac{\partial\Phi}{\partial I_2}\frac{\partial I_2}{\partial\boldsymbol{C}} + \frac{\partial\Phi}{\partial I_3}\frac{\partial I_3}{\partial\boldsymbol{C}} . \tag{6.58}$$

The derivatives of $I_i$ $(i = 1, 2, 3)$ with respect to $\boldsymbol{C}$ are given by

$$\frac{\partial I_1}{\partial\boldsymbol{C}} = \boldsymbol{I} , \qquad \frac{\partial I_2}{\partial\boldsymbol{C}} = I_1\boldsymbol{I} - \boldsymbol{C} , \qquad \frac{\partial I_3}{\partial\boldsymbol{C}} = I_3\boldsymbol{C}^{-1} . \tag{6.59}$$

EXAMPLE 6.2
For example, to prove the third equality, we proceed as follows. From relations (1.144) and (2.110), we write the successive relations

$$\begin{aligned}\frac{\partial I_3}{\partial \boldsymbol{C}} &= \sum_1^3 \frac{\partial(\lambda_1^2\lambda_2^2\lambda_3^2)}{\partial \lambda_i^2}(\boldsymbol{A}_i \otimes \boldsymbol{A}_i)\\ &= \lambda_2^2\lambda_3^2(\boldsymbol{A}_1 \otimes \boldsymbol{A}_1) + \lambda_1^2\lambda_3^2(\boldsymbol{A}_2 \otimes \boldsymbol{A}_2) + \lambda_2^2\lambda_1^2(\boldsymbol{A}_3 \otimes \boldsymbol{A}_3)\\ &= \lambda_1^2\lambda_2^2\lambda_3^2\lambda_1^{-2}(\boldsymbol{A}_1 \otimes \boldsymbol{A}_1) + \lambda_1^2\lambda_2^2\lambda_3^2\lambda_2^{-2}(\boldsymbol{A}_2 \otimes \boldsymbol{A}_2) + \lambda_1^2\lambda_2^2\lambda_3^2\lambda_3^{-2}(\boldsymbol{A}_3 \otimes \boldsymbol{A}_3)\\ &= I_3 \sum_1^3 \lambda_i^{-2}(\boldsymbol{A}_i \otimes \boldsymbol{A}_i) = I_3 \boldsymbol{C}^{-1}.\end{aligned}$$

By substituting (6.59) in (6.58), we have

$$\frac{\partial \Phi(\boldsymbol{C})}{\partial \boldsymbol{C}} = I_3 \frac{\partial \Phi}{\partial I_3}\boldsymbol{C}^{-1} + \left(\frac{\partial \Phi}{\partial I_1} + I_1 \frac{\partial \Phi}{\partial I_2}\right)\boldsymbol{I} - \frac{\partial \Phi}{\partial I_2}\boldsymbol{C}. \tag{6.60}$$

And inserting (6.60) in (6.51), the general form of the constitutive equation of a hyperelastic material becomes

$$\boldsymbol{S} = 2\left(I_3 \frac{\partial \Phi}{\partial I_3}\boldsymbol{C}^{-1} + \left(\frac{\partial \Phi}{\partial I_1} + I_1 \frac{\partial \Phi}{\partial I_2}\right)\boldsymbol{I} - \frac{\partial \Phi}{\partial I_2}\boldsymbol{C}\right). \tag{6.61}$$

Note that when the deformation is zero, $\boldsymbol{S} = \boldsymbol{0}$. In this case, $\boldsymbol{C} = \boldsymbol{I}$, or $\lambda_1 = \lambda_2 = \lambda_3 = 1$, or even, using the invariants of the tensor $\boldsymbol{C}$, $I_1 = 3$, $I_2 = 3$, $I_3 = 1$. In addition, for $\boldsymbol{S} = \boldsymbol{0}$ and $\boldsymbol{C} = \boldsymbol{I}$, the constitutive equation (6.61) leads to the following condition for the partial derivatives

$$\frac{\partial \Phi}{\partial I_1} + 2\frac{\partial \Phi}{\partial I_2} + \frac{\partial \Phi}{\partial I_3} = 0. \tag{6.62}$$

Combining (3.152) and (6.61), we obtain the the constitutive equation for the Cauchy stress

$$\boldsymbol{\sigma} = 2J^{-1}\left(I_3(\boldsymbol{c})\frac{\partial \Phi}{\partial I_3(\boldsymbol{c})}\boldsymbol{I} + \left(\frac{\partial \Phi}{\partial I_1(\boldsymbol{c})} + I_1(\boldsymbol{c})\frac{\partial \Phi}{\partial I_2(\boldsymbol{c})}\right)\boldsymbol{c} - \frac{\partial \Phi}{\partial I_2(\boldsymbol{c})}\boldsymbol{c}^2\right), \tag{6.63}$$

where $\boldsymbol{c}$ is the Cauchy-Green deformation tensor (2.89). We recall that the tensors $\boldsymbol{C}$ and $\boldsymbol{c}$ have the same ***principal stretches*** $\lambda_i^2\,(i = 1, 2, 3)$. Thus the corresponding invariants are also equal.

Consequently, when the energy function of a certain hyperelastic material is known, its constitutive relation is established either by (6.61) or by (6.63).

For an isotropic material, the strain energy (6.56) can also be written as a symmetric function of the principal stretches $\lambda_i$ $(i = 1, 2, 3)$

$$\widehat{\mathcal{W}}(\boldsymbol{C}) = \phi(\lambda_1, \lambda_2, \lambda_3)\,. \tag{6.64}$$

Differentiating (6.64), we obtain

$$\frac{\partial \widehat{\mathcal{W}}(\boldsymbol{C})}{\partial \boldsymbol{C}} = \frac{\partial \phi}{\partial \lambda_i^2} \frac{\partial \lambda_i^2}{\partial \boldsymbol{C}} = \frac{1}{2\lambda_i} \frac{\partial \phi}{\partial \lambda_i} \frac{\partial \lambda_i^2}{\partial \boldsymbol{C}}\,. \tag{6.65}$$

From the tensor analysis relation

$$\frac{\partial \lambda_i^2}{\partial \boldsymbol{C}} = \boldsymbol{A}_i \otimes \boldsymbol{A}_i\,, \tag{6.66}$$

where $\lambda_i^2$ are the ***principal values*** of $\boldsymbol{C}$, and $\boldsymbol{A}_i$ are the corresponding ***principal directions***, we obtain

$$\frac{\partial \widehat{\mathcal{W}}(\boldsymbol{C})}{\partial \boldsymbol{C}} = \sum_{i=1}^{3} \frac{1}{2\lambda_i} \frac{\partial \phi}{\partial \lambda_i} \boldsymbol{A}_i \otimes \boldsymbol{A}_i\,. \tag{6.67}$$

Combining this last relation with (6.51), we have

$$\boldsymbol{S} = \sum_{i=1}^{3} \frac{1}{\lambda_i} \frac{\partial \phi}{\partial \lambda_i} \boldsymbol{A}_i \otimes \boldsymbol{A}_i\,. \tag{6.68}$$

From the spectral decomposition of a tensor (1.125), we have for the principal values of the second Piola-Kirchhoff stress tensor $S_i$ $(i = 1, 2, 3)$,

$$S_i = \frac{1}{\lambda_i} \frac{\partial \phi}{\partial \lambda_i}\,. \tag{6.69}$$

Recalling the relation between $\boldsymbol{S}$ and $\boldsymbol{P}$ and the developments in section 2.7.3, the corresponding constitutive equation using the principal values of the first Piola-Kirchoff tensor is obtained as follows:

$$\begin{aligned} \boldsymbol{P} = \boldsymbol{F}\boldsymbol{S} &= \boldsymbol{F}\left(\sum_{i=1}^{3} \frac{1}{\lambda_i} \frac{\partial \phi}{\partial \lambda_i} \boldsymbol{A}_i \otimes \boldsymbol{A}_i\right) \\ &= \sum_{i=1}^{3} \frac{1}{\lambda_i} \frac{\partial \phi}{\partial \lambda_i} (\boldsymbol{F}\boldsymbol{A})_i \otimes \boldsymbol{A}_i = \sum_{i=1}^{3} \frac{\partial \phi}{\partial \lambda_i} \boldsymbol{b}_i \otimes \boldsymbol{A}_i\,. \end{aligned} \tag{6.70}$$

Thus,

$$P_i = \frac{\partial \phi}{\partial \lambda_i}\,. \tag{6.71}$$

The principal values $\sigma_i$ of the Cauchy stress tensor are obtained using (3.141), (2.113), the property (1.70), and $\boldsymbol{F}\boldsymbol{A}_i = \lambda_i \boldsymbol{b}_i$ (see exercise 2.11),

$$\begin{aligned} \boldsymbol{\sigma} = J^{-1}\boldsymbol{F}\boldsymbol{P}^T &= J^{-1}\boldsymbol{F}\left(\sum_{i=1}^{3} \frac{\partial \phi}{\partial \lambda_i} (\boldsymbol{b}_i \otimes \boldsymbol{A}_i)^T\right) \\ &= J^{-1}\left(\sum_{i=1}^{3} \frac{\partial \phi}{\partial \lambda_i} \boldsymbol{F}\boldsymbol{A}_i \otimes \boldsymbol{b}_i\right) = J^{-1}\left(\sum_{i=1}^{3} \lambda_i \frac{\partial \phi}{\partial \lambda_i} \boldsymbol{b}_i \otimes \boldsymbol{b}_i\right)\,. \end{aligned} \tag{6.72}$$

Then,

$$\sigma_i = J^{-1}\lambda_i \frac{\partial \phi}{\partial \lambda_i}\,. \tag{6.73}$$

The constitutive equations (6.61) and (6.63) are valid for an arbitrary hyperelastic material. However, hyperelastic materials for which the behavior in strain is quasi ***incompressible*** do exist. This means that their volume remains almost unchanged during a deformation (i.e., isochoric motion). Such materials include rubber and rubber-like materials as well as soft biological tissues. For these materials we extract from (3.38) and (3.7)

$$J = \frac{dv}{dV} = \lambda_1\lambda_2\lambda_3 = 1\,. \tag{6.74}$$

The ***incompressibility condition*** also introduces the following relation:

$$I_3 = {\lambda_1}^2{\lambda_2}^2{\lambda_3}^2 = 1\,. \tag{6.75}$$

In this case, the constitutive equation (6.61) becomes

$$\boldsymbol{S} = -p\boldsymbol{C}^{-1} + 2\left(\frac{\partial \Phi}{\partial I_1} + I_1\frac{\partial \Phi}{\partial I_2}\right)\boldsymbol{I} - 2\frac{\partial \Phi}{\partial I_2}\boldsymbol{C}\,, \tag{6.76}$$

where $I_3\,\partial\Phi/\partial I_3 = \partial\Phi/\partial I_3$ is replaced by $-p/2$ with $p$ being a parameter similar to pressure. When the Cauchy stress tensor is used for an incompressible material, relation (6.63) becomes

$$\boldsymbol{\sigma} = -p\boldsymbol{I} + 2\left(\frac{\partial \Phi}{\partial I_1} + I_1\frac{\partial \Phi}{\partial I_2}\right)\boldsymbol{c} - 2\frac{\partial \Phi}{\partial I_2}\boldsymbol{c}^2\,. \tag{6.77}$$

In addition, when the strain energy function is expressed as a function of the principal stretches, the constitutive relations (6.69), (6.71), and (6.73) take the form

$$S_i = -\frac{p}{{\lambda_i}^2} + \frac{1}{\lambda_i}\frac{\partial \phi}{\partial \lambda_i} \qquad P_i = -\frac{p}{\lambda_i} + \frac{\partial \phi}{\partial \lambda_i} \qquad \text{and} \quad \sigma_i = -p + \lambda_i\frac{\partial \phi}{\partial \lambda_i}\,. \tag{6.78}$$

The parameter $p$ is a constant which produces no work during the motion. It is generally associated with a ***hydrostatic pressure*** and is calculated from the equilibrium conditions and the boundary conditions.

It is often very useful to express the principal stresses as functions of the principal stretches. This is easy to do since for an isotropic material the directions of the principal stresses and stretches coincide. Consequently, as the principal values of the Cauchy-Green deformation tensor $\boldsymbol{c}$ are $\lambda_1^2, \lambda_2^2, \lambda_3^2$, the principal stresses resulting from (6.77) are

$$\sigma_i = -p + 2\left(\frac{\partial \Phi}{\partial I_1} + (\lambda_1^2+\lambda_2^2+\lambda_3^2)\frac{\partial \Phi}{\partial I_2}\right){\lambda_i}^2 - 2\frac{\partial \Phi}{\partial I_2}\lambda_i^4 \quad i = 1,2,3\,. \tag{6.79}$$

### 6.5.2 Forms of the Strain Energy Function

The constitutive equation is specified once the energy function has been identified. The mathematical conditions imposed until now are based on objectivity and isotropy. Other requirements can come from the type of boundary value problem, the experimental configuration, and the uniqueness of the solution. In general, the explicit definition of the energy function is based on methodological developments, experimental data, and/or the material microstructure.

We have already shown that the strain energy of an isotropic material can be expressed as a function of the three invariants $I_i$ $(i = 1, 2, 3)$ or as a symmetric function of the principal stretches $\lambda_i^2$ $(i = 1, 2, 3)$ of $\boldsymbol{C}$. Assuming that the function is of class $C^\infty$, (6.56) can be expressed as an infinite series of powers of $I_1 - 3$, $I_2 - 3$, $I_3 - 1$

$$\Phi(I_1, I_2, I_3) = \sum_{i,\,j,\,k=0}^{\infty} C_{ijk}(I_1 - 3)^i (I_2 - 3)^j (I_3 - 1)^k\,, \qquad (6.80)$$

where $C_{ijk}$ are material parameters, independent of the strain. In the reference configuration, that is, in the case of no applied stress, $I_1 = I_2 = 3$, $I_3 = 1$, and $\Phi(3,3,1) = C_{000} = 0$. In addition, from a physical point of view, the energy function should increase with the strain such that $\Phi(I_1, I_2, I_3) \geq 0$. Alternatively, (6.80) can be written as

$$\Phi(I_1, I_2, I_3) = \sum_{i,j=0}^{\infty} C_{ij0}(I_1 - 3)^i (I_2 - 3)^j + \sum_{k=1}^{\infty} (I_3 - 1)^k\,. \qquad (6.81)$$

In practice, only a limited number of terms are necessary in (6.80) or (6.81) in order to correctly express the strain response of a particular material. For incompressible materials, $I_3 = 1$, and (6.80) or (6.81) becomes a function of the first two invariants

$$\Phi(I_1, I_2) = \sum_{i,j=0}^{\infty} C_{ij}(I_1 - 3)^i (I_2 - 3)^j\,. \qquad (6.82)$$

To obtain a state without stress for zero strain, the first coefficient $C_{00}$ must be zero. Note that the material parameters are necessarily evaluated by a detailed experiment with specific identification procedures, a process which becomes more difficult when the number of parameters in the energy function increases.

In the recent past a certain number of ***energy functions*** have been proposed. Among them, we retain a few for incompressible materials. The simplest is the ***neo-Hookean model***, which results from (6.82) with $(i, j) = (1, 0)$

$$\Phi(I_1) = C_{10}(I_1 - 3)\,. \qquad (6.83)$$

This model has its origin in the statistical theory of rubber elasticity and gives satisfactory results for stretching ratios less than 2. The constant is expressed by $C_{10} = nk_BT$ where $n$ is the number of chains per unit volume,

$k_B = 1.381 \times 10^{-23}\,\mathrm{J\,K^{-1}}$ is Boltzmann's constant, and $T$ is the absolute temperature. This constant is normally related to the shear modulus of the material.

Another often-used model for rubber elasticity is known as the Mooney or ***Mooney-Rivlin*** strain energy function. From (6.82) for $(i,j) = (1,0)$ and $(i,j) = (0,1)$, we obtain

$$\Phi(I_1, I_2) = C_{10}(I_1 - 3) + C_{01}(I_2 - 3)\,. \tag{6.84}$$

This model played an important role in the development of non-linear elasticity. It can be used for stretch ratios up to 4. For larger stretches, the model becomes inaccurate.

The model proposed by Valanis and Landel assumes that the strain energy function can be written as the sum of three parts, each of which is a function of a single stretch:

$$\phi(\lambda_1, \lambda_2, \lambda_3) = \omega(\lambda_1) + \omega(\lambda_2) + \omega(\lambda_3)\,. \tag{6.85}$$

Here, $\omega(\lambda_i)$ indicates a function of $\lambda_i$ $(i = 1,2,3)$. The decomposition (6.85) corresponds to the ***Valanis-Landel hypothesis***.

A general form of the energy function was proposed by Ogden [36]. It is expressed as a function of the *principal stretches* as follows:

$$\phi(\lambda_1, \lambda_2, \lambda_3) = \sum_{i=1}^{N} \frac{\mu_i}{\alpha_i}\left(\lambda_1^{\alpha_i} + \lambda_2^{\alpha_i} + \lambda_3^{\alpha_i} - 3\right), \tag{6.86}$$

where $\alpha_i$ and $\mu_i$ are the material parameters. This model has been tested for simple and biaxial tension and simple shear of rubber. Excellent results are obtained for a large range of stretch ratios with the choice of $N = 3$. Note that the models mentioned above, as well as many others available in the literature, are special cases of (6.86). It reduces to the neo-Hookean model with $N = 1, \alpha_1 = 2$, $C_{10} = \mu_1/2$ and using the expression for the first invariant (6.57). We obtain the Mooney-Rivlin form (6.84) with $N = 2$, $\alpha_1 = 2$, $\alpha_2 = -2$, $C_{10} = \mu_1/2$, $C_{01} = -\mu_2/2$ and using the expressions for the first and second invariants (6.57). Note that with the Valanis-Landel hypothesis, the strain energy (6.86) takes the form

$$\phi(\lambda_1, \lambda_2, \lambda_3) = \sum_{k=1}^{3} \omega(\lambda_k) \qquad \text{and} \qquad \omega(\lambda_k) = \sum_{i=1}^{N} \frac{\mu_i}{\alpha_i}\left(\lambda_k^{\alpha_i} - 1\right). \tag{6.87}$$

For additional models of incompressible and compressible hyperplastic materials the reader is referred to [4, 20, 36]. We finish this section with the definition of the Saint-Venant-Kirchhoff model.

From the representation theorem (1.140) and with (6.51), we can write

$$\boldsymbol{S} = \beta_0 \boldsymbol{I} + \beta_1 \boldsymbol{C} + \beta_2 \boldsymbol{C}^2\,, \tag{6.88}$$

where the parameters $\beta_0, \beta_1, \beta_2$ are scalar functions of the invariants (6.57). Assuming that $\boldsymbol{S} = \boldsymbol{0}$ in the reference configuration for which $\boldsymbol{C} = \boldsymbol{I}$, we can show that near the reference configuration

$$\boldsymbol{S} = \lambda \operatorname{tr} \boldsymbol{E} \boldsymbol{I} + 2\mu \boldsymbol{E} + \mathrm{o}(\boldsymbol{E}), \tag{6.89}$$

where $\boldsymbol{E}$ is the strain tensor (2.82) and $\mathrm{o}(\boldsymbol{E})$ are the higher order terms that are negligible. We see that there are two parameters, $\lambda$, $\mu$, in the approximation of order 1; they are the ***Lamé coefficients*** or parameters.

Gabriel Lamé (1795–1870) was born in Tours. He was sent with Clapeyron to Saint Petersburg to teach applied mathematics and physics to the students at the School of Public Works. Returning to Paris, he was named professor at the École Polytechnique and subsequently at the Sorbonne. His main contributions were in applied mathematics and elasticity. He notably wrote *Leçons sur la théorie mathématique de l'élasticité des corps solides.*

**Fig. 6.2** Gabriel Lamé

Note that we should not confuse this constant $\lambda$ with the stretch parameter since in (6.89), we use the strain tensor $\boldsymbol{E}$ and not the invariants (6.57). When we take the term $\mathrm{o}(\boldsymbol{E})$ to zero, we obtain

$$\boldsymbol{S} = \lambda \operatorname{tr} \boldsymbol{E} \boldsymbol{I} + 2\mu \boldsymbol{E}\ , \tag{6.90}$$

which is the constitutive equation of a ***Saint-Venant-Kirchhoff material***. Relation (6.90) represents the classical non-linear model for compressible hyperelastic materials. It is adequate for the analysis of relatively small strains of isotropic homogeneous elastic materials. It can also be applied in non-linear analysis of large displacements with the displacement-strain relations given by (2.86). Further discussions on the suitability of this model can be found in [7, 20]. It can easily be shown that such a material is hyperelastic with the strain energy function given by

$$\widehat{\mathcal{W}}(\boldsymbol{E}) = \frac{\lambda}{2} \left(\operatorname{tr} \boldsymbol{E}\right)^2 + \mu \operatorname{tr} \boldsymbol{E}^2 . \tag{6.91}$$

We can also express (6.90) as a function of the first Piola-Kirchhoff tensor using (3.152)

$$\boldsymbol{P} = \boldsymbol{F}\boldsymbol{S} = \lambda(\operatorname{tr} \boldsymbol{E})\boldsymbol{F} + 2\mu \boldsymbol{F}\boldsymbol{E} . \tag{6.92}$$

### 6.5.3 Reduction to Simple Stress States

In this section we express the Cauchy stress tensor (6.79) for some simple cases of loads applied to isotropic, incompressible materials. The reader is referred to [4, 20, 36] for examples of compressible materials.

**Biaxial Stretch**

This type of motion is encountered in the case of thin plates with a plane load in two orthogonal directions; it is defined such that we have two independent principal stretches $\lambda_1$, $\lambda_2$, and, by (6.74), $\lambda_3 = \lambda_1^{-1}\lambda_2^{-1}$. The corresponding stresses are $\sigma_1$, $\sigma_2 \neq 0$ and $\sigma_3 = 0$. Setting $\sigma_3 = 0$ in (6.79), we obtain for the parameter $p$

$$p = 2\frac{1}{\lambda_1^2\lambda_2^2}\frac{\partial\Phi}{\partial I_1} + 2\frac{\lambda_1^2+\lambda_2^2}{\lambda_1^2\lambda_2^2}\frac{\partial\Phi}{\partial I_2}. \tag{6.93}$$

Introducing (6.93) in (6.79) and after a few algebraic manipulations, we can write the following expressions for the stresses:

$$\begin{aligned} \sigma_1 &= 2\left(\lambda_1^2 - \frac{1}{\lambda_1^2\lambda_2^2}\right)\left(\frac{\partial\Phi}{\partial I_1} + \lambda_2^2\frac{\partial\Phi}{\partial I_2}\right) \\ \sigma_2 &= 2\left(\lambda_2^2 - \frac{1}{\lambda_1^2\lambda_2^2}\right)\left(\frac{\partial\Phi}{\partial I_1} + \lambda_1^2\frac{\partial\Phi}{\partial I_2}\right). \end{aligned} \tag{6.94}$$

**Equibiaxial Stretch**

This type of motion is a special case of the preceding case with $\sigma_1 = \sigma_2$ and $\sigma_3 = 0$. Consequently, $\lambda_1 = \lambda_2 = \lambda$ and $\sigma_1 = \sigma_2 = \sigma$. Using (6.94), the stress $\sigma$ becomes

$$\sigma = 2\left(\lambda^2 - \frac{1}{\lambda^4}\right)\left(\frac{\partial\Phi}{\partial I_1} + \lambda^2\frac{\partial\Phi}{\partial I_2}\right). \tag{6.95}$$

Such loads are found in spherical shells under pressure where the two stresses tangent to the middle plane of the shell are equal and the third, normal to the shell surface, is considered to be zero.

**Uniaxial Stretch**

When the material is loaded in only one direction, we have $\lambda_1 = \lambda$ and from the incompressibility condition $\lambda_2 = \lambda_3 = \lambda^{-1/2}$. The stresses are $\sigma_1 = \sigma$ and $\sigma_2 = \sigma_3 = 0$. With these values of the parameters and (6.94), we obtain

$$\sigma = 2\left(\lambda^2 - \frac{1}{\lambda}\right)\left(\frac{\partial\Phi}{\partial I_1} + \frac{1}{\lambda}\frac{\partial\Phi}{\partial I_2}\right). \tag{6.96}$$

EXAMPLE 6.3

**Inflation of a Balloon**

As an example of hyperelasticity, we will present the case of inflation of a balloon made of a rubber material. We will use here the neo-Hookean and Mooney-Rivlin models to describe the pressure in a spherical balloon as a function of the stretch and assume that the material is isotropic and incompressible. The spherical balloon has an initial thickness $e_i$ and a radius $R$ such that $R \gg e_i$. In the deformed configuration, the thickness and the radius are $e$ and $r$, respectively as we can see in figure 6.3.

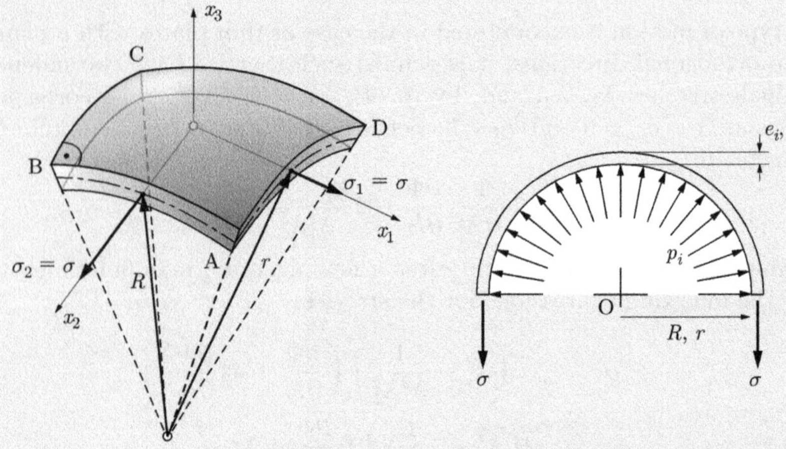


**Fig. 6.3** Inflation of a balloon: (a) geometry and (b) boundary conditions

Because of the spherical symmetry of the load and geometry, the two principal stresses are equal, $\sigma_1 = \sigma_2 = \sigma$, while the third, $\sigma_3 = 0$. Thus the state of stress is equibiaxial. To relate the internal pressure, $p_i$, to the stress, $\sigma$, consider the equilibrium of a hemisphere in the deformed configuration. The projection of the force due to the pressure onto the plane through the center is in equilibrium with the stresses in the thickness of the balloon, $\pi r^2 p_i = 2\pi r e \sigma$, from which we obtain

$$p_i = 2\frac{e}{r}\sigma \,. \tag{6.97}$$

To describe the deformation of the balloon, we define the stretch ratio by $\lambda = r/R$. The incompressibility condition is expressed by setting the material volume equal in the deformed and undeformed conditions, that is, $4\pi r^2 e = 4\pi R^2 e_i$. Thus

$$\frac{e}{e_i} = \frac{1}{\lambda^2} \,. \tag{6.98}$$

Combining the above expressions we obtain for the neo-Hookean model

$$\sigma = 2\left(\lambda^2 - \frac{1}{\lambda^4}\right)\frac{\partial \Phi}{\partial I_1} = 2C_{10}\left(\lambda^2 - \frac{1}{\lambda^4}\right) \tag{6.99}$$

and the pressure

$$p_i(\lambda) = 4C_{10}\frac{e_i}{R}\frac{1}{\lambda}\left(1 - \frac{1}{\lambda^6}\right) . \tag{6.100}$$

Taking $C_{10} = \partial\Phi/\partial I_1$ and $C_{01} = \partial\Phi/\partial I_2$ in the Mooney-Rivlin model, we obtain for the stress

$$\sigma = 2\left(\lambda^2 - \frac{1}{\lambda^4}\right)\left(C_{10} + \lambda^2 C_{01}\right) . \tag{6.101}$$

Using (6.101) and (6.98) in (6.97), we obtain

$$p_i(\lambda) = 4C_{10}\frac{e_i}{R}\frac{1}{\lambda}\left(1 - \frac{1}{\lambda^6}\right)\left(1 + \eta\lambda^2\right) , \qquad (6.102)$$

with $\eta = C_{01}/C_{10}$. Setting $e_i/R = 0.01$, the normalized pressure $p_i(\lambda)/C_{10}$ according to (6.102) is shown in figure 6.4 as a function of stretch ratio $\lambda$ for different values of $\eta$. The curve corresponding to $\eta = 0$ represents the neo-Hookean model (6.100).

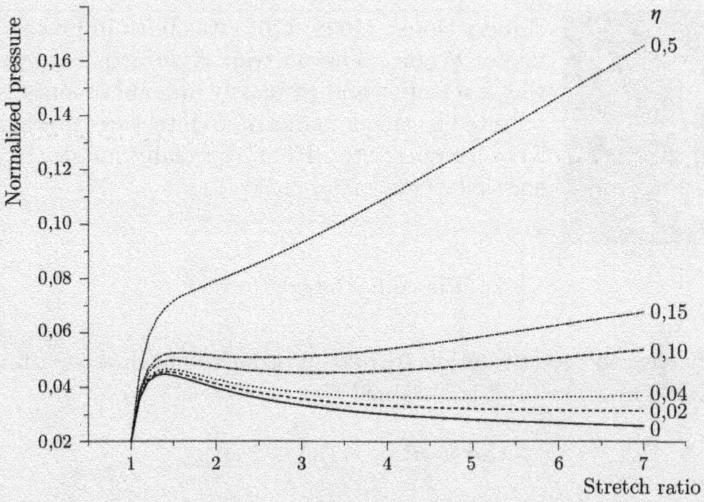


**Fig. 6.4** Normalized pressure in a balloon as a function of different values of $\eta$

## 6.6 Linear Infinitesimal Elasticity

Since in many cases the displacements and strains of elastic solids are small, we use the linearized theory developed for infinitesimal displacements and strains introduced in section 2.9.

Classical elasticity is thus a theory linearized around the natural state of a material considered to be homogeneous and isotropic. For this case we showed that the difference between the Green-Lagrange and Euler-Almansi tensors is proportional to the terms of order 2, which are considered negligible in the linearization. We thus use the *infinitesimal strain tensor* $\boldsymbol{\varepsilon}$ defined by (2.150), such that

$$\varepsilon_{ij} = \frac{1}{2}\left(\frac{\partial U_i}{\partial X_j} + \frac{\partial U_j}{\partial X_i}\right) = \frac{1}{2}\left(\frac{\partial u_i}{\partial x_j} + \frac{\partial u_j}{\partial x_i}\right) . \qquad (6.103)$$

With this approximation, the Piola-Kirchhoff tensors reduce to the Cauchy tensor $\boldsymbol{\sigma}$ (sec. 3.9). In addition, the principle of objectivity is satisfied (sec. 2.11). Also with these linearizations, the Saint-Venant-Kirchhoff equations become

$$\boldsymbol{\sigma} = \lambda \operatorname{tr} \boldsymbol{\varepsilon} \boldsymbol{I} + 2\mu\boldsymbol{\varepsilon}\,, \tag{6.104}$$

where the scalar coefficients $\lambda$ and $\mu$ are the *Lamé coefficients of elasticity*, whose dimensions are force per unit surface (Pa). Thus, linear elasticity relates the stress to the infinitesimal strain with a linear equation known as ***Hooke's law***.

Robert Hooke (1635–1703) was born in Freshwater on the Isle of Wight. The portrait is an artist's impression. He was a scientist and primarily an experimentalist. He contributed to the domains of architecture, mechanics, chemistry, physics, etc. He is especially known for his law of elasticity (*ut tensio, sic vis*).

**Fig. 6.5** Robert Hooke

It is very easy to invert relation (6.104) in order to obtain $\boldsymbol{\varepsilon}$ as a function of $\boldsymbol{\sigma}$ as follows. The trace of $\boldsymbol{\sigma}$ is obtained by

$$\operatorname{tr} \boldsymbol{\sigma} = \sigma_{mm} = (3\lambda + 2\mu)\varepsilon_{mm}\,. \tag{6.105}$$

With (6.104) and (6.105), we have

$$\boldsymbol{\varepsilon} = \frac{-\lambda}{2\mu(3\lambda + 2\mu)} \operatorname{tr} \boldsymbol{\sigma} \boldsymbol{I} + \frac{\boldsymbol{\sigma}}{2\mu}\,. \tag{6.106}$$

This relation exists only if

$$3\lambda + 2\mu \neq 0 \qquad \text{and} \qquad \mu \neq 0\,. \tag{6.107}$$

Equation (6.106) can also be written as

$$\boldsymbol{\varepsilon} = -\frac{\nu}{E} \operatorname{tr} \boldsymbol{\sigma}\, \boldsymbol{I} + \frac{(1+\nu)}{E}\boldsymbol{\sigma}\,, \tag{6.108}$$

where $E$ is *Young's modulus* (Pa) and $\nu$ is *Poisson's dimensionless ratio*. These coefficients are related to Lamé's by

$$E = \frac{\mu(3\lambda + 2\mu)}{\lambda + \mu} \qquad \nu = \frac{\lambda}{2(\lambda + \mu)}\,. \tag{6.109}$$

Table 6.2 shows values of Young's modulus, Poisson's ratio, and the density of a few elastic materials used in engineering.

**Table 6.2** Material constants for elastic solids.

| | $E$ (Pa) | $\nu$ | $\rho$ (kg/m$^3$) |
|---|---|---|---|
| steel | $200\,10^9$ | 0.27 | 7 850 |
| glass | $69\,10^9$ | 0.19 | 2 500 |
| rubber | $0.05\,10^9$ | 0.50 | 850 |

We will offer an interpretation of the elastic coefficients by considering a few simple cases. The first example is that of *simple traction* (fig. 6.6). A bar is subjected to traction in direction $x_1$. We assume that the lateral surfaces of the bar are free with no contact forces acting on them.

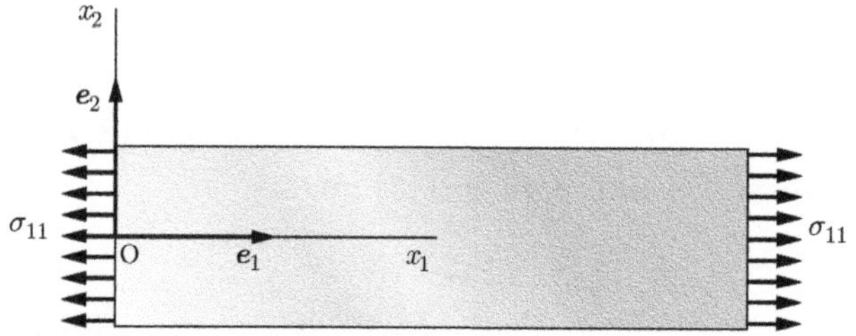


**Fig. 6.6** Simple traction

The stress tensor has only a single non-zero component, namely $\sigma_{11}$. To calculate the strains, we use equations (6.106). We obtain

$$\begin{aligned}\varepsilon_{11} &= \frac{\lambda+\mu}{\mu(3\lambda+2\mu)}\sigma_{11}\\ \varepsilon_{22} = \varepsilon_{33} &= -\frac{\lambda}{2\mu(3\lambda+2\mu)}\sigma_{11} = -\frac{\lambda}{2(\lambda+\mu)}\frac{\lambda+\mu}{\mu(3\lambda+2\mu)}\sigma_{11}\\ \varepsilon_{12} = \varepsilon_{23} &= \varepsilon_{31} = 0\,.\end{aligned} \tag{6.110}$$

With (6.109), equations (6.110) become

$$\varepsilon_{11} = \frac{1}{E}\sigma_{11} \qquad \varepsilon_{22} = \varepsilon_{33} = -\frac{\nu}{E}\sigma_{11} = -\nu\varepsilon_{11}\,. \tag{6.111}$$

Poisson's ratio thus corresponds to the lateral contraction of the sample under traction. We can express the Lamé coefficients as functions of $E$ and $\nu$. This follows by inverting (6.109)

$$\lambda = \frac{E\nu}{(1+\nu)(1-2\nu)} \qquad \mu = \frac{E}{2(1+\nu)}\,. \tag{6.112}$$

The modulus $E$ is positive because experience shows that if $\sigma_{11} > 0$, then $\varepsilon_{11}$ is also; the part stretches under traction. From experience we also see that $\nu$ is also positive.

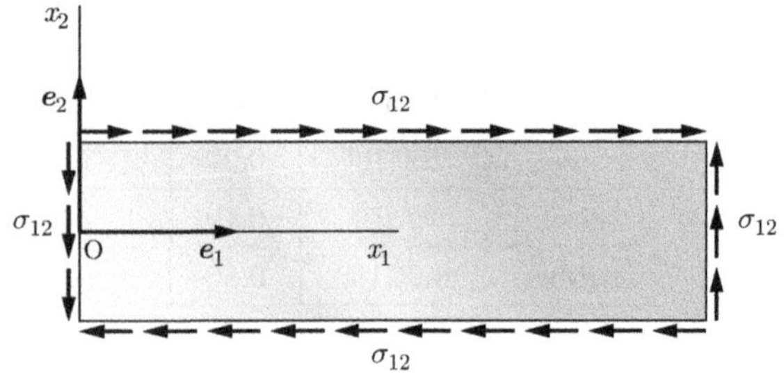


**Fig. 6.7** Simple shear

The second example is that of *shear* applied to a block (fig. 6.7) such that the components of the stress tensor are written

$$\sigma_{12} = \sigma_{21} \tag{6.113}$$

with

$$\sigma_{ij} = 0 \qquad \forall(i,j) \neq (1,2) \text{ and } (2,1)\,.$$

The strain tensor (6.106) with (6.113) gives

$$\varepsilon_{12} = \varepsilon_{21} = \frac{\sigma_{12}}{2\mu}\,, \tag{6.114}$$

the other components being zero. As the component $\varepsilon_{12}$ is half the complement of the angle formed, after deformation, by directions initially oriented along the directions $x_1$ and $x_2$ by (2.158), we have

$$\frac{\sigma_{12}}{\mu} = \phi_{12}\,. \tag{6.115}$$

We call the Lamé coefficient $\mu$ the ***shear modulus***, or also the ***modulus of rigidity in shear***.

The third example is that of uniform local contraction such that the stress tensor is diagonal (or spherical) with

$$\sigma_{ij} = -p\,\delta_{ij}\,, \tag{6.116}$$

where $p$ denotes the pressure. Then by (6.106), we calculate

$$\varepsilon_{ij} = -\frac{p}{(3\lambda + 2\mu)}\,\delta_{ij} \tag{6.117}$$

or

$$p = -\frac{1}{3}\,(3\lambda + 2\mu)\varepsilon_{kk} = -K\varepsilon_{kk}\,. \tag{6.118}$$

The coefficient defined by the relation

$$K = \frac{(3\lambda + 2\mu)}{3} = \frac{1}{3}\,\frac{E}{1 - 2\nu} \tag{6.119}$$

is the ***bulk modulus***. For a given value of pressure, the volume variation $\varepsilon_{kk}$ will be smaller when $K$ is larger. Experience shows that $K$ is positive, which leads to $\nu \leq 0.5$. The special case $\nu = 1/2$ is that for incompressible elastic materials for which $K \to \infty$. An example of this type of material is rubber, which we take to be incompressible.

## 6.7 Heat Conduction

We have seen in section 5.4.1 that the heat flux is written as

$$\boldsymbol{q}(\boldsymbol{X},t) = \boldsymbol{\mathcal{Q}}(\chi, T, \boldsymbol{X}, t). \tag{6.120}$$

Applying the general principles of the constitutive laws, we arrive at the equation

$$\boldsymbol{q}(\boldsymbol{X},t) = \boldsymbol{\mathcal{Q}}\left(\boldsymbol{F}, T, \frac{\partial T}{\partial \boldsymbol{X}}, \boldsymbol{X}\right). \tag{6.121}$$

We note that the heat flux depends on the temperature gradient. When we study conductive heat transfer in fluids and solids, we can show that $\boldsymbol{q}$ depends mostly on the temperature gradient $\partial T/\partial \boldsymbol{X}$ and only weakly on $\boldsymbol{F}$. This is confirmed by experience. We should thus write, in the Eulerian representation,

$$q_i = K_{il} \frac{\partial T}{\partial x_l}. \tag{6.122}$$

Furthermore, we can specialize equation (6.122) for the case where the heat flux is given by ***Fourier's law*** with $K_{il} = -k\,\delta_{il}$

$$\boldsymbol{q} = -k(T)\,\boldsymbol{\nabla} T \qquad q_i = -k \frac{\partial T}{\partial x_i}. \tag{6.123}$$

Parameter $k$ is the coefficient of thermal conductivity. Its SI units are $\mathrm{W\,m^{-1}\,K^{-1}}$. Fourier's law is valid for both fluids and solids.

Joseph Fourier (1768–1830) was born in Auxerre. A brilliant student at the École Polytechnique, he became a professor at the age of 16. He participated in the Egyptian campaign with Champollion. After returning to France, Napoleon named him prefect of Isère. Elected to the Académie des Sciences in 1817, he became its perpetual secretary in 1822. Fourier wrote *Théorie analytique de la chaleur*, where he introduced the partial differential equation for heat diffusion. He solved it with the series of periodic functions which are named after him.

**Fig. 6.8** Joseph Fourier

## 6.8 Second Principle of Thermodynamics for Viscous Fluids

Inequality (4.83) is applied to Newtonian viscous fluids by introducing the forms (6.14) for the stress field and (6.123) for the conductive heat flux. We also add forms for the internal energy per unit mass, $u$, and the entropy per unit mass,

$s$, taken as functions of only temperature $T$ and density $\rho$. We will establish the state equation:

$$\rho(du - T\,ds) - \frac{p}{\rho}\,d\rho = 0\,. \tag{6.124}$$

We also propose to show that the three coefficients $\kappa$, $\mu$, and $k$ must always be positive, with the coefficient $\kappa$ defined by

$$\kappa = \frac{1}{3}\,(3\lambda + 2\mu)\,. \tag{6.125}$$

To start, we state the following postulate.

POSTULATE *The Clausius-Duhem inequality (4.83) is satisfied at all times, for arbitrary histories and independently of the temperature, density, deviatoric strain rate,* $\boldsymbol{d}^d$, *and the thermal gradient,* $(\partial T/\partial \boldsymbol{x})$.

These quantities are called the thermodynamic model variables. Their history, that is, their values as functions of time, for a given material point, is called a ***thermodynamic process***. We note that the constitutive equations express $\sigma_{ij}$, $q_i$, $u$, and $s$ as functions of the thermodynamic process (the history of $\rho$ giving that of $\operatorname{tr}\boldsymbol{d}$ by the conservation of mass). First, consider the following relations, in index form:

$$d^d_{ij} = d_{ij} - \frac{1}{3}\,d_{mm}\,\delta_{ij} \tag{6.126}$$

$$\frac{\partial v_i}{\partial x_i} = d_{ii} = -\frac{1}{\rho}\frac{D\rho}{Dt} \tag{6.127}$$

$$d^d_{ij}\,d^d_{ij} = d^d_{ij}\left(d_{ij} - \frac{1}{3}\,d_{mm}\,\delta_{ij}\right) = d^d_{ij}\,d_{ij}\ ,$$

since $\operatorname{tr}\boldsymbol{d}^d = 0$. The constitutive equation (6.14) can be rewritten with (6.125) and (6.126):

$$\sigma_{ij} = -p\,\delta_{ij} + \kappa\,d_{kk}\,\delta_{ij} + 2\mu\,d^d_{ij}\ . \tag{6.128}$$

We can then extract the following development from (4.24), (6.128), and (6.123):

$$\begin{aligned}\sigma_{ji}\frac{\partial v_i}{\partial x_j} &- \frac{q_i}{T}\frac{\partial T}{\partial x_i}\\ &= -p\,d_{ii} + \kappa\,(d_{mm})^2 + 2\mu\,d^d_{ij}\,d^d_{ij} + \frac{k}{T}\left(\frac{\partial T}{\partial x_i}\right)\left(\frac{\partial T}{\partial x_i}\right)\\ &= \frac{p}{\rho}\frac{D\rho}{Dt} + \kappa\,(d_{mm})^2 + 2\mu\,d^d_{ij}\,d^d_{ij} + \frac{k}{T}\left(\frac{\partial T}{\partial x_i}\right)\left(\frac{\partial T}{\partial x_i}\right).\end{aligned} \tag{6.129}$$

The Clausius-Duhem inequality (4.83) is thus written in the following form, valid for any Newtonian viscous fluid:

$$\rho\left(\frac{Du}{Dt} - T\frac{Ds}{Dt}\right) - \frac{p}{\rho}\frac{D\rho}{Dt} \le \kappa\,(d_{mm})^2 + 2\mu\,d^d_{ij}\,d^d_{ij} + \frac{k}{T}\left(\frac{\partial T}{\partial x_i}\right)\left(\frac{\partial T}{\partial x_i}\right). \tag{6.130}$$

This inequality must be satisfied for an arbitrary thermodynamic process (within the limits of the model's applicability). Hence, we can also require that it be valid at any arbitrary time, regardless of the thermodynamic variables ($T$, $\rho$, $d^d_{ij}$, $\partial T/\partial x_i$) and their material derivatives (eqns. (6.1) and (6.6)). Let $\widehat{T}$, $\widehat{\rho}$, $\widehat{d}^d_{ij}$, $\widehat{T}_{,i}$ be the values of the thermodynamic variables and $\widehat{\dot{T}}$, $\widehat{\dot{\rho}}$, ..., be those of their material derivatives. Then, by Taylor series expansion in time about the values of the terms in inequality (6.130), we must have

$$\begin{aligned} E \equiv \widehat{\rho}\left(\frac{\partial u}{\partial \rho} - T\frac{\partial s}{\partial \rho}\right)_{(\widehat{T},\widehat{\rho})} \widehat{\dot{\rho}} \\ + \widehat{\rho}\left(\frac{\partial u}{\partial T} - T\frac{\partial s}{\partial T}\right)_{(\widehat{T},\widehat{\rho})} \widehat{\dot{T}} - \frac{p(\widehat{T},\widehat{\rho})}{\widehat{\rho}}\widehat{\dot{\rho}} \\ \leq \kappa\left(-\frac{\widehat{\dot{\rho}}}{\widehat{\rho}}\right)^2 + 2\mu\, \widehat{d}^d_{ij}\, \widehat{d}^d_{ij} + \frac{k}{\widehat{T}}\widehat{T}_{,i}\widehat{T}_{,i}\,. \end{aligned} \tag{6.131}$$

According to the adopted postulate, we can, without changing $\widehat{T}$, $\widehat{\rho}$, $\widehat{\dot{T}}$, and $\widehat{\dot{\rho}}$, let the arbitrary values $\widehat{d}^d_{ij}$ and $\widehat{T}_{,i}$ be zero in (6.131). This amounts to considering an identical thermodynamic process, but without heat flux nor a deviatoric rate of deformation at the material point being followed. We thus have the following inequality:

$$E \leq \kappa\left(-\frac{\widehat{\dot{\rho}}}{\widehat{\rho}}\right)^2 . \tag{6.132}$$

We can then, following the same postulate, multiply $\widehat{\dot{T}}$ and $\widehat{\dot{\rho}}$ by an arbitrary quantity $\varepsilon$, either positive or negative, without changing $\widehat{T}$ nor $\widehat{\rho}$ (which amounts to slowing the process at the instant under study or to considering a slow process in the opposite sense). We thereby obtain, for any $\varepsilon \in \mathbb{R}$, the inequality

$$\varepsilon E \leq \varepsilon^2 \kappa\left(-\frac{\widehat{\dot{\rho}}}{\widehat{\rho}}\right)^2 ,$$

and thus inevitably the equality

$$E = \left(\rho\left(\frac{Du}{Dt} - T\frac{Ds}{Dt}\right) - \frac{p}{\rho}\frac{D\rho}{Dt}\right)_{(\widehat{T},\widehat{\rho},\widehat{\dot{T}},\widehat{\dot{\rho}})} = 0\,. \tag{6.133}$$

By writing (6.133) in differential form, we find the relation (6.124) we sought. As soon as (6.124) is satisfied, the left-hand side of (6.131) becomes identically zero. For the right-hand side to remain positive, it is necessary and sufficient that the coefficients $\kappa$, $\mu$, and $k$ be positive. We observe then that we have a linear combination of squares of independent expressions which must be a positive definite quadratic form, and thus with positive coefficients.

The right-hand side of inequality (6.130),

$$\kappa\,(d_{mm})^2 + 2\mu\, d^d_{ij}\, d^d_{ij} + \frac{k}{T}\left(\frac{\partial T}{\partial x_i}\right)\left(\frac{\partial T}{\partial x_i}\right), \tag{6.134}$$

measures the ***local irreversibility*** of the process being studied.

## 6.9 Ideal Gas Thermodynamics

In this section we establish the relation between the thermodynamics of continuous mechanics and classical thermodynamics. For the latter, we refer to the book by [35].

We introduce the definition of the ***enthalpy*** per unit mass

$$h = u + \frac{p}{\rho} \tag{6.135}$$

for a Newtonian viscous fluid.

The constitutive equation (6.12), coupled with the hypothesis that the internal energy per unit mass does not depend on the density, and thus $u = u(T)$, yields the ***ideal gas*** model. This gas satisfies Boyle's law, or the Boyle-Mariotte law, that is, "at constant temperature, the product of the pressure $p$ and the volume $V$ is constant". We deduce from (6.135) that the enthalpy per unit mass does not depend on density either, so that we have the equations

$$p = \rho R T \tag{6.136}$$
$$u = u(T) \tag{6.137}$$
$$h = h(T)\,. \tag{6.138}$$

If we introduce the notion of specific volume $v = 1/\rho$, then the state equation, (6.136), corresponds to $pv =$ cnst in an isothermal process. Differentiating, we define the specific heat capacities at constant volume and pressure

$$du = c_v(T)\, dT \qquad\qquad dh = c_p(T)\, dT\,. \tag{6.139}$$

Also, differentiation of (6.135), taking into account (6.136) and (6.137), leads to

$$dh = du + R\, dT\,, \tag{6.140}$$

from which we obtain

$$c_p(T) - c_v(T) = R\,. \tag{6.141}$$

$R$ is the ideal gas constant expressed in $\mathrm{J\,kg^{-1}\,K^{-1}}$. Various expressions of $ds$ are found by combining (6.124) and (6.137):

$$ds = \begin{cases} c_v\, d(\log T) - R\, d(\log \rho) \\ c_p\, d(\log T) - R\, d(\log p) \\ c_v\, d(\log p) - c_p\, d(\log \rho)\,. \end{cases} \tag{6.142}$$

Note that physical observations reveal that for an ideal gas, the coefficients of viscosity and thermal conduction generally depend only on the absolute temperature.

Important simplifications can be made if we assume that, for a certain temperature range, the heat capacities are constant. In this case we can write, within an additive constant, in the given temperature range

$$u = c_v T \tag{6.143}$$

$$h = c_p T \tag{6.144}$$

$$s = c_v \log p - c_p \log \rho \,.$$

An ***isentropic flow*** is one for which

$$\frac{p}{\rho^\gamma} = \text{cnst} \,, \tag{6.145}$$

with the definition of the heat capacity ratio:

$$\gamma = \frac{c_p}{c_v} \,. \tag{6.146}$$

A fluid in isentropic flow is a ***barotropic*** fluid for which the density is only a function of pressure such that $\rho = \rho(p)$. In this case, we easily show that

$$\frac{1}{\rho(p)} \boldsymbol{\nabla} p = \boldsymbol{\nabla} \int \frac{dp}{\rho(p)} \,. \tag{6.147}$$

So, for a given function $f(p)$, we have

$$\boldsymbol{\nabla} f(p) = \frac{df}{dp} \boldsymbol{\nabla} p \,. \tag{6.148}$$

Setting

$$f(p) = \int \frac{dp}{\rho(p)} \,, \tag{6.149}$$

then it follows that

$$\frac{df}{dp} = \frac{1}{\rho(p)} \,. \tag{6.150}$$

Finally, within a constant, we can write

$$h = \frac{c_p}{R} \frac{p}{\rho} = \frac{\gamma}{\gamma - 1} \frac{p}{\rho} \,. \tag{6.151}$$

The speed of sound $a$ is defined by the relation

$$a^2 = \left. \frac{\partial p}{\partial \rho} \right|_s \,. \tag{6.152}$$

In the special case of an ideal gas, (6.152) takes the form, with use of (6.145)

$$a^2 = \gamma \frac{p}{\rho} \,, \tag{6.153}$$

so that with (6.151) we obtain, to within an additive constant,

$$h = \frac{a^2}{\gamma - 1} . \tag{6.154}$$

For air considered as an ideal gas, we have the following constants at 300°K: $R_{air} = 287$ J kg$^{-1}$ K$^{-1}$, $\gamma_{air} = 1.401$, $c_p = 1006$ J kg$^{-1}$ K$^{-1}$. Air conducts heat as predicted by Fourier's law, (6.123), with $k = 0.0262$ W m$^{-1}$ K$^{-1}$. The speed of sound in air is 340 m s$^{-1}$ at standard temperature (288 degrees K). For comparison, the speed of sound in sea water is about 1500 m s$^{-1}$.

## 6.10 Second Principle of Thermodynamics for Classical Elastic Media

We examine an elastic solid subjected to small displacements and infinitesimal strains (sec. 2.9). Thus we express all quantities in terms of their material description and use lowercase symbols for convenience. We also assume that geometric changes are negligible and that the deformation process takes place slowly enough so that thermodynamic equilibrium is maintained in the entire body at all times. In this case, the internal energy and free energy defined by equation (4.84) take the forms

$$u = u(\boldsymbol{\varepsilon}, T) \qquad f = f(\boldsymbol{\varepsilon}, T) . \tag{6.155}$$

Classical elasticity assumes the reversibility of the thermodynamic phenomena, given that the solid is not subject to permanent deformation. Then, the Clausius-Duhem inequality (4.83) becomes an equality, that is,

$$\rho\left(\frac{Du}{Dt} - T\frac{Ds}{Dt}\right) = \boldsymbol{\sigma} : \boldsymbol{d} - \frac{1}{T}\boldsymbol{q} \cdot \boldsymbol{\nabla} T . \tag{6.156}$$

Assume an adiabatic process ($\boldsymbol{q} = 0$). Using (4.84) and since for small deformations $\boldsymbol{d} = \dot{\boldsymbol{\varepsilon}}$, relation (6.156) becomes

$$\rho\left(\frac{Du}{Dt} - T\frac{Ds}{Dt}\right) = \boldsymbol{\sigma} : \dot{\boldsymbol{\varepsilon}} = \rho\left(\frac{Df}{Dt} + s\frac{DT}{Dt}\right) . \tag{6.157}$$

From the second equality of (6.157), we obtain

$$\frac{1}{\rho}\boldsymbol{\sigma}\, d\boldsymbol{\varepsilon} - s\, dT = df . \tag{6.158}$$

Accounting for (6.155), it follows in index notation

$$\frac{1}{\rho}\sigma_{ij} = \frac{\partial f}{\partial \varepsilon_{ij}} \qquad s = -\frac{\partial f}{\partial T} . \tag{6.159}$$

The ***free energy*** is thus a potential for the tensor $\boldsymbol{\sigma}/\rho$ and also for the entropy, $s$. If we reduce the dependence of $f$ to be only on $\varepsilon_{ij}$, by considering only isothermal processes, then we can expand $f$ in the neighborhood of the natural state of the elastic material. We have

$$f = f_0 + \left.\frac{\partial f}{\partial \varepsilon_{ij}}\right|_0 \varepsilon_{ij} + \cdots , \qquad (6.160)$$

where the index zero denotes the unstressed natural state. From $(6.159)_1$, the coefficients of the linear term are zero ($\sigma_{ij}|_0 = 0$). Thus, $(f - f_0)$ is quadratic in $\varepsilon_{ij}$, and, in the identity

$$\frac{\partial}{\partial \varepsilon_{ij}}\left(\rho(f - f_0)\right) = \rho\frac{\partial f}{\partial \varepsilon_{ij}} + (f - f_0)\frac{\partial \rho}{\partial \varepsilon_{ij}} , \qquad (6.161)$$

the second term on the right-hand side is negligible by the linear theory. In fact, if we combine (2.147) and (3.37) we obtain $\rho = \rho_0(1 + O(\varepsilon)) \approx \rho_0$ (since there is no distinction between the material and spatial description when the strains are infinitesimal). Thus $\rho \approx \rho_0$ and density can be considered constant during infinitesimal deformation. The first relation of (6.159) can be written as

$$\sigma_{ij} = \frac{\partial W}{\partial \varepsilon_{ij}} , \qquad (6.162)$$

with

$$W = \rho(f - f_0) \qquad (6.163)$$

being the ***strain energy*** per unit volume with $\sigma_{ij}$ and $\varepsilon_{ij}$ as corresponding conjugate parameters. With equation (6.104), the potential energy $W$ can be set in the form (compare with (6.91))

$$W = \frac{\lambda}{2}\varepsilon_{ii}\varepsilon_{kk} + \mu\varepsilon_{ij}\varepsilon_{ij} . \qquad (6.164)$$

## 6.11 Thermoelasticity

We make the assumption of small strains and displacements and consider small deviations from a reference temperature $T_0$. Since we are developing an approximate theory, we will expand $f(\boldsymbol{\varepsilon}, T)$ in a Taylor series in the neighborhood of $\boldsymbol{\varepsilon} = 0$ and $T = T_0$ which we truncate after the quadratic terms. As (6.163) gives zero stress for $\boldsymbol{\varepsilon} = 0$ and $T = T_0$, the expansion has no linear term in only $\boldsymbol{\varepsilon}$. It is appropriate to work with $\rho f$ rather than with $f$ [27]. We have

$$\begin{aligned}\rho f = {} & \rho f_0 - \rho s_0(T - T_0) + \frac{\lambda}{2}\varepsilon_{ii}\varepsilon_{kk} \\ & + \mu\varepsilon_{ij}\varepsilon_{ij} + \varepsilon_{ij}c_{ij}(T - T_0) - \frac{\rho c}{2T_0}(T - T_0)^2 ,\end{aligned} \qquad (6.165)$$

an expression for which the coefficients $f_0$, $s_0$, $c_{ij}$, and $c$ are still to be determined, and in which the factors $\rho$ and $\rho/T_0$ have been added to simplify subsequent steps. If we apply (6.165) to the case where $\varepsilon = 0$ and $T = T_0$, we observe that $f_0$ is the free energy of the natural state.

If we impose isotropy on the elastic material, the tensor $c_{ij}$ must be isotropic and of the form $a\,\delta_{ij}$ with $a$ a scalar. Taking this scalar as $a = -(3\lambda + 2\mu)\alpha$, with $\alpha$ yet to be determined, we obtain

$$c_{ij}\varepsilon_{ij}(T - T_0) = -(3\lambda + 2\mu)\alpha\varepsilon_{kk}(T - T_0)\,. \tag{6.166}$$

We can easily calculate $\sigma_{ij}$ with the relation (6.159) applied to (6.165) and accounting for (6.166) we obtain

$$\sigma_{ij} = \rho\frac{\partial f}{\partial \varepsilon_{ij}} = \lambda\varepsilon_{kk}\,\delta_{ij} + 2\mu\varepsilon_{ij} - (3\lambda + 2\mu)\alpha(T - T_0)\,\delta_{ij}\,. \tag{6.167}$$

This is the generalization of *Hooke's law* (6.104). We can invert the relation to obtain $\boldsymbol{\varepsilon}$ as a function of $\boldsymbol{\sigma}$

$$\boldsymbol{\varepsilon} = \frac{1}{2\mu}\left(\boldsymbol{\sigma} + \left(2\mu\alpha(T - T_0) - \frac{\lambda}{3\lambda + 2\mu}\,\mathrm{tr}\,\boldsymbol{\sigma}\right)\boldsymbol{I}\right). \tag{6.168}$$

This equation resembles (6.106) except for the additional term $\alpha(T - T_0)\,\delta_{ij}$ which comes from the thermal effects. It corresponds to a uniform extension $\alpha(T - T_0)$ in all directions, that is, to a dilation of volume $3\alpha(T - T_0)$. The coefficient $\alpha$ is the ***thermal expansion coefficient***. It has dimensions of an inverse temperature.

If we now apply $(6.159)_2$ to (6.165) and (6.166), we can evaluate the entropy

$$\rho s = -\rho\frac{\partial f}{\partial T} = \rho s_0 + (3\lambda + 2\mu)\alpha\varepsilon_{kk} + \frac{\rho c}{T_0}\,(T - T_0)\,. \tag{6.169}$$

The quantity $s_0$ is the entropy of the natural state.

Having evaluated $f$ and $s$, respectively, with (6.165) and (6.169), we can obtain the internal energy

$$\begin{aligned}\rho u &= \rho(f + Ts)\\ &= \rho u_0 + \frac{\lambda}{2}\,(\mathrm{tr}\;\boldsymbol{\varepsilon})^2 + \mu\boldsymbol{\varepsilon} : \boldsymbol{\varepsilon} + (3\lambda + 2\mu)\alpha T_0\,\mathrm{tr}\;\boldsymbol{\varepsilon} + \frac{\rho c}{2T_0}\,(T^2 - T_0^2)\end{aligned} \tag{6.170}$$

with $u_0$ being the internal energy in the natural state. Equation (6.170) shows that the internal energy cannot be obtained simply by a linear combination of the strain energy (first three terms of the right-hand side of (6.170)) and a thermal energy (last term of (6.170)). Its structure is more complicated implying a coupling between mechanical and thermal effects. Calculating the partial derivative of $u$ with respect to the temperature, we obtain

$$\left.\frac{\partial u}{\partial T}\right|_{T_0} = c\,, \tag{6.171}$$

which is the *heat per unit mass* at the temperature $T = T_0$.

## 6.12 Exercises

**6.1** Prove that $\dot{\boldsymbol{C}} = 2\boldsymbol{d}$, from equations (6.2), is valid for a simple fluid.

**6.2** Express the energy equation (4.23) for the case of a Newtonian viscous fluid (6.14).

Simplify this expression for the case of a perfect fluid. If the fluid is an ideal gas, what happens to the energy equation?

**6.3** Express the energy equation (4.23) for the case of an incompressible Newtonian viscous fluid (6.16).

Simplify this expression for the case of a perfect fluid.

**6.4** Prove that

1) the tensors $\boldsymbol{U}, \boldsymbol{C}$, and $\boldsymbol{S}$ have the same principal directions;
2) the tensors $\boldsymbol{V}, \boldsymbol{c}$, and $\boldsymbol{\sigma}$ have the same principal directions.

**6.5** From (6.61), (3.152), and relations (2.88)–(2.90), prove relation (6.63).

**6.6** Use the Cayley-Hamilton equation, (1.123), to show that the stress relations (6.61) and (6.63) can be written as

$$\boldsymbol{S} = 2\left(\left(\frac{\partial \Phi}{\partial I_1} + I_1\frac{\partial \Phi}{\partial I_2} + I_2\frac{\partial \Phi}{\partial I_3}\right)\boldsymbol{I} - \left(I_1\frac{\partial \Phi}{\partial I_3} + \frac{\partial \Phi}{\partial I_2}\right)\boldsymbol{C} + \frac{\partial \Phi}{\partial I_3}\boldsymbol{C}^2\right) \tag{6.172}$$

$$\boldsymbol{\sigma} = 2J^{-1}\left(\left(I_2\frac{\partial \Phi}{\partial I_2} + I_3\frac{\partial \Phi}{\partial I_3}\right)\boldsymbol{I} + \frac{\partial \Phi}{\partial I_1}\boldsymbol{c} - I_3\frac{\partial \Phi}{\partial I_2}\boldsymbol{c}^{-1}\right). \tag{6.173}$$

**6.7** Using (6.80), prove that the energy is zero in the reference configuration when $C_{000} = 0$ and that it is stress free when the coefficients satisfy the condition $C_{100} + 2C_{010} + C_{001} = 0$.

**6.8** Using the expressions of the invariants (6.57) and relation (1.144), prove the first two equalities of (6.59).

**6.9** For the neo-Hookean model, find the stretch ratio for which the maximum pressure is obtained in the case of a balloon under internal pressure (6.100).

**6.10** Use the Ogden model (6.86) to express the stress components for the cases of uniaxial, biaxial, and equibiaxial stretching of an incompressible material. With the choice $N = 3$ and the material parameters $\alpha_1 = 1.3$, $\alpha_2 = 5$, $\alpha_3 = -2$, $\mu_1 = 0.63$ MPa, $\mu_2 = 0.0012$ MPa, $\mu_3 = -0.01$ MPa, sketch the stress components as functions of the corresponding stretches for uniaxial and equibiaxial stretching.

**6.11** Calculate the free energy in a bar of linear elastic material subject to a simple tension load $\sigma_{11}$.

**6.12** Calculate the induced strain in a rectilinear bar of length $L$ along the axis $x_1$ in a temperature field $T = T_0 + (T_1 - T_0)x_1/L$.

**6.13** Given Fourier's law of conduction and the ideal gas state equation for a perfect compressible fluid, write the conservation of energy equation with temperature as the principal unknown of the problem.

**6.14** Written in index form, Hooke's law (6.104) is,

$$\sigma_{ij} = \lambda\,\varepsilon_{kk}\,\delta_{ij} + 2\mu\,\varepsilon_{ij}\,. \tag{6.174}$$

1) Decompose the stress and strain tensors into their hydrostatic and deviatoric parts

$$\sigma_{ij} = \sigma^d_{ij} + \sigma_0\delta_{ij} \quad \sigma_0 = \frac{1}{3}\sigma_{kk} \quad \sigma^d_{ij} = \sigma_{ij} - \frac{1}{3}\sigma_{kk}\delta_{ij} \tag{6.175}$$

$$\varepsilon_{ij} = \varepsilon^d_{ij} + \varepsilon_0\delta_{ij} \quad \varepsilon_0 = \frac{1}{3}\varepsilon_{kk} \quad \varepsilon^d_{ij} = \varepsilon_{ij} - \frac{1}{3}\varepsilon_{kk}\delta_{ij}\,. \tag{6.176}$$

Show that (6.174) is equivalent to

$$\sigma^d_{ij} = 2\mu\varepsilon^d_{ij} \qquad \sigma_0 = 3K\varepsilon_0\,, \tag{6.177}$$

where the bulk modulus $K$ is defined in (6.119).

2) Show that the principal axes of the stress and strain tensors coincide.

3) Show that the potential strain energy is

$$W(\boldsymbol{\varepsilon}) = \frac{1}{2}\lambda(\varepsilon_{kk})^2 + \mu\varepsilon_{ij}\varepsilon_{ij} = \frac{9}{2}K(\varepsilon_0)^2 + \mu\varepsilon^d_{ij}\varepsilon^d_{ij}\,. \tag{6.178}$$

4) Show that the stability condition $W(\boldsymbol{\varepsilon}) > 0\ \forall\boldsymbol{\varepsilon} \neq \mathbf{0}$ amounts to imposing the conditions $K > 0$ and $\mu > 0$.

**6.15** Consider a Hookean solid for which the stress-strain relation is given by (6.104) and its inverse by (6.106).

1) If the state of stress is that of hydrostatic compression, that is,

$$\sigma_{ij} = \sigma\,\delta_{ij}\,, \tag{6.179}$$

show that the corresponding deformed state is given by

$$\varepsilon_{ij} = \varepsilon\,\delta_{ij} \qquad \varepsilon = \frac{\sigma}{3K}\,, \tag{6.180}$$

with $K$ defined in (6.119).

2) If the state of strain is that of simple shear, that is,

$$\varepsilon_{ij} = \frac{1}{2}\gamma(m_i n_j + m_j n_i) \quad m_i m_i = n_i n_i = 1 \quad m_i n_i = 0\,, \tag{6.181}$$

show that the corresponding stress state is given by

$$\sigma_{ji} = \tau(m_i n_j + m_j n_i) \qquad \tau = \mu\gamma\,. \tag{6.182}$$

Thus $\mu$ is also called the modulus of rigidity.

3) If the stress state is that of simple tension, that is,

$$\sigma_{ij} = \sigma n_i n_j \qquad n_i n_i = 1\,, \tag{6.183}$$

show that the corresponding state of strain is given by

$$\varepsilon_{ij} = \varepsilon_n n_i n_j + \varepsilon_T(\delta_{ij} - n_i n_j) \quad \varepsilon_n = \frac{\sigma}{E} \quad \varepsilon_T = -\nu\varepsilon_n\,, \tag{6.184}$$

where Young's modulus and Poisson's ratio are found as defined in (6.109).

CHAPTER 7

# Introduction to Solid Mechanics

## 7.1 Introduction

In a typical solid mechanics problem we are interested in the calculation of the displacement, stress, and strain (which, in general, are functions of time) for every point in the body. Often, for several materials, the stress-strain behavior is non-linear, inelastic, and anisotropic and the corresponding mathematical formulation is difficult. Many theories have been developed and are currently used in engineering. Among these, we will mention linear and non-linear elasticity, viscoelasticity, plasticity, and viscoplasticity. Their development has been stimulated by the use of new materials. Each of these approaches aims to model certain specific aspects of the material behavior. In solid mechanics one of the simplest stress-strain relations is the case of linear dependence between stress and strain. Such linear relations can be applied to all materials for small loads or displacements and often yield satisfactory results. Also linear elastic theory is the basis for the study of solid mechanics. For a large number of materials, such as metals and ceramics, the strain remains small and obeys Hooke's law when the applied loads are not too large. In addition, the study of linear elasticity is justified as a first step to the study of dissipative phenomena such as viscoelastic and elastoplastic. In this chapter, we present the elements of ***linear elastic theory*** of a homogeneous and isotropic medium subjected to simple loads. The discussion is focused firstly on solids subjected to static loads (i.e., independent of time) and secondly on the description of waves in linear elastic media subjected to time dependent excitations. Representative examples are given in both cases.

Additional reading can be found in the following texts: [6, 7, 30, 43, 47, 48, 54].

## 7.2 Fundamental Equations of Static Linear Elasticity

In this section we describe the deformation of a solid subjected to volume forces as well as loads imposed on its boundary. We only consider isothermal and stationary (static or elastostatic) problems.

### 7.2.1 Static Linear Elastic Field Equations

In the context of static linear elasticity, the strains and the stresses are governed by the system of equations composed as follows:

1) six equations that define the strain–displacement (2.150)

$$\boldsymbol{\varepsilon} = \frac{1}{2}\left(\nabla \boldsymbol{u} + (\nabla \boldsymbol{u})^T\right) ; \tag{7.1}$$

2) three equilibrium equations (3.125)

$$\mathbf{div}\, \boldsymbol{\sigma} + \boldsymbol{f} = 0 \ , \tag{7.2}$$

with $\boldsymbol{f} = \rho \boldsymbol{b}$ being the volume force;

3) six equations that define the homogeneous isotropic constitutive law (6.104)

$$\boldsymbol{\sigma} = \lambda \operatorname{tr} \boldsymbol{\varepsilon} \boldsymbol{I} + 2\mu \boldsymbol{\varepsilon} = \frac{\nu E}{(1+\nu)(1-2\nu)} \operatorname{tr} \boldsymbol{\varepsilon}\, \boldsymbol{I} + \frac{E}{1+\nu}\, \boldsymbol{\varepsilon} \tag{7.3}$$

or its inverse (6.108)

$$\boldsymbol{\varepsilon} = \frac{-\lambda}{2\mu(3\lambda+2\mu)} \operatorname{tr} \boldsymbol{\sigma} \boldsymbol{I} + \frac{\boldsymbol{\sigma}}{2\mu} = -\frac{\nu}{E} \operatorname{tr} \boldsymbol{\sigma} \boldsymbol{I} + \frac{(1+\nu)}{E}\, \boldsymbol{\sigma} \ , \tag{7.4}$$

where the elasticity coefficients $\lambda$, $\mu$, $E$ and $\nu$ are independent of position.

A simple count shows that there are fifteen unknowns (three displacement components $u_i$, six strain components $\varepsilon_{ij}$, and six stress components $\sigma_{ij}$) and fifteen equations; the problem is thus well posed. We have shown in section 6.10 that a linear elastic solid satisfies the second principle of thermodynamics and that there exists a potential function which, in this case, has a quadratic form in the strains (6.164) or the stresses.

There are two ways to combine the preceding fifteen scalar equations. Taking first the three displacement components $u_i$ as the primary unknowns and introducing (7.1) in (7.3), we obtain

$$\sigma_{ij} = \lambda u_{k,k}\, \delta_{ij} + \mu(u_{i,j} + u_{j,i}) \tag{7.5}$$

and by substituting into (7.2)

$$(\lambda + \mu) u_{k,ki} + \mu u_{i,jj} + f_i = 0 \, . \tag{7.6}$$

Relations (7.6) are ***Navier's equations*** which we can also write in the form

$$(\lambda + \mu)\nabla(\operatorname{div} \boldsymbol{u}) + \mu \Delta \boldsymbol{u} + \boldsymbol{f} = 0 \ , \tag{7.7}$$

where the displacement must be continuously differentiable at least two times. These equations can be solved when we impose the boundary conditions in terms either of the displacement or the contact forces expressed as functions

Claude Louis Marie Henri Navier (1785–1836) was born in Dijon and died in Paris. Navier graduated from the École Polytechnique and became an ordinary engineer for the Ponts et Chaussées in 1808. His expertise was the resistance of structures and bridges, for which he established the equations of elasticity. His memoir was read at the Académie des Sciences in May 1821 and published in 1827. Following this work, he made the synthesis of inviscid fluid dynamics proposed by Euler and with viscous effects. In 1822 at the Académie, he presented the equations for the flows of viscous incompressible fluids. This memoir was also published in 1827.

**Fig. 7.1** Claude Louis Marie Henri Navier

of displacement with (7.5). Once the displacements are known, the strains can be obtained by (7.1) and the stresses by (7.3).

We can also consider the six stress components $\sigma_{ij}$ as unknowns. Then, substituting relations (7.4) in the six compatibility equations (2.174)

$$\varepsilon_{ij,kl} + \varepsilon_{kl,ij} - \varepsilon_{jl,ik} - \varepsilon_{ik,jl} = 0\,, \tag{7.8}$$

we obtain

$$(1+\nu)\sigma_{ij,kk} - \nu\sigma_{mm,nn}\,\delta_{ij} + \sigma_{pp,ij} - (1+\nu)(\sigma_{iq,qj} + \sigma_{jr,ri}) = 0\,. \tag{7.9}$$

At the same time, from the equilibrium equation (7.2) we have the result that

$$\sigma_{iq,qj} + \sigma_{jr,ri} = -f_{i,j} - f_{j,i}\,. \tag{7.10}$$

Thus, (7.9) becomes

$$(1+\nu)\sigma_{ij,kk} - \nu\sigma_{mm,nn}\,\delta_{ij} + \sigma_{pp,ij} + (1+\nu)(f_{i,j} + f_{j,i}) = 0\,. \tag{7.11}$$

Taking the trace of this equation, leads to

$$(1-\nu)\sigma_{mm,nn} = -(1+\nu)f_{k,k} \tag{7.12}$$

which allows the simplification of relation (7.11) knowing that $\nu \neq 1$

$$\sigma_{ij,kk} + \frac{1}{1+\nu}\,\sigma_{mm,ij} + f_{i,j} + f_{j,i} + \frac{\nu}{1-\nu}\,f_{n,n}\delta_{ij} = 0\,. \tag{7.13}$$

These expressions are the ***Beltrami-Michell compatibility equations***. If the volume forces are constant, (7.13) reduces to

$$\sigma_{ij,kk} + \frac{1}{1+\nu}\sigma_{mm,ij} = 0\,. \tag{7.14}$$

Note that equations (7.14) are trivially satisfied when the components $\sigma_{ij}$ are affine functions of $\boldsymbol{x}$. Subsequently, the stress field inside the body must satisfy the three equilibrium equations, the Beltrami-Michell equations, and the boundary conditions of the problem. When the stresses are known, the strains are calculated with (7.4) and the displacements with (7.1).

### 7.2.2 Boundary Conditions

The preceding system of equations can only be solved if the appropriate boundary conditions are imposed. Let a solid occupy a domain $\Omega$ in $\mathbb{R}^3$ with boundary $\partial\Omega$. In general, we can divide the surface $\partial\Omega$ into two parts: $\partial\Omega = S_u \cup S_t$ with $S_u \cap S_t = \emptyset$, where

- $S_u$ represents the part of $\partial\Omega$ on which the displacement components $u_i$ are imposed, that is,

$$u_i = \overline{u_i} \qquad \text{on } S_u\,, \tag{7.15}$$

- $S_t$ represent the part of $\partial\Omega$ on which components of the stress vector $t_i$ are specified, that is,

$$t_i = \sigma_{ij} n_j = \bar{t}_i \qquad \text{on } S_t\,, \tag{7.16}$$

  where $n_j$ are the outgoing unit normal vector components on $S_t$.

We can classify elastostatic problems into three types according to the imposed boundary conditions:

- type I, where we only have displacement boundary conditions (7.15) and $S_u$ is not empty;
- type II, where we only have stress boundary conditions (7.16) and $S_t$ is not empty;
- type III, or mixed, where the boundary conditions specify both displacements and stresses with $S_u$ and $S_t$ not empty at the same time.

Note that stress and displacement cannot both be imposed at the same location.

Having formulated the boundary conditions, the questions of existence and uniqueness of the solution to the linear elastic problem are posed. A discussion on this subject can be found in [47].

#### Saint-Venant's Principle

Although it is relatively easy to define the boundary conditions and their type, it is often more difficult to specify them precisely, especially when we consider surface forces. This is because information on the exact distribution of contact forces is not easily known. In order to overcome this difficulty, the elasticity boundary problem is replaced with another for the same body but with the substitution of boundary conditions by statically equivalent conditions. According to Saint-Venant's principle, the effects from this replacement of the real conditions by statically equivalent conditions are local and sufficiently removed from the boundary, and so the solution to the original problem is practically identical to that of the equivalent problem. The distance at which the differences are no longer significant depends on the characteristic linear scale of the structure under consideration. This principle has proven to be very useful in many problems of practical interest.

### 7.2.3 Superposition Principle

It is worth noting that in linear elastic theory, the fifteen equations (7.1)–(7.4), as well as the boundary conditions, are linear. This leads to the formulation of the superposition principle, which is stated as follows for a type II problem. Let a body occupying the domain $\Omega$ of $\mathbb{R}^3$ be subject to forces $\overline{t_i}^{(1)}$ on $\partial\Omega$ and to volume forces $\overline{f_i}^{(1)}$. The stress field produced in this body by these forces is denoted $\sigma_{ij}^{(1)}$. The same body subject to surface forces $\overline{t_i}^{(2)}$ and volume forces $\overline{f_i}^{(2)}$ leads to the stress field ${\sigma_{ij}}^{(2)}$. The simultaneous application of the surface forces $(\overline{t_i}^{(1)} + \overline{t_i}^{(2)})$ and the volume forces $(\overline{f_i}^{(1)} + \overline{f_i}^{(2)})$ results in the stress field $({\sigma_{ij}}^{(1)} + {\sigma_{ij}}^{(2)})$. Consequently, the strains in the body are obtained from equations (7.4) by inserting the stresses $({\sigma_{ij}}^{(1)} + {\sigma_{ij}}^{(2)})$.

The superposition principle applies equally to type I and III problems.

## 7.3 Plane Isotropic Linear Elasticity

Many important practical problems do not require the solution of the three-dimensional problem for the state of stress and strain. Because of the particular geometry of the solid and the form of the loads, the elasticity equations can be considered as functions of only two spatial variables. The problem is then reduced to a plane problem.

In this section, two important cases of plane linear elasticity are defined. They are the cases of plane strain and plane stress.

### 7.3.1 Plane Strain States

Let a long prismatic bar be subject to lateral forces (fig. 7.2). We assume that the volume force component along $x_3$ is zero while the components in

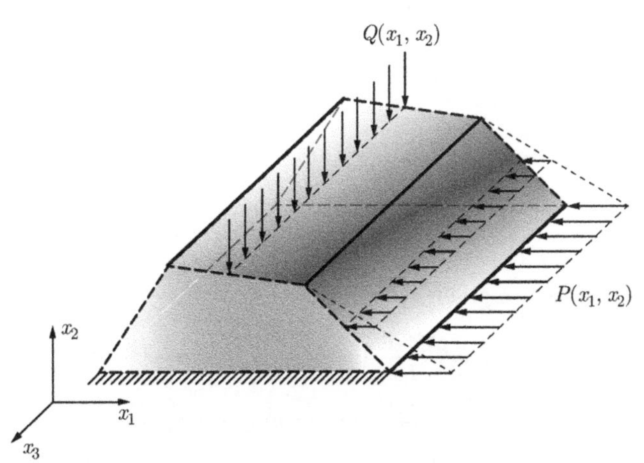


**Fig. 7.2** The case of a long prismatic structure loaded in plane strain

directions $x_1$ and $x_2$ are functions of $x_1$ and $x_2$. Because of the long length of the bar along the axis $x_3$, we can assume that the displacement $u_3$ at a certain distance from the ends is a function only of the coordinate $x_3$ and that the displacements $u_1$ and $u_2$ only depend on $x_1$ and $x_2$

$$u_1 = u_1(x_1, x_2) \qquad u_2 = u_2(x_1, x_2) \qquad u_3 = u_3(x_3)\,. \tag{7.17}$$

For a prismatic structure of infinite length or when its ends are fixed, we can assume in addition that $u_3 = 0$ in each section. The strain components are then given by

$$\varepsilon_{11} = \frac{\partial u_1}{\partial x_1} \qquad \varepsilon_{22} = \frac{\partial u_2}{\partial x_2} \qquad \varepsilon_{12} = \frac{1}{2}\left(\frac{\partial u_1}{\partial x_2} + \frac{\partial u_2}{\partial x_1}\right) \tag{7.18}$$

and

$$\begin{aligned} \varepsilon_{33} &= \frac{\partial u_3}{\partial x_3} = 0 \\ \varepsilon_{13} &= \frac{1}{2}\left(\frac{\partial u_1}{\partial x_3} + \frac{\partial u_3}{\partial x_1}\right) = 0 \\ \varepsilon_{23} &= \frac{1}{2}\left(\frac{\partial u_2}{\partial x_3} + \frac{\partial u_3}{\partial x_2}\right) = 0\,. \end{aligned} \tag{7.19}$$

The strain state so defined is called plane strain. Using Hooke's law (7.3), we obtain that the stresses $\sigma_{11}, \sigma_{22}, \sigma_{33}$, and $\sigma_{12}$ are functions only of $x_1$ and $x_2$, and that $\sigma_{23}$ and $\sigma_{31}$ are zero everywhere. Consequently, the equilibrium equations (7.2) become

$$\frac{\partial \sigma_{11}}{\partial x_1} + \frac{\partial \sigma_{12}}{\partial x_2} + f_1 = 0 \qquad \frac{\partial \sigma_{21}}{\partial x_1} + \frac{\partial \sigma_{22}}{\partial x_2} + f_2 = 0\,. \tag{7.20}$$

Knowing the strains (equations (7.18) and (7.19)), we can employ the stress-strain relations (7.3) to calculate the stresses as follows:

$$\sigma_{11} = \frac{E}{(1+\nu)(1-2\nu)}\left(\varepsilon_{11}(1-\nu) + \nu\varepsilon_{22}\right) \tag{7.21}$$

$$\sigma_{22} = \frac{E}{(1+\nu)(1-2\nu)}\left(\varepsilon_{22}(1-\nu) + \nu\varepsilon_{11}\right) \tag{7.22}$$

$$\sigma_{12} = \frac{E}{(1+\nu)}\varepsilon_{12} \tag{7.23}$$

$$\sigma_{33} = \frac{E}{(1+\nu)(1-2\nu)}\nu(\varepsilon_{11} + \varepsilon_{22})\,. \tag{7.24}$$

Inversely, the strains are given by

$$\varepsilon_{11} = \frac{1+\nu}{E}\left((1-\nu)\sigma_{11} - \nu\sigma_{22}\right) \tag{7.25}$$

$$\varepsilon_{22} = \frac{1+\nu}{E}\left((1-\nu)\sigma_{22} - \nu\sigma_{11}\right) \tag{7.26}$$

$$\varepsilon_{12} = \frac{1+\nu}{E}\sigma_{12}\,. \tag{7.27}$$

We impose the same restrictions on the surface forces (fig. 7.3). The surface forces $\bar{t}_1$ and $\bar{t}_2$ must be functions of only $x_1$ and $x_2$, with $\bar{t}_3 = 0$, in order that the strain be truly plane. Thus, for conditions of type II, we write

$$\bar{t}_1 = \sigma_{11} n_1 + \sigma_{12} n_2 \qquad \bar{t}_2 = \sigma_{12} n_1 + \sigma_{22} n_2 \,, \tag{7.28}$$

where $n_1$ and $n_2$ are the components of outgoing unit normal $\boldsymbol{n}$ on $\partial\Omega$. When the stresses are chosen as unknowns, the compatibility equations must be used. Under the plane strain assumption, the only compatibility equation that is not automatically satisfied is

$$\frac{\partial^2 \varepsilon_{11}}{\partial {x_2}^2} + \frac{\partial^2 \varepsilon_{22}}{\partial {x_1}^2} = 2 \frac{\partial^2 \varepsilon_{12}}{\partial x_1 \partial x_2} \,. \tag{7.29}$$

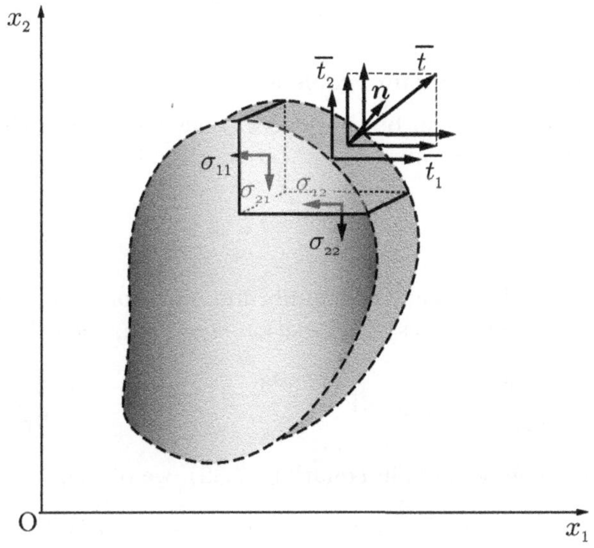


**Fig. 7.3** Boundary conditions for plane elasticity

Thus, in the case of plane strain, eight quantities, $\varepsilon_{11}$, $\varepsilon_{22}$, $\varepsilon_{12}$, $\sigma_{11}$, $\sigma_{22}$, $\sigma_{12}$, $u_1$, and $u_2$, must be determined to satisfy equations (7.18), (7.20), and (7.25)–(7.27) as well as the boundary conditions, (7.28). The eight preceding scalar equations can be reduced to three in the following way.

1) Introducing equations (7.25)–(7.27) in (7.29), it follows that

$$\frac{\partial^2}{\partial x_2^2}\big((1-\nu)\sigma_{11} - \nu\sigma_{22}\big) + \frac{\partial^2}{\partial x_1^2}\big((1-\nu)\sigma_{22} - \nu\sigma_{11}\big) = 2\frac{\partial^2 \sigma_{12}}{\partial x_1 \partial x_2} \,. \tag{7.30}$$

2) Differentiating the first and second equations of (7.20) with respect to $x_1$ and $x_2$, respectively, and adding the two resulting equations, we obtain

$$-2\frac{\partial^2 \sigma_{12}}{\partial x_1 \partial x_2} = \left(\frac{\partial^2 \sigma_{11}}{\partial {x_1}^2} + \frac{\partial^2 \sigma_{22}}{\partial {x_2}^2}\right) + \left(\frac{\partial f_1}{\partial x_1} + \frac{\partial f_2}{\partial x_2}\right) . \tag{7.31}$$

3) Substiting (7.31) in (7.30), we have

$$\left(\frac{\partial^2}{\partial x_1^2}+\frac{\partial^2}{\partial x_2^2}\right)(\sigma_{11}+\sigma_{22})=-\frac{1}{1-\nu}\left(\frac{\partial f_1}{\partial x_1}+\frac{\partial f_2}{\partial x_2}\right) , \qquad (7.32)$$

which is the compatibility equation as a function of the stress components.

After the preceding analysis, we now have a set of three equations: two equilibrium equations (7.20) and a compatibility equation (7.32). These three equations have $\sigma_{11}, \sigma_{22}$, and $\sigma_{12}$ as unknowns. This system of equations, with boundary conditions (7.28), can be used to obtain a solution for a plane strain problem. We note that the solution satisfying this system is unique. After having determined the stress components, the strains are calculated with equations (7.25)–(7.27) and the displacements with (7.18).

**Stress Function for Plane Strain Problems**

The plane strain problem can be simplified to one equation with a single variable. If we assume that the volume forces are derived from a potential $V(x_1, x_2)$

$$f_i=-\frac{\partial V}{\partial x_i}, \quad i=1,2 , \qquad (7.33)$$

it is not difficult to show that the equilibrium equations are met if the stress components are the derivatives of a function $\Phi(x_1, x_2)$, such that

$$\sigma_{11}=V+\frac{\partial^2\Phi}{\partial {x_2}^2} \qquad \sigma_{22}=V+\frac{\partial^2\Phi}{\partial {x_1}^2} \qquad \sigma_{12}=-\frac{\partial^2\Phi}{\partial x_1\partial x_2} . \qquad (7.34)$$

Introducing these components in equation (7.32), we obtain

$$\frac{\partial^4\Phi}{\partial x_1^4}+2\frac{\partial^4\Phi}{\partial x_1^2\partial x_2^2}+\frac{\partial^4\Phi}{\partial x_2^4}+\frac{1-2\nu}{1-\nu}\left(\frac{\partial^2 V}{\partial {x_1}^2}+\frac{\partial^2 V}{\partial {x_2}^2}\right)=0 \qquad (7.35)$$

or

$$\triangle\triangle\Phi+\frac{1-2\nu}{1-\nu}\triangle V=0 . \qquad (7.36)$$

When the volume forces are negligible, the stresses are given by

$$\sigma_{11}=\frac{\partial^2\Phi}{\partial {x_2}^2} \qquad \sigma_{22}=\frac{\partial^2\Phi}{\partial {x_1}^2} \qquad \sigma_{12}=-\frac{\partial^2\Phi}{\partial x_1\partial x_2} \qquad (7.37)$$

and equation (7.36) becomes

$$\triangle\triangle\Phi=0 , \qquad (7.38)$$

which is a ***biharmonic equation***. The plane strain problem in linear elasticity is thus reduced to seeking a function $\Phi$, called an ***Airy stress function***, satisfying (7.38) for the appropriate boundary conditions. The knowledge of this function permits the determination of the stresses with (7.34), the strains with (7.25)–(7.27), and the displacements with (7.18).

### 7.3.2 Plane Stress States

Now we consider figure 7.4 where we have the opposite of the long bar of figure 7.2. We assume that the body has one dimension, along $x_3$, that is very small with respect to the dimensions in the plane $x_1x_2$. We also suppose that the surface forces are applied parallel to the plane $x_1x_2$. The volume forces along $x_3$ are zero while in the directions $x_1$ and $x_2$, they are functions only of $x_1$ and $x_2$. In view of the geometry and the applied loads, we can assume that the stress components $\sigma_{33}$, $\sigma_{13}$, and $\sigma_{23}$ are zero everywhere and that the other components $\sigma_{11}$, $\sigma_{22}$, and $\sigma_{12}$ remain almost constant across the thickness of the plate. Such a stress state is called plane stress and is written as

$$\sigma_{11} = \sigma_{11}(x_1, x_2) \qquad \sigma_{22} = \sigma_{22}(x_1, x_2) \qquad \sigma_{12} = \sigma_{12}(x_1, x_2) \tag{7.39}$$

$$\sigma_{33} = \sigma_{13} = \sigma_{23} = 0\,. \tag{7.40}$$

The equilibrium equations are given by (7.20) and the boundary conditions by (7.28). To obtain the stress-strain relations, we use relation (7.4) which reduces to

$$\begin{aligned} \varepsilon_{11} &= \frac{1}{E}\left(\sigma_{11} - \nu\sigma_{22}\right) \qquad \varepsilon_{22} = \frac{1}{E}\left(\sigma_{22} - \nu\sigma_{11}\right) \\ \varepsilon_{12} &= \frac{1+\nu}{E}\sigma_{12} \end{aligned} \tag{7.41}$$

$$\varepsilon_{13} = \varepsilon_{23} = 0 \qquad \varepsilon_{33} = -\frac{\nu}{E}\left(\sigma_{11} + \sigma_{22}\right). \tag{7.42}$$

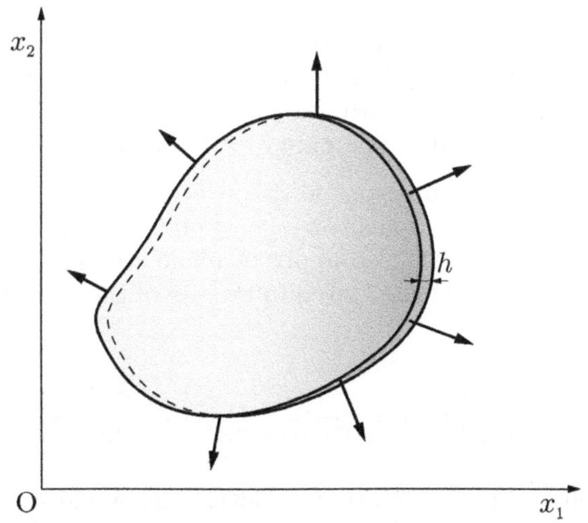


**Fig. 7.4** Thin plate with plane loads

Inverting these relations, we have the stresses

$$\begin{aligned}\sigma_{11} &= \frac{E}{1-\nu^2}\left(\varepsilon_{11}+\nu\varepsilon_{22}\right)\\ \sigma_{22} &= \frac{E}{1-\nu^2}\left(\varepsilon_{22}+\nu\varepsilon_{11}\right)\\ \sigma_{12} &= \frac{E}{1+\nu}\,\varepsilon_{12}\,.\end{aligned} \tag{7.43}$$

Extracting $(\sigma_{11}+\sigma_{22})$ from equations (7.41) and inserting the result in the last equation of (7.42), we obtain

$$\varepsilon_{33} = -\frac{\nu}{1-\nu}\left(\varepsilon_{11}+\varepsilon_{22}\right). \tag{7.44}$$

This equation gives the normal strain "out of the plane" as a function of the tangent strains "in the plane". Note that $\varepsilon_{33}$ is not among the quantities that characterize the plane stress. Nonetheless, we can obtain it independently using the last equation. We can naturally obtain $u_3$ using $\varepsilon_{33} = \partial u_3/\partial x_3$. The displacements $u_1$ and $u_2$ are independent of $x_3$ and the strain-displacement relations are given by (7.18).

As for the compatibility equation, we have at our disposal relation (7.29) and the following equations which come from the non-zero component $\varepsilon_{33}$

$$\frac{\partial\varepsilon_{33}}{\partial x_1} = \frac{\partial\varepsilon_{33}}{\partial x_2} = \frac{\partial^2\varepsilon_{33}}{\partial x_1\partial x_2} = 0\,. \tag{7.45}$$

The integration of the last relation imposes the following condition on $\varepsilon_{33}$

$$\varepsilon_{33} = A_0 + A_1x_1 + A_2x_2\,. \tag{7.46}$$

In the solution of plane stress problems, this condition is generally too restrictive and is not satisfied; only equation (7.29) is considered. Although the resulting solutions are approximate, they are satisfactory as long as the plate thickness remains very small as compared to the planar dimensions.

As in plane strain, the equations for plane stress reduce to three equations in which the three stress components $\sigma_{11}$, $\sigma_{22}$, and $\sigma_{12}$ appear. This is evident as equations (7.18) and (7.29) also apply to plane stress. The substitution of the strain components (7.41) in (7.29) and the use of the equilibrium equations (7.20) yield

$$\left(\frac{\partial^2}{\partial x_1^2}+\frac{\partial^2}{\partial x_2^2}\right)\left(\sigma_{11}+\sigma_{22}\right) = -(1+\nu)\left(\frac{\partial f_1}{\partial x_1}+\frac{\partial f_2}{\partial x_2}\right). \tag{7.47}$$

This last equation and the equilibrium equations form a system of three equations in three unknowns. We also note that for a given problem, the solution to this system is unique.

**Stress Function for Plane Stress Problems**

Substituting equations (7.37) in equation (7.47) and neglecting the volume forces, we again end up with a stress function satisfying

$$\triangle\triangle\Phi = 0\,. \tag{7.48}$$

To summarize, we have shown that one single differential equation governs the stress function for the two cases of two-dimensional linear elasticity: plane strain and plane stress. The distinction between these two cases appears after the stress calculation, when the solution for the strains with equations (7.25)–(7.27) and (7.41)–(7.42), respectively, for the states of plane strain and plane stress is carried out.

Using the appropriate combinations of elastic constants, the plane strain equations can be converted into the plane stress equations. That is:

*From plane strain to plane stress:*

the stress-strain relations (7.21)–(7.24) for the plane strain case can be converted into those for plane stress (7.43) if we replace Young's modulus $E$ in (7.43) with $E(1+2\nu)/(1+\nu)^2$ and Poisson's ratio $\nu$ by $\nu/(1+\nu)$.

*From plane stress to plane strain:*

similarly, the stress-strain relations (7.43) for the plane stress case can be converted into those for plane strain (7.21)–(7.24) if we replace Young's modulus $E$ in (7.43) with $E/(1-\nu^2)$ and Poisson's ratio $\nu$ with $\nu/(1-\nu)$.

Then, the solution of a plane stress problem can be determined from the corresponding problem in plane deformation, and vice-versa.

## 7.4 Solution Methods in Linear Elasticity

We have shown in section 7.2 that the solution of a three-dimensional isotropic linear elasticity problem requires the treatment of fifteen equations in fifteen unknowns satisfying the specified boundary conditions. These fifteen equations are combined such that: (a) three unknowns are the displacement components, solutions of equations (7.6), (b) six unknowns are the stress components, solutions of equations (7.13). Plane elastic problems, either plane strain or plane stress, can be reduced to eight equations in eight unknowns. The number of equations and unknowns can also be reduced in a way similar to the three-dimensional case.

Directly obtaining analytical solutions in elasticity problems is not easy, and often, is not even possible. Consequently, methods based on the rigorous use of applied mathematics are proposed to handle the different classes of problems, while other techniques permit approximate solutions based on intuition and experience. Below, we present a list of the methods most often used in linear elasticity.

- *Inverse Method.* In this method, the displacement or stress field is assigned to the body and we determine all the other quantities, including external forces. Although the solution to inverse problems pose no particular difficulties, it is not always possible to find a solution that the engineer finds interesting [6, 54].
- *Method of Potential.* To simplify the elasticity equations, we introduce potential functions. Potentials for the displacements yield the solution to Navier's equations and those for stress yield systems of stress in equilibrium [15, 47, 50, 65].
- *Semi-Inverse Method.* In this method, part of the stress and displacement fields are specified. Then, knowing these elements and applying the elasticity theory, the equations which must be satisfied by the remaining stresses and displacements are determined. These equations are normally easy to integrate, and combined with the portion of data originally specified, they yield a complete and precise solution for many interesting problems in engineering. Saint-Venant applied this method to the bending and torsion of prismatic bars [6, 54, 61].
- *Complex Variable Methods.* This method uses analytic functions defined in the complex plane to solve elasticity problems. It can only be applied to plane problems [47].
- *Variational Methods.* These methods are based on the fact that the elasticity equations can be obtained by minimizing a generalized energy function; for example, see [15, 47].
- *Others.* Other methods include integral transform methods and numerical approaches such as the finite element method; for example, see [15, 21, 48].

In this chapter we present the method of potential and semi-inverse method for the solution of representative problems with the principal objective of highlighting the classical formulations of elasticity theory. The application of other methods for the solution of various problems is abundantly treated in the literature [15, 47, 48, 54, 61].

In the preceding sections, we have shown that an elasticity problem can be formulated in terms of displacements with Navier's equations (7.6) as field equations. Another formulation is based on the stresses, for which the compatibility equations (7.8) and the equilibrium equations (7.2) constitute a system of nine equations. In this section, we develop a general context where the displacement or stress functions are introduced in order to satisfy Navier's equations (7.6) or the Beltrami-Michell compatibility equations (7.13) and the equilibrium equations (7.2) respectively. We show that such functions provide the solution to certain elasticity problems. For simplicity, we only consider the case where there are no volume forces. When volume forces are taken into consideration, the methodology becomes more difficult and is beyond the scope of this text. The reader can find more advanced and pertinent complements in references [15, 47].

### 7.4.1 Displacement Functions

To solve Navier's equations, potential or displacement functions are introduced in such a way that the displacement vector in Navier's equations is obtained from the derivatives of these functions. These potential functions are governed by Laplace's equation or the biharmonic equation, well known in mathematical physics. To advance further in that sense, we introduce Helmholtz' decomposition theorem; see [8] for a proof.

**Helmholtz' Theorem**

*A finite and continuous vector field* $\boldsymbol{a}$*, that is zero at infinity, can be represented as the sum of an irrotational field* $\boldsymbol{b}$ *and a solenoidal field* $\mathbf{c}$

$$\boldsymbol{a} = \boldsymbol{b} + \mathbf{c} \tag{7.49}$$

*with*

$$\boldsymbol{\nabla} \times \boldsymbol{b} = 0 \quad \textit{and} \quad \operatorname{div} \mathbf{c} = 0 \; . \tag{7.50}$$

To formulate the solution to Navier's equations in terms of potential functions, we state the following definitions.

- For an irrotational field, there exists a scalar potential $\varphi$ such that $\boldsymbol{b} = \boldsymbol{\nabla}\varphi$. Since the gradient operator only involves first derivatives, the function $\varphi$ is determined to within an arbitrary additive constant.
- For a solenoidal field, there always exists a potential vector $\boldsymbol{\Psi}$ such that $\boldsymbol{c} = \boldsymbol{\nabla} \times \boldsymbol{\Psi}$. This potential is determined to within an additive vector function.

Thus a finite, continuous displacement field $\boldsymbol{u}$, that goes to zero at infinity, can be represented, following Helmholtz' theorem, by the sum

$$\boldsymbol{u} = \boldsymbol{\nabla}\varphi + \boldsymbol{\nabla} \times \boldsymbol{\Psi} \tag{7.51}$$

with the conditions that $\boldsymbol{\nabla} \times \boldsymbol{\nabla}\varphi = \mathbf{0}$ and $\operatorname{div}(\boldsymbol{\nabla} \times \boldsymbol{\Psi}) = 0$. Note that $\boldsymbol{u}$ has three scalar components while $\varphi$ and $\boldsymbol{\Psi}$ together have four. We can thus impose the following condition without loss of generality:

$$\operatorname{div} \boldsymbol{\Psi} = 0 \; . \tag{7.52}$$

It is interesting to examine the divergence and the curl of the displacement expressed in (7.51). Using (7.52), (1.179), (1.188), and (1.190), we obtain

$$\operatorname{div} \boldsymbol{u} = \operatorname{div} \boldsymbol{\nabla}\varphi + \operatorname{div}(\boldsymbol{\nabla} \times \boldsymbol{\Psi}) = \operatorname{div} \boldsymbol{\nabla}\varphi + 0 = \nabla^2\varphi \tag{7.53}$$

$$\begin{aligned} \boldsymbol{\nabla} \times \boldsymbol{u} &= \boldsymbol{\nabla} \times \boldsymbol{\nabla}\varphi + \boldsymbol{\nabla} \times (\boldsymbol{\nabla} \times \boldsymbol{\Psi}) = \mathbf{0} + \boldsymbol{\nabla} \times (\boldsymbol{\nabla} \times \boldsymbol{\Psi}) \\ &= \nabla(\operatorname{div} \boldsymbol{\Psi}) - \nabla^2\boldsymbol{\Psi} = -\nabla^2\boldsymbol{\Psi} \; . \end{aligned} \tag{7.54}$$

Note that with (2.163), $\operatorname{div} \boldsymbol{u} = \varepsilon_{ii}$ and thus $\nabla^2\varphi = \varepsilon_{ii}$. The curl of the displacement vector, i.e., $\boldsymbol{\nabla} \times \boldsymbol{u}$, is related to the body rotation vector, whose components are those of the antisymmetric infinitesimal rotation tensor $\boldsymbol{\omega}$, multiplied by the factor 2 (see eqn. (2.168)).

As was previously justified, we assume that $\boldsymbol{f} = \mathbf{0}$ in (7.7). Introducing (7.51) in (7.7) and with (7.53), also the vector identities (1.180), (1.188), and relations (1.236), (1.237), we obtain

$$(\lambda + 2\mu)\boldsymbol{\nabla}(\nabla^2\varphi) + \mu\boldsymbol{\nabla} \times (\nabla^2\boldsymbol{\Psi}) = 0 \ . \tag{7.55}$$

Then, every pair of functions $\varphi$ and $\boldsymbol{\Psi}$ satisfying (7.55) produces a displacement field, given by (7.51), that is a solution to Navier's equations. Inversely, for any displacement $\boldsymbol{u}$ that satisfies Navier's equations, there exists at least one set of functions $\varphi$ and $\boldsymbol{\Psi}$ satisfying (7.51); for example, see [6, 15, 65].

**Lamé Strain Potential**

Particular solutions of (7.55) are generated from the two equations

$$\nabla^2\varphi = \text{cnst} \quad \text{and} \quad \nabla^2\boldsymbol{\Psi} = \text{cnst} \ . \tag{7.56}$$

When

$$\nabla^2\varphi = \text{cnst} \quad \text{and} \quad \boldsymbol{\Psi} = 0 \ , \tag{7.57}$$

the function $\varphi$ is called the Lamé strain potential and the displacement is obtained from

$$\boldsymbol{u} = \boldsymbol{\nabla}\varphi \ , \tag{7.58}$$

which satisfies Navier's equation. To simplify the solution in the applications, it is very common to write (7.58) in the form

$$\boldsymbol{u} = \frac{1}{2\mu}\boldsymbol{\nabla}\varphi \ . \tag{7.59}$$

Thus any function that satisfies Poisson's equation (7.57) can serve as a strain potential. When $\varphi$ is known, the displacement vector is obtained with (7.59), the strains with (7.1), and the stresses with (7.3). Note that all these quantities are expressed as first and second derivatives of $\varphi$. As examples, we have

$$u_i = \frac{1}{2\mu}\varphi_{,i} \tag{7.60}$$

$$\varepsilon_{ij} = \frac{1}{2}\left(u_{i,j} + u_{j,i}\right) = \frac{1}{2\mu}\varphi_{,ij} \tag{7.61}$$

$$\varepsilon_{kk} = u_{k,k} = \frac{1}{2\mu}\varphi_{,kk} \tag{7.62}$$

$$\sigma_{ij} = \lambda\delta_{ij}\varepsilon_{kk} + 2\mu\varepsilon_{ij} = \frac{\lambda}{2\mu}\varphi_{,kk}\delta_{ij} + \varphi_{,ij} \ . \tag{7.63}$$

In many practical elasticity problems, the objective is not to obtain a general solution, but a particular solution. Thus, for simplicity, consider

$$\nabla^2\varphi = 0 \ , \tag{7.64}$$

which is Laplace's equation, and $\varphi$ a harmonic function. Below we list a few useful harmonic functions for the solution of certain practical problems:

$$\varphi(r,\theta) = Cr^n \cos n\theta, \quad r^2 = x_1^2 + x_2^2, \tag{7.65}$$

$$\varphi(r) = C \ln \frac{r}{K}, \quad r^2 = x_1^2 + x_2^2, \tag{7.66}$$

$$\varphi(\theta) = C\theta, \quad \theta = \tan^{-1} \frac{x_2}{x_1}, \tag{7.67}$$

$$\varphi(R) = \frac{C}{R}, \quad R^2 = x_1^2 + x_2^2 + x_3^2 \,. \tag{7.68}$$

When we use spherical coordinates from Appendix B in the following examples, we replace $r$ with the symbol $R$. The two Poisson type functions below are useful for the solution of the hollow sphere and cylinder subject to internal and external pressure when combined with (7.66) and (7.68):

$$\varphi(R) = DR^2, \quad R^2 = x_1^2 + x_2^2 + x_3^2, \tag{7.69}$$

$$\varphi(r) = Dr^2, \quad r^2 = x_1^2 + x_2^2 \,. \tag{7.70}$$

**Hollow Sphere Subject to Internal and External Pressure**

Let a hollow sphere with internal and external radii $r_i$ and $r_e$ (fig. 7.5) be subject to internal and external pressures $P_i$ and $P_e$, respectively. The volume forces are considered negligible.

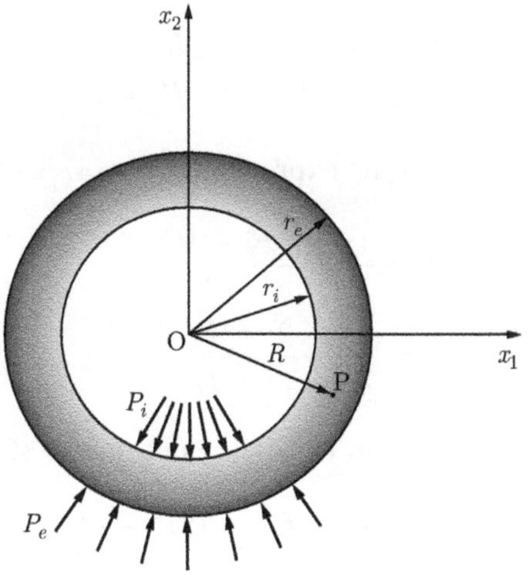


**Fig. 7.5** Hollow sphere subject to internal and external pressure loads

Because of spherical symmetry, we use spherical coordinates $(R, \theta, \varphi)$. In this system all the shear stresses and strains are zero, and of the three components of the vector displacement $u_R, u_\varphi, u_\theta$, only $u_R$ is non-zero.

The solution to this problem can be obtained by combining the two potentials (7.68) and (7.69)

$$\varphi(R) = \frac{C}{R} + DR^2 \,. \tag{7.71}$$

This potential satisfies Poisson's equation $(7.56)_1$ and yields the stresses and strains that fulfill all the geometric characteristics of the problem. With (7.60) and the equations in Appendix B, the displacement components are

$$u_R = \frac{1}{2\mu}\left(-\frac{C}{R^2} + 2DR\right), \quad u_\theta = u_\varphi = 0\ . \tag{7.72}$$

With the displacement (7.72), the strains are obtained using relations (B.22)–(B.24) from Appendix B

$$\begin{aligned} \varepsilon_{RR} &= \frac{1}{2\mu}\left(\frac{2C}{R^3} + 2D\right), \quad \varepsilon_{\varphi\varphi} = \varepsilon_{\theta\theta} = \frac{1}{2\mu}\left(-\frac{C}{R^3} + 2D\right) \\ \varepsilon_{\theta\varphi} &= \varepsilon_{\theta R} = \varepsilon_{\varphi R} = 0\ . \end{aligned} \tag{7.73}$$

Inserting these components into Hooke's law, (7.3), we obtain the stresses

$$\begin{aligned} \sigma_{RR} &= \frac{2C}{R^3} + 2\frac{1+\nu}{1-2\nu}D, \quad \sigma_{\varphi\varphi} = \sigma_{\theta\theta} = -\frac{C}{R^3} + 2\frac{1+\nu}{1-2\nu}D \\ \sigma_{\theta\varphi} &= \sigma_{\theta R} = \sigma_{\varphi R} = 0\ . \end{aligned} \tag{7.74}$$

The constants $C$ and $D$ in (7.74) are determined by imposing the boundary conditions

$$\begin{aligned} \sigma_{RR} &= -P_i \quad \text{at} \quad R = r_i \\ \sigma_{RR} &= -P_e \quad \text{at} \quad R = r_e\ . \end{aligned} \tag{7.75}$$

Applying these conditions to the expression for $\sigma_{RR}$, leads to

$$\begin{aligned} C &= \frac{1}{2}\frac{r_e^3 r_i^3 (P_e - P_i)}{r_e^3 - r_i^3} \\ D &= \frac{1}{2}\frac{1-2\nu}{1+\nu}\frac{r_i^3 P_i - r_e^3 P_e}{r_e^3 - r_i^3}\ . \end{aligned} \tag{7.76}$$

Substituting these expressions in (7.74), we obtain the stresses

$$\begin{aligned} \sigma_{RR} &= \frac{1}{R^3}\frac{r_e^3 r_i^3 (P_e - P_i)}{r_e^3 - r_i^3} + \frac{r_i^3 P_i - r_e^3 P_e}{r_e^3 - r_i^3} \\ &= -\frac{P_i\left(\frac{r_e^3}{R^3} - 1\right)}{\left(\frac{r_e^3}{r_i^3} - 1\right)} - \frac{P_e\left(1 - \frac{r_i^3}{R^3}\right)}{\left(1 - \frac{r_i^3}{r_e^3}\right)} \end{aligned} \tag{7.77}$$

$$\begin{aligned} \sigma_{\varphi\varphi} &= \sigma_{\theta\theta} = -\frac{1}{2R^3}\frac{r_e^3 r_i^3 (P_e - P_i)}{r_e^3 - r_i^3} + \frac{r_i^3 P_i - r_e^3 P_e}{r_e^3 - r_i^3} \\ &= \frac{1}{2}\left(\frac{P_i\left(\frac{r_e^3}{R^3} + 2\right)}{\left(\frac{r_e^3}{r_i^3} - 1\right)} - \frac{P_e\left(\frac{r_i^3}{R^3} + 2\right)}{\left(1 - \frac{r_i^3}{r_e^3}\right)}\right)\ . \end{aligned} \tag{7.78}$$

The non-zero component of the displacement becomes

$$\begin{aligned} u_R &= \frac{R}{2\mu}\left(-\frac{1}{2R^3}\frac{r_e^3 r_i^3(P_e-P_i)}{r_e^3-r_i^3}+\frac{1-2\nu}{1+\nu}\frac{r_i^3P_i-r_e^3P_e}{r_e^3-r_i^3}\right) \\ &= \frac{R}{2\mu}\left(P_i\frac{\frac{1}{2}\frac{r_e^3}{R^3}+\frac{1-2\nu}{1+\nu}}{\frac{r_e^3}{r_i^3}-1}-P_e\frac{\frac{1}{2}\frac{r_i^3}{R^3}+\frac{1-2\nu}{1+\nu}}{1-\frac{r_i^3}{r_e^3}}\right) . \end{aligned} \tag{7.79}$$

It is interesting to notice that if $r_e \gg r_i$, the stresses and displacements are approximated by

$$\sigma_{RR} = -P_i\frac{r_i^3}{R^3}-P_e\left(1-\frac{r_i^3}{R^3}\right) \tag{7.80}$$

$$\sigma_{\theta\theta} = \sigma_{\varphi\varphi} = \frac{P_i}{2}\frac{r_i^3}{R^3}-\frac{P_e}{2}\left(\frac{r_i^3}{R^3}+2\right) \tag{7.81}$$

$$u_R = \frac{R}{2\mu}\left(\frac{P_i}{2}\frac{r_i^3}{R^3}-P_e\left(\frac{1-2\nu}{1+\nu}+\frac{1}{2}\frac{r_i^3}{R^3}\right)\right) . \tag{7.82}$$

On the internal surface of the sphere, $R = r_i$, the stresses and displacements become

$$\sigma_{\theta\theta}|_{R=r_i} = \sigma_{\varphi\varphi}|_{R=r_i} = \frac{P_i}{2}-\frac{3P_e}{2}, \quad \sigma_{RR}|_{R=r_i} = -P_i \tag{7.83}$$

$$u_R|_{R=r_i} = \frac{R}{2\mu}\left(\frac{P_i}{2}-\frac{3P_e}{2}\frac{1-\nu}{1+\nu}\right) . \tag{7.84}$$

When $R \to \infty$, $r_i/R \to 0$ and equations (7.80)–(7.82) simplify to

$$\sigma_{RR} = \sigma_{\theta\theta} = \sigma_{\varphi\varphi} = -P_e \quad \text{and} \quad u_R = -\frac{RP_e}{2\mu}.\frac{1-2\nu}{1+\nu} \tag{7.85}$$

**Hollow Cylinder with Fixed Ends under Internal and External Pressure**

A hollow cylinder with internal and external radii $r_i$ and $r_e$ (fig. 7.6) is subject to internal and external pressures $P_i$ and $P_e$, respectively. The volume forces are assumed to be zero.

Because of the cylindrical symmetry, it is in this case useful to use cylindrical coordinates $(r, \theta, z)$ (Appendix A). In this system, all the shear stresses and strains are zero, and of the three displacement vector components, only the component $u_r$ is non-zero. Here we are in the case of a plane strain problem, since no deformation is allowed in the direction of the axis of the cylinder, subject to the boundary conditions

$$\sigma_{rr} = -P_i, \quad \sigma_{r\theta} = 0 \quad \text{at} \quad r = r_i \tag{7.86}$$

$$\sigma_{rr} = -P_e, \quad \sigma_{r\theta} = 0 \quad \text{at} \quad r = r_e . \tag{7.87}$$

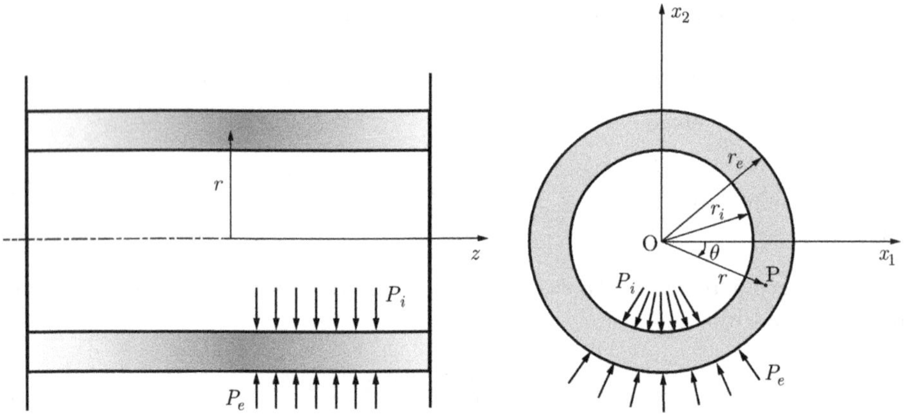


**Fig. 7.6** Hollow cylinder subject to internal and external pressure loads

The solution to this problem can be obtained by combining the two potentials (7.66) and (7.70)

$$\varphi(r) = C_1 \ln \frac{r}{K} + C_2 r^2 \tag{7.88}$$

where $C_1, K$, and $C_2$ are constants to determine from the boundary conditions. Applying the procedure from the preceding example, the stress and displacement components are

$$\begin{aligned} \sigma_{rr} &= \frac{1}{r_e^2 - r_i^2}\left(r_i^2 P_i - r_e^2 P_e + \frac{r_i^2 r_e^2}{r^2}(P_e - P_i)\right) \\ \sigma_{\theta\theta} &= \frac{1}{r_e^2 - r_i^2}\left(r_i^2 P_i - r_e^2 P_e - \frac{r_i^2 r_e^2}{r^2}(P_e - P_i)\right) \end{aligned} \tag{7.89}$$

$$\sigma_{zz} = 2\nu \frac{r_i^2 P_i - r_e^2 P_e}{r_e^2 - r_i^2} \tag{7.90}$$

$$\sigma_{r\theta} = \sigma_{rz} = \sigma_{\theta z} = 0 \tag{7.91}$$

$$u_r = \frac{r}{2\mu}\left(-\frac{1}{r^2}\frac{r_e^2 r_i^2 (P_e - P_i)}{r_e^2 - r_i^2} + \frac{r_i^2 P_i - r_e^2 P_e}{r_e^2 - r_i^2}(1 - 2\nu)\right) . \tag{7.92}$$

This problem is also solved with two different methods, example 7.4, for the case of a plane stress state.

### Galerkin Vector

The displacement vector (7.51) is represented by a sum of first derivatives, via the differential operator $\boldsymbol{\nabla}(\bullet) = \partial(\bullet)/\partial x_i$, of a scalar function $\varphi$ and a vector function $\boldsymbol{\Psi}$. To be able to construct solutions devoted to general applications, the use of second order differential operators is more appropriate. The operators of this type are the Laplacian $\nabla^2$ (sec. 1.4.8) and $\boldsymbol{\nabla}(\text{div}(\bullet)) = \frac{\partial}{\partial x_i}\left(\frac{\partial(\bullet)}{\partial x_j}\right)$. These operators can be expressed in an arbitrary coordinate system and are applied to a vector function.

Let $\boldsymbol{V}$ be a vector function related to the displacement $\boldsymbol{u}$ by the expression

$$2\mu\boldsymbol{u} = 2(1-\nu)\nabla^2\boldsymbol{V} - \boldsymbol{\nabla}(\text{div}\,\boldsymbol{V})\ . \tag{7.93}$$

The factor $2\mu$ is introduced for simplicity in the applications. The vector $\boldsymbol{V}$ is the Galerkin vector which yields a general solution of Navier's equations. Introducing (7.93) in (7.7), using vector identities (1.188), (1.191), and (1.236), and bearing in mind that $2(1-\nu) = (\lambda+2\mu)/(\lambda+\mu)$, we obtain

$$\nabla^2\left(\nabla^2\boldsymbol{V}\right) = 0\ . \tag{7.94}$$

Consequently, any biharmonic vector function can serve as a Galerkin vector, and the displacement $\boldsymbol{u}$ in (7.93) will satisfy (7.7). Thus relations (7.93) and (7.94) are equivalent to Navier's equations. The comparison of (7.93) and (7.51) allows us to write

$$\varphi = -\frac{1}{2\mu}\,\text{div}\,\boldsymbol{V} \tag{7.95}$$

$$\boldsymbol{\nabla}\times\boldsymbol{\Psi} = \frac{2(1-\nu)}{2\mu}\nabla^2\boldsymbol{V}\ . \tag{7.96}$$

If we also impose the condition that $\boldsymbol{V}$ be harmonic, i.e., $\nabla^2\boldsymbol{V} = 0$, then (7.96) leads to $\boldsymbol{\nabla}\times\boldsymbol{\Psi} = 0$. In addition, because of vector identity (1.191), it follows from (7.95) that $\varphi$ is a harmonic function, $\nabla^2\varphi = 0$. Thus, $\varphi$ is a Lamé strain potential, defined earlier.

### Love's Strain Function

A particular case of the Galerkin vector appears when $\boldsymbol{V} = V_3\boldsymbol{e}_3$. Then, we have Love's strain function. Condition (7.94) becomes

$$\nabla^2\left(\nabla^2 V_3\right) = 0\ , \tag{7.97}$$

and (7.93) is written as

$$2\mu\boldsymbol{u} = 2(1-\nu)(\nabla^2 V_3)\boldsymbol{e}_3 - \boldsymbol{\nabla}(\frac{\partial V_3}{\partial x_3})\ . \tag{7.98}$$

The three displacement components are easily expressed in Cartesian coordinates

$$2\mu u_1 = -\frac{\partial^2 V_3}{\partial x_1\partial x_3},\quad 2\mu u_2 = -\frac{\partial^2 V_3}{\partial x_2\partial x_3},\quad 2\mu u_3 = 2(1-\nu)\nabla^2 V_3 - \frac{\partial^2 V_3}{\partial x_3^2}\ . \tag{7.99}$$

Love introduced this particular form of the vector for the study of solids of revolution under symmetric loads. In such cases, the displacement components are expressed in terms of this function in cylindrical coordinates

$$2\mu u_r = -\frac{\partial^2 V_z}{\partial r \partial z}, \quad 2\mu u_\theta = -\frac{1}{r}\frac{\partial^2 V_z}{\partial \theta \partial z}, \quad 2\mu u_z = 2(1-\nu)\nabla^2 V_z - \frac{\partial^2 V_z}{\partial z^2} \; . \quad (7.100)$$

**Kelvin's Problem: Concentrated Force Inside an Infinite Body**

An application of this strain potential is the problem of a single force concentrated inside an infinite body. This application is known as Kelvin's problem, defined in figure 7.7(a) [15].

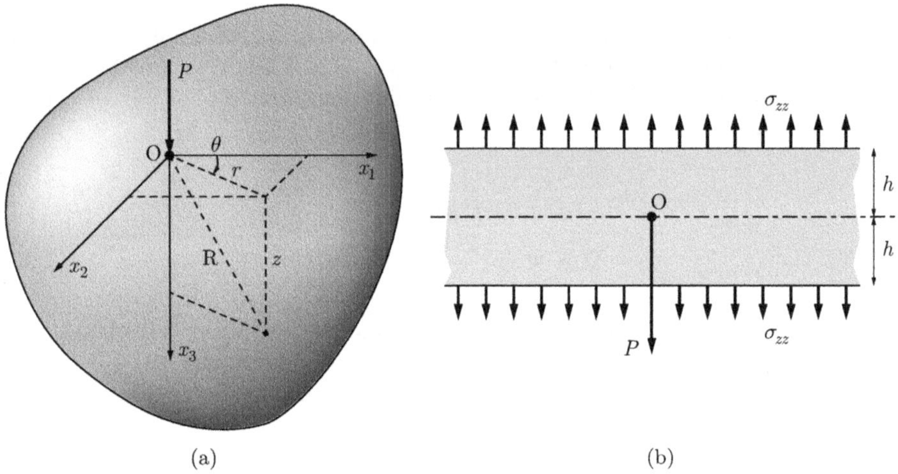


**Fig. 7.7** (a) Infinite solid subject to a concentrated force (b) View of a section

A force $P$ is applied to point O parallel to the axis $x_3$. It satisfies the following boundary conditions:

- all the stresses are zero at infinity;
- the singularity at the origin is equivalent to the applied force $P$. Thus the concentrated force can be considered as the limit of a system of forces that are applied to the surface of a small cavity situated at the origin.

The solution to this problem is obtained in cylindrical coordinates. Then, because of angular symmetry, Love's strain potential is independent of $\theta$, i.e.,

$$V_z = V_z(r, z) \; . \quad (7.101)$$

Using the strain–displacement and stress–strain relations (see Appendix A), the stress components are expressed by the relations

$$\sigma_{rr} = \frac{\partial}{\partial z}\left(\nu\nabla^2 V_z - \frac{\partial^2 V_z}{\partial r^2}\right) \tag{7.102}$$

$$\sigma_{\theta\theta} = \frac{\partial}{\partial z}\left(\nu\nabla^2 V_z - \frac{1}{r}\frac{\partial V_z}{\partial r} - \frac{1}{r^2}\frac{\partial^2 V_z}{\partial \theta^2}\right) \tag{7.103}$$

$$\sigma_{zz} = \frac{\partial}{\partial z}\left((2-\nu)\nabla^2 V_z - \frac{\partial^2 V_z}{\partial r^2}\right) \tag{7.104}$$

$$\sigma_{r\theta} = -\frac{\partial^3}{\partial r\partial\theta\partial z}\left(\frac{V_z}{r}\right) \tag{7.105}$$

$$\sigma_{\theta z} = \frac{1}{r}\frac{\partial}{\partial \theta}\left((1-\nu)\nabla^2 V_z - \frac{\partial^2 V_z}{\partial z^2}\right) \tag{7.106}$$

$$\sigma_{zr} = \frac{\partial}{\partial r}\left((1-\nu)\nabla^2 V_z - \frac{\partial^2 V_z}{\partial z^2}\right) . \tag{7.107}$$

The particular form of function (7.101) must satisfy (7.97) and its third derivatives, which appear in the stress components (eqns. (7.102)–(7.107)), with the latter going to zero at infinity with a singularity at the origin. A function that meets these requirements is given by

$$V_z = K(r^2 + z^2)^{1/2} . \tag{7.108}$$

Using (7.108) in (7.100) and (7.102)–(7.107), we have

$$2\mu u_r = \frac{Krz}{(r^2+z^2)^{3/2}}, \quad 2\mu u_\theta = 0, \tag{7.109}$$

$$2\mu u_z = K\left[\frac{2(1-2\nu)}{(r^2+z^2)^{1/2}} + \frac{1}{(r^2+z^2)^{1/2}} + \frac{z^2}{(r^2+z^2)^{3/2}}\right]$$

$$\sigma_{rr} = K\left[\frac{(1-2\nu)z}{(r^2+z^2)^{3/2}} - \frac{3r^2 z}{(r^2+z^2)^{5/2}}\right] \tag{7.110}$$

$$\sigma_{\theta\theta} = \frac{(1-2\nu)Kz}{(r^2+z^2)^{3/2}} \tag{7.111}$$

$$\sigma_{zz} = -K\left[\frac{(1-2\nu)z}{(r^2+z^2)^{3/2}} + \frac{3z^3}{(r^2+z^2)^{5/2}}\right] \tag{7.112}$$

$$\sigma_{rz} = -K\left[\frac{(1-2\nu)r}{(r^2+z^2)^{3/2}} + \frac{3rz^2}{(r^2+z^2)^{5/2}}\right] \tag{7.113}$$

$$\sigma_{r\theta} = \sigma_{\theta z} = 0 . \tag{7.114}$$

Note that the stresses are undefined at the origin where there is a singularity, and that they go to zero at infinity. To determine the constant $K$, it is necessary to consider the force equilibrium in the vertical direction. Consider a symmetric horizontal band of height $\pm h$ that contains the horizontal axis and the point O (fig. 7.7(b)). The equilibrium of the forces is written as

$$P = \int_0^\infty 2\pi\, r dr \sigma_{zz}|_{z=-h} - \int_0^\infty 2\pi\, r dr \sigma_{zz}|_{z=+h} . \tag{7.115}$$

Inserting (7.112) in (7.115) and integrating, we find the value of the constant

$$K = \frac{P}{8\pi(1-\nu)} \; . \tag{7.116}$$

This parameter is substituted in (7.109)–(7.113) to obtain the displacement and the stress components, respectively.

**Cerruti's Problem: Tangential Force at the End of a Semi-Infinite Body**

Using the potential method, we can solve certain problems by combining a Lamé strain potential and a Galerkin vector. A well-known problem which uses such an approach is Cerruti's problem, where a tangential force $P$ acts on the surface of a semi-infinite solid body as exhibited in figure 7.8 [15].

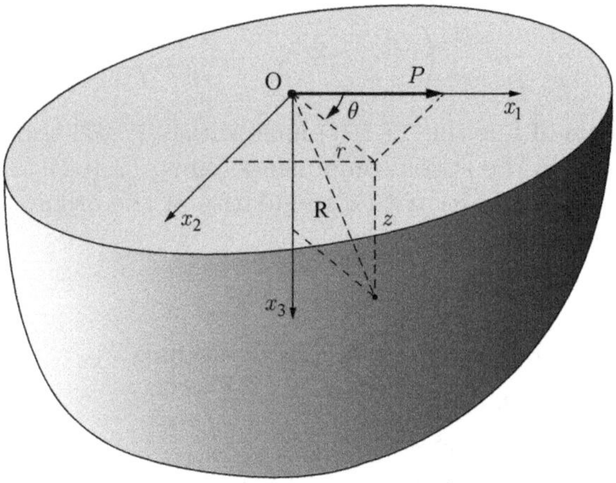


**Fig. 7.8** Semi-infinite solid body subject to a tangential surface force

This problem can be solved with the combination of a Galerkin vector with components

$$V_1 = AR, \quad V_2 = 0, \quad V_3 = Bx_1 \ln(R + x_3) \tag{7.117}$$

and a Lamé strain potential

$$\varphi = \frac{Cx_1}{R + x_3} \; , \tag{7.118}$$

where the coefficients $A, B$, and $C$ are constants and $R^2 = x_1^2 + x_2^2 + x_3^2$. In this case, the displacement vector is given by the superposition of equations (7.59) and (7.93)

$$2\mu \boldsymbol{u} = \boldsymbol{\nabla}\varphi + 2(1-\nu)\nabla^2 \boldsymbol{V} - \boldsymbol{\nabla}(\text{div}\, \boldsymbol{V}) \; . \tag{7.119}$$

The three constants $A, B$, and $C$ are determined by the boundary conditions, which are (fig. 7.8)

- on $x_3 = 0$, $\sigma_{33} = \sigma_{23} = 0$,
- along $x_1$ and $\forall x_3 > 0$ the sum of the forces is zero: $P + \int_{-\infty}^{+\infty} \int_{-\infty}^{+\infty} \sigma_{13}\, dx_1\, dx_2 = 0$.

These conditions yield

$$A = \frac{P}{4\pi(1-\nu)}, \quad B = \frac{P(1-2\nu)}{4\pi(1-\nu)}, \quad C = \frac{P(1-2\nu)}{2\pi} . \tag{7.120}$$

These constants can be inserted in (7.117) and (7.118) to calculate the displacement. The strains come from (7.1) and the stresses from (7.3).

### The Papkovich-Neuber Presentation

Note that equation (7.55) is of third order, while that which governs the Galerkin vector, (7.94), is of fourth order. In the formulation we will now present, we propose a system of equations of second order that is equivalent to Navier's equations. More precisely, the displacement vector is expressed by a combination of harmonic functions [6, 13, 65]

$$2\mu \boldsymbol{u} = \boldsymbol{\alpha} - \nabla \left[ \beta + \frac{\boldsymbol{\alpha} \cdot \boldsymbol{x}}{4(1-\nu)} \right] , \tag{7.121}$$

where $\boldsymbol{\alpha}$ and $\beta$ are vector and scalar fields, respectively, and $\boldsymbol{x}$ is the position vector. Introducing $\boldsymbol{u}$ with $\boldsymbol{f} = \boldsymbol{0}$ in (7.7), and using vector identities (1.188), (1.236), and (1.240) as well as the relation between elastic constants (6.109), we obtain the following equation:

$$\mu \nabla^2 \boldsymbol{\alpha} - (\lambda + 2\mu)\nabla(\nabla^2 \beta) - \left( \frac{\lambda + \mu}{2} \right) \nabla(\boldsymbol{x} \cdot \nabla^2 \boldsymbol{\alpha}) = 0 . \tag{7.122}$$

This last equation is satisfied when

$$\nabla^2 \boldsymbol{\alpha} = 0, \quad \nabla^2 \beta = 0 . \tag{7.123}$$

These equations are of second order, not higher as were equations (7.55) and (7.94). It should be noted that the four scalar functions in (7.123) are not independent. It can be shown that, for all convex domains, the number of independent functions is reduced to 3 [13]. In addition, the vector $\boldsymbol{\alpha}$ and the scalar $\beta$ are related to the Galerkin vector as follows:

$$\boldsymbol{\alpha} = 2(1-\nu)\nabla^2 \boldsymbol{V} \tag{7.124}$$

$$\beta = \boldsymbol{\nabla} \cdot \boldsymbol{V} - \frac{\boldsymbol{\alpha} \cdot \boldsymbol{x}}{4(1-\nu)} . \tag{7.125}$$

A special case for these four functions, in cylindrical coordinates, is given by relations

$$\alpha_r = \alpha_\theta = 0, \quad \alpha_z = \alpha_z(r, z), \quad \beta = \beta(r, z) , \tag{7.126}$$

which we will use to solve the following, very important, elasticity problem.

**Boussinesq's Problem: Vertical Force at the End of a Semi-Infinite Body**

Let a semi-infinite solid body be subject to a force $P$ acting in the direction of the vertical axis, normal to the surface as is shown in figure 7.9.

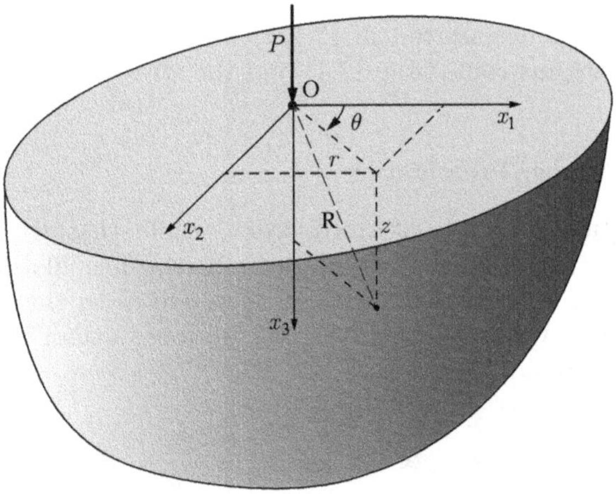


**Fig. 7.9** Semi-infinite solid body subject to a concentrated vertical force on its surface

The boundary conditions for this problem are

- $\sigma_{rz} = 0$ everywhere on the surface,
- the resulting vertical force on a horizontal plane due to $\sigma_{zz}$ at depth $z$ is equal to the applied force $P$. Note that $\sigma_{zz}$ is undefined at the origin.

Referring to figure 7.9, we define the distance $R^2 = r^2 + z^2$. The solution to this problem is obtained by considering the following functions:

$$\begin{aligned} \alpha_r &= \alpha_\theta = 0 \\ \alpha_z &= 4(1-\nu)\frac{K}{R} \\ \beta &= C\ln(R+z)\ . \end{aligned} \tag{7.127}$$

The substitution of (7.127) in (7.121) gives

$$\boldsymbol{u} = \frac{4(1-\nu)}{2\mu}\frac{K}{R}\boldsymbol{e}_z - \frac{1}{2\mu}\boldsymbol{\nabla}\left(C\ln(R+z) + \frac{Kz}{R}\right)\ , \tag{7.128}$$

for which, in cylindrical components, the displacements are

$$u_r = -\frac{Cr}{2\mu R(R+z)} + \frac{Kzr}{2\mu R^3},\ u_\theta = 0,\ u_z = \frac{(3-4\nu)K - C}{2\mu R} + \frac{Kz^2}{2\mu R^3}\ . \tag{7.129}$$

Inserting (7.129) in the strain–displacement relations, (A.24)–(A.26), and the resulting strains in the corresponding stress–strain relations (7.3), the stresses that are necessary to apply the boundary conditions are

$$\sigma_{rz} = \frac{r}{R^3}\left(C - K(1-2\nu) - \frac{3Kz^2}{R^3}\right) \tag{7.130}$$

$$\sigma_{zz} = -\frac{3Kz^3}{R^5} \ . \tag{7.131}$$

The first boundary condition stated above leads to

$$C = K(1-2\nu) \ . \tag{7.132}$$

In order to determine $K$, the accumulated force at depth $z$ due to $\sigma_{zz}$ is equal to the applied force $P$

$$P = \int_{r=0}^{r=\infty} \frac{3Kz^3}{R^5} 2\pi\, r dr \ . \tag{7.133}$$

The integration of (7.133) yields

$$K = P/2\pi \ , \tag{7.134}$$

and relation (7.132) leads to

$$C = P(1-2\nu)/2\pi \ . \tag{7.135}$$

Inserting (7.134) and (7.135) in (7.129), the displacement components are

$$u_r = \frac{P}{4\pi\mu R}\left(\frac{zr}{R^2} - \frac{(1-2\nu)r}{R+z}\right), \quad u_\theta = 0, \quad u_z = \frac{P}{4\pi\mu R}\left(2(1-\nu) + \frac{z^2}{R^2}\right) \ . \tag{7.136}$$

With the displacements known, we can calculate the strain using (A24)–(A26). These strains can be inserted in (7.3) to express the non-zero stress components as follows:

$$\sigma_{rr} = \frac{P}{2\pi R^2}\left(-\frac{3r^2 z}{R^3} + \frac{R(1-2\nu)}{R+z}\right) \tag{7.137}$$

$$\sigma_{\theta\theta} = \frac{(1-2\nu)P}{2\pi R^2}\left(\frac{z}{R} - \frac{R}{R+z}\right) \tag{7.138}$$

$$\sigma_{zz} = -\frac{3Pz^3}{2\pi R^5}, \quad \sigma_{rz} = -\frac{3Prz^3}{2\pi R^5} \ . \tag{7.139}$$

To end this section, it is necessary to mention that many important practical problems (for example, contact of solid bodies) imply the analysis of stresses and strains in semi-infinite domains subject to loads applied on free surfaces. The solutions to this type of problems are obtained by integration of Boussinesq's and Cerruti's solutions as presented in this section. The reader can find such solutions in the literature [23].

### 7.4.2 Stress Functions and Airy Solutions for Plane Problems

In section 7.2, we showed that the stress field at every point in a body in equilibrium is governed by equations (7.2), the Beltrami-Michell compatibility equations (7.14), and the boundary conditions. In a manner similar to the study of displacement functions, we proposed functions that yield stress fields satisfying the systems of equations mentioned above. However, since the stress is a second order tensor, the function we seek should also be a tensor. For simplicity in the following exposition, we ignore the volume forces.

We introduce a tensor stress function $\boldsymbol{\Phi}(\boldsymbol{x})$ that is symmetric and that expresses the six stress components as follows:

$$\sigma_{11} = \frac{\partial^2\Phi_{22}}{\partial x_3^2} + \frac{\partial^2\Phi_{33}}{\partial x_2^2} - 2\frac{\partial^2\Phi_{23}}{\partial x_2\partial x_3} \tag{7.140}$$

$$\sigma_{22} = \frac{\partial^2\Phi_{33}}{\partial x_1^2} + \frac{\partial^2\Phi_{11}}{\partial x_3^2} - 2\frac{\partial^2\Phi_{31}}{\partial x_3\partial x_1} \tag{7.141}$$

$$\sigma_{33} = \frac{\partial^2\Phi_{11}}{\partial x_2^2} + \frac{\partial^2\Phi_{22}}{\partial x_1^2} - 2\frac{\partial^2\Phi_{12}}{\partial x_1\partial x_2} \tag{7.142}$$

$$\sigma_{12} = \frac{\partial^2\Phi_{23}}{\partial x_3\partial x_1} + \frac{\partial^2\Phi_{31}}{\partial x_3\partial x_2} - \frac{\partial^2\Phi_{33}}{\partial x_1\partial x_2} - \frac{\partial^2\Phi_{12}}{\partial x_3^2} \tag{7.143}$$

$$\sigma_{23} = \frac{\partial^2\Phi_{31}}{\partial x_1\partial x_2} + \frac{\partial^2\Phi_{12}}{\partial x_1\partial x_3} - \frac{\partial^2\Phi_{11}}{\partial x_2\partial x_3} - \frac{\partial^2\Phi_{23}}{\partial x_1^2} \tag{7.144}$$

$$\sigma_{31} = \frac{\partial^2\Phi_{12}}{\partial x_2\partial x_3} + \frac{\partial^2\Phi_{23}}{\partial x_2\partial x_1} - \frac{\partial^2\Phi_{22}}{\partial x_3\partial x_1} - \frac{\partial^2\Phi_{31}}{\partial x_2^2} \,. \tag{7.145}$$

It is easy to verify that the equilibrium equations (7.2) are satisfied with these stress components (7.140)–(7.145) when there are no volume forces ($\boldsymbol{f} = 0$). Two alternatives have been proposed to generate complete solutions from the stress functions [6, 30]. These are the Maxwell and Morera functions. More precisely, if we keep only the diagonal components $\Phi_{ii}$, we define a Maxwell system; in the case where we keep the off diagonal terms, we define a Morera system. Each of these sets of stress functions is complete, in the sense that for every stress distribution that satisfies the equilibrium equations there exists a set of Maxwell functions and a set of Morera functions. In this section, we will only discuss Maxwell functions.

We point out that if the component $\Phi_{33}$ is the only non-zero component in the Maxwell formulation, then we have the Airy function for plane problems. Since in plane problems we make the distinction between plane stress and plane strain (sec. 7.3), we examine these two problems in terms of stress functions. Starting from the single component $\Phi_{33} = \Phi_{33}(x_1, x_2)$, independent of $x_3$, equations (7.140)–(7.145) yield

$$\sigma_{11} = \frac{\partial^2\Phi_{33}}{\partial x_2^2}, \quad \sigma_{22} = \frac{\partial^2\Phi_{33}}{\partial x_1^2}, \quad \sigma_{12} = -\frac{\partial^2\Phi_{33}}{\partial x_1\partial x_2} \tag{7.146}$$

$$\sigma_{33} = 0 \tag{7.147}$$

$$\sigma_{23} = \sigma_{31} = 0 \,, \tag{7.148}$$

which corresponds to the case of plane stress since $\sigma_{33} = 0$. In order to further examine the nature of $\Phi_{33}$, we must turn to the Beltrami-Michell compatibility equations. The six equations (7.14) are all written explicitly as

$$\nabla^2\sigma_{11} + \frac{1}{1+\nu}\frac{\partial^2\sigma_{kk}}{\partial x_1^2} = 0 \tag{7.149}$$

$$\nabla^2\sigma_{22} + \frac{1}{1+\nu}\frac{\partial^2\sigma_{kk}}{\partial x_2^2} = 0 \tag{7.150}$$

$$\nabla^2\sigma_{33} + \frac{1}{1+\nu}\frac{\partial^2\sigma_{kk}}{\partial x_3^2} = 0 \tag{7.151}$$

$$\nabla^2\sigma_{12} + \frac{1}{1+\nu}\frac{\partial^2\sigma_{kk}}{\partial x_1 \partial x_2} = 0 \tag{7.152}$$

$$\nabla^2\sigma_{23} + \frac{1}{1+\nu}\frac{\partial^2\sigma_{kk}}{\partial x_2 \partial x_3} = 0 \tag{7.153}$$

$$\nabla^2\sigma_{31} + \frac{1}{1+\nu}\frac{\partial^2\sigma_{kk}}{\partial x_3 \partial x_1} = 0\ . \tag{7.154}$$

Note that, using (7.146), equations (7.151), (7.153), and (7.154) are satisfied, while (7.149), (7.150), and (7.152) are not. This is because of the approximate nature of the plane stress problem (sec. 7.3.2). Nonetheless, inserting (7.146) in (7.149) and (7.150) and then adding them, we get easily that $\Phi_{33}$ satisfies the biharmonic equation

$$\Delta\Delta\Phi_{33} = \frac{\partial^4\Phi_{33}}{\partial x_1^4} + 2\frac{\partial^4\Phi_{33}}{\partial x_1^2 \partial x_2^2} + \frac{\partial^4\Phi_{33}}{\partial x_2^4} = 0\ . \tag{7.155}$$

For plane strain problems, the stress component $\sigma_{33}$ is related to components $\sigma_{11}$ and $\sigma_{22}$ with

$$\sigma_{33} = \nu(\sigma_{11} + \sigma_{22})\ . \tag{7.156}$$

To satisfy this requirement in a Maxwell formulation, it is necessary to include components $\Phi_{11}$ and $\Phi_{22}$ in addition to $\Phi_{33}$ and to impose the conditions

$$\Phi_{11} = \Phi_{22} = \nu\Phi_{33} \quad \text{with} \quad \Phi_{12} = \Phi_{23} = \Phi_{31} = 0\ . \tag{7.157}$$

In this case, each compatibility equation (7.149)–(7.151) leads to the biharmonic equation (7.155), while satisfying relations (7.152)–(7.154).

Comparing the results of this section with those in section 7.3, we notice that the stress function $\Phi(x_1, x_2)$ defined in section 7.3 is a special case of the Maxwell formulation.

Thus, for plane problems, when $\Phi_{33}$ is known and satisfies the biharmonic equation (7.155), the stress components for plane stress (7.146) and those of (7.146) and (7.156) for plane strain satisfy the equilibrium equations. We consider that such a stress state is a solution to the problem if it satisfies the boundary conditions.

Before presenting a few examples, we note that it is relatively easy to find a stress function that satisfies (7.155). However, satisfying the boundary conditions is not always simple. In general, we should be guided by our intuition and experience as to the nature of the necessary function. A common practice consists of using polynomial forms and finding the appropriate combinations which satisfy the boundary conditions. Replacing $\Phi_{33}(x_1, x_2)$ with $\Phi(x_1, x_2)$, an appropriate polynomial function can be written as

$$\begin{aligned}\Phi(x_1, x_2) &= a_1 x_1^2 + a_2 x_1 x_2 + a_3 x_2^2 + b_1 x_1^3 + b_2 x_1^2 x_2 + b_3 x_1 x_2^2 + b_4 x_2^3 \\ &\quad + c_1 x_1^4 + c_2 x_1^3 x_2 + c_3 x_1^2 x_2^2 + c_4 x_1 x_2^3 + c_5 x_2^4 + \cdots \, . \qquad (7.158)\end{aligned}$$

Note that all polynomial terms of degree less than or equal to three satisfy (7.155). Terms of higher order should not be considered, but if they must be included, their coefficients should be chosen with care to satisfy the biharmonic equation. This approach proves to be effective in many problems with rectangular domains. However, polynomial functions cannot easily describe discontinuities in the geometry and load distribution. Thus Saint-Venant's principle is often used to replace the real boundary conditions with statically equivalent conditions. Note that the solution method that is based on a stress function satisfying (7.155) is a semi-inverse method, since the polynomial function is given and we seek a problem that can be solved with this function.

Because of the rotational symmetry of many practical problems, the Airy functions are generally presented in cylindrical coordinates. The function proposed by Michell [48] offers a solution to (7.155) for plane problems:

$$\begin{aligned}\Phi(r, \theta) &= A_0 + A_1 \ln r + A_2 r^2 + A_3 r^2 \ln r \\ &\quad + (A_4 + A_5 \ln r + A_6 r^2 + A_7 r^2 \ln r)\theta \\ &\quad + \left( A_{11} r + A_{12} r \ln r + \frac{A_{13}}{r} + A_{14} r^3 + A_{15} r\theta + A_{16} r\theta \ln r \right) \cos\theta \\ &\quad + \left( B_{11} r + B_{12} r \ln r + \frac{B_{13}}{r} + B_{14} r^3 + B_{15} r\theta + B_{16} r\theta \ln r \right) \sin\theta \\ &\quad + \sum_{n=2}^{\infty} \left( A_{n1} r^n + A_{n2} r^{2+n} + A_{n3} r^{-n} + A_{n4} r^{2-n} \right) \cos n\theta \\ &\quad + \sum_{n=2}^{\infty} \left( B_{n1} r^n + B_{n2} r^{2+n} + B_{n3} r^{-n} + B_{n4} r^{2-n} \right) \sin n\theta \, . \qquad (7.159)\end{aligned}$$

Here, $\Phi_{33}(r, \theta)$ has been replaced with $\Phi(r, \theta)$. The coefficients $A_0, \ldots, A_7$; $A_{11}, \ldots, A_{16}$; $B_{11}, \ldots, B_{16}$; $A_{n1}, \ldots, A_{n4}$; $B_{n1}, \ldots, B_{n4}$ are constants and $n$ is an integer. We choose various terms in (7.159) to solve many different problems in polar coordinates. A few examples are given below.

**Normal Linear Load on the Flat Edge of a Semi-Infinite Plate**

Consider a plate of unit thickness subject to a load $P$ distributed along a line across its thickness as shown in figure 7.10.

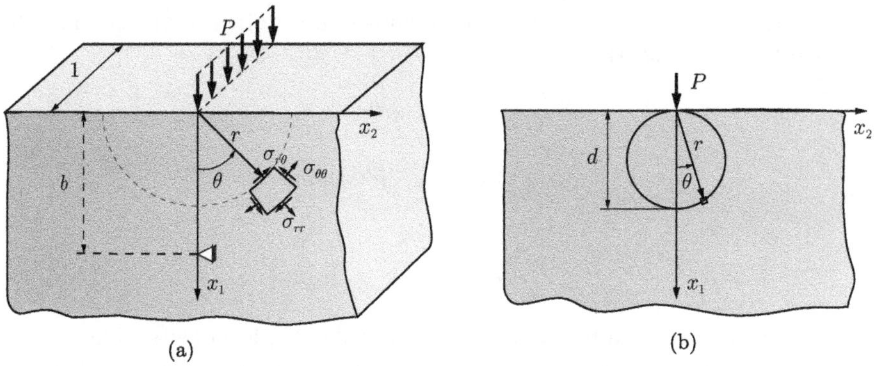


**Fig. 7.10** (a) Semi-infinite plate subject to a concentrated vertical force on its edge surface (b) Circle of diameter $d$ for each point of which the stress is the same

The plane stress problem has stress components in cylindrical coordinates $\sigma_{rr}, \sigma_{\theta\theta}, \sigma_{r\theta}$. The boundary conditions for this problem are

- the stress components $\sigma_{\theta\theta} = \sigma_{r\theta} = 0$ for $\theta = \pm\pi/2$;
- the vertical force is in equilibrium with the vertical component of the radial stress at a distance $r$.

The solution to the problem is obtained by setting the following Airy function:

$$\Phi(r, \theta) = Cr\theta \sin\theta \; . \tag{7.160}$$

Knowing $\Phi$, the stress components obtained from relations (7.146) in cylindrical coordinates (A.28)–(A.30) are

$$\sigma_{rr} = \frac{2C\cos\theta}{r}, \quad \sigma_{\theta\theta} = \sigma_{r\theta} = 0 \; . \tag{7.161}$$

Applying the second boundary condition, the constant $C$ is determined from the relation

$$P + \int_{-\pi/2}^{+\pi/2} \sigma_{rr} \cos\theta (r d\theta) = P + 2C \int_{-\pi/2}^{+\pi/2} \cos^2\theta d\theta = 0 \quad \text{and} \quad C = -\frac{P}{\pi} \; . \tag{7.162}$$

Thus, the stress components are

$$\sigma_{rr} = -\frac{2P}{\pi}\frac{\cos\theta}{r}, \quad \sigma_{\theta\theta} = \sigma_{r\theta} = 0 \; . \tag{7.163}$$

We note for the circle of diameter $d$ tangent to the surface at the origin, whose center is on the vertical axis, that $r = d\cos\theta$, and that the stress (fig. 7.10(b))

$$\sigma_{rr} = -\frac{2P}{\pi}\frac{1}{d} \tag{7.164}$$

is the same for all points on the circle.

Once the stresses are known, the strains are obtained from Hooke's law (7.4) (see (A.24)–(A.26))

$$\begin{aligned}
\varepsilon_{rr} &= \frac{\partial u_r}{\partial r} = -\frac{2P}{\pi E}\frac{\cos\theta}{r} \\
\varepsilon_{\theta\theta} &= \frac{u_r}{r} + \frac{1}{r}\frac{\partial u_\theta}{\partial \theta} = \frac{2P\nu}{\pi E}\frac{\cos\theta}{r} \\
\varepsilon_{r\theta} &= \frac{1}{2}\left(\frac{1}{r}\frac{\partial u_r}{\partial \theta} + \frac{\partial u_\theta}{\partial r} - \frac{u_\theta}{r}\right) = 0 \,.
\end{aligned} \tag{7.165}$$

These equations can be integrated to calculate the displacements. The integration constants are then evaluated in order to eliminate the rigid body motion of the plate. This is done by imposing $u_\theta(r,\theta)|_{\theta=0} = 0$ and $u_r(r,\theta)|_{\theta=0,r=b} = 0$, where $b$ is an arbitrary distance along axis $x_1$ (fig. 7.10(a))

$$\begin{aligned}
u_r &= \frac{2P}{\pi E}\cos\theta \ln\frac{b}{r} - \frac{(1-\nu)P}{\pi E}\theta\sin\theta \\
u_\theta &= \frac{(1+\nu)P}{\pi E}\sin\theta - \frac{2P}{\pi E}\sin\theta\ln\frac{b}{r} - \frac{(1-\nu)P}{\pi E}\theta\cos\theta \,.
\end{aligned} \tag{7.166}$$

**Hollow Cylinder with Free Ends Subject to Internal and External Pressure**

Given the cylindrical geometry of the body and applied loads, and by taking the ends of the cylinder to be free, we can show that $\sigma_{zz} = 0$. Thus it is a plane stress problem. The boundary conditions are independent of $\theta$; in addition, as the stress distribution is symmetric with respect to the axis $x_3$, it is implied that $\sigma_{r\theta} = 0$. The boundary conditions of the problem are given by (7.86) and (7.87).

This problem is solved by two methods. In the first method, we use Navier's equations (7.6). In the second method, we define an Airy stress function appropriate for the problem, and we use it to calculate the stress, strain, and displacement components.

In the first method, we consider that an element of the cylinder cannot move axially because of the symmetry of the load and the geometry. Thus the only non-zero component of displacement is $u_r$, and the strain–displacement relations in cylindrical coordinates become

$$\varepsilon_{rr} = \frac{du_r}{dr} \tag{7.167}$$

$$\varepsilon_{\theta\theta} = \frac{u_r}{r} \tag{7.168}$$

$$\varepsilon_{r\theta} = \frac{1}{2}\left(\frac{du_\theta}{dr} - \frac{u_\theta}{r}\right) = 0 \tag{7.169}$$

$$\varepsilon_{zz} = \frac{-\nu}{1-\nu}\left(\varepsilon_{\theta\theta} + \varepsilon_{rr}\right) \,. \tag{7.170}$$

In the case of plane stress, the stress–strain relations (7.41) are

$$\varepsilon_{rr} = \frac{1}{E}\left(\sigma_{rr} - \nu\sigma_{\theta\theta}\right) \qquad \varepsilon_{\theta\theta} = \frac{1}{E}\left(\sigma_{\theta\theta} - \nu\sigma_{rr}\right). \tag{7.171}$$

Solving these two relations for the stresses and using (7.167) and (7.168), we have

$$\sigma_{rr} = \frac{E}{1-\nu^2}\left(\frac{du_r}{dr} + \nu\frac{u_r}{r}\right) \tag{7.172}$$

$$\sigma_{\theta\theta} = \frac{E}{1-\nu^2}\left(\frac{u_r}{r} + \nu\frac{du_r}{dr}\right). \tag{7.173}$$

With $u_\theta = u_z = 0$, and $u_r \neq 0$, only one of Navier's equations is not satisfied. Assuming no volume force, this one is written as

$$\frac{d^2u_r}{dr^2} + \frac{1}{r}\frac{du_r}{dr} - \frac{u_r}{r^2} = 0\,, \tag{7.174}$$

the solution of which is given by

$$u_r = C_1 r + \frac{C_2}{r}\,. \tag{7.175}$$

Using (7.175) in (7.172) and (7.173), leads to

$$\sigma_{rr} = \frac{E}{1-\nu^2}\left(C_1(1+\nu) - C_2\frac{1-\nu}{r^2}\right) \tag{7.176}$$

$$\sigma_{\theta\theta} = \frac{E}{1-\nu^2}\left(C_1(1+\nu) + C_2\frac{1-\nu}{r^2}\right). \tag{7.177}$$

We determine the constants $C_1$ and $C_2$ with the boundary conditions (7.86) and (7.87), which yield

$$C_1 = \frac{1-\nu}{E}\frac{r_i^2P_i - r_e^2P_e}{r_e^2 - r_i^2} \qquad C_2 = \frac{1+\nu}{E}\frac{r_i^2r_e^2(P_e - P_i)}{r_e^2 - r_i^2}\,. \tag{7.178}$$

Finally, the stresses and displacements take the form

$$\begin{aligned}\sigma_{rr} &= \frac{1}{r_e^2 - r_i^2}\left(r_i^2P_i - r_e^2P_e + \frac{r_i^2r_e^2}{r^2}\left(P_e - P_i\right)\right)\\ \sigma_{\theta\theta} &= \frac{1}{r_e^2 - r_i^2}\left(r_i^2P_i - r_e^2P_e - \frac{r_i^2r_e^2}{r^2}\left(P_e - P_i\right)\right)\end{aligned} \tag{7.179}$$

$$u_r = \frac{1-\nu}{E}\frac{r_i^2P_i - r_e^2P_e}{r_e^2 - r_i^2}r - \frac{1+\nu}{E}\frac{(P_e - P_i)}{r_e^2 - r_i^2}\frac{r_i^2r_e^2}{r}\,. \tag{7.180}$$

Now, let us verify the assumption $\sigma_{zz} = 0$. If the ends of the cylinder are free, then $\varepsilon_{zz} = \text{cnst}$. Consequently, the stress–strain relations give

$$\sigma_{zz} = \nu(\sigma_{rr} + \sigma_{\theta\theta}) + E\varepsilon_{zz} = c\,. \tag{7.181}$$

The constant $c$ is determined by imposing that the total normal force at the end of the cylinder is zero

$$\int_{r_i}^{r_e} \sigma_{zz} 2\pi r \, dr = \pi c\left(r_e^2 - r_i^2\right) = 0 \Longrightarrow c = 0 \qquad \text{and} \qquad \sigma_{zz} = 0\,. \tag{7.182}$$

In the second method, we define an appropriate stress function. Taking into account the symmetry of the load and the geometry, the stress function $\Phi$ is independent of $\theta$ and is then only a function of $r$. Therefore, a stress function is taken in the form

$$\Phi(r) = A \ln r + Br^2 + Cr^2 \ln r + D\,. \tag{7.183}$$

Although this function is a general solution of the biharmonic equation (7.38), (7.48), or also (7.155), the analysis of the radial displacement $u_r$ leads to the conclusion that $C = 0$. The constant $D$ does not affect the components of stress. Thus, we only retain the two first terms of (7.183) in the following. With this function, the biharmonic equation (7.38) in cylindrical coordinates (A.27),

$$\frac{d^4\Phi}{dr^4} + \frac{2}{r}\frac{d^3\Phi}{dr^3} - \frac{1}{r^2}\frac{d^2\Phi}{dr^2} + \frac{1}{r^3}\frac{d\Phi}{dr} = 0\,, \tag{7.184}$$

is automatically satisfied and the stress components are

$$\begin{aligned}
\sigma_{rr} &= \frac{1}{r}\frac{d\Phi}{dr} = \frac{A}{r^2} + 2B \\
\sigma_{\theta\theta} &= \frac{d^2\Phi}{dr^2} = -\frac{A}{r^2} + 2B \\
\sigma_{r\theta} &= 0\,.
\end{aligned} \tag{7.185}$$

With the boundary conditions (7.86) and (7.87), we obtain the constants

$$A = \frac{r_i^2 r_e^2}{r_e^2 - r_i^2}\left(P_e - P_i\right) \qquad B = \frac{r_i^2 P_i - r_e^2 P_e}{r_e^2 - r_i^2}\,. \tag{7.186}$$

Then the stresses are expressed

$$\begin{aligned}
\sigma_{rr} &= \frac{1}{r_e^2 - r_i^2}\left(r_i^2 P_i - r_e^2 P_e + \frac{r_i^2 r_e^2}{r^2}\left(P_e - P_i\right)\right) \\
\sigma_{\theta\theta} &= \frac{1}{r_e^2 - r_i^2}\left(r_i^2 P_i - r_e^2 P_e - \frac{r_i^2 r_e^2}{r^2}\left(P_e - P_i\right)\right)\,,
\end{aligned} \tag{7.187}$$

which are the same as seen in expressions (7.179 ) and (7.89). The reason that the stresses are the same in the two problems comes from the fact that the Airy stress function is identical for plane stress and plane strain problems. The difference only appears in the stress–strain relations and the displacements. The stresses known, the strains are evaluated with (7.171).

The integration of the latter directly yields the displacement $u_r$, (7.180).

In the case of a thin-walled container under pressure with thickness $e = r_e - r_i$ and $e \ll r_i$, we can make the following approximations:

$$\begin{aligned} &r_e^2 - r_i^2 = (r_e - r_i)(r_e + r_i) \approx 2er_i \\ &r_i^2 P_i - r_e^2 P_e \approx r_i^2 (P_i - P_e) \\ &r_e^2 \approx r_i^2 \qquad\qquad r^2 \approx r_i^2 \,. \end{aligned} \tag{7.188}$$

Taking these approximations into account, the stresses (7.187) reduce to the expressions

$$\sigma_{rr} \approx 0 \tag{7.189}$$

$$\sigma_{\theta\theta} \approx \frac{r_i(P_i - P_e)}{e} \,. \tag{7.190}$$

In the solutions given in texts on introductory material properties [54], it is supposed that $\sigma_{rr}$ is zero, because $e \ll r_i$, while $\sigma_{\theta\theta}$ is obtained by equilibrium of an element of the container.

We must point out that the solution given in this example is only valid for sections sufficiently far from the ends of the cylinder.

**Long, Thin Beam with a Uniform Load $q$**

A long, thin beam with length $l$, height $h$, and thickness $d = 1$ is subject to a uniformly distributed load $q$ (N/m$^2$) (fig. 7.11). We propose to determine the stresses, strains, and displacements when the beam has simple supports. The thickness $d$ of the beam is assumed to be small relative to its plane dimensions, (that is, $h, l \gg d$) and the load $q$ is parallel to the plane $(x_1, x_2)$. In this plane stress problem we neglect the weight of the beam.

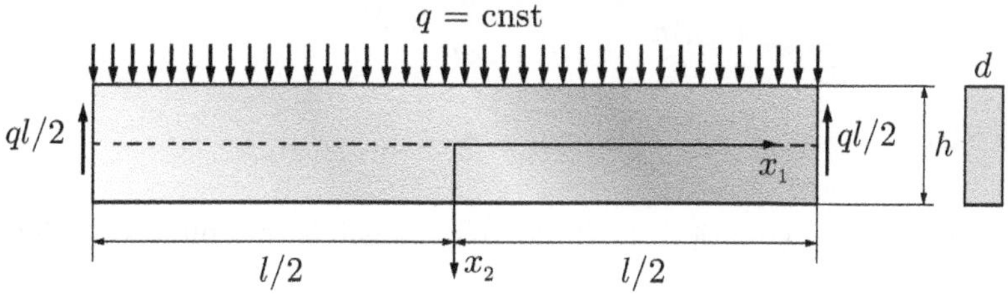


**Fig. 7.11** A long, thin beam with a uniform load

The boundary conditions are

$$\begin{aligned} &x_2 = -\frac{h}{2} \qquad \sigma_{22} = -q \qquad \sigma_{12} = 0 \\ &x_2 = \frac{h}{2} \qquad \sigma_{22} = 0 \qquad \sigma_{12} = 0 \,. \end{aligned} \tag{7.191}$$

The evaluation of the axial force $N_1$ (along $x_1$), the total moment $M_3$ (with respect to axis $x_3$) and the tangential force $N_2$ parallel to the section (along

$x_2$) at the ends $x_1 = \pm l/2$ of the beam, leads to

$$\begin{aligned} N_1 &= \int_{-h/2}^{h/2} \sigma_{11}\, dx_2 = 0 \\ M_3 &= \int_{-h/2}^{h/2} \sigma_{11} x_2\, dx_2 = 0 \\ N_2 &= \int_{-h/2}^{h/2} \sigma_{21}\, dx_2 = -\frac{ql}{2}\,. \end{aligned} \tag{7.192}$$

A stress function that yields a solution to the problem is written as

$$\Phi(x_1, x_2) = A x_2^3 \left( x_1^2 - \frac{x_2^2}{5} \right) + B x_1^2 x_2 + C x_2^3 + D x_1^2\,, \tag{7.193}$$

where $A, B, C$, and $D$ are constants which are determined later from the boundary conditions (7.191) and (7.192). First we verify that this function satisfies the biharmonic equation (7.38). Using equations (7.37), the three stress field components are given by

$$\begin{aligned} \sigma_{11} &= 6 A x_2 x_1^2 - 4 A x_2^3 + 6 C x_2 \\ \sigma_{22} &= 2 A x_2^3 + 2 B x_2 + 2D \\ \sigma_{12} &= -6 A x_2^2 x_1 - 2 B x_1\,. \end{aligned} \tag{7.194}$$

With boundary conditions (7.191), we obtain

$$\begin{aligned} &-\frac{Ah^3}{4} - Bh + 2D = -q \\ &\frac{Ah^3}{4} + Bh + 2D = 0 \\ &\frac{3Ah^2}{2} + 2B = 0\,. \end{aligned} \tag{7.195}$$

This sytem of three equations in three unknowns has the following solution:

$$A = -\frac{q}{h^3} \qquad B = \frac{3q}{4h} \qquad D = -\frac{q}{4}\,. \tag{7.196}$$

The constant $C$ is obtained from the condition $M_3 = 0$ ($N_1 = 0$ is satisfied for all values of the constants)

$$C = \frac{q}{24 I_3} \left( \frac{l^2}{2} - \frac{h^2}{5} \right), \tag{7.197}$$

where $I_3 = h^3/12$ is the moment of inertia of the section with respect to $x_3$. It can easily be shown that the two remaining integrals of (7.192) are verified. Inserting these values of the constants in (7.194), the stresses in the beam become

$$\begin{aligned}\sigma_{11} &= \frac{q}{2I_3}x_2\left(\frac{l^2}{4}-\frac{x_1^2}{2}\right)+\frac{q}{2I_3}x_2\left(\frac{2}{3}x_2^2-\frac{h^2}{20}\right)\\ \sigma_{22} &= -\frac{q}{2I_3}\left(\frac{x_2^3}{3}-\frac{h^2x_2}{4}+\frac{h^3}{12}\right)\\ \sigma_{12} &= -\frac{q}{2I_3}x_1\left(\frac{h^2}{4}-x_2^2\right).\end{aligned} \tag{7.198}$$

The first term (of the equation for $\sigma_{11}$) results from elementary beam theory. The second is an additional term resulting from taking into consideration $\sigma_{22}$, which does not depend on $x_1$ and which becomes negligible when $l \gg h$. Note that the solution is only valid for sections sufficiently far from the supports.

With the stresses (7.198), the strains are given by (7.41). The displacements are calculated by integrating relations (7.18) with the following boundary conditions:

- at $x_1$ and $x_2 = 0$,

$$u_1 = 0 \qquad u_2 = f \qquad \frac{\partial u_2}{\partial x_1} = 0\,;$$

- at $x_1 = \pm l/2$ and $x_2 = 0$,

$$u_2 = 0\,, \tag{7.199}$$

  where $f$ is the *maximum deflection* at the center of the beam which we determine next.

We have

$$\begin{aligned}u_1 &= \frac{q}{2EI_3}\left(\left(\frac{l^2x_1}{4}-\frac{x_1^3}{3}\right)x_2+\left(\frac{2x_2^3}{3}-\frac{h^2x_2}{10}\right)x_1\right.\\ &\quad\left.+\nu\left(\frac{x_2^3}{3}-\frac{h^2x_2}{4}+\frac{h^3}{12}\right)x_1\right)\\ u_2 &= -\frac{q}{2EI_3}\left(\frac{x_2^4}{12}-\frac{h^2x_2^2}{8}+\frac{h^3x_2}{12}+\nu\left(\left(\frac{l_2}{4}-x_1^2\right)\frac{x_2^2}{2}+\frac{x_2^4}{6}-\frac{h^2x_2^2}{20}\right)\right)\\ &\quad-\frac{q}{2EI_3}\left(\frac{l^2x_1^2}{8}-\frac{x_1^4}{12}-\frac{h^2x_1^2}{20}+\left(1+\frac{1}{2}\nu\right)\frac{h^2x_1^2}{4}\right)+f\,.\end{aligned} \tag{7.200}$$

The deflection $f$ at the center of the beam is obtained by using the expression obtained for $u_2$ in the first condition (7.199)

$$f = \frac{5}{384}\frac{ql^4}{EI_3}\left(1+\frac{12}{5}\frac{h^2}{l^2}\left(\frac{4}{5}+\frac{\nu}{2}\right)\right). \tag{7.201}$$

Note that the first term of (7.201) is the deflection resulting from elementary beam theory. The second term appears because we have taken $\sigma_{22}$ into account along $x_2$. This term is especially important for short beams (that is $l \sim h$). For long, thin beams, we have $l \gg h$ and this term becomes negligible.

## 7.5 Wave Propagation in a Linear Elastic Medium

In this chapter up until now, we have presented static elastic problems. That is, the solid is considered to be at rest under the loads which were applied sufficiently slowly, such that the dynamic effects could be ignored. Such an approach is justified for many practical elasticity problems; this is known as "linear elastostatic" analysis. However, there are many problems in solid mechanics where we take into account the dynamic effects, that is, the inertial forces. These emerge when the external loads are applied at high rates, including vibrations, impacts, and explosions. Sudden displacements also create dynamic effects such as the slipping of a seismic fault. Such dynamic loads produce stress and strain waves which are transmitted across the body at different velocities dependent on the deformation mode.

To understand and analyze the dynamic response of an elastic medium, the static equilibrium equations (7.7) must be replaced by motion equations thus defining the linear elastodynamic problem. In this section we present the general three-dimensional equations of motion for a linear elastic solid as well as the wave propagation solutions in simple structural elements.

The reader can also consult [16, 54] for complementary information.

### 7.5.1 Shear and Dilatation Waves

The motion equation in terms of displacements can be obtained from Navier's equations (7.6) by adding an inertial force component and by taking the displacements as functions of $x_i$ and $t$, $u_i = u_i(x_i, t)$. Assuming, as in the case of static analysis (sec. 7.4), that there are no volume forces, the motion equations (7.7) are

$$(\lambda + \mu)\,\boldsymbol{\nabla}(\mathrm{div}\ \boldsymbol{u}) + \mu\boldsymbol{\nabla}^2\boldsymbol{u} = \rho\boldsymbol{a}\ , \tag{7.202}$$

which in index form is

$$(\lambda + \mu)\,u_{k,ki} + \mu u_{i,jj} = \rho a_i\ . \tag{7.203}$$

The acceleration can be expressed in terms of the displacement vector as $\boldsymbol{a} = \partial^2\boldsymbol{u}/\partial t^2$ or $a_i = \partial^2 u_i/\partial t^2$. Note that because of the hypothesis of small displacements in section 2.9, the non-linear acceleration term in (2.33) is of second order in $\varepsilon$ (eqn. (2.139)) and can thus be neglected. Consequently, relation (7.203) becomes

$$(\lambda + \mu)\,u_{k,ki} + \mu u_{i,jj} = \rho\frac{\partial^2 u_i}{\partial t^2}\ . \tag{7.204}$$

Since the deformations are small, the motions examined in this model are small elastic oscillations or elastic waves. According to the type of deformation, we can distinguish two types of waves as follows. First, assume that the load produces waves which are associated with no volume changes. Then $\varepsilon_{ii} = \mathrm{div}\ \boldsymbol{u} = 0$ and (7.204) can be replaced by

$$\mu u_{i,jj} = \rho\frac{\partial^2 u_i}{\partial t^2}\ . \tag{7.205}$$

The waves described by this last equation are called ***shear waves*** or ***distortion waves***. Next we assume that the deformation produced by the applied load is irrotational. In other words, the rotation tensor (2.166) is zero, or

$$\omega_{32} = \frac{1}{2}\left(\frac{\partial u_3}{\partial x_2} - \frac{\partial u_2}{\partial x_3}\right) = 0, \quad \omega_{13} = \frac{1}{2}\left(\frac{\partial u_1}{\partial x_3} - \frac{\partial u_3}{\partial x_1}\right) = 0 \ ,$$
$$\omega_{21} = \frac{1}{2}\left(\frac{\partial u_2}{\partial x_1} - \frac{\partial u_1}{\partial x_2}\right) = 0,$$

or finally

$$\frac{\partial u_2}{\partial x_1} - \frac{\partial u_1}{\partial x_2} = 0, \quad \frac{\partial u_1}{\partial x_3} - \frac{\partial u_3}{\partial x_1} = 0, \quad \frac{\partial u_3}{\partial x_2} - \frac{\partial u_2}{\partial x_3} = 0 \ . \tag{7.206}$$

These relations imply that $\boldsymbol{u}$ can be derived from a potential function $\phi(x_i, t)$

$$\boldsymbol{u} = \boldsymbol{\nabla}\phi \ . \tag{7.207}$$

We can easily see using (1.236) that relation (7.207) leads to the expressions

$$\text{div}\ \boldsymbol{u} = \frac{\partial u_i}{\partial x_i} = \frac{\partial^2 \phi}{\partial x_i \partial x_i} = \nabla^2\phi, \quad \text{and} \quad \boldsymbol{\nabla}(\text{div}\ \boldsymbol{u}) = \nabla^2\boldsymbol{u} = \boldsymbol{\nabla}\nabla^2\phi \ . \tag{7.208}$$

Using these expressions in (7.202), we obtain the ***irrotational*** or ***dilatational wave*** equation

$$(\lambda + 2\mu)\,\nabla^2\boldsymbol{u} = \rho\frac{\partial^2 \boldsymbol{u}}{\partial t^2} \ , \tag{7.209}$$

or

$$(\lambda + 2\mu)\, u_{i,jj} = \rho\frac{\partial^2 u_i}{\partial t^2} \ . \tag{7.210}$$

It is interesting to express equations (7.205) and (7.209) in a similar form

$$c^2\nabla^2\boldsymbol{u}' = \frac{\partial^2 \boldsymbol{u}'}{\partial t^2} \ , \tag{7.211}$$

where we have

$$c = c_1 = \sqrt{\frac{\lambda + 2\mu}{\rho}} = \sqrt{\frac{E(1-\nu)}{\rho(1-2\nu)(1+\nu)}} \ , \tag{7.212}$$

for dilatational waves, and

$$c = c_2 = \sqrt{\frac{\mu}{\rho}} = \sqrt{\frac{E}{2\rho(1+\nu)}} \ , \tag{7.213}$$

for shear waves. Note that $c_1$ and $c_2$ have dimensions of a velocity, since $(MLT^{-2}L^{-2}M^{-1}L^3)^{1/2} = (L^2T^{-2})^{1/2}$, and $c_1 > c_2$, thus showing that elastic dilatation waves travel at a higher speed than elastic shear waves. Note that the general case of wave propagation in a linear elastic medium can be obtained by the superposition of dilatation and shear waves.

To express the motion equations in terms of scalar and vector potentials $\varphi, \boldsymbol{\Psi}$, it is necessary to insert (7.51) in (7.202). To do that, we deduce the second time derivative of $\boldsymbol{u}$ and its divergence as follows:

$$\ddot{\boldsymbol{u}} = \boldsymbol{\nabla}\ddot{\varphi} + \boldsymbol{\nabla} \times \ddot{\boldsymbol{\Psi}} \tag{7.214}$$

$$\text{div}\ \boldsymbol{u} = \text{div}(\boldsymbol{\nabla}\varphi) + \text{div}(\boldsymbol{\nabla} \times \boldsymbol{\Psi}) = \nabla^2\varphi \tag{7.215}$$

$$\begin{aligned}\nabla^2\boldsymbol{u} &= \nabla^2(\boldsymbol{\nabla}\varphi + \boldsymbol{\nabla} \times \boldsymbol{\Psi}) = \nabla^2(\boldsymbol{\nabla}\varphi) + \nabla^2(\boldsymbol{\nabla} \times \boldsymbol{\Psi}) \\ &= \boldsymbol{\nabla}(\nabla^2\varphi) + \boldsymbol{\nabla} \times (\nabla^2\boldsymbol{\Psi})\ .\end{aligned} \tag{7.216}$$

We used identities (1.188) and (1.180) to deduce (7.215), and relations (1.236) and (1.237) to obtain (7.216). Thus the motion equation (7.202) becomes

$$\begin{aligned}&(\lambda + \mu)\,\boldsymbol{\nabla}(\nabla^2\varphi) + \mu\left(\boldsymbol{\nabla}(\nabla^2\varphi) + \boldsymbol{\nabla} \times (\nabla^2\boldsymbol{\Psi})\right) \\ &= \rho(\boldsymbol{\nabla}\ddot{\varphi} + \boldsymbol{\nabla} \times \ddot{\boldsymbol{\Psi}})\ ,\end{aligned} \tag{7.217}$$

which we can rewrite in the form

$$\boldsymbol{\nabla}\left((\lambda + 2\mu)\nabla^2\varphi - \rho\ddot{\varphi}\right) + \boldsymbol{\nabla} \times \left(\mu\nabla^2\boldsymbol{\Psi} - \rho\ddot{\boldsymbol{\Psi}}\right) = 0\ . \tag{7.218}$$

Equality (7.218) is satisfied if

$$(\lambda + 2\mu)\nabla^2\varphi - \rho\ddot{\varphi} = 0 \tag{7.219}$$

$$\mu\nabla^2\boldsymbol{\Psi} - \rho\ddot{\boldsymbol{\Psi}} = \boldsymbol{0}\ . \tag{7.220}$$

Finally, we can write

$$(\lambda + 2\mu)\nabla^2\varphi = \rho\frac{\partial^2\varphi}{\partial t^2} \tag{7.221}$$

$$\mu\nabla^2\boldsymbol{\Psi} = \rho\frac{\partial^2\boldsymbol{\Psi}}{\partial t^2}\ . \tag{7.222}$$

It is interesting to note that (7.221) and (7.222) resemble relations (7.210) and (7.205), respectively (see also exercises 7.5 and 7.6). Furthermore, using representation (7.51) for the displacement field, the elastodynamic problem reduces to the resolution of the wave equations (7.221) and (7.222).

### 7.5.2 Rayleigh Surface Waves

In the previous section, we examined the case of wave propagation in an infinite, isotropic linear elastic body. Often, however, we must treat waves along free surfaces or interfaces between two bodies. In this case, wave propagation becomes more complicated. Surface waves were analyzed by Rayleigh (fig. 7.12) and involve longitudinal and transversal wave types at the same time. Surface waves also appear after earthquakes, explosions, and impacts. In this section, we summarize the essential equations for waves of this type. For a detailed analysis, the reader is referred to more elaborate treatments of the subject [16].

Lord Rayleigh (John William Strutt), born in 1842, studied mathematics at Trinity College in Cambridge. He inherited the title of Lord upon the death of his father in 1872 and devoted part of his time to the management of the domain while being at the same time an active scientist. In 1879, he became director of Cavendish Laboratory at Cambridge, succeeding Maxwell. He died in 1919. His contributions are numerous in the study of sound, vibrations, electrodynamics, electromagnetism, and fluid and solid mechanics. His text *The Theory of Sound* which appeared in 1877 constitutes a classic reference in the domain. His most important discovery was that of argon in 1894, for which he was awarded the Nobel prize for physics in 1904.

**Fig. 7.12** Lord Rayleigh

The schematic of a half space is shown in figure 7.13. The wave propagates in the $x_1$ direction, such that the displacement field is given by $u_1(x_1, x_3, t), u_3(x_1, x_3, t), u_2 = 0$. To solve this problem using representation (7.51), we assume that

$$\varphi = \varphi(x_1, x_3, t) \quad \text{and} \quad \boldsymbol{\Psi} = -\Psi_2(x_1, x_3, t)\boldsymbol{e}_2 \ . \tag{7.223}$$

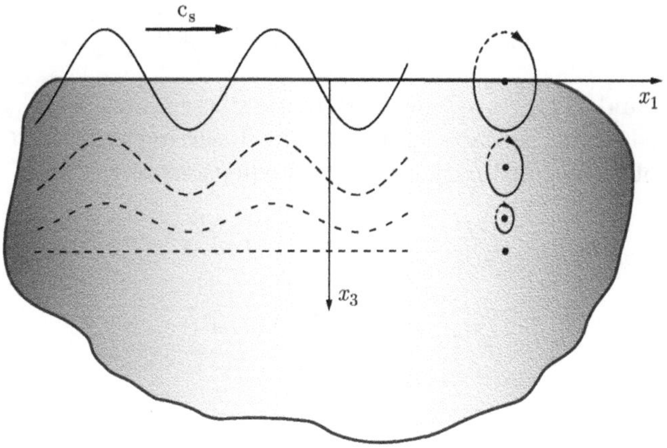


**Fig. 7.13** Propagation of a Rayleigh wave in direction $x_1$

Based on these expressions and (7.51), we deduce the displacement components

$$u_1(x_1, x_3, t) = \frac{\partial \varphi}{\partial x_1} + \frac{\partial \Psi_2}{\partial x_3}, \quad u_3(x_1, x_3, t) = \frac{\partial \varphi}{\partial x_3} - \frac{\partial \Psi_2}{\partial x_1} \tag{7.224}$$

$$\text{div } \boldsymbol{u} = \frac{\partial u_1}{\partial x_1} + \frac{\partial u_3}{\partial x_3} = \nabla^2 \varphi \ , \tag{7.225}$$

as well as the components of the infinitesimal rotation tensor

$$\begin{aligned}\omega_{13} &= \frac{1}{2}\left(\frac{\partial u_1}{\partial x_3} - \frac{\partial u_3}{\partial x_1}\right) = -\frac{1}{2}\nabla^2\Psi_2 \\ \omega_{21} &= \omega_{32} = 0\,.\end{aligned} \tag{7.226}$$

For this plane problem, the motion equations (7.203) reduce to

$$\begin{aligned}(\lambda+\mu)\frac{\partial}{\partial x_1}\left(\frac{\partial u_1}{\partial x_1} + \frac{\partial u_3}{\partial x_3}\right) + \mu\left(\frac{\partial^2 u_1}{\partial x_1^2} + \frac{\partial^2 u_1}{\partial x_3^2}\right) &= \rho\frac{\partial^2 u_1}{\partial t^2} \\ (\lambda+\mu)\frac{\partial}{\partial x_3}\left(\frac{\partial u_1}{\partial x_1} + \frac{\partial u_3}{\partial x_3}\right) + \mu\left(\frac{\partial^2 u_3}{\partial x_1^2} + \frac{\partial^2 u_3}{\partial x_3^2}\right) &= \rho\frac{\partial^2 u_3}{\partial t^2}\,.\end{aligned} \tag{7.227}$$

Using (7.224)–(7.226) in (7.227), leads to

$$\begin{aligned}(\lambda+2\mu)\frac{\partial}{\partial x_1}\nabla^2\varphi + \mu\frac{\partial}{\partial x_3}(\nabla^2\Psi_2) &= \rho\left(\frac{\partial}{\partial x_1}\frac{\partial^2\varphi}{\partial t^2} + \frac{\partial}{\partial x_3}\frac{\partial^2\Psi_2}{\partial t^2}\right) \\ (\lambda+2\mu)\frac{\partial}{\partial x_3}\nabla^2\varphi - \mu\frac{\partial}{\partial x_1}(\nabla^2\Psi_2) &= \rho\left(\frac{\partial}{\partial x_3}\frac{\partial^2\varphi}{\partial t^2} - \frac{\partial}{\partial x_1}\frac{\partial^2\Psi_2}{\partial t^2}\right).\end{aligned} \tag{7.228}$$

Note that these last equations are satisfied with (7.221) and (7.222). To continue further, we consider only harmonic forms of $\varphi(x_1, x_3)$ and $\Psi_2(x_1, x_3)$ with a wave propagating in direction $x_1$

$$\begin{aligned}\varphi(x_1, x_3, t) &= H(x_3)e^{i(kx_1-\omega t)} \\ \Psi_2(x_1, x_3, t) &= G(x_3)e^{i(kx_1-\omega t)}\,,\end{aligned} \tag{7.229}$$

where $H(x_3)$ and $G(x_3)$ are functions to be determined, $k = \omega/c_s$ with $c_s$ the wave speed on the surface, and $i^2 = -1$. Inserting (7.229) in (7.221) and (7.222), we obtain two ordinary differential equations of the form

$$\frac{d^2H}{dx_3^2} - \left(k^2 - \frac{\omega^2}{c_1^2}\right)H = 0 \tag{7.230}$$

$$\frac{d^2G}{dx_3^2} - \left(k^2 - \frac{\omega^2}{c_2^2}\right)G = 0\,. \tag{7.231}$$

Defining the symbols

$$\zeta_1^2 = \left(k^2 - \frac{\omega^2}{c_1^2}\right), \quad \zeta_2^2 = \left(k^2 - \frac{\omega^2}{c_2^2}\right)\,, \tag{7.232}$$

the solutions of (7.230) and (7.231) are written as follows:

$$H(x_3) = A_1e^{-\zeta_1 x_3} + B_1e^{\zeta_1 x_3} \tag{7.233}$$

$$G(x_3) = A_2e^{-\zeta_2 x_3} + B_2e^{\zeta_2 x_3}\,. \tag{7.234}$$

Reasoning about the phenomena from a physical viewpoint, the terms with a positive exponent yield waves with unlimited amplitude, which is unrealistic.

Thus the corresponding constants are zero: $B_1 = B_2 = 0$. Using (7.233) and (7.234) in (7.229), the solution takes the form

$$\begin{aligned} \varphi(x_1, x_3, t) &= A_1 e^{-\zeta_1 x_3} e^{i(kx_1 - \omega t)} \\ \Psi_2(x_1, x_3, t) &= A_2 e^{-\zeta_2 x_3} e^{i(kx_1 - \omega t)} . \end{aligned} \tag{7.235}$$

For this problem, the following boundary conditions must be satisfied on the free surface

$$\sigma_{33} = \sigma_{31} = \sigma_{32} = 0 \quad \text{at} \quad x_3 = 0 . \tag{7.236}$$

To apply these conditions, we need to express the stresses (7.3) in terms of the functions (7.51). Using the displacements (7.51), we calculate the strains and inserting them in (7.3), we obtain

$$\begin{aligned} \sigma_{33} &= \lambda \nabla^2 \varphi + 2\mu \frac{\partial^2 \varphi}{\partial x_3^2} - 2\mu \frac{\partial^2 \Psi_2}{\partial x_3 \partial x_1} \\ \sigma_{31} &= \mu \left( 2 \frac{\partial^2 \varphi}{\partial x_3 \partial x_1} - \frac{\partial^2 \Psi_2}{\partial x_1^2} + \frac{\partial^2 \Psi_2}{\partial x_3^2} \right) . \end{aligned} \tag{7.237}$$

The component $\sigma_{32}$ is zero, since the displacement field is independent of $x_2$, which eliminates all the derivatives with respect to $x_2$. With assumptions (7.223) and solution (7.235) known, (7.237) for $x_3 = 0$ yields two homogeneous equations

$$A_1 \left[ (\lambda + 2\mu)\zeta_1^2 - \lambda k^2 \right] + 2iA_2 \mu \zeta_2 k = 0 \tag{7.238}$$

$$-2iA_1 \zeta_1 k + A_2 \left[ \zeta_2^2 + k^2 \right] = 0 . \tag{7.239}$$

A non-trivial solution for $A_1, A_2$, requires that the determinant of the system of equations be zero, which leads to the following characteristic equation:

$$\left( \frac{c_s}{c_2} \right)^6 - 8 \left( \frac{c_s}{c_2} \right)^4 + (24 - 16\kappa^{-2}) \left( \frac{c_s}{c_2} \right)^2 + 16(\kappa^{-2} - 1) = 0 , \tag{7.240}$$

with $\kappa^{-2} = c_2^2/c_1^2 = \mu/(\lambda + 2\mu) = (1 - 2\nu)/(1 - \nu)$. Thus, the wave speed $c_s$ depends on the material via Poisson's coefficient. The polynomial equation (7.240) is treated as a reduced cubic equation with $(c_s/c_2)^2$ as unknown. In the interest of simplicity, we consider a material with $\nu = 1/4$. In this case $\kappa^{-2} = 1/3$, and the roots of (7.240) are

$$c_s^2/c_2^2 = 4,\ 2 + 2/\sqrt{2},\ 2 - 2/\sqrt{2}. \tag{7.241}$$

Among these three roots, two are not realistic, as they lead to imaginary values for the parameters $\zeta_1$ and $\zeta_2$. Therefore, we retain the third root, which yields $c_s/c_2 = 0.9194$ or

$$c_s = 0.9194 \sqrt{\frac{\mu}{\rho}} . \tag{7.242}$$

For the case where $\nu = 0.5$, corresponding to the largest value of Poisson's coefficient, we obtain $c_s = 0.9553\sqrt{\mu/\rho}$. Thus the speed of a surface wave is slightly smaller than the speed of the shear waves (7.213).

The next important parameters to calculate are the displacement components (7.224). Knowing the solution (7.235), we can easily express the displacement as follows:

$$\begin{aligned} u_1(x_1, x_3, t) &= \frac{\partial \varphi}{\partial x_1} + \frac{\partial \Psi_2}{\partial x_3} \\ &= \left(iA_1 k e^{-\zeta_1 x_3} - A_2 \zeta_2 k e^{-\zeta_2 x_3}\right) e^{i(kx_3 - \omega t)} . \end{aligned} \quad (7.243)$$

Using (7.239) to eliminate $A_2$ and Euler's formula to replace the exponentials, we have

$$u_1(x_1, x_3, t) = -A_1 k \left( e^{-\zeta_1 x_3} - \frac{2\zeta_1 \zeta_2}{\zeta_2^2 + k^2} e^{-\zeta_2 x_3} \right) \sin(kx_1 - \omega t) . \quad (7.244)$$

Similarly, we obtain a complex function for $u_3(x_1, x_3, t)$ of which the real part is

$$\begin{aligned} u_3(x_1, x_3, t) &= \frac{\partial \varphi}{\partial x_3} - \frac{\partial \Psi_2}{\partial x_1} \\ &= -A_1 \zeta_1 \left( e^{-\zeta_1 x_3} - \frac{2k^2}{\zeta_2^2 + k^2} e^{-\zeta_2 x_3} \right) \cos(kx_1 - \omega t) . \end{aligned} \quad (7.245)$$

The motion we have calculated describes the displacement of a particle in the vertical plane $Ox_1x_3$. We display in figure 7.13 schematics of these motions for particles along the vertical axis. Note that the motion of a particle traces an ellipse with the long axis normal to $x_1$ and the short axis normal to $x_3$. Recall that the parametric equations of an ellipse are expressed as $u_1 = C_1 \sin\theta$ and $u_3 = C_3 \cos\theta$, where $C_1$ and $C_3$ are the ellipse's semi-axes. This is due to the fact that dilatation waves and shear waves are both present in surface wave propagation. In addition, the solution expressed by (7.244) and (7.245) shows that the amplitude of the Rayleigh wave diminishes very rapidly along the axis $x_3$. The rate of this decrease depends on the values of $\zeta_1$ and $\zeta_2$ defined by (7.232).

### 7.5.3 One-Dimensional Elastic Plane Waves

When a dynamic perturbation occurs (impact load, earthquake, explosion, etc), the waves propagate in all directions. At relatively large distances from the perturbation, we can consider the generated waves to propagate in a plane. Thus the material particle is displaced either in the propagation direction or perpendicular to it. These waves are called ***longitudinal and transversal waves***, respectively, and correspond to the dilatation and shear waves previously defined.

Take a longitudinal wave travelling in direction $x_1$. In this case, $u_2 = u_3 = 0$ and $u_1$ is function only of $x_1$ and time $t$, and (7.209) reduces to

$$c_1^2 \frac{\partial^2 u_1}{\partial x_1^2} = \frac{\partial^2 u_1}{\partial t^2} . \quad (7.246)$$

This last equation can be solved by the well-known method of separation of variables which is presented in the next section. For this particular equation however, there exists a special method called d'Alembert's solution which is described below.

We assume that there exists a function $f$ with continuous first and second derivatives. We obtain the first and second derivatives using the chain rule

$$\frac{\partial f(x_1 - c_1 t)}{\partial t} = -c_1 \frac{\partial f(x_1 - c_1 t)}{\partial x_1} , \tag{7.247}$$

$$\frac{\partial^2 f(x_1 - c_1 t)}{\partial t^2} = c_1^2 \frac{\partial^2 f(x_1 - c_1 t)}{\partial x_1^2} . \tag{7.248}$$

Obviously, from the second equation, $f$ satisfies the wave equation (7.246). Similarly, we can show that for a function $g$, with continuous first and second derivatives, $g(x_1 + c_1 t)$ is a solution of (7.246). Given that (7.246) is a linear equation, the sum of $f$ and $g$,

$$u_1(x_1, t) = f(x_1 - c_1 t) + g(x_1 + c_1 t), \tag{7.249}$$

must also be a solution of (7.246). The solution (7.249) represents the propagation of perturbations, to the right (function $f(x_1 - c_1 t)$) and to the left (function $g(x_1 + c_1 t)$). For example, consider a cord stretched from $-\infty$ to $+\infty$. The function $f(x_1 - c_1 t)$ is constant when $x_1 - c_1 t =$ constant. Thus an increase in time is necessary to compensate for the increase of $x_1$ to maintain the function constant as shown in figure 7.14. This behavior represents propagation of an undeformed perturbation to the right along the cord as time increases. Similarly, $g(x_1 + c_1 t)$ represents a perturbation travelling to the left. To continue further in this analysis, it is necessary to define the functions $f$ and $g$.

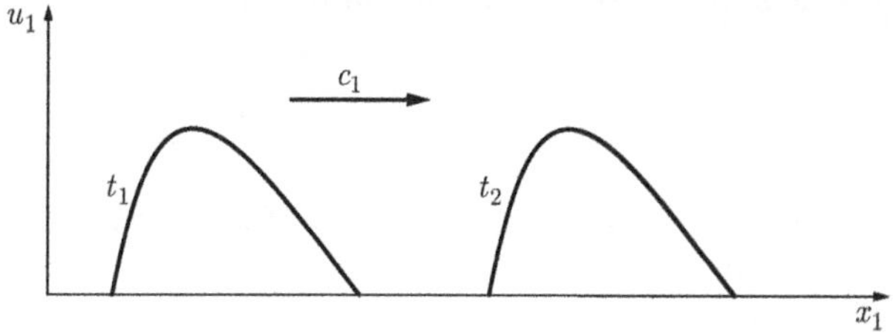


**Fig. 7.14** Propagation of a perturbation given by $f(x_1 - c_1 t)$ in (7.249)

The specific forms of $f$ and $g$ are determined by the initial displacement described by the function $\phi(x_1)$ and the initial velocity of the cord, $\theta(x_1)$, at every point $x_1$. With these two functions and (7.249), we obtain

$$\begin{aligned} u_1(x_1, 0) = \phi(x_1) &= f(x_1 - c_1 t)\big|_{t=0} + g(x_1 + c_1 t)\big|_{t=0} \\ &= f(x_1) + g(x_1) , \end{aligned} \tag{7.250}$$

$$\begin{aligned} \frac{\partial u_1(x_1, t)}{\partial t}\Big|_{x_1, t=0} &= \theta(x_1) = -c_1 \frac{\partial f(x_1 - c_1 t)}{\partial x_1}\Big|_{t=0} + c_1 \frac{\partial g(x_1 + c_1 t)}{\partial x_1}\Big|_{t=0} \\ &= -c_1 \frac{\partial f(x_1)}{\partial x_1} + c_1 \frac{\partial g(x_1)}{\partial x_1} . \end{aligned} \tag{7.251}$$

Integrating (7.251) with respect to $x_1$, leads to

$$-f(x_1) + g(x_1) = \frac{1}{c_1} \int_{x_0}^{x_1} \theta(x_1')dx_1' \,. \tag{7.252}$$

Combining (7.252) with (7.250), we find

$$f(x_1) = \frac{1}{2}\left[\phi(x_1) - \frac{1}{c_1}\int_{x_0}^{x_1} \theta(x_1')dx_1'\right] \tag{7.253}$$

$$g(x_1) = \frac{1}{2}\left[\phi(x_1) + \frac{1}{c_1}\int_{x_0}^{x_1} \theta(x_1')dx_1'\right] \,. \tag{7.254}$$

With the forms of $f$ and $g$ known, the complete solution is written as

$$\begin{aligned}
u_1(x_1,t) &= f(x_1 - c_1t) + g(x_1 + c_1t) \\
&= \left[\frac{\phi(x_1 - c_1t)}{2} - \frac{1}{2c_1}\int_{x_0}^{x_1-c_1t} \theta(x_1')dx_1'\right] \\
&+ \left[\frac{\phi(x_1 + c_1t)}{2} + \frac{1}{2c_1}\int_{x_0}^{x_1+c_1t} \theta(x_1')dx_1'\right] \\
&= \frac{\phi(x_1 + c_1t) + \phi(x_1 - c_1t)}{2} + \frac{1}{2c_1}\int_{x_1-c_1t}^{x_1+c_1t} \theta(x_1')dx_1' \,.
\end{aligned} \tag{7.255}$$

EXAMPLE 7.1

An infinite cord is subject to the initial conditions

$$\phi(x_1) = \frac{0.02}{1+9x_1^2}, \quad \theta(x_1) = 0 \,. \tag{7.256}$$

Find the expression of the subsequent motion of the cord. We assume that the wave speed is $220\,\mathrm{ms}^{-1}$.
Using (7.256) in (7.255), we obtain

$$\begin{aligned}
u_1(x_1,t) &= \frac{1}{2}\left[\phi(x_1 + c_1t) + \phi(x_1 - c_1t)\right] \\
&= \frac{1}{2}\frac{0.02}{1+9(x_1-c_1t)^2} + \frac{1}{2}\frac{0.02}{1+9(x_1+c_1t)^2} \,.
\end{aligned} \tag{7.257}$$

The motion is shown schematically in figure 7.15. The configuration at time $t_0 = 0$ shows the initial perturbation (7.256). Then, it splits into two waves that propagate to the left and right as indicated in solution (7.257) and shown in figure 7.15 for $t_0 < t_1 < t_2 < t_3$.

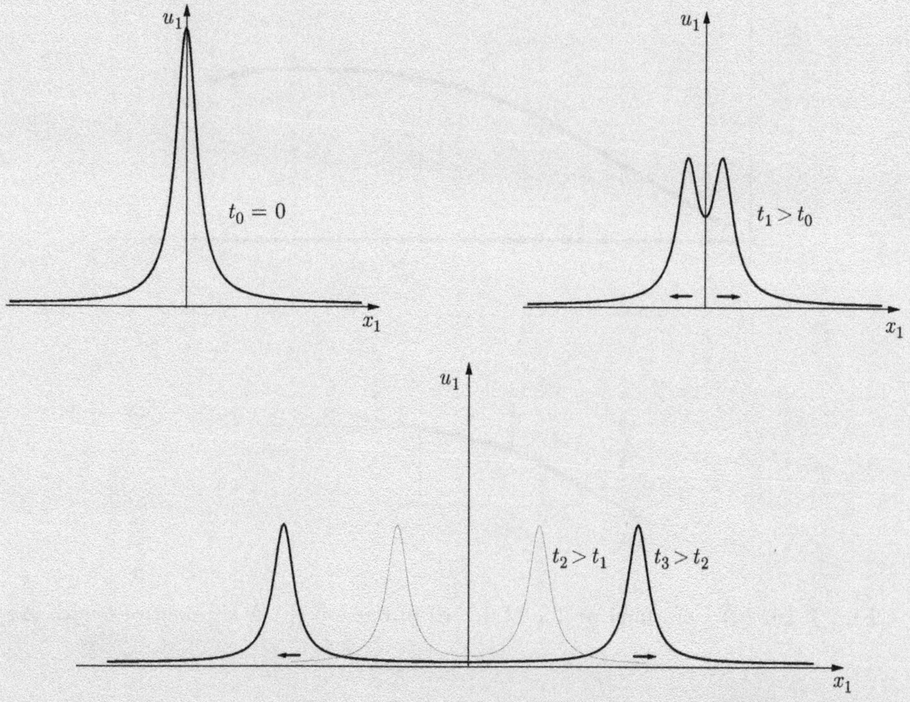


**Fig. 7.15** Propagation of a perturbation along an infinite length cord in the initial configuration ($t_0 = 0$) and later times ($t_1, t_2, t_3$)

### 7.5.4 Propagation of a Wave in an Elastic Cord

The vibration of a cord is a physical problem which illustrates many aspects of wave propagation. It allows us to examine d'Alembert's solution (7.249) and has many applications in the study of musical instruments. In addition, the mathematical analysis of a cord in tension has applications in the study of high tension power transmission lines.

Consider an elastic cord subject to tension $\mathcal{T}$ between two points on the axis as shown in figure 7.16(a). In the following formulation of the problem, we make several hypotheses.

- The motion is entirely in a plane, and every particle of the cord moves at a right angle with respect to the position of the cord in equilibrium. In this case $u_1 = u_3 = 0$ and $u_2$ is only a function of $x_1$ and time $t$. For simplicity in the expression of the equations, we set $u_2(x_1, t) = v(x_1, t)$.
- The cord can only transmit force in the direction of its length.
- The slopes of the deformed cord are small.
- The deflections of the cord are supposed small so that they do not significantly affect the tension $\mathcal{T}$ and there is no energy dissipation.

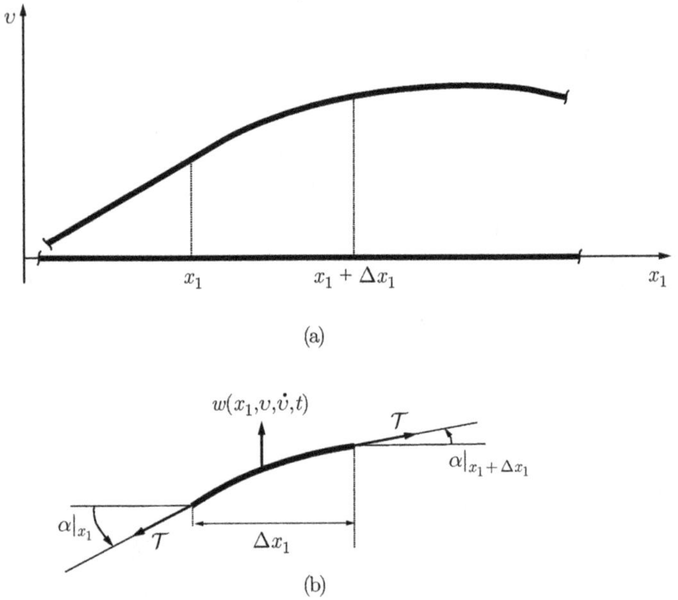


**Fig. 7.16** (a) Deformed profile of a cord in tension (b) A segment of length $\Delta x_1$

The mass of the cord per unit length is a known function $m'(x_1)$. In addition to the elastic and inertial forces inherent in the system, the cord can be subject to a distributed load $w(x_1, v, \dot{v}, t)$. Now consider a segment $\Delta x_1$ (fig. 7.16(b)). Applying Newton's law to the segment in figure 7.16(b), we can write

$$m'\Delta x_1 \frac{\partial^2 v}{\partial t^2} = \mathcal{T} \sin\alpha\Big|_{x_1+\Delta x_1} - \mathcal{T}\sin\alpha\Big|_{x_1} + w\Delta x_1 \ . \tag{7.258}$$

The third hypothesis implies that $\sin\alpha\big|_{x_1} \approx \tan\alpha\big|_{x_1}$ and $\sin\alpha\big|_{x_1+\Delta x_1} \approx \tan\alpha\big|_{x_1+\Delta x_1}$. Inserting these approximations in (7.258) and dividing by $\Delta x_1$ we have

$$m'\frac{\partial^2 v}{\partial t^2} = \mathcal{T}\frac{\tan\alpha\big|_{x_1+\Delta x_1} - \tan\alpha\big|_{x_1}}{\Delta x_1} + w \ . \tag{7.259}$$

Then, by taking the limit $\Delta x_1 \to 0$ and by observing that $\tan\alpha\big|_{x_1} = \frac{\partial v}{\partial x_1}$ and that $\tan\alpha\big|_{x_1+\Delta x_1} = \frac{\partial v}{\partial x_1} + \frac{\partial^2 v}{\partial x_1^2}dx_1$, relation (7.259) reduces to the following differential equation:

$$\frac{\partial^2 v}{\partial t^2} = \frac{\mathcal{T}}{m'}\frac{\partial^2 v}{\partial x_1^2} + \frac{w}{m'} \ . \tag{7.260}$$

In the majority of practical problems, the external forces are negligible and $m'(x_1)$ is constant along the cord. Thus $w(x_1, v, \dot{v}, t)$ can be considered to be zero and $m'(x_1)$ is replaced by a constant $m'_0$. In these conditions, (7.260) further reduces to

$$\frac{\partial^2 v}{\partial t^2} = a^2\frac{\partial^2 v}{\partial x_1^2}, \quad a^2 = \frac{\mathcal{T}}{m'_0} \ , \tag{7.261}$$

where $a$ has the dimensions of a velocity, since $(MLT^{-2}M^{-1}L)^{1/2} = (L^2T^{-2})^{1/2}$. Thus (7.261) is the equation of wave propagation (7.246) for which the solution

$$v(x_1, t) = f(x_1 - at) + g(x_1 + at) \tag{7.262}$$

represents two waves propagating along the cord at speed $a$, one to the right, and one to the left as seen in figure 7.15.

It is worth noticing that in addition to describing the waves on a cord, equation (7.261), or (7.246), is applicable to different types of wave propagation. In particular, a sound wave $v(x_i, t)$ represents the displacement of a gas where a wave propagates. In the case of electromagnetic waves, $v(x_1, t)$ represents the electric or magnetic field component.

Now consider the solution of the partial differential equation by the method of separation of variables. The solution by this method allows us to directly treat the boundary problem encountered in many engineering and physics applications. According to this method, the solution of (7.261) is expressed in the form

$$v(x_1, t) = X(x_1)T(t) \ . \tag{7.263}$$

By substitution of (7.263) in (7.261), we obtain

$$a^2 \frac{d^2X/dx_1^2}{X} = \frac{d^2T/dt^2}{T} = \gamma \ . \tag{7.264}$$

Then, from (7.264) we have two ordinary differential equations

$$\frac{d^2X}{dx_1^2} - \frac{\gamma}{a^2}X = 0 \tag{7.265}$$

$$\frac{d^2T}{dt^2} - \gamma T = 0 \ . \tag{7.266}$$

The solution of these equations depends on the positive, negative, or zero value of the parameter $\gamma$. If $\gamma > 0$ or $\gamma = 0$, the solution of (7.261) is not periodic and cannot describe the undamped vibration of a cord. The only values that produce a periodically vibrating cord are those corresponding to $\gamma < 0$.

Since $\gamma$ is negative, it is customary to define $\gamma = -\omega^2$. Then (7.265) and (7.266) take the forms

$$\frac{d^2X}{dx_1^2} + \left(\frac{\omega}{a}\right)^2 X = 0, \quad \frac{d^2T}{dt^2} + \omega^2 T = 0 \ , \tag{7.267}$$

for which the solutions are

$$X(x_1) = A\cos\frac{\omega}{a}x_1 + B\sin\frac{\omega}{a}x_1 \tag{7.268}$$

$$T(t) = C\cos\omega t + D\sin\omega t \ , \tag{7.269}$$

such that

$$v(x_1, t) = \left(A\cos\frac{\omega}{a}x_1 + B\sin\frac{\omega}{a}x_1\right)(C\cos\omega t + D\sin\omega t) \ , \tag{7.270}$$

where $A, B, C$, and $D$ are arbitrary constants and $\omega$ can be interpreted as a ***circular frequency*** that depends on the boundary conditions and which must be evaluated. Note that the solution (7.270) is periodic since it is the same with an increment of time by a factor $2\pi/\omega$. The first part of the expression defines the form of the cord, the second, its motion.

Equation (7.270) can be rewritten in the following form

$$\begin{aligned} v(x_1, t) &= AC \cos \frac{\omega}{a} x_1 \cos \omega t + AD \cos \frac{\omega}{a} x_1 \sin \omega t \\ &+ BC \sin \frac{\omega}{a} x_1 \cos \omega t + BD \sin \frac{\omega}{a} x_1 \sin \omega t \, . \end{aligned} \tag{7.271}$$

Using elementary trigonometric identities, we express (7.271) as

$$\begin{aligned} v(x_1, t) &= B_1 \sin \left( \frac{\omega}{a} x_1 + \omega t \right) + B_2 \sin \left( \frac{\omega}{a} x_1 - \omega t \right) \\ &+ B_3 \cos \left( \frac{\omega}{a} x_1 + \omega t \right) + B_4 \cos \left( \frac{\omega}{a} x_1 - \omega t \right) \, , \end{aligned} \tag{7.272}$$

where $B_i (i = 1, 2, 3, 4)$ are arbitrary constants. It is interesting to note that this last equation resembles d'Alembert's solution (7.249) and expresses harmonic wave propagation along a cord in tension. For example, one term of this solution

$$v(x_1, t) = B_4 \cos \left( \frac{\omega}{a} x_1 - \omega t \right) \tag{7.273}$$

represents a wave propagating in direction $x_1$ at speed $a$ as we show in figure 7.17. It is clear that the other terms of (7.272) can be similarly interpreted.

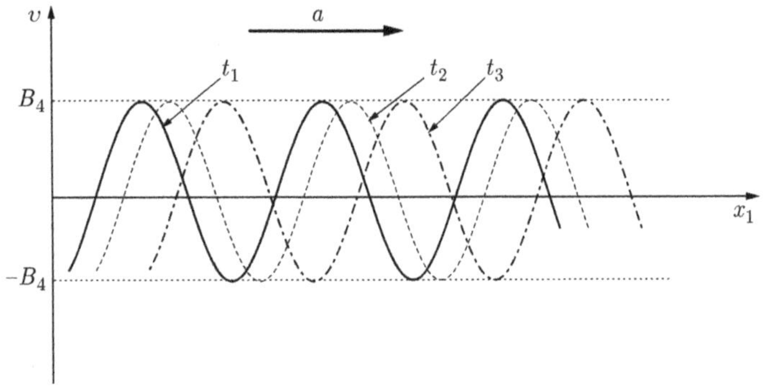


**Fig. 7.17** Transverse deviations at successive time intervals as result of wave propagation (7.273)

Now let us examine the energy components during wave propagation. Assuming that there is no energy dissipation, the cord contains kinetic and potential energies of deformation. From (7.273), the speed and deformation are expressed as

$$\frac{\partial v}{\partial t} = -B_4 \omega \sin \left( \frac{\omega}{a} x_1 - \omega t \right) \tag{7.274}$$

$$\varepsilon_{12} = \frac{1}{2} \frac{\partial v}{\partial x_1} = -B_4 \frac{1}{2} \frac{\omega}{a} \sin \left( \frac{\omega}{a} x_1 - \omega t \right) \, . \tag{7.275}$$

Let $S$ be the area of a transverse section of the cord. Using (7.273), the kinetic and potential energies of an element $dx_1$ are

$$dE = \frac{1}{2}\rho S dx_1 \left(\frac{\partial v}{\partial t}\right)^2 = \frac{1}{2}\rho (B_4\omega)^2 \sin^2\left(\frac{\omega}{a}x_1 - \omega t\right) S dx_1 \qquad (7.276)$$

$$dU = 2\mu S \varepsilon_{12}^2 dx_1 = \mu\frac{1}{2}\left(B_4\frac{\omega}{a}\right)^2 \sin^2\left(\frac{\omega}{a}x_1 - \omega t\right) S dx_1 \,. \qquad (7.277)$$

Eliminating the parameter for the wave speed $a$ with (7.213), it is obvious that $dE = dU$.

EXAMPLE 7.2

**Vibration of an Elastic Cord**

Now consider an elastic cord stretched between two points separated by a distance $\ell$ (fig. 7.18). The general solution to this problem is given by (7.270).

To obtain the solution for the cord in figure 7.18, we need to determine the four constants $A, B, C$, and $D$ and the parameter $\omega$ for the boundary conditions $v(0,t) = v(\ell,t) = 0$ and the initial condition. In order to do so, it is useful to choose the velocity of the cord as zero at time $t = 0$ and to assume that at $t = 0$ the cord in its initial form is in a normal, or principal, mode. We will define such a mode later in this section.

The two boundary conditions specified above lead to

$$\begin{aligned} 0 &= A\left(D\sin\omega t + C\cos\omega t\right) \,, \\ 0 &= \left(A\cos\frac{\omega}{a}\ell + B\sin\frac{\omega}{a}\ell\right)\left(C\cos\omega t + D\sin\omega t\right) \,. \end{aligned} \qquad (7.278)$$

These must be satisfied at all times, thus

$$A = 0, \quad B\sin\frac{\omega}{a}\ell = 0 \,. \qquad (7.279)$$

For a non-trivial solution, $B$ cannot also be zero, thus $\sin\frac{\omega}{a}\ell = 0$, from which we obtain

$$\frac{\omega}{a}\ell = n\pi \,, \qquad (7.280)$$

with $n = 1, 2, 3, \ldots$. Therefore $\omega$ is given by

$$\omega = \frac{n\pi}{\ell}a \,, \qquad (7.281)$$

and the ***frequency*** and ***period*** of the wave are

$$f = \frac{\omega}{2\pi} = \frac{n}{2\ell}a, \quad T = \frac{1}{f} = \frac{1}{a}\frac{2\ell}{n} \,. \qquad (7.282)$$

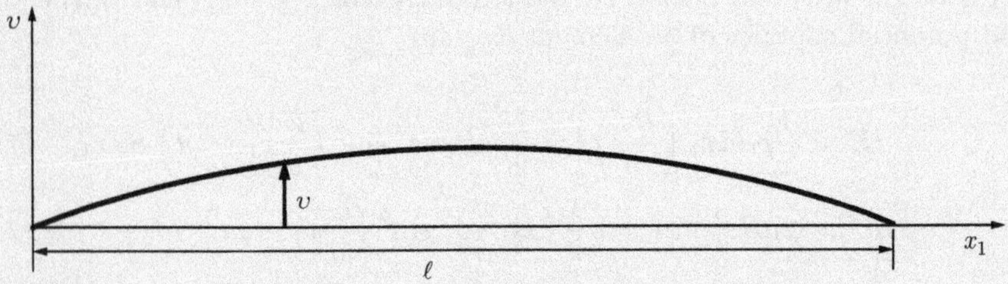


**Fig. 7.18** Deformed elastic cord between two points separated by a distance $\ell$

We will study two initial conditons in the following. Differentiating (7.270) with respect to time, we find for the velocity

$$\frac{\partial v}{\partial t} = \left(B \sin \frac{\omega}{a} x_1\right) (D \cos \omega t - C \sin \omega t)\omega \ . \tag{7.283}$$

By imposing zero velocity at $t = 0$, we obtain $D = 0$ since $B \neq 0$. Consequently the solution reduces to

$$v(x_1, t) = BC \sin \frac{\omega}{a} x_1 \cos \omega t \ . \tag{7.284}$$

At $t = 0$, this last equation yields

$$v(x_1, 0) = BC \sin \frac{\omega}{a} x_1 \ , \tag{7.285}$$

which defines a ***principal mode*** configuration for the initial condition. Finally, the solution to the vibrating cord is expressed by

$$v(x_1, t) = \mathcal{A} \sin \frac{\omega}{a} x_1 \cos \omega t = \mathcal{A} \sin \frac{n\pi}{\ell} x_1 \cos \frac{n\pi a}{\ell} t \ , \tag{7.286}$$

where $\mathcal{A}$ represents the maximum value (or amplitude) of the deviation $v(x_1, t)$ and $n = 1, 2, 3, \ldots$. Note that the first term in the product (7.286) represents the configuration, and the second, the motion of the cord. The ***wavelength*** $x_\lambda$ is defined by the length of a sinusoidal wave for $\sin \frac{\omega}{a} x_1$, or

$$\frac{\omega}{a} x_1 = 2\pi \quad \text{or} \quad x_\lambda = \frac{2\ell}{n} \ . \tag{7.287}$$

Figure 7.19 shows the forms of the three principal vibration modes according to (7.286). They are

$$\text{(a)} \quad n = 1,\ x_\lambda = 2\ell,\ \omega = \frac{\pi}{\ell} a \, ;$$

$$\text{(b)} \quad n = 2,\ x_\lambda = \ell,\ \omega = \frac{2\pi}{\ell} a \, ;$$

$$\text{(c)} \quad n = 3,\ x_\lambda = \frac{2}{3}\ell,\ \omega = \frac{3\pi}{\ell} a \ .$$

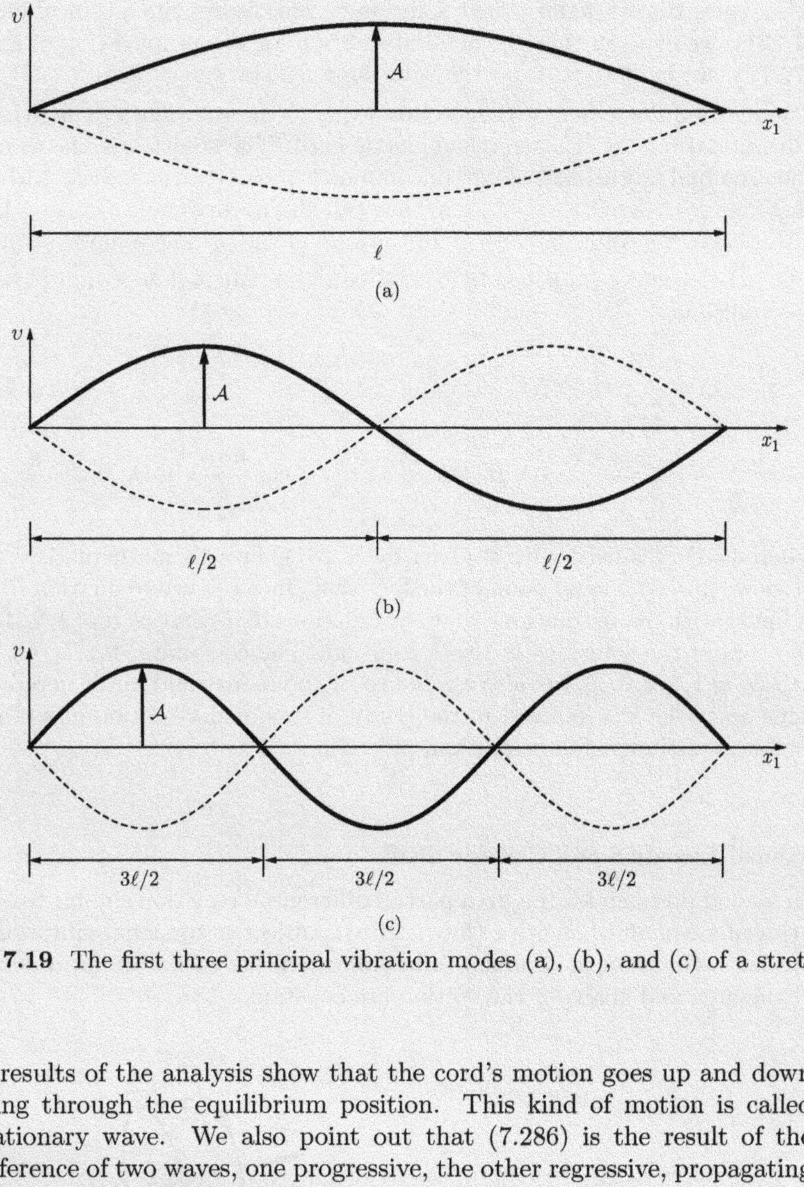


**Fig. 7.19** The first three principal vibration modes (a), (b), and (c) of a stretched cord

The results of the analysis show that the cord's motion goes up and down passing through the equilibrium position. This kind of motion is called a stationary wave. We also point out that (7.286) is the result of the interference of two waves, one progressive, the other regressive, propagating in opposite directions. To show this, we can use trigonometric identities to rewrite (7.286) as follows:

$$
\begin{aligned}
v(x_1, t) &= \frac{\mathcal{A}}{2}\Big[\left(\sin\frac{\omega}{a}x_1\cos\omega t + \cos\frac{\omega}{a}x_1\sin\omega t\right) \\
&+ \left(\sin\frac{\omega}{a}x_1\cos\omega t - \cos\frac{\omega}{a}x_1\sin\omega t\right)\Big] \\
&= \mathcal{A}\left[\sin\frac{\omega}{a}(x_1 + at) + \sin\frac{\omega}{a}(x_1 - at)\right] . \qquad (7.288)
\end{aligned}
$$

Also, since the stationary wave solution (7.286) represents a typical term of (7.271), we can say that the general solution for the propagation of a wave (7.272) can be derived from the solution for a stationary wave (7.271).

The method described in this section leads to the solution for the principal vibrational modes of a stretched elastic cord. The solution is the same for the principal vibrational modes of common types of elastic bodies. Although these modes can exist in isolation, they can also occur simultaneously. In the latter case, the solution consists of the sum of the principal mode solutions.

Thus the elastic cord solution (7.270) can be expressed as a sum over $n$ of the solutions

$$\begin{aligned} v(x_1,t) &= \sum_{n=1}^{\infty} u_{1n}(x_1,t) \\ &= \sum_{n=1}^{\infty} \left(A_n \cos\frac{n\pi}{\ell}x_1 + B_n \sin\frac{n\pi}{\ell}x_1\right)\left(C_n \cos\frac{n\pi a}{\ell}t + D_n \sin\frac{n\pi a}{\ell}t\right) , \end{aligned} \tag{7.289}$$

where (7.270) is used with $\omega$ given by (7.281). From a mathematical point of view, this is an expression of the fact that since the wave equation (7.261) is linear with many solutions that are functions of $n$ (that is, $n = 1, 2, 3, \ldots$), the sum of the solutions is also a solution. The constants $A_n, B_n, C_n$, and $D_n, (n = 1, 2, 3, \ldots)$, are also related to the boundary and initial conditions. This approach is explained in the study of torsional vibration of a circular elastic shaft in the following example.

**Torsional Vibration of a Circular Shaft**

As a second problem leading to a partial differential equation similar to (7.261), we consider a shaft of length $\ell$ (fig. 7.20(a)) subject to torsional vibration. The material in the shaft is assumed to be homogeneous with a uniform density $\rho$, and the area and shape of the section are constant.

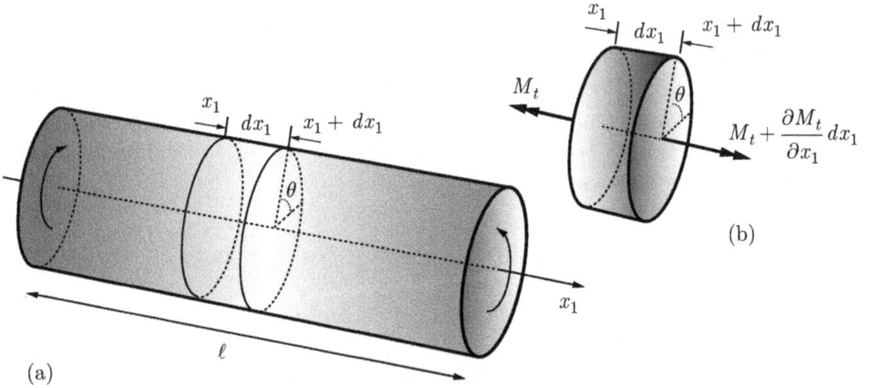


**Fig. 7.20** (a) A circular shaft subject to torque (b) An infinitesimal slice $dx_1$

In addition to the elastic and inertial forces inherent in the system, the shaft could be subject to a torque distributed by unit length $w(x_1, \theta, \dot{\theta}, t)$. In the following formulation of the problem, we assume that

- the transversal sections of the shaft remain plane during deformation,
- a typical transverse section rotates around its center of gravity,
- the rotation of the shaft is small,
- there is no energy dissipation.

From elementary solid mechanics, we know that, for static conditions, the variation of the torsion angle $\theta(x_1)$ along the shaft axis $d\theta/dx_1$ is expressed as follows:

$$\frac{d\theta(x_1)}{dx_1} = \frac{M_t(x_1)}{\mu I_p} , \tag{7.290}$$

where $M_t(x_1)$ is the applied torque, $\mu$ the shear modulus, and $I_p$ the polar moment of inertia of the transverse circular section. For the problem treated here, the torsion angle is a function of the time and space variables. Then (7.290) is rewritten as

$$\frac{\partial\theta(x_1,t)}{\partial x_1} = \frac{M_t(x_1,t)}{\mu I_p} . \tag{7.291}$$

The partial derivative of (7.291) with respect to $x_1$ yields

$$\frac{\partial M_t(x_1,t)}{\partial x_1} = \mu I_p \frac{\partial^2\theta(x_1,t)}{\partial x_1^2} . \tag{7.292}$$

In order to establish the differential equation of motion, we consider an infinitesimal segment of the shaft, between two transversal sections separated by distance $dx_1$ (fig. 7.20(b)). The moment of inertia of this slice is

$$dJ = \int r^2 dm = \int \rho r^2 dS dx_1 = \rho dx_1 \int r^2 dS = \rho I_p dx_1 , \tag{7.293}$$

where $dm$ is the mass of an element at a distance $r$ from the center of the section. Applying Newton's law in a torsional form to the infinitesimal segment $dx_1$ (fig. 7.20(b)), we have

$$(I_p\, \rho\, dx_1)\, \frac{\partial^2\theta(x_1,t)}{\partial t^2} = \left(M_t(x_1,t) + \frac{\partial M_t(x_1,t)}{\partial x_1} dx_1\right) - M_t(x_1,t) + w\, dx_1$$
$$= \frac{\partial M_t(x_1,t)}{\partial x_1} dx_1 + w\, dx_1 .$$

Dividing by $dx_1$ and using (7.292) leads to

$$\frac{\partial^2\theta(x_1,t)}{\partial t^2} = a^2 \frac{\partial^2\theta(x_1,t)}{\partial x_1^2} + w . \tag{7.294}$$

In many interesting cases, the external torque $w$ can be neglected, that is, $w(x_1, \theta, \dot{\theta}, t) = 0$. Then, relation (7.294) has the same form as the wave equation

(7.246) or (7.261) for an elastic cord. Note that $a = \sqrt{\mu/\rho}$ has the dimensions of a velocity.

By separation of variables, the solution is of the form

$$\theta(x_1, t) = X(x_1)T(t) \ . \tag{7.295}$$

Following the same procedure as for the cord problem treated in the preceding section, the solution is expressed as

$$\theta(x_1, t) = \left(A\cos\frac{\omega}{a}x_1 + B\sin\frac{\omega}{a}x_1\right)(C\cos\omega t + D\sin\omega t) \ . \tag{7.296}$$

The ends of the shaft can be fixed or free with respect to rotation. In this section we study the case of a shaft with both ends free. To determine the constants in (7.296), the boundary and initial conditions must be specified. Given the case of free shaft ends, the torque there must be zero. Taking into account (7.291), these conditions yield

$$\frac{\partial\theta(0,t)}{\partial x_1} = \frac{\partial\theta(\ell,t)}{\partial x_1} = 0 \ . \tag{7.297}$$

From (7.296), we find

$$\frac{\partial\theta}{\partial x_1} = \left(-A\frac{\omega}{a}\sin\frac{\omega}{a}x_1 + B\frac{\omega}{a}\cos\frac{\omega}{a}x_1\right)(C\cos\omega t + D\sin\omega t) \ . \tag{7.298}$$

The first condition of (7.297) leads to

$$B\frac{\omega}{a}(C\cos\omega t + D\sin\omega t) = 0, \forall\, t \ . \tag{7.299}$$

Thus $B = 0$. And similarly, imposing the second condition (7.297), we have

$$-A\frac{\omega}{a}\sin\frac{\omega}{a}\ell(C\cos\omega t + D\sin\omega t) = 0, \forall\, t \ . \tag{7.300}$$

For a non-trivial solution, $A$ cannot be zero, so we must have

$$\sin\frac{\omega}{a}\ell = 0, \quad \text{or} \quad \frac{\omega}{a}\ell = n\pi \ . \tag{7.301}$$

As for the case of the elastic cord, $\omega$ takes the following values

$$\omega_n = \frac{n\pi a}{\ell}, \qquad n = 1, 2, 3, \dots \ . \tag{7.302}$$

It is clear that we have an infinite number of solutions. The $n^{th}$ solution of the problem is

$$\theta_n(x_1, t) = \cos\frac{\omega_n}{a}x_1\,(C_n\cos\omega_n t + D_n\sin\omega_n t) \ . \tag{7.303}$$

Note that the constant $A_n$ has been absorbed into the constants $C_n$ and $D_n$. Since the wave equation is linear, the sum over $n$ of all the solutions is also a solution

$$\theta(x_1, t) = \sum_{n=1}^{\infty}\theta_n(x_1, t) = \sum_{n=1}^{\infty}\cos\frac{n\pi x_1}{\ell}\left(C_n\cos\frac{n\pi a}{\ell}t + D_n\sin\frac{n\pi a}{\ell}t\right) \ . \tag{7.304}$$

Relation (7.304) must satisfy the following initial conditions

$$\theta(x_1, 0) = f(x_1) \quad \text{and} \quad \frac{\partial \theta(x_1, 0)}{\partial t} = g(x_1) \, , \tag{7.305}$$

where the two functions $f(x_1)$ and $g(x_1)$ are known. To satisfy these conditions, we must have

$$\theta(x_1, 0) = f(x_1) = \sum_{n=1}^{\infty} C_n \cos \frac{n\pi x_1}{\ell} \, . \tag{7.306}$$

This last result implies that $C_n$ are the coefficients in the half-range cosine expansion of $f(x_1)$ over the interval $(0, \ell)$

$$C_n = \frac{2}{\ell} \int_0^{\ell} f(x_1) \cos \frac{n\pi x_1}{\ell} dx_1 \, . \tag{7.307}$$

The second initial condition imposes

$$\left. \frac{\partial \theta(x_1, t)}{\partial t} \right|_{t=0} = g(x_1) = \sum_{n=1}^{\infty} D_n \frac{n\pi a}{\ell} \cos \frac{n\pi x_1}{\ell} \, , \tag{7.308}$$

such that $D_n \frac{n\pi a}{\ell}$ are the coefficients in the half-range cosine expansion of $g(x_1)$ over the interval $(0, \ell)$

$$D_n \frac{n\pi a}{\ell} = \frac{2}{\ell} \int_0^{\ell} g(x_1) \cos \frac{n\pi x_1}{\ell} dx_1 \quad \text{or} \quad D_n = \frac{2}{n\pi a} \int_0^{\ell} g(x_1) \cos \frac{n\pi x_1}{\ell} dx_1 \, . \tag{7.309}$$

Note that the same analysis can be carried out for different boundary conditions, that is, when both ends of the shaft are fixed or when one is fixed and the other is free in rotation.

**Longitudinal Vibration of a Prismatic Beam**

Using elementary beam theory, and following the same procedure as for torsion of the circular bar treated above, the equation for longitudinal waves in a beam (fig. 7.21) is

$$\frac{\partial^2 u_1(x_1, t)}{\partial t^2} = a^2 \frac{\partial^2 u_1(x_1, t)}{\partial x_1^2} \, , \tag{7.310}$$

where

$$a = \sqrt{\frac{E}{\rho}} \tag{7.311}$$

is the longitudinal wave speed.

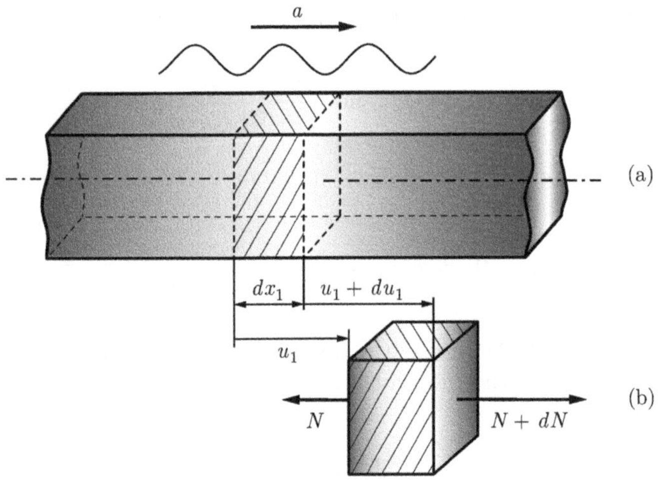


**Fig. 7.21** (a) A thin beam subject to longitudinal vibrations (b) A slice of length $dx_1$

The derivation of (7.310) is left as an exercise for the reader.

We conclude this section with values of the wave speed in typical elastic bodies. Using the properties from table 6.2, we obtain the values in the following table.

**Table 7.1** Wave speed in elastic solids

| Wave speed ($m/s$) | Steel | Glass | Rubber |
|---|---|---|---|
| Dilatation | 5240 | 5505 | $\to \infty$ |
| Longitudinal | 5047 | 5253 | 242 |
| Shear or transversal | 3169 | 3405 | 140 |

The first line for the wave speed corresponds to relation (7.212) and the second to (7.311). The transversal wave speed is from (7.213). Note that the dilatation wave speed for rubber is not defined, since we assume incompressibility and $\nu = 0.5$. Using expression (7.311), the longitudinal wave speed has a finite value. For the other materials, the longitudinal speed is smaller by about 4%, because the lateral effects induced by Poisson's coefficient are neglected in elementary beam theory.

## 7.6 Exercises

**7.1** Show that in the case of plane strain, the equilibrium equations can be written in terms of the displacement in the form

$$\mu\left(\frac{\partial^2 u_1}{\partial x_1{}^2}+\frac{\partial^2 u_1}{\partial x_2{}^2}\right)+(\lambda+\mu)\frac{\partial}{\partial x_1}\left(\frac{\partial u_1}{\partial x_1}+\frac{\partial u_2}{\partial x_2}\right)+f_1=0 \tag{7.312}$$

$$\mu\left(\frac{\partial^2 u_2}{\partial x_1{}^2}+\frac{\partial^2 u_2}{\partial x_2{}^2}\right)+(\lambda+\mu)\frac{\partial}{\partial x_2}\left(\frac{\partial u_1}{\partial x_1}+\frac{\partial u_2}{\partial x_2}\right)+f_2=0\,, \tag{7.313}$$

where $\mu = E/2(1+\nu)$ is the shear modulus and $\lambda$ is related to $E$ and $\nu$ by relation (6.112).

**7.2** Show that in the case of plane stress, the equilibrium equations can be expressed in terms of the displacement in the form

$$\mu\left(\frac{\partial^2 u_1}{\partial x_1{}^2}+\frac{\partial^2 u_1}{\partial x_2{}^2}\right)+\frac{E}{2(1-\nu)}\frac{\partial}{\partial x_1}\left(\frac{\partial u_1}{\partial x_1}+\frac{\partial u_2}{\partial x_2}\right)+f_1=0 \tag{7.314}$$

$$\mu\left(\frac{\partial^2 u_2}{\partial x_1{}^2}+\frac{\partial^2 u_2}{\partial x_2{}^2}\right)+\frac{E}{2(1-\nu)}\frac{\partial}{\partial x_2}\left(\frac{\partial u_1}{\partial x_1}+\frac{\partial u_2}{\partial x_2}\right)+f_2=0\,. \tag{7.315}$$

**7.3** Show that if a vector field $\boldsymbol{g}(\boldsymbol{x})$ is such that

$$g_{i,mmnn}=0\,, \tag{7.316}$$

then the displacement field defined by

$$u_i=\frac{\lambda+2\mu}{\mu(\lambda+\mu)}g_{i,mm}-\frac{1}{\mu}g_{n,ni} \tag{7.317}$$

satisfies Navier's equations without volume forces.

**7.4** Prove that Navier's equations (7.7) with $\boldsymbol{f}=\boldsymbol{0}$ are equivalent to the following equations:

1) $(\lambda+2\mu)\nabla^2\boldsymbol{u}+(\lambda+\mu)\boldsymbol{\nabla}\times\boldsymbol{\nabla}\times\boldsymbol{u}=0\,,$
2) $(1-2\nu)\nabla^2\boldsymbol{u}+\boldsymbol{\nabla}\cdot\boldsymbol{\nabla}\boldsymbol{u}=0\,,$
3) $(\lambda+2\mu)\nabla\boldsymbol{\nabla}\cdot\boldsymbol{u}-\mu\boldsymbol{\nabla}\times\boldsymbol{\nabla}\times\boldsymbol{u}=0\,.$

**7.5** Express the wave equation (7.209) in terms of div $\boldsymbol{u}$ as follows:

$$(\lambda+2\mu)\nabla^2(\text{div }\boldsymbol{u})=\rho\frac{\partial^2(\text{div }\boldsymbol{u})}{\partial t^2}\,. \tag{7.318}$$

**7.6** Express the wave equation (7.205) in terms of the rotation components of the displacement field $\nabla\times\boldsymbol{u}$ as follows:

$$\mu\nabla^2(\boldsymbol{\nabla}\times\boldsymbol{u})=\rho\frac{\partial^2(\boldsymbol{\nabla}\times\boldsymbol{u})}{\partial t^2}\,. \tag{7.319}$$

**7.7** For the corner of angle $2\alpha$ and unit thickness, with a moment $M$ applied as shown in figure 7.22, determine whether a stress function of the form

$$\Phi(r,\theta) = A\theta + B\sin 2\theta \tag{7.320}$$

is a solution of the problem. Then show that $\sigma_{rr}$ is given by

$$\sigma_{rr} = -\frac{2C}{r^2}\sin 2\theta\ , \tag{7.321}$$

with $C = M/(\sin 2\alpha - 2\alpha\cos 2\alpha)$.

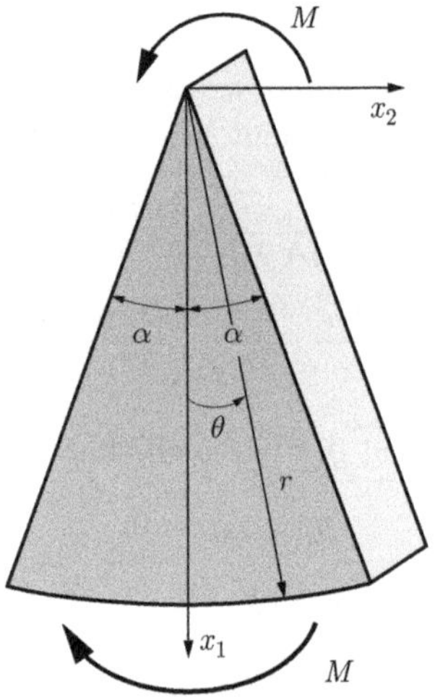


**Fig. 7.22** Corner subject to a moment $M$

**7.8** For a corner with angle $2\alpha$ and unit thickness, with the load $P$ as indicated in figure 7.23.

(a) Prove that the stress function

$$\Phi(r,\theta) = Cr\theta\sin\theta \tag{7.322}$$

provides a solution for the stresses of the problem. Then express the components of the stress and determine the constant $C$.

(b) How can the stresses be found for a semi-infinite unit thickness plate with a linear force load as in the previous case?

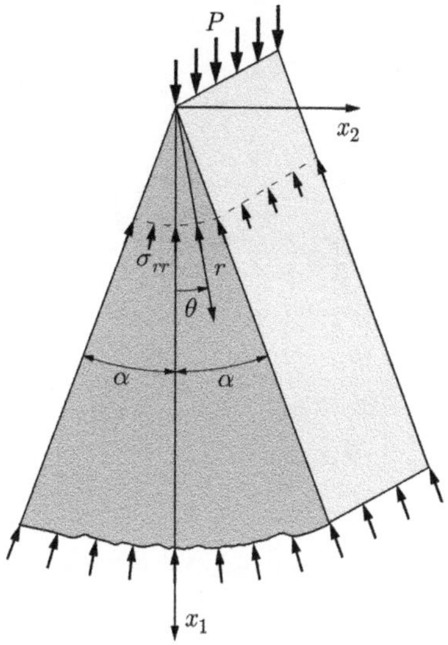


**Fig. 7.23** Corner subject to a linear force $P$

CHAPTER 8

# Introduction to Newtonian Fluid Mechanics

## 8.1 Introduction

Fluids are omnipresent in nature, technology, and everyday life, for example, in the water for washing one's hands and coffee or tea, the air we breathe, the blood circulating in our vascular system, etc. The two most common fluids, air and water, are typical examples of the two major categories: compressible and incompressible fluids.

We have all encountered a flat tire on our bicycle and have had to pump air into the tube. Then we experimentally discover that air can be compressed, thus the pressure increased, which is very useful in this particular case. As for water, we generally consider it to be an incompressible fluid. It is an idealization of reality, since the speed of sound in water is 1500 $ms^{-1}$, that is, around five times the speed of sound in air.

The effects of compressibility result in characteristic physical phenomena such as the propagation of acoustic waves or the presence of shock waves in supersonic flow. Nonetheless, we can treat air as an incompressible fluid when the Mach number is small. This is common practice in automobile aerodynamics. We can also treat water as a compressible fluid, if we are interested in the propagation of acoustic waves, for example, in the oceans.

Another classification of fluids is made on the basis of their Newtonian or non-Newtonian character. Air and water are Newtonian fluids. Molten polymers, blood, mud, agro-alimentary liquids, paints, toothpaste, etc. are non-Newtonian.

Couette flow between two circular cylinders is an adequate experiment to discriminate these two fluid categories. We have two vertical, coaxial cylinders. The outer one, for example, can be fixed, while the inner one is forced to rotate at a constant speed of around a dozen revolutions per minute. The annular space between them is filled with liquid fluid, up to a certain height. At the free surface the liquid is in contact with air. In the case of a Newtonian fluid (such as water), we observe that the free surface takes the form of a paraboloid of revolution under the action of centrifugal force. A non-Newtonian fluid, however, accumulates around the rotating inner cylinder. This is the Weissenberg effect.

The Navier-Stokes equations describe the dynamics of viscous flow. They are derived starting from the conservation and constitutive equations. In this chapter, we choose the constitutive equations of Newtonian viscous fluids, for which the tensor $\boldsymbol{\sigma}$ is a linear function of the tensor $\boldsymbol{d}$.

In some industrial processes we encounter materials whose behavior in the fluid state departs from Newtonian character. In these cases we turn to concepts from rheology to study the constitutive equation which best represents the phenomena associated with fluid flow. The reader is referred to the monograph [9] for more information.

Nonetheless, the Navier-Stokes equations constitute a sufficiently rich model to be applied in a very large number of cases. We have chosen the Eulerian representation of the conservation equations, as fluids generally experience very large motions, but also because the common problems are defined in spatial (as opposed to material) coordinate systems.

The physics of Newtonian fluids is characterized by the non-dimensional Reynolds number defined by the relation

$$Re = \frac{UL}{\nu} , \tag{8.1}$$

where $U$ and $L$ are, respectively, a reference velocity and length for the flow under consideration, and $\nu$ is the kinematic viscosity of the fluid. This number can take values from zero to several million. When $Re$ is near zero, the flows are laminar. Their geometric configuration and dynamics are relatively simple. Their interpretation with analytic solutions permits a profound understanding of the associated physics. As $Re$ increases, the laminar flows experience instabilities which gradually lead to fully developed turbulence. It is the latter condition that we experience during a flight when an aircraft is violently shaken by atmospheric turbulence. Understanding turbulence still remains one of the great challenges of physics.

The reader can find additional information in the following texts: [2, 3, 25, 26, 28, 56, 64].

## 8.2 Physical Considerations for Laminar and Turbulent Incompressible Flows

In this section we only consider incompressible fluid flows as a function of the Reynolds number (eqn. (8.1)). As we have already stated, the physics of flow changes drastically, going from creeping flows at a very low Reynolds number to those for which $Re \sim 10^6 \sim 10^7$.

Creeping fluid flows are laminar (from the Latin word *laminae*: thin layers); they are very often stationary, and the streamlines follow the contours of the obstacles placed in the flow. These flows present effects that are easy to predict and to interpret, as they correspond to the linear Stokes equation. They can be found in lava, terrestrial magma, molten glass, and very viscous polymers.

As $Re$ increases, the non-linear terms of the Navier-Stokes equations become preponderant and for values of a few dozen, the laminar flows become unstable and secondary flows are produced. These are called transitional flows.

**Fig. 8.1** Flow around a cylinder at $Re \simeq 0$

An excellent example is the uniform, parallel flow upstream of a horizontal circular cylinder. The Reynolds number is defined by $U$, the uniform upstream velocity, $L = D$, the diameter of the cylinder, and $\nu$, the kinematic viscosity of the fluid. Figure 8.1[(1)] shows the flow at $Re \simeq 0$ for which the streamlines are symmetric with respect to the horizontal, vertical, and diagonal directions.

As $Re$ grows, for the values 13.1 and 26 shown in figure 8.2, it is seen that the flow is stationary and symmetric with respect to the horizontal axis. However, two counter-rotating recirculation zones appear behind the cylinder. The

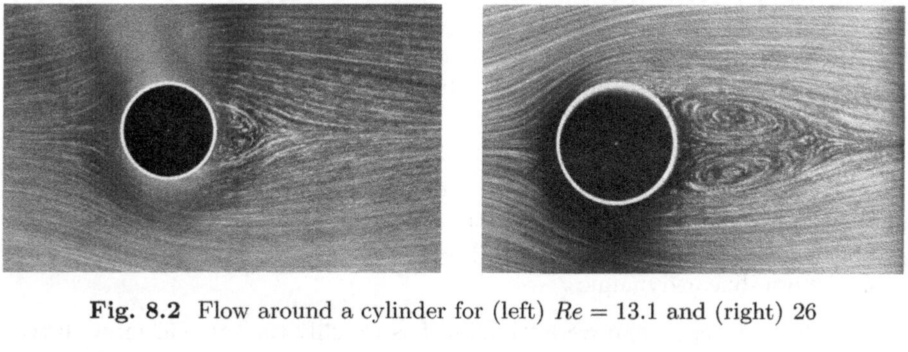

**Fig. 8.2** Flow around a cylinder for (left) $Re = 13.1$ and (right) 26

[(1)]Figures 8.1–8.4, 8.7, and 8.11 are taken from text [62]. Attempts to identify the copyright owner have not as yet succeeded, and he or she is invited to contact the publisher.

length of the recirculation zone increases linearly with $Re$ while the distance separating the centers of the vortices grows as $\sqrt{Re}$.

At $Re = 47.5$, the first critical Reynolds number is reached, at which point the physical phenomena become unstable. A von Kármán vortex street is produced behind the cylinder with vortices alternately shed above and below. A similar vortex street is shown in figure 8.3 for $Re = 140$, taken from [62]. The shed vortices are regularly produced at a frequency corresponding to a limit cycle in phase space: a Hopf bifurcation. This frequency, denoted $f$, leads to the definition of the Strouhal number, $St$

$$St = \frac{fD}{U} \,. \tag{8.2}$$

For values of $Re$ around one hundred, $St$ is 0.13.

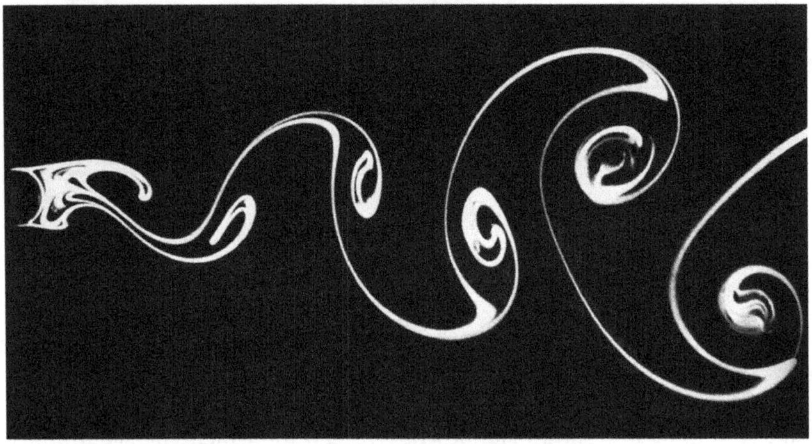

**Fig. 8.3** von Kármán vortex street for $Re = 140$

Stability analyses are based on the Ginzburg-Landau equation ([10]) which determines the non-linear development of perturbations superimposed on an underlying flow. This theory extends over a vast domain that this book cannot cover. We refer the reader to specialized texts, for example, [5, 11, 44]. If the Reynolds number is further increased, the flow passes through transitional regimes before finally attaining the turbulent state. An excellent synthesis of the dynamics of the wakes of circular cylinders is that of Williamson [66].

Figure 8.4 shows the flow pattern for weak turbulence. The boundary layer, where viscous effects are of the same order of magnitude as inertial effects, is laminar in front of the cylinder, develops around it, undergoes a separation, and produces a turbulent wake. It is still possible to observe two vortices resulting from the non-linear dynamics.

At $Re = 10^4$ as in figure 8.5, the flow has roughly the same form, with two identifiable vortices.

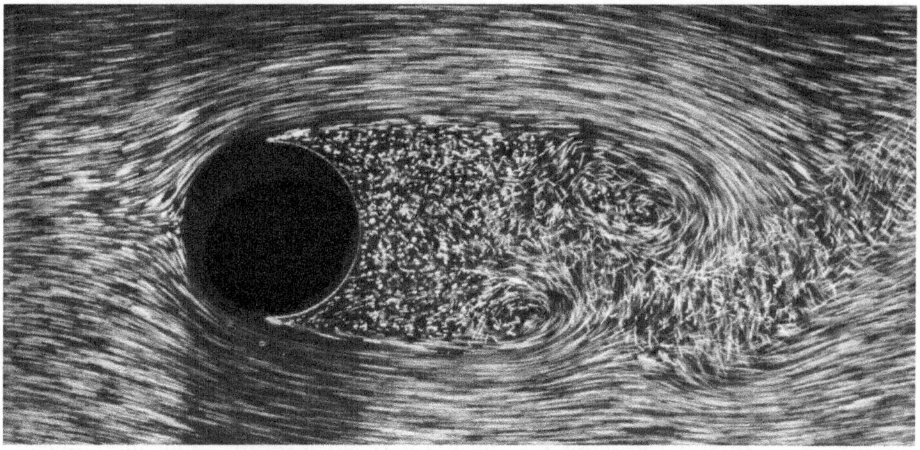

**Fig. 8.4** von Kármán vortex street for $Re = 2000$

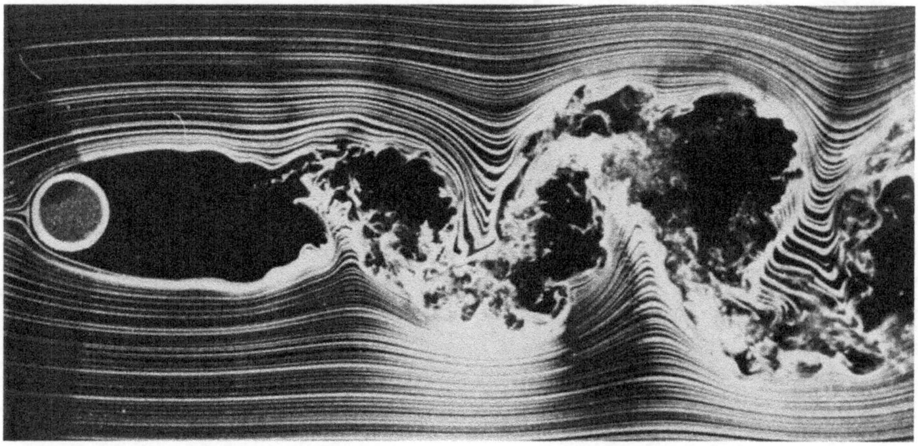

**Fig. 8.5** von Kármán vortex street for $Re = 10^4$

When the Reynolds number reaches a few tens of thousands, or millions, the physics of the flow presents a multitude of spatial and temporal scales; fully developed turbulence is present. Turbulence exists in the majority of flows in nature. Everyone has certainly experienced it during a flight: the chaotic and random effects of turbulence which correspond to dynamics with very rapid variation. However, understanding of turbulence is one of the rare challenges in modern physics that has not been completely attained.

Non-linear equations as a rule are very difficult to solve analytically, and the Navier-Stokes equations do not escape this rule. It is one of the reasons why numerical simulation has come to dominate as the only way of performing an in-depth analysis of such complex phenomena. The volume and finite element methods constitute a pertinent choice to perform this type of calculation [40].

## 8.3 Physical Considerations for Compressible Fluid Flows

Compressible fluids, such as air or gases in general, present phenomena that are complex and very interesting for scientists and engineers.

Compressible flows and their effects need to be taken into account for high speed flows in particular, which we refer to as of gas dynamics. In this case, the values of the Reynolds number are very large. For example, as the kinematic viscosity of air at room temperature is $\nu_{air} = 1.45\,10^{-5}\,\mathrm{m}^2\mathrm{s}^{-1}$, with speeds of the order of hundreds of meters per second, the Reynolds number is in the millions. The flows are obviously turbulent. In aerodynamics, viscous effects are present in zones near the body or obstacles in the flow. This is one reason for which we can neglect turbulence and viscous effects and only take into account pressure as an important variable. It is the pressure that will produce the necessary lift on the wing for the flight of an aircraft. The modeling of these problems thus uses the Euler equations for a perfect fluid. If viscous and turbulent effects are taken into account, the Navier-Stokes equations are needed to compute the ***drag***, that is, the force acting in the opposite direction of a rigid body in steady translation in a fluid at rest at infinity.

Compressible flows are characterized by the global ***Mach number***

$$M = \frac{U}{a}\,, \tag{8.3}$$

where $U$ is a reference velocity such as that of the flow upstream of a body, and $a$ is the speed of sound defined by (6.152). The Mach number typically is between 0 and 8. The case $M = 0$ is that of an incompressible fluid corresponding to infinite speed of sound as $\rho = \text{cnst}$.

### 8.3.1 Subsonic, Supersonic and Hypersonic Flows

Consider a gas in uniform stationary flow with velocity $\boldsymbol{v} = U\boldsymbol{e}_1$. The presence of a fixed object in this flow at a point P generates a perturbation, or sound wave, which propagates in space at the speed of sound (for the gas or air). This perturbation produces pressure and density fluctuations. The velocity at which the perturbation propagates with respect to a fixed coordinate system is thus the sum of the velocity of the gas plus the speed of sound $a$ in all directions, characterized by a unit vector $\boldsymbol{n}$. The resulting velocities, $\boldsymbol{v} + a\boldsymbol{n}$, of the perturbation from P depend on the various directions of $\boldsymbol{n}$ in space. Graphically, they can be obtained by drawing a horizontal vector $\boldsymbol{v}$ from P; at its end, construct the sphere of radius $a$. All vectors from P to a point on the sphere are possible solutions for the speed and direction of the propagation.

For $0 < M < 1$, we have subsonic flow with $U < a$. Referring to figure 8.6(a), we note that the resulting velocity propagates into all space as the sphere encloses the point P. The flow can also be interpreted in the following way. As the flow is moving from left to right at velocity $\|\boldsymbol{v}\| < a$, the emitted wave at the initial time $t = t_0$ is found at time $t = t_1 > t_0$ on the sphere of radius $a(t_1 - t_0)$.

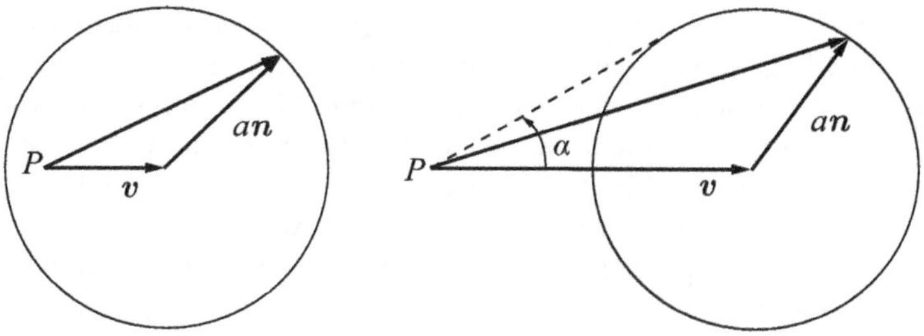


(a) (b)

**Fig. 8.6** Propagation of perturbations in a gas: (a) subsonic case and (b) supersonic case

During this time the flow has moved a distance $(t_1 - t_0)\boldsymbol{v}$. As $\|\boldsymbol{v}\| < a$, the point P remains inside the sphere created by the initial perturbation.

If $M > 1$, the flow is supersonic. Figure 8.6(b) shows that the resulting velocity is contained within a cone that has its vertex at P and is tangent to the sphere centered at the end of $\boldsymbol{v}$. This cone has a half angle of $\alpha$ at the apex such that

$$\sin\alpha = \frac{a}{\|\boldsymbol{v}\|} = \frac{1}{M} , \tag{8.4}$$

and the sine is inversely proportional to the Mach number. Note that the Mach number in (8.4) is based on the velocity $\boldsymbol{v}$; it is thus a local Mach number which varies with the position. In a supersonic flow, all perturbations propagate *downstream* in a cone whose angle decreases as $M$ increases. The angle $\alpha$ defined by (8.4) is the ***Mach angle***.

In summary, subsonic flow around a body affects all space in front of and behind the body. The amplitude of the perturbation diminishes with distance. For supersonic flow, the perturbation is produced when the flow reaches the obstacle and only propagates downstream. It cannot "advance" upstream. This phenomenological difference is explained by the mathematical model associated with propagation of acoustic waves. For the subsonic case the equation is elliptic, while for the supersonic case the equation becomes hyperbolic.

The special case $M = 1$ corresponds to a sonic flow and the Mach angle is 90°. All the spherically propagating perturbations are tangent to a plane perpendicular to $\boldsymbol{v}$. The small (infinitesimal) perturbations accumulate to create a finite amplitude perturbation: the ***sound barrier***.

Flow for $M > 5$ is called hypersonic. In this case, the air molecules dis-associate and the gas becomes ionized. Then one must take into account the chemical reactions between the ionized gas components, and the thermodynamic effects become dominant. These flows are encountered around missiles or reentry vehicles, as, for example, space shuttles.

### 8.3.2 Shock Waves

When a flying object is not small (e.g., not a slender body), the generated perturbations are no longer infinitesimal, and the separation between the zone of silence and the zone where the perturbations propagate becomes a curved surface across which the pressure, density, and velocity are subjected to sudden changes of finite amplitude. This abrupt change of the physical quantities is called a ***shock*** and the associated surface is the ***shock wave***. Note that the shock wave is a compression wave. In reality, a shock wave has a certain thickness of the order of a few millimeters. However, when the body is a complete airplane, we approximate the shock wave as a discontinuous surface which simplifies the mathematical treatment.

The following photographs were obtained by shadowgraphy, which accentuates the index of refraction variations due to the variations of density. Light is viewed after passing through the flow, on the opposite side; the source itself is blocked in a focal plane, leaving only the deviated light to form an image. The highlights thus correspond to strong refraction in the gas and sketch the geometric configuration of the flow. These photographs were published by von Kármán [63].

Going from $M = 0.84$ to $M = 0.971$, we see in figures 8.7 and 8.8 the configuration of the shock waves that are produced by a projectile in free flight through air with an incidence angle less than $1.5°$. We can also recognize the presence of a turbulent wake behind the body. Note that the nose of the projectile has a half angle equal to $20°$.

Nearing the speed of sound, the configuration of the shock waves extends laterally over greater distances. Figures 8.9 and 8.10 show all the complexity of the waves and their interactions.

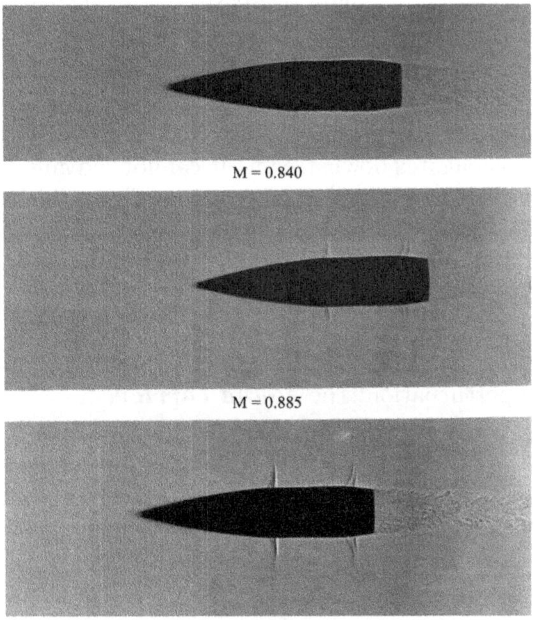


**Fig. 8.7** Subsonic flow around a projectile at $M = 0.84, 0.885$, and $0.9$

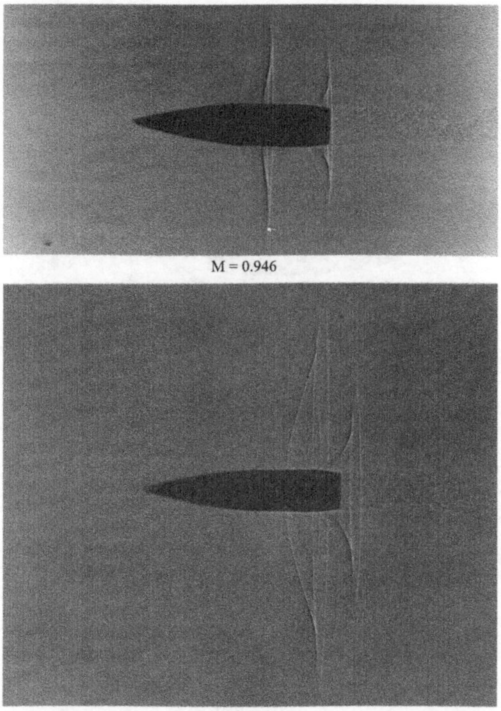


**Fig. 8.8** Subsonic flow around a projectile at $M = 0.946$ and 0.971

M = 0.978

**Fig. 8.9** Flow near the speed of sound, $M = 0.978$

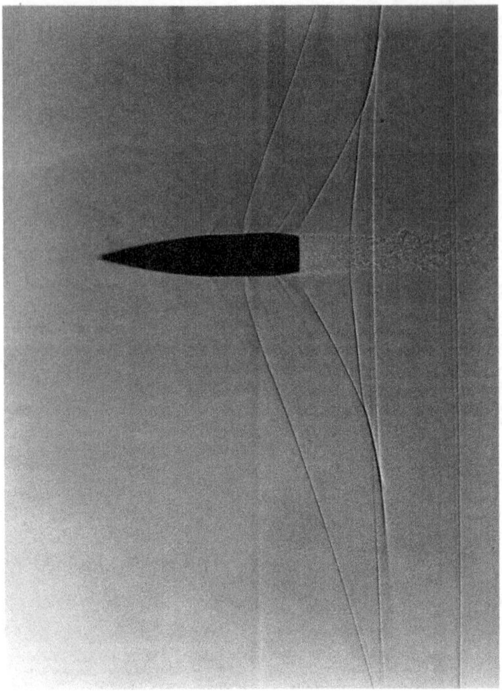

M = 0.990

**Fig. 8.10** Flow near the speed of sound, $M = 0.99$

Finally, in figure 8.11 we have a supersonic flow around a sphere of diameter 1.27 cm in free flight in air. The Mach number is $M = 1.53$. The shock wave produced is curved and detached from the body taking a position in front of it. Behind the shock, the flow returns to subsonic and covers the spherical surface to around 45°, where 0° is the horizontal, upstream pole of the sphere. At an angle around 90°, the laminar boundary layer separates with an oblique shock and becomes turbulent. The wake downstream of the sphere produces a system of weak perturbations which result in a second shock wave.

## 8.4 Navier-Stokes Equations

In this section we formulate the Navier-Stokes equations for compressible Newtonian fluids, and then for an incompressible Newtonian fluid.

### 8.4.1 Navier-Stokes Equations for an Ideal Gas with Constant Heat Capacity

Let us write the Navier-Stokes equations for the special case of a compressible ideal gas with constant heat capacities.

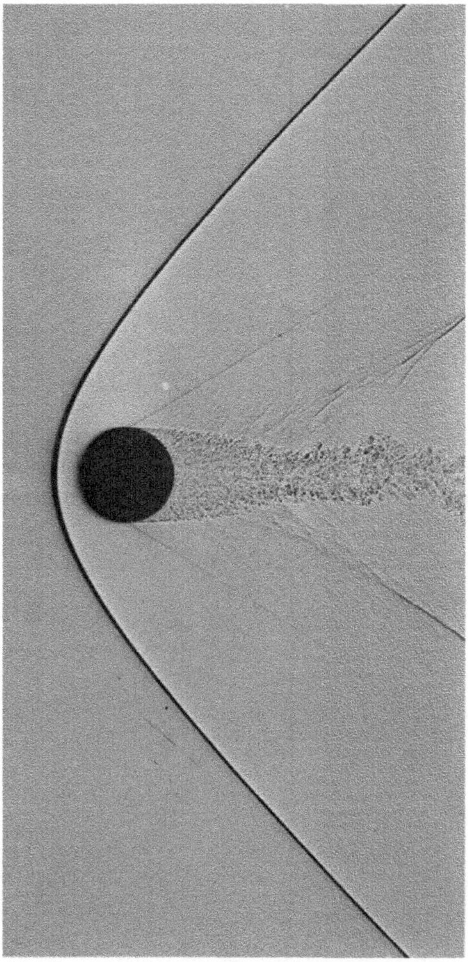

**Fig. 8.11** Supersonic flow around a sphere at $M = 1.53$

With relation (6.143), the energy equation (4.23) can be put in the form

$$\rho c_v \frac{DT}{Dt} = \boldsymbol{\sigma} : \boldsymbol{\nabla} \boldsymbol{v} - \operatorname{div} \boldsymbol{q} + r \,. \tag{8.5}$$

The expression $\boldsymbol{\sigma} : \boldsymbol{\nabla} \boldsymbol{v}$ can be written as

$$\begin{aligned} \sigma_{ij} \frac{\partial v_i}{\partial x_j} &= \sigma_{ij}\, d_{ij} = -p\, \delta_{ij}\, d_{ij} + \lambda\, d_{kk}\, d_{ij}\, \delta_{ij} + 2\mu (d_{ij})^2 \\ &= -p\, d_{ii} + \lambda (d_{ii})^2 + 2\mu (d_{ij})^2 \,, \end{aligned}$$

or by using (4.25)

$$\boldsymbol{\sigma} : \boldsymbol{L} = \boldsymbol{\sigma} : \boldsymbol{d} = -p \operatorname{tr} \boldsymbol{d} + \lambda (\operatorname{tr} \boldsymbol{d})^2 + 2\mu (\boldsymbol{d} : \boldsymbol{d}) \,. \tag{8.6}$$

From the equation of conservation of mass (3.41), we have the equality

$$\operatorname{tr}\boldsymbol{d} = -\frac{1}{\rho}\frac{D\rho}{Dt}\,.$$

Then, the energy equation (8.5) becomes

$$\rho c_v \frac{DT}{Dt} = \frac{p}{\rho}\frac{D\rho}{Dt} + \lambda\,(\operatorname{tr}\boldsymbol{d})^2 + 2\mu\,\boldsymbol{d} : \boldsymbol{d} - \operatorname{div}\boldsymbol{q} + r\,. \tag{8.7}$$

Using the equation of state (6.136), we transform (8.7), which becomes

$$\rho c_v \frac{DT}{Dt} = \frac{Dp}{Dt} - \rho R\frac{DT}{Dt} + \lambda(\operatorname{tr}\boldsymbol{d})^2 + 2\mu\,\boldsymbol{d} : \boldsymbol{d} - \operatorname{div}\boldsymbol{q} + r\,. \tag{8.8}$$

In order to obtain the full set of equations, we use the conservation of mass (3.41), insert the constitutive equation (6.14) in the conservation of momentum (3.96), and finally, we modify the energy equation (8.8) with Fourier's law (6.123) and relation (6.141). The Navier-Stokes system of equations is then written in index form as

$$\frac{\partial\rho}{\partial t} + \frac{\partial\rho v_i}{\partial x_i} = 0 \tag{8.9}$$

$$\rho\frac{Dv_i}{Dt} = -\frac{\partial p}{\partial x_i} + \frac{\partial}{\partial x_i}\,(\lambda d_{kk}) + \frac{\partial}{\partial x_j}\,(2\mu\,d_{ij}) + \rho b_i \tag{8.10}$$

$$\rho c_p\left(\frac{DT}{Dt} - \frac{\gamma-1}{\gamma}\frac{T}{p}\frac{Dp}{Dt}\right) = \frac{\partial}{\partial x_j}\left(k\frac{\partial T}{\partial x_j}\right) + \lambda\left(\frac{\partial v_i}{\partial x_i}\right)^2 + 2\mu\,d_{ij}\,d_{ij} + r \tag{8.11}$$

$$p = \rho RT\,. \tag{8.12}$$

Note that $\rho c_p \frac{\gamma-1}{\gamma}\frac{T}{p} = 1$.

Equations (8.10) and (8.11) are simplified when $\lambda$, $\mu$, and $k$ are constants. In addition, we use ***Stokes' hypothesis***

$$3\lambda + 2\mu = 0\,. \tag{8.13}$$

This relation has been established based on reasoning from the kinetic theory of gases. Although this hypothesis is valid for monatomic gases, it is not valid for polyatomic gases. Nonetheless it is widely used in aerodynamics applications.

Equations (8.10) and (8.11) then become

$$\rho\frac{Dv_i}{Dt} = -\frac{\partial p}{\partial x_i} + \mu\frac{\partial^2 v_i}{\partial x_j \partial x_j} + \frac{\mu}{3}\frac{\partial}{\partial x_i}(d_{kk}) + \rho b_i \tag{8.14}$$

$$\rho c_p\left(\frac{DT}{Dt} - \frac{\gamma-1}{\gamma}\frac{T}{p}\frac{Dp}{Dt}\right) = k\frac{\partial^2 T}{\partial x_j \partial x_j} - \frac{2}{3}\,\mu(d_{kk})^2 + 2\mu\,d_{ij}\,d_{ij} + r\,. \tag{8.15}$$

### 8.4.2 Navier-Stokes Equations for an Incompressible Fluid in Isothermal Flow

For an isothermal flow, $T = \text{cnst}$; the principle of conservation of energy is trivially satisfied. Taking into account the incompressibility ($\rho = \text{cnst}$), the preceding equations are simplified as

$$\operatorname{div} \boldsymbol{v} = 0 \tag{8.16}$$

$$\rho \frac{D\boldsymbol{v}}{Dt} = -\boldsymbol{\nabla} p + \mu \Delta \boldsymbol{v} + \rho \boldsymbol{b} \,. \tag{8.17}$$

Equation (8.17) is a non-linear second-order partial differential equation. It states that acceleration is produced by the actions of the pressure gradient, the viscous forces, and the body forces.

## 8.5 Non-Dimensional Form of the Navier-Stokes Equations

### 8.5.1 Compressible Fluid Case

Denote the reference values of length, speed, pressure, density, and temperature that characterize the flow under consideration by $L$, $U$, $p_0$, $\rho_0$, and $T_0$. The time scale is $L/U$ and the scale for inertial forces is $U^2/L$. Now we introduce non-dimensional variables and functions (denoted with primes) with relations

$$x_i = L x_i' \qquad t = \frac{L}{U} t' \qquad v_i = U v_i' \qquad p = p_0 p'$$

$$\rho = \rho_0 \rho' \qquad T = T_0 T' \qquad b_i = U^2 \frac{b_i'}{L} \,.$$

Furthermore, for the sake of simplicity, we make the hypothesis that $r = 0$.

We reformulate equations (8.9), (8.12), (8.14), and (8.15) with non-dimensional values, including constant characteristic values $\mu_0$ and $k_0$ estimated at the temperature $T_0$, as well as $c_p$, $\gamma$, and $R$:

$$\frac{\partial \rho'}{\partial t'} + v_j' \frac{\partial \rho'}{\partial x_j'} + \rho' \frac{\partial v_j'}{\partial x_j'} = 0 \tag{8.18}$$

$$\frac{\partial v_i'}{\partial t'} + v_k' \frac{\partial v_i'}{\partial x_k'}$$

$$= -\frac{p_0}{\rho_0 U^2} \frac{1}{\rho'} \frac{\partial p'}{\partial x_i'} + \frac{\mu_0}{U L \rho_0} \frac{1}{\rho'} \left( \frac{\partial^2 v_i'}{\partial {x_j'}^2} + \frac{1}{3} \frac{\partial}{\partial x_i'} (d_{kk}') \right) + b_i' \tag{8.19}$$

$$\rho' \left( \frac{DT'}{Dt'} - \frac{\gamma - 1}{\gamma} \frac{T'}{p'} \frac{Dp'}{Dt'} \right)$$

$$= \frac{k_0}{\mu_0 c_p} \frac{\mu_0}{\rho_0 U L} \frac{\partial^2 T'}{\partial {x_j'}^2} - \frac{\mu_0}{\rho_0 U L} \frac{U^2}{c_p T_0} \left( \frac{2}{3} (d_{kk}')^2 - \frac{1}{2} \left( \frac{\partial v_i'}{\partial x_j'} + \frac{\partial v_j'}{\partial x_i'} \right)^2 \right) \tag{8.20}$$

$$p' = \rho' T' \,, \tag{8.21}$$

if we set $p_0 = \rho_0 \, R \, T_0$.

In relations (8.18)–(8.20) three non-dimensional numbers appear:

- the ***Reynolds number***

$$Re = \rho_0 \frac{UL}{\mu_0} = \frac{UL}{\nu_0} \, ;$$

- the ***Prandtl number***

$$Pr = \frac{c_p \mu_0}{k_0} = \frac{\nu_0}{\Lambda} \, ;$$

- the ***Mach number***

$$M = \frac{U}{a_0} \, ,$$

which appear also in the group

$$\frac{p_0}{\rho_0 U^2} = \frac{RT_0}{U^2} = \frac{a_0^2}{\gamma U^2} = \frac{1}{\gamma M^2} \, .$$

The denominator of the Mach number $a_0$ is the characteristic speed of sound (eqn. (6.153)). The coefficient $\Lambda$ defined by relation

$$\Lambda = \frac{k_0}{\rho_0 c_p}$$

appearing in the Prandtl number is called the ***thermal diffusivity***.

The Reynolds number expresses the relative importance of the inertial forces with respect to the viscous forces. It takes values from zero up to several million. For $Re = 0$, the Navier-Stokes equations reduce to the Stokes equation. They govern the dynamics of slow or creeping laminar flows. For $Re \sim 10^6$, the flow is turbulent. The Prandtl number estimates the relative importance of the viscous and thermal diffusion phenomena ($Pr = 0.71$ for room temperature air). The Mach number characterizes the compressibility effects. Its value is $M = 0$ for incompressible fluids. It is between $0 < M < 1$ for subsonic flows and $M > 1$ for supersonic flows.

The Navier-Stokes equations take the non-dimensional form

$$\frac{D\rho'}{Dt'} + \rho' \operatorname{div} \boldsymbol{v}' = 0 \tag{8.22}$$

$$\rho' \frac{D\boldsymbol{v}'}{Dt'} = -\frac{1}{\gamma M^2} \boldsymbol{\nabla} p' + \frac{1}{Re} \left( \boldsymbol{\nabla}^2 \boldsymbol{v}' + \frac{1}{3} \boldsymbol{\nabla} (\operatorname{div} \boldsymbol{v}') \right) + \rho' \boldsymbol{b}' \tag{8.23}$$

$$\rho' \left( \frac{DT'}{Dt'} - \frac{\gamma - 1}{\gamma} \frac{T'}{\rho'} \frac{Dp'}{Dt'} \right)$$

$$= \frac{1}{Pr\,Re} \boldsymbol{\nabla}^2 T' - (\gamma - 1) \frac{M^2}{Re} \left( \frac{2}{3} (\operatorname{div} \boldsymbol{v}')^2 - \frac{1}{2} \left( \frac{\partial v_i'}{\partial x_j} + \frac{\partial v_j'}{\partial x_i} \right)^2 \right) \tag{8.24}$$

$$p' = \rho' T' \, . \tag{8.25}$$

If we fix the coordinates $x_i$, time $t$, and all the parameters $M$, $Pr$, $\gamma$, and take $Re \to \infty$, then the system (8.22)–(8.25) leads to the Euler system of equations for perfect (inviscid) fluids. Taking the limit where the Mach number goes to zero, with all the other parameters fixed, should lead to the Navier-Stokes equations for an incompressible fluid.

However, examination of the system (8.22)–(8.25) shows that this is not so, and that the term $-(1/\gamma M^2)\boldsymbol{\nabla} p$ becomes dominant. This behavior is due to the choice of the non-dimensional pressure $p' = p/p_0$, which was made by considering pressure to be a thermodynamic variable. The motion equation reveals that pressure is also a dynamic variable. It is more natural to choose

$$p^* = \frac{p - p_0}{\rho_0 U^2}$$

for the non-dimensional pressure.

In this case, equation (8.23) becomes

$$\rho' \frac{D\boldsymbol{v}'}{Dt'} = -\boldsymbol{\nabla} p^* + \frac{1}{Re}\left(\boldsymbol{\nabla}^2 \boldsymbol{v}' + \frac{1}{3}\boldsymbol{\nabla}(\operatorname{div} \boldsymbol{v}')\right) + \rho' \boldsymbol{b}' . \tag{8.26}$$

The limiting case of equations (8.18), (8.26), (8.20), and (8.21) when the Mach number goes to zero, yields the relations

$$\frac{D\rho'}{Dt'} + \rho' \operatorname{div} \boldsymbol{v}' = 0 \tag{8.27}$$

$$\rho' \frac{D\boldsymbol{v}'}{Dt'} = -\boldsymbol{\nabla} p^* + \frac{1}{Re}\left(\boldsymbol{\nabla}^2 \boldsymbol{v}' + \frac{1}{3}\boldsymbol{\nabla}(\operatorname{div} \boldsymbol{v}')\right) + \rho' \boldsymbol{b}' \tag{8.28}$$

$$\rho' \frac{DT'}{Dt'} = \frac{1}{Pr\, Re} \boldsymbol{\nabla}^2 T' \tag{8.29}$$

$$\rho' T' = 1 , \tag{8.30}$$

which are those for an ***incompressible fluid***, but which still may experience thermal expansion.

To obtain (8.29), we calculate

$$\frac{Dp'}{Dt'} = \frac{1}{p_0}\frac{Dp}{Dt'} = \frac{\rho_0 U^2}{p_0}\frac{Dp^*}{Dt'} = \frac{U^2}{RT_0}\frac{Dp^*}{Dt'} = \gamma M^2 \frac{Dp^*}{Dt'} .$$

Equation (8.30) comes from the following evaluation:

$$p' = \rho' T' = \frac{p}{p_0} = 1 + p^* \frac{U^2}{RT_0} = 1 + \gamma M^2 p^* .$$

If, in addition, we assume that at the domain wall $T' = 1$, then equations (8.29) and (8.30) as well as the boundary conditions on $T'$ are satisfied by

$$\rho' = 1 \tag{8.31}$$

$$T' = 1 . \tag{8.32}$$

Consequently, in this case, equations (8.27) and (8.28) reduce to the equations of an isothermal, incompressible flow.

### 8.5.2 Case of an Incompressible Fluid in Isothermal Flow

In a first case, we examine the aerodynamic point of view. We state the hypothesis that the body force is that due to gravity: $\boldsymbol{b} = \boldsymbol{g}$. We use the same scales of time, length, and speed as in the section above. For the pressure, we set

$$p' = \frac{p - p_0}{\rho U^2}$$

and for the gravity force, we introduce

$$\boldsymbol{g}' = \frac{\boldsymbol{g}}{g},$$

where $g = \|\boldsymbol{g}\|$ is the gravitational acceleration. Equation (8.17), in reduced form, becomes

$$\frac{D\boldsymbol{v}'}{Dt'} = -\boldsymbol{\nabla} p' + \frac{1}{Re}\boldsymbol{\nabla}^2 \boldsymbol{v}' + \frac{1}{\mathrm{Fr}}\boldsymbol{g}'. \tag{8.33}$$

Another non-dimensional quantity appears, the *Froude number*

$$\mathrm{Fr} = \frac{U^2}{Lg}.$$

This number compares the inertial forces to the gravitational forces.

In a second case, we develop the point of view of rheologists for whom the flow phenomena are dominated by viscous effects. Going back to equation (8.17), but normalizing the reduced forms of time and pressure by the viscosity

$$t' = \frac{\nu t}{L^2} \qquad \text{and} \qquad p' = \frac{p - p_0}{\frac{\mu U}{L}},$$

the ***reduced form of the Navier-Stokes equations*** for an incompressible fluid is written as

$$\frac{\partial v_i'}{\partial t'} + Re\left(v_k' \frac{\partial v_i'}{\partial x_k'}\right) = -\frac{\partial p'}{\partial x_i'} + \boldsymbol{\nabla}^2 v_i' + \frac{Re}{Fr}\boldsymbol{g}'. \tag{8.34}$$

Equations (8.33) and (8.34) are different because the normalization of time is, on one hand, by the advection time (inertial term) $L/U$, and on the other, by the characteristic time of molecular diffusion $L^2/\nu$.

Consequently, the limiting form of the Navier-Stokes equations is obtained from (8.33) as $Re \to \infty$

$$\frac{D\boldsymbol{v}'}{Dt'} = -\boldsymbol{\nabla} p' + \frac{1}{Fr}\boldsymbol{g}'. \tag{8.35}$$

These are the Euler equations in non-dimensional form. The dimensional form is

$$\rho\frac{D\boldsymbol{v}}{Dt} = -\boldsymbol{\nabla} p + \rho\boldsymbol{g}. \tag{8.36}$$

Inversely, when $Re \to 0$, equation (8.34) simplifies to

$$\frac{\partial \boldsymbol{v}'}{\partial t'} = -\boldsymbol{\nabla} p' + \boldsymbol{\nabla}^2 \boldsymbol{v}' . \tag{8.37}$$

This is the non-dimensional ***Stokes equation***. This equation is linear, unlike the Navier-Stokes equations which are non-linear. In dimensional form, it is written as

$$\rho \frac{\partial \boldsymbol{v}}{\partial t} = -\boldsymbol{\nabla} p + \mu \boldsymbol{\nabla}^2 \boldsymbol{v} . \tag{8.38}$$

Recall that the kinematic viscosity of water is $\nu_{water} = 10^{-6} m^2 s^{-1}$. Thus if $U$ and $L$ are both of order 1, the Reynolds number will be $O(10^6)$. This value is typical for turbulence, a subject which itself merits an entire book. The reader is referred to the specialized literature [9, 39, 53].

## 8.6 Boundary and Initial Conditions

### 8.6.1 Viscous Fluid

A viscous fluid in contact with a rigid wall will adhere to the wall due to the effects of viscosity. The no-slip condition can therefore be written as

$$\boldsymbol{v}_{fluid} = \boldsymbol{v}_{wall} . \tag{8.39}$$

Otherwise, if the fluid (liquid) is in contact with a gaseous medium, we assume that the contact forces are in equilibrium on both sides of the contact surface and we write

$$n_i \sigma_{ij}^{fluid} = n_i \sigma_{ij}^{gas} .$$

If the gas is inviscid, this becomes

$$n_i \sigma_{ij}^{fluid} = -n_j p^{gas} . \tag{8.40}$$

Projecting (8.40) onto the interface normal, we obtain

$$n_i n_j \sigma_{ij}^{fluid} = -p^{gas} . \tag{8.41}$$

For the tangential component, we have

$$n_i \tau_j \sigma_{ij}^{fluid} = -n_j \tau_j p^{gas} = 0 , \tag{8.42}$$

since $\boldsymbol{n} \cdot \boldsymbol{\tau} = 0$, where $\boldsymbol{\tau}$ denotes the unit tangent vector at the contact surface.

Relations (8.41) and (8.42) express the conditions which are said to be on a "free surface". They imply that we know the form of the surface for their application. However, the form of the surface is itself part of the solution of the problem at hand. We find then that free surface problems constitute one of the biggest difficulties in fluid mechanics as they are intrinsically non-linear.

For certain fluids, condition (8.41) needs to be extended to take into account surface tension. Then we have

$$n_i n_j \sigma_{ij}^{fluid} = -p^{gas} + \sigma \left( \frac{1}{R_1} + \frac{1}{R_2} \right) , \qquad (8.43)$$

in which $R_1$ and $R_2$ are the principal radii of curvature of the surface, and $\sigma$ is the liquid-gas surface tension coefficient expressed in $N/m$.

In practice, we generally limit the study to a part of the zone occupied by the fluid. In this case, it is necessary to add the conditions on the entry section, where the velocity vector is typically imposed, and the exit surface, where contact forces are usually imposed. The latter are most often taken to be zero, which corresponds to a situation where the fluid is allowed to exit at its own speed.

For the case of a transient problem, the initial conditions are the velocities, which are often zero at the start.

### 8.6.2 Perfect Fluid

As viscosity plays no role here, the fluid can slip along a wall. The adherence condition no longer applies. We impose that the normal component of velocity of the fluid be zero with respect to the wall with which it is in contact. The slip condition is then written as

$$\boldsymbol{v}_{fluid} \cdot \boldsymbol{n} = \boldsymbol{v}_{wall} \cdot \boldsymbol{n} \, . \qquad (8.44)$$

Similarly, we impose the value of the normal component of the fluid velocity for the entry section and the pressure on the exit section. For transient flows, we proceed as for viscous flows.

Finally, in aerodynamics (external flows, for example, the flow around a wing profile or an airfoil), we very often find conditions to impose on an immaterial boundary (which may be at infinity). The typical example is a finite obstacle placed in an unconfined flow. In this case we impose the condition that the flow is uniform at infinity.

## 8.7 Exact Solutions of the Navier-Stokes Equations

### 8.7.1 Plane Stationary Flows

In this section, we examine some exact solutions of the Navier-Stokes equations for plane stationary flows.

**Plane Couette Flow of an Incompressible Fluid**

We consider the two-dimensional stationary flow of an incompressible viscous fluid between parallel plates.

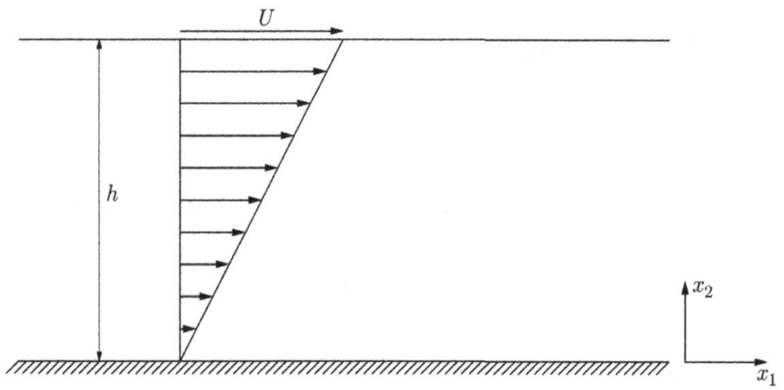


**Fig. 8.12** Plane Couette flow

Figure 8.12 shows the flow domain. We see that the lower boundary is fixed while the upper boundary moves in its own plane at a given constant velocity $U$ in the direction $x_1$.

Since the flow is two-dimensional, the vector $\boldsymbol{v}$ reduces to two components, $\boldsymbol{v} = (v_1, v_2, 0)$. We assume that the flow is developed, that is, that the transient effects and those from the upstream edges of the plates are negligible. Then, we can have $v_1$ as a function only of $x_2$. The incompressibility condition (8.16) becomes

$$\frac{\partial v_2}{\partial x_2} = 0 \tag{8.45}$$

indicating that $v_2$ is not a function of $x_2$; it is thus a function of $x_1$. However, since at the two boundaries $v_2$ is zero for all $x_1$, we conclude that $v_2 = 0$ everywhere. We write the two-dimensional Navier-Stokes equation (8.17) for the velocity component $v_1$ as

$$\rho(\frac{\partial v_1}{\partial t} + v_1 \frac{\partial v_1}{\partial x_1} + v_2 \frac{\partial v_1}{\partial x_2}) = -\frac{\partial p}{\partial x_1} + \mu \Delta v_1 + \rho b_1 . \tag{8.46}$$

As the gravitational force is oriented along the negative direction of the axis $x_2$, $b_1 = 0$. In addition, the problem is stationary, thus $\partial v_1/\partial t = 0$. The term $v_1 \partial v_1/\partial x_1$ is zero as $v_1 = v_1(x_2)$. Finally $v_2 \partial v_1/\partial x_2$ is also zero since $v_2 = 0$. We can assume that the horizontal component of the pressure gradient is zero as the flow is forced kinematically by the motion of the upper plate. Thus, we are left with

$$\mu \frac{d^2 v_1}{d{x_2}^2} = 0 . \tag{8.47}$$

Integrating (8.47) once, we obtain

$$\mu \frac{d v_1}{d x_2} = C . \tag{8.48}$$

This relation shows that the shear stress is constant across the height of the channel. Integrating again leads to

$$v_1 = Ax_2 + B. \tag{8.49}$$

The no-slip boundary conditions

$$v_1(x_2 = 0) = 0, \; v_1(x_2 = h) = U \tag{8.50}$$

permit us to determine the integration constants; we obtain a linear velocity profile

$$v_1 = \frac{Ux_2}{h}. \tag{8.51}$$

This is the profile that was used in example 6.1.

The shear stress component (6.16) obtained with (8.51) is a constant

$$\sigma_{12} = \mu \frac{dv_1}{dx_2} = \mu \frac{U}{h} \, . \tag{8.52}$$

If we examine the second Navier-Stokes equation, in direction $x_2$, we have

$$0 = -\frac{\partial p}{\partial x_2} - \rho g, \tag{8.53}$$

with $g$ the gravitational acceleration. Integrating this relation and taking into account the independence of $p$ with respect to $x_1$, leads to

$$p = -\rho g x_2 + C. \tag{8.54}$$

As the pressure in an incompressible fluid is only known to an arbitrary constant, we choose it by imposing $p(x_2 = h) = 0$ which yields $C = \rho g h$. The pressure is in hydrostatic equilibrium

$$p = \rho g(h - x_2) \, . \tag{8.55}$$

**Plane Poiseuille Flow of an Incompressible Fluid**

Consider the two-dimensional stationary flow of a viscous incompressible fluid in a channel formed by two fixed walls. Figure 8.13 shows the geometric configuration of the domain. In this case, a longitudinal pressure gradient, along direction $x_1$, is established. We assume that the flow is developed and that the fluid particles move on paths parallel to the walls. Reasoning as for Couette flow, we can write $v_1 = v_1(x_2)$, $v_2 = 0$.

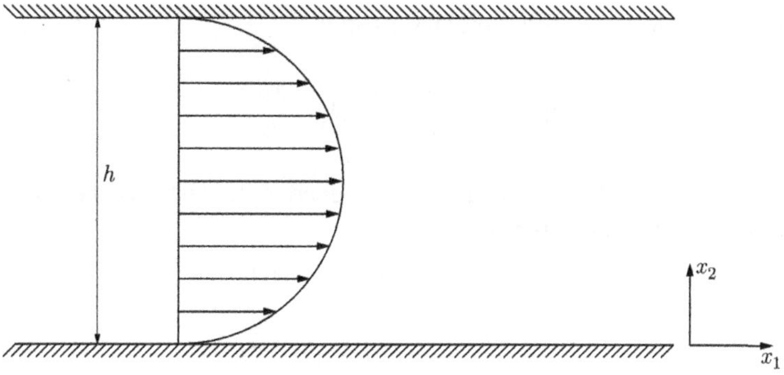


**Fig. 8.13** Plane Poiseuille flow

The dynamic equation for velocity $v_1$ is relation (8.17), which for Poiseuille flow reduces to

$$0 = -\frac{\partial p}{\partial x_1} + \mu \frac{\partial^2 v_1}{\partial {x_2}^2} \,. \tag{8.56}$$

As for Couette flow, the pressure in the vertical direction is in hydrostatic equilibrium

$$0 = -\frac{\partial p}{\partial x_2} - \rho g \,. \tag{8.57}$$

Integrating this relation, we obtain

$$p = -\rho g x_2 + P(x_1) \,. \tag{8.58}$$

The term introduced by integration, $P(x_1)$, is the pressure on the lower wall, for $x_2 = 0$. The pressure gradient in direction $x_1$ can be written as

$$\frac{\partial p}{\partial x_1} = \frac{dP}{dx_1} \,, \tag{8.59}$$

as it is a function only of $x_1$. Equation (8.56) yields

$$\frac{d^2 v_1}{d{x_2}^2} = \frac{1}{\mu}\frac{dP}{dx_1} = C. \tag{8.60}$$

We see that the first term is a function of $x_2$ while the second is a function of $x_1$. It follows that these two terms must be equal to the same constant $C$. The integration of (8.60) gives us

$$v_1 = \frac{1}{\mu}\frac{dP}{dx_1}\frac{{x_2}^2}{2} + Ax_2 + B. \tag{8.61}$$

Imposing the boundary conditions

$$v_1(x_2 = 0) = v_1(x_2 = h) = 0, \tag{8.62}$$

yields the parabolic Poiseuille velocity profile

$$v_1 = -\frac{h^2}{2\mu}\frac{dP}{dx_1}\frac{x_2}{h}(1-\frac{x_2}{h}) . \qquad (8.63)$$

As the pressure in the channel diminishes linearly with distance $x_1$, $dP/dx_1 < 0$, and the flow is in the positive direction of the axis $x_1$.

The shear stress component obtained from (8.63) is

$$\sigma_{12} = \mu\frac{dv_1}{dx_2} = -\frac{h}{2}\frac{dP}{dx_1}(1-\frac{2x_2}{h}) . \qquad (8.64)$$

We note that the shear (8.64) is zero on the axis of symmetry of the channel, $x_2 = h/2$, and the absolute value is at a maximum on the two walls.

We can calculate the volume flux or flow rate through the section $S$ of the channel. The general definition of the volume flux is given by the relation

$$Q = \int_S \boldsymbol{v}\cdot\boldsymbol{n}\, dS . \qquad (8.65)$$

Considering a unit surface for direction $x_3$, the flow rate in the two-dimensional channel is written

$$Q = \int_0^h v_1\, dx_2 = -\frac{h^3}{12\mu}\frac{dP}{dx_1} = \frac{h^3}{12\mu}\frac{\Delta p}{L} , \qquad (8.66)$$

with $\Delta P$ the pressure difference observed at two points with the same ordinate, $x_2$, separated by a distance $L$ in direction $x_1$. We define the average velocity by $Q = v_{avg}\, h$, from which we have

$$v_{avg} = \frac{h^2}{12\mu}\frac{\Delta P}{L} . \qquad (8.67)$$

As the maximum velocity, $v_{max}$, is attained on the axis of symmetry of the channel, at $x_2/h = 1/2$, it follows that

$$v_{max} = \frac{h^2}{8\mu}\frac{\Delta P}{L} \qquad (8.68)$$

and, consequently

$$v_{avg} = \frac{2}{3}v_{max} . \qquad (8.69)$$

In the case where the two-dimensional channel is replaced by a pipe with circular section (see sec. 8.7.2), we obtain the average velocity equal to half the maximum velocity. This shows that the zone of high velocity constitutes a smaller fraction of the section.

**Flow of an Incompressible Fluid on an Inclined Plane**

We have a stationary, two-dimensional flow of a Newtonian viscous fluid on a plane inclined at angle $\alpha$ to the vertical (fig. 8.14). The thickness of the fluid layer is uniform and equal to $h$. At the free surface, the fluid is in contact with ambient air which we consider to be a perfect fluid at atmospheric pressure $p_a$. We assume that the flow in the air does not affect the flow of the viscous fluid. The flow is parallel because the trajectories of the fluid particles are parallel to the inclined plane. Then, $\boldsymbol{v} = (v_1, 0, 0)$. From incompressibility we obtain

$$\frac{\partial v_1}{\partial x_1} = 0, \tag{8.70}$$

therefore we deduce that $v_1 = v_1(x_2)$. The only components of the stress tensor are $\sigma_{12}$ or $\sigma_{21}$. As the pressure is uniform at the free surface, the pressure in the viscous fluid can not depend on direction $x_1$, but only on $x_2$.

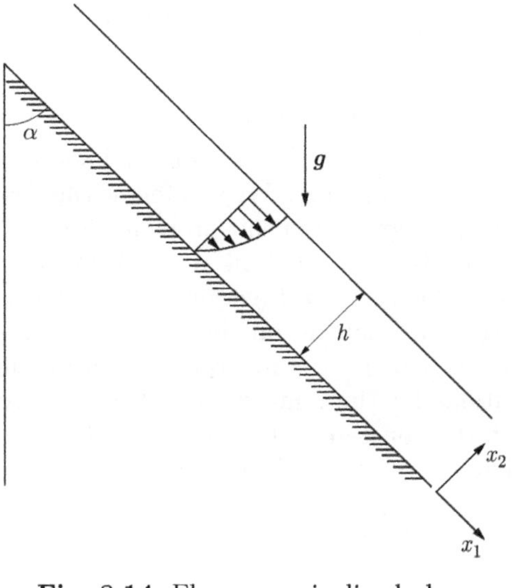


**Fig. 8.14** Flow on an inclined plane

From the motion equation (3.96), written in direction $x_1$, it follows that

$$\frac{\partial \sigma_{12}}{\partial x_2} + \rho b_1 = \frac{\partial \sigma_{12}}{\partial x_2} + \rho g cos\alpha = 0 \ . \tag{8.71}$$

Integrating this relation, we have

$$\sigma_{12} = -\rho g \, x_2 cos\alpha + C \ . \tag{8.72}$$

At the free surface, $x_2 = h$, the shear stress should be zero. We obtain

$$\sigma_{12} = \rho g cos\alpha (h - x_2) \ . \tag{8.73}$$

As $\sigma_{12} = \mu\, dv_1/dx_2$, we can evaluate the component $v_1$ by integrating with respect to $x_2$, taking into account the boundary condition $v_1(x_2 = 0) = 0$. The velocity profile is given by the relation

$$v_1 = \frac{\rho g cos\alpha}{2\mu} x_2(2h - x_2) \ . \tag{8.74}$$

The Navier-Stokes equation for direction $x_2$ yields the relation

$$-\frac{\partial p}{\partial x_2} + \rho b_2 = -\frac{\partial p}{\partial x_2} - \rho g sin\alpha = 0 \ . \tag{8.75}$$

Integrating with respect to $x_2$ and taking into account the condition on the free surface $p(x_2 = h) = p_a$, we can write

$$p = p_a - (\rho\, g sin\alpha)(x_2 - h) \ . \tag{8.76}$$

The flow rate per unit length in direction $x_3$ is obtained from

$$Q = \int_0^h v_1\, dx_2 = \frac{\rho g cos\alpha\, h^3}{2\mu}. \tag{8.77}$$

**Plane Couette Flow of a Compressible Fluid**

Consider the stationary, developed, two-dimensional flow of a viscous compressible fluid between two parallel planes. We will follow the development proposed by Panton [38]. We neglect gravity in this problem. Figure 8.15 shows the flow domain. The lower wall is fixed, and the upper wall moves in its own plane at the given constant velocity $U$ in direction $x_1$. Since the compressible fluid has material properties that depend on the density and the temperature (see eqn. (6.15)), the energy equation is incorporated in our model and we must give boundary conditions for the temperature. Here, the temperature is fixed on the upper wall at the constant value $T_0$, while the condition on the lower wall is adiabatic, that is, $q_2 = -k\, \partial T/\partial x_2 = 0$. To simplify the problem, we assume that no material property depends on $x_1$, $x_3$, or $t$. The unknowns of the problem are then such that

$$\boldsymbol{v} = \big(v_1(x_2),\, v_2(x_2),\, 0\big) \qquad \rho = \rho(x_2) \qquad T = T(x_2)\,. \tag{8.78}$$

The equation for the conservation of mass (8.9) becomes

$$\frac{d(\rho v_2)}{dx_2} = 0\,. \tag{8.79}$$

By integration, $\rho v_2$ is a constant. Since the vertical velocity component $v_2$ is zero at the two walls, we must have $v_2 = 0$ everywhere. The Navier-Stokes equations (8.10) become

$$0 = \frac{d}{dx_2}\left(\mu \frac{dv_1}{dx_2}\right), \tag{8.80}$$

$$0 = -\frac{dp}{dx_2}\,. \tag{8.81}$$

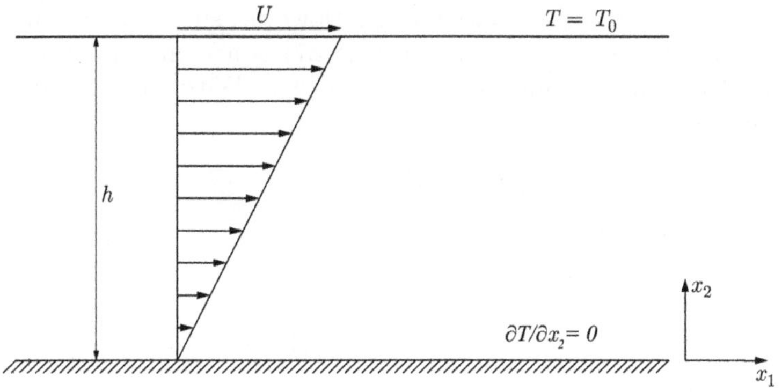


**Fig. 8.15** Plane Couette flow for a compressible fluid

Relation (8.81) shows that the pressure is constant everywhere. Consequently, we can assume that the material properties are functions only of temperature, since $p = \rho RT = \text{cnst}$. The integration of (8.80) yields

$$\mu \frac{dv_1}{dx_2} = \sigma_0 \ . \tag{8.82}$$

with the convention that $\sigma_0$ represents a constant shear stress.

The heat flux has only one non-zero component: $q_2$, given our assumption for the temperature field. In this case, the energy equation (8.11) is

$$0 = \frac{d}{dx_2}\left(k\frac{dT}{dx_2}\right) + \mu\left(\frac{dv_1}{dx_2}\right)^2 = -\frac{d\,q_2}{dx_2} + \sigma_0\frac{dv_1}{dx_2} \, . \tag{8.83}$$

Integration of (8.83) yields

$$-q_2 + \sigma_0 v_1 = C \, . \tag{8.84}$$

Since we have an adiabatic condition on the fixed lower wall where $v_1(x_2 = 0) = 0$, the imposition of the boundary conditions on (8.84) gives $C = 0$, and we obtain

$$k\frac{dT}{dx_2} + \sigma_0 v_1 = k\frac{dT}{dx_2} + \mu v_1 \frac{dv_1}{dx_2} = 0 \, . \tag{8.85}$$

Writing (8.85) in the form

$$\frac{d}{dx_2}\left(\frac{1}{2}v_1^2\right) = -\frac{k}{\mu}\frac{dT}{dx_2} \tag{8.86}$$

and taking into account relations $k = k(T)$ and $\mu = \mu(T)$, the integration from a point with ordinate $x_2$ to the upper wall gives

$$\frac{1}{2}(U^2 - v_1^2) = -\int_T^{T_0} \frac{k(T')}{\mu(T')}\, dT' \, . \tag{8.87}$$

This equation yields $v_1$ as a function of $T$. However, since $k$ and $\mu$ are positive functions of $T$ (sec. 6.8), the integral in (8.87) is a monotonic function of $T$ and we can thus find the inverse function $T(v_1)$. With this inverse function, equation (8.82) becomes

$$\mu\big(T(v_1)\big)\, dv_1 = \sigma_0\, dx_2 \ , \tag{8.88}$$

and integrating from the lower wall, where $v_1 = 0$, to an arbitrarily chosen point with ordinate $x_2$, we obtain

$$x_2 = \frac{1}{\sigma_0} \int_0^{v_1} \mu\big(T(v_1')\big)\, dv_1{}' \, . \tag{8.89}$$

This equation is the velocity profile given in inverse form $x_2 = x_2(v_1)$.

We will illustrate this theory with results calculated with power law models of viscosity and thermal conductivity

$$\mu = \mu_0 \left(\frac{T}{T_0}\right)^n \qquad k = k_0 \left(\frac{T}{T_0}\right)^n . \tag{8.90}$$

The exponent $n$ of the power law is obtained by a polynomial approximation of experimental data (by least squares) in the range of temperatures concerned in the problem under consideration. The reference quantities $\mu_0$ and $k_0$ are the values corresponding to the reference temperature $T_0$ of the upper wall. Combining (8.87) and (8.90), leads to

$$\frac{1}{2}\,(U^2 - v_1^2) = -\frac{k_0}{\mu_0} \int_T^{T_0} dT' \tag{8.91}$$

and thus

$$T = T_0 + \frac{\mu_0}{2k_0}(U^2 - v_1^2) \, . \tag{8.92}$$

The inverse form of the velocity profile is evaluated by inserting (8.90) in (8.89):

$$\begin{aligned} x_2 &= \frac{\mu_0}{\sigma_0} \int_0^{v_1} \left(\frac{T}{T_0}\right)^n dv_1' \\ &= \frac{\mu_0}{\sigma_0} \int_0^{v_1} \left(1 + \frac{\mu_0}{2k_0 T_0}\left(U^2 - {v_1'}^2\right)\right)^n dv_1' \, . \end{aligned} \tag{8.93}$$

Consider two cases: $n = 0$ for constant properties and $n = 1$ which is close to the behavior of an ideal gas.

When $n = 0$, relation (8.93) produces

$$x_2 = \frac{\mu_0}{\sigma_0}\, v_1 \, . \tag{8.94}$$

Evaluating $\sigma_0$ at the upper wall, we again find the Couette profile for an incompressible fluid (8.51). The temperature profile (8.92) becomes

$$T = T_0 + \frac{\mu_0 U^2}{2k_0}\left(1 - \left(\frac{x_2}{h}\right)^2\right) . \tag{8.95}$$

The case $n = 1$ leads to the inverse form of the velocity profile

$$\frac{x_2}{h} = \frac{v_1}{U} \frac{1 + \dfrac{\mu_0 U^2}{2k_0 T_0}\left(1 - \dfrac{1}{3}\left(\dfrac{v_1}{U}\right)^2\right)}{1 + \dfrac{\mu_0 U^2}{3k_0 T_0}} . \tag{8.96}$$

Relation (8.96) cannot be explicitly inverted to obtain the solution $v_1$. We thus can not obtain an explicit form of the temperature that always satisfies equation (8.92).

### 8.7.2 Stationary Axisymmetric Flows

In this section we consider exact solutions of the Navier-Stokes equations for stationary flows in axisymmetric geometries of revolution. We integrate the Navier-Stokes equations expressed in a cylindrical coordinate system. The vector velocity has components $v_r, v_\theta$, and $v_z$ which we call the radial, azimuthal, and axial velocities, respectively.

#### Circular Couette Flow

Consider the stationary flow of an incompressible Newtonian viscous fluid between two concentric cylinders supposed to be of infinite axial length. We denote by $R_1$ and $R_2$ the radii of the internal and external cylinders, respectively, and $\omega_1$ and $\omega_2$ their respective rates of angular rotation, as shown in figure 8.16. We want to calculate the azimuthal velocity $v_\theta$. This flow is known by the name of circular Couette flow. We neglect the effects of the volume forces. The flow has no axial velocity since there is no pressure gradient in that direction. In addition, due to the symmetry of revolution, it also does not depend on the azimuthal coordinate, thus $\partial(\bullet)/\partial\theta = 0$. The two velocity components $v_r$ and $v_\theta$, stationary, thus independent of time, are functions uniquely of the radial coordinate, $r$, so $v_r = v_r(r)$ and $v_\theta = v_\theta(r)$. Applying adherence to the wall, the boundary conditions are

$$v_r(R_1) = v_r(R_2) = 0, \quad v_\theta(R_1) = \omega_1 R_1,\ v_\theta(R_2) = \omega_2 R_2 . \tag{8.97}$$

With these assumptions about the velocity profile, the continuity equation (A.31) becomes

$$\frac{1}{r}\frac{d}{dr}(r\, v_r) = 0 . \tag{8.98}$$

Taking into account the condition that $v_r$ is zero at the boundaries (8.97), the solution is

$$v_r = 0 . \tag{8.99}$$

In this case the Navier-Stokes equations (A.32)–(A.33) reduce to

$$\frac{1}{\rho}\frac{\partial p}{\partial r} = \frac{v_\theta^2}{r} , \tag{8.100}$$

$$\frac{1}{r}\frac{\partial}{\partial r}\left(r\frac{\partial v_\theta}{\partial r}\right) - \frac{v_\theta}{r^2} = 0 . \tag{8.101}$$

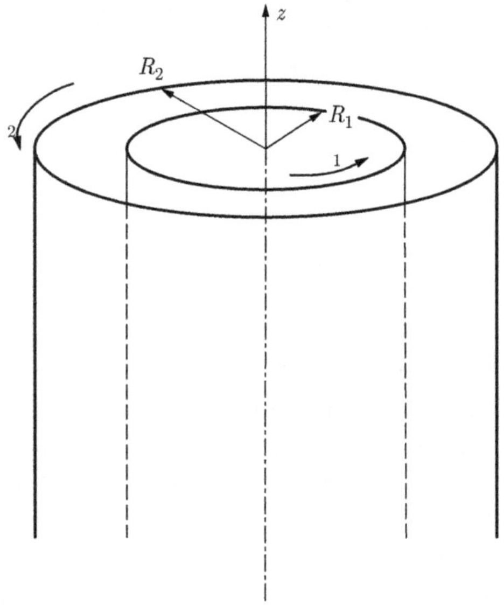


**Fig. 8.16** Circular Couette flow

The solution for the component $v_\theta$ is in the form $v_\theta = \sum_{-\infty}^{+\infty} a_n r^n$. Plugging this series into (8.101), we easily find that $n = \pm 1$. Imposing the boundary conditions leads to

$$v_\theta = Ar + \frac{B}{r} = \frac{\omega_2 R_2^2 - \omega_1 R_1^2}{R_2^2 - R_1^2}\, r - \frac{(\omega_2 - \omega_1) R_1^2 R_2^2}{R_2^2 - R_1^2}\, \frac{1}{r} \tag{8.102}$$

after solving for the constants $A$ and $B$. The first term on the right-hand side corresponds to rotation of all the fluid around the central axis. If $\omega_1 = \omega_2 = \omega$, the velocity becomes $v_\theta = \omega r$, which shows that the fluid rotates as a rigid body around the axis. The second term on the right-hand side corresponds to a deformation of the particles over time. If $R_2 \to \infty$ and $\omega_2 = 0$, then we have the case of a cylinder in an infinite fluid. The velocity $v_\theta = \omega_1 R_1^2 / r$ gives circular streamlines around the cylinder, and the velocity distribution is irrotational, that is, $\mathbf{curl}\ \boldsymbol{v} = \mathbf{0}$.

A tangential shear stress $\sigma_{\theta r}$ acts on a surface element with a radial normal, which is expressed by (A.5)

$$\sigma_{\theta r} = \mu \left( \frac{\partial v_\theta}{\partial r} - \frac{v_\theta}{r} + \frac{1}{r}\frac{\partial v_r}{\partial \theta} \right) = \mu \left( \frac{\partial v_\theta}{\partial r} - \frac{v_\theta}{r} \right) = \mu\, r \frac{\partial}{\partial r} \left( \frac{v_\theta}{r} \right) . \tag{8.103}$$

Combining (8.102) and (8.103), we obtain

$$\sigma_{\theta r} = - \frac{2B\mu}{r^2} . \tag{8.104}$$

Next we calculate the viscous moment, $M$, that acts on the interior cylinder per unit axial length. This moment is equal to the component $\sigma_{\theta r}$ evaluated at

$r = R_1$ and the area, $2\pi R_1$, on which this stress acts, multiplied by the lever arm, $R_1$, the distance between the axis and the point where the force acts. We have

$$M = -2\pi R_1^2 \frac{2B\mu}{R_1^2} = 4\pi\mu \frac{(\omega_2 - \omega_1)R_1^2 R_2^2}{R_2^2 - R_1^2} \ . \tag{8.105}$$

This last relation indicates that we can measure the viscosity $\mu$ of a fluid in a Couette viscometer where the drive motor imposes a torque on one of the cylinders and we measure the resulting rotation speed of the other one.

**Circular Poiseuille Flow in a Cylindrical Pipe**

Poiseuille flow in a circular pipe with radius $R$ is subject to the action of an imposed pressure gradient in direction $z$ (fig. 8.17). The flow is stationary. From the Navier-Stokes equations in cylindrical coordinates, we show first that the only non-zero component of the velocity is $v_z$.

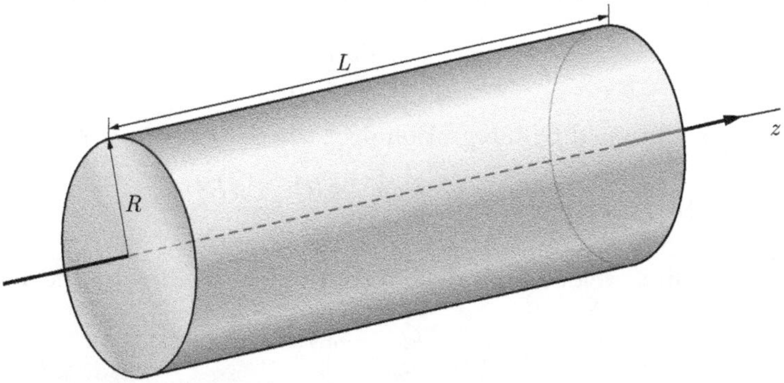


**Fig. 8.17** Poiseuille flow in a circular section cylindrical pipe

Given the hypotheses of axial symmetry and stationary flow, $v_\theta = 0$ and the only two components of velocity, $v_r$ and $v_z$, are functions only of $r$. The continuity equation (A.31) is then

$$\frac{1}{r}\frac{\partial}{\partial r}(r\, v_r) = 0 \ . \tag{8.106}$$

Integration yields

$$r v_r = f(z) \ .$$

But, since $v_r = 0$ at the wall, $r = R$, we conclude that $f(z) = 0$ and thus that $v_r$ is zero everywhere in the flow. The Navier-Stokes equation for the radial component of velocity (A.32) reduces to $\partial p/\partial r = 0$. The pressure depends only on $z$ and not on $r$. The equation for the velocity component $v_z$ (A.34) yields

$$-\frac{dp}{dz} + \mu \left( \frac{\partial^2 v_z}{\partial r^2} + \frac{1}{r}\frac{\partial v_z}{\partial r} \right) = 0$$

or

$$\frac{dp}{dz} = \frac{\mu}{r}\frac{d}{dr}\left(r\frac{dv_z}{dr}\right) .$$

The left-hand side term only depends on $z$; on the right-hand side there is only dependence on $r$. Thus the two terms must be equal to a constant. Integrating, we obtain

$$v_z = \left(\frac{dp}{dz}\right)\frac{1}{\mu}\left(\frac{r^2}{4} + A\ln r + B\right) .$$

The velocity must be finite on the axis $r = 0$. This leads to $A \equiv 0$. Taking into account the condition $v_z(R) = 0$, we have

$$v_z = -\left(\frac{dp}{dz}\right)\frac{R^2}{4\mu}\left(1 - \left(\frac{r}{R}\right)^2\right) .$$

In Poiseuille flow, the velocity profile is parabolic. The maximum velocity at the center is

$$v_{max} = -\left(\frac{dp}{dz}\right)\frac{R^2}{4\mu} . \tag{8.107}$$

The flow rate is obtained by integration over the section of the pipe. We have

$$Q = 2\pi\int_0^R v_z(r)\, r\, dr = -\left(\frac{dp}{dz}\right)\frac{\pi R^4}{8\mu} = \frac{\pi R^2 v_{max}}{2} . \tag{8.108}$$

The average, or flux, velocity obtained from the flux divided by the area of the section $S$ is

$$v_{avg} = \frac{Q}{S} = \frac{v_{max}}{2} . \tag{8.109}$$

The maximum velocity is thus equal to twice the average velocity. The shear stress at the cylinder wall, which we denote $\tau_w$, is given by the component $\sigma_{zr}$ evaluated at $r = R$

$$\tau_w = -\mu\frac{dv_z}{dr}\,|_{r=R} = -\left(\frac{dp}{dz}\right)\frac{R}{2} = \frac{2\mu v_{max}}{R} = \frac{4\mu v_{avg}}{R} . \tag{8.110}$$

The sign change between $\tau_w$ and $\sigma_{zr}$ comes from the fact that $\tau_w$ represents the shear force exercised on the wall by the fluid. The friction coefficient is defined by the ratio of the stress at the wall to the average dynamic pressure

$$C_f = \frac{\tau_w}{\frac{\rho v_{avg}^2}{2}} = \frac{8\mu}{\rho R v_{avg}} = \frac{8\nu}{R v_{avg}} = \frac{16}{Re_D} , \tag{8.111}$$

with $Re_D$ being the Reynolds number based on the average velocity and the diameter of the section. It is common to define the head loss coefficient $\lambda$ by the relation

$$-\left(\frac{dp}{dz}\right) = \frac{\rho v_{avg}^2}{2}\frac{\lambda}{D} . \tag{8.112}$$

Thus it follows that

$$\lambda = 4C_f = \frac{64}{Re_D} . \tag{8.113}$$

### 8.7.3 Plane Non-Stationary Flows

In this section we turn our attention to plane flows that depend on time. This situation leads to partial differential equations with independent variables of space and time. In order to arrive at an analytic solution of the problem, we use a change of variables to obtain an ordinary differential equation that is easier to solve.

**Transient Flow in a Semi-Infinite Space**

An incompressible Newtonian viscous fluid occupies a half space ($x_2 \geq 0$), and is at rest for $t < 0$ (fig. 8.18). At time $t = 0$, the rigid plane which limits the half space is instantaneously put into motion at the constant velocity $U$ in the positive direction of axis $x_1$. The motion is two-dimensional such that $v_3 = 0$.

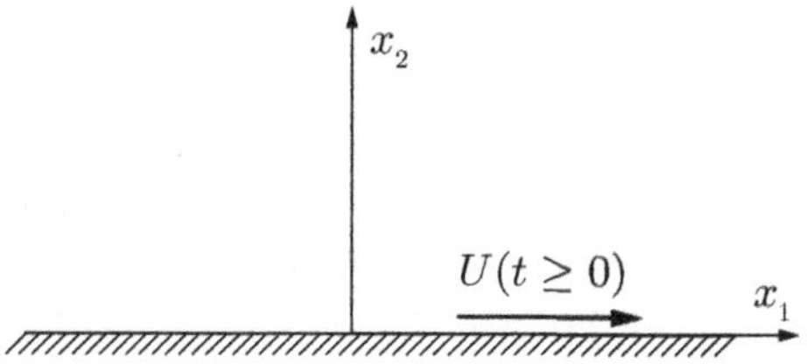


**Fig. 8.18** Unsteady flow in an infinite half space

The boundary and initial conditions are given by

$$t < 0, \quad v_1 = v_2 = 0, \quad \forall x_1, x_2 \tag{8.114}$$

$$t \geq 0, \quad v_1 = U, v_2 = 0, \text{ for } x_2 = 0, \tag{8.115}$$

$$v_1 = v_2 = 0, \text{ for } x_2 \to \infty. \tag{8.116}$$

We assume that $v_1$ and $v_2$ are functions of $x_2$ and $t$

$$v_1 = v_1(x_2, t),\ v_2 = v_2(x_2, t)\ , \tag{8.117}$$

and that the pressure $p$ is a function only of $x_2$ (there is no horizontal pressure gradient; the flow is generated entirely by the motion of the moving wall). The conservation of mass becomes

$$\frac{\partial v_2(x_2, t)}{\partial x_2} = 0\ . \tag{8.118}$$

The component $v_2$ only depends on time, and with conditions (8.115) and (8.116), it is identically zero for all $t$. The Navier-Stokes equations become

$$\rho \frac{\partial v_1}{\partial t} = \mu \frac{\partial^2 v_1}{\partial x_2^2}\ , \tag{8.119}$$

$$\frac{\partial p}{\partial x_2} = 0\ . \tag{8.120}$$

The pressure $p$ is constant.

We can, if we wish, include the effect of gravity in the pressure calculation, by writing

$$\frac{\partial p}{\partial x_2} = -\rho g x_2 \ . \tag{8.121}$$

Integration of this relation leads to the calculation of the hydrostatic pressure, where the pressure at a point is equal to the weight of the column of fluid located above that position. The hydrostatic pressure, as its name suggests, does not participate in the dynamics of the flow.

The motion equation (8.119) is a diffusion equation, of the same type as the "heat equation". We can transform this partial differential equation into an ordinary differential equation with a variable change that we obtain from dimensional analysis. Since the problem has no spatial scale other than the variable $x_2$ nor time scale other than that of $t$ itself, we combine them to form the non-dimensional group

$$\eta = \frac{x_2}{2\sqrt{\nu t}} \ . \tag{8.122}$$

This permits us to obtain an ordinary differential equation for which the solution is a function of $\eta$. It is called a self similar solution because the velocity profile with respect to the variable $x_2$ is similar for all times $t$.

Setting

$$v_1 = U\, f(\eta) \ , \tag{8.123}$$

relation (8.119) becomes

$$f'' + 2\eta f' = 0 \ , \tag{8.124}$$

with conditions

$$\eta = 0,\ f = 1; \quad \eta \to \infty, f = 0 \ . \tag{8.125}$$

Integrating (8.124), we obtain

$$f = A \int_0^{\eta} e^{-\eta'^2} d\eta' + B \ . \tag{8.126}$$

Taking into account conditions (8.125), we have for $\eta = 0$, $B = 1$ and for $\eta = \infty$, $A = -2/\sqrt{\pi}$ where we introduced the error function $\mathrm{erf}(x)$ defined by [1]

$$\mathrm{erf}\,(x) = \frac{2}{\sqrt{\pi}} \int_0^{x} e^{-\tau^2} d\tau \ , \tag{8.127}$$

such that $\mathrm{erf}(\infty) = 1$. Then

$$f = 1 - \mathrm{erf}\,\eta \ , \tag{8.128}$$

and from (8.123) the velocity of the fluid for $t > 0$ is

$$v_1 = U[1 - \mathrm{erf}(\frac{x_2}{2\sqrt{\nu t}})] \ . \tag{8.129}$$

The velocity profile $v_1/U$ as a function of $\eta$ is shown in figure 8.19. For a fixed value of $t$, the variable $\eta$ is proportional to $x_2$. Then, we can deduce the

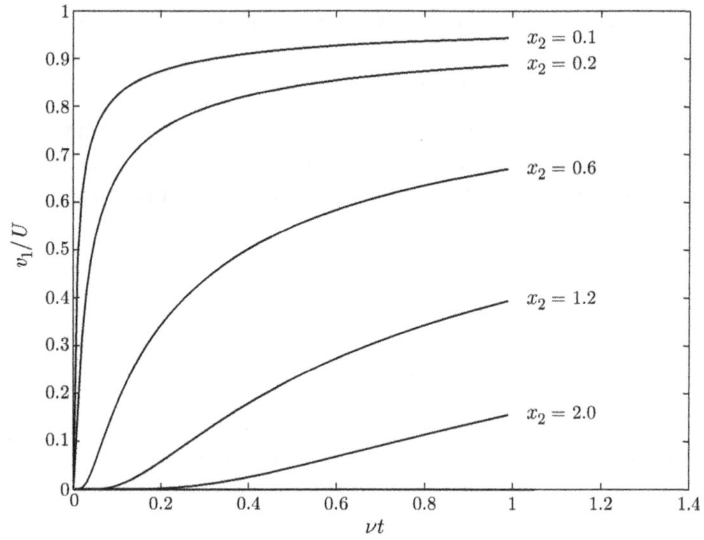


**Fig. 8.19** Transient flow in an infinite half space

velocity profile at every instant as a function of the distance from the wall. An interesting question is to know the depth of the penetration of the wall motion into the semi-infinite space. More precisely, for a given $t$, what is the distance at which the velocity attains, for example, one per cent of the value of $U$? Examining the function erf, $1 - \text{erf}$ has the value 0.01 for $\eta \sim 2$. Defined as such, the penetration depth $\delta$ is given by

$$\eta_\delta = \frac{\delta}{2\sqrt{\nu t}} \simeq 2, \quad \delta \simeq 4\sqrt{\nu t}\ , \tag{8.130}$$

which is proportional to the square root of the kinematic viscosity and time. Thus, if the viscosity is very small, the fluid "sticks" less to the wall and it has a weak effect. If $t$ tends to infinity, the velocity at every point in the half space goes to $U$.

### Flow on an Oscillating Plane

Consider the flow produced by the periodic horizontal oscillation of a plate in its plane (fig. 8.20).

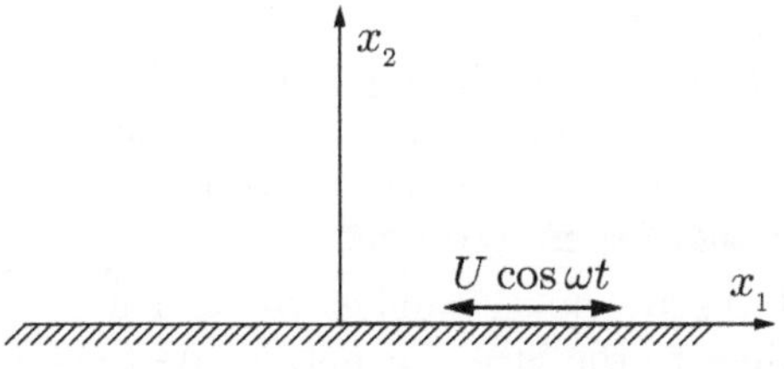


**Fig. 8.20** Unsteady flow on an oscillating plane

Equation (8.119) still applies, and we must solve it with the boundary conditions

$$v_1 = U \cos\omega t \quad \text{for} \quad x_2 = 0 \, . \tag{8.131}$$

After the initial transient phenomena, the fluid velocity gradually becomes a periodic function of time at the same frequency as the oscillation of the plate. Here we examine this periodic regime. Assume that solution $v_1$ is of the form

$$v_1 = \Re\left(e^{i\omega t} f(x_2)\right) \, . \tag{8.132}$$

The combination of (8.119) and (8.132) yields

$$i\omega f = \nu \frac{d^2 f}{dx_2^2} \, .$$

Thus the only solution that remains finite as $x_2 \to \infty$ is

$$f = A \exp\left(-(1+i)(\omega/2\nu)^{1/2} x_2\right) \, .$$

The imposition of the boundary condition (8.131) leads to $A = U$ and the solution becomes

$$v_1 = U \exp\left(-(\omega/2\nu)^{1/2} x_2\right) \cos\left(\omega t - (\omega/2\nu)^{1/2} x_2\right) \, . \tag{8.133}$$

The velocity profile represents a damped harmonic oscillation of amplitude $Ue^{-x_2\sqrt{\omega/2\nu}}$ in a fluid where a layer at distance $x_2$ has a phase lag of $x_2\sqrt{\omega/2\nu}$ with respect to the motion at the wall. Two layers of fluid separated by the distance $2\pi(2\nu/\omega)^{1/2}$ oscillate in phase. This distance constitutes an estimation of the length of the motion and is called the viscous wave penetration depth.

## 8.8 Stokes Flow

In this section, consider first the Stokes equation, valid for very slow flows that we conventionally call creeping flows or Stokes flows. These flows are dominated by viscous forces which are much larger than inertial forces. Examples come from technologies in such diverse domains as convection currents in high temperature glass melting furnaces, lubricants in bearings, and the flow of oils and mud (although the latter may have pronounced non-Newtonian behavior). In nature (another source of interesting cases), we find convection in terrestrial magma, the flow of lava, the swimming of fish, the propulsion of microorganisms, and the squirming of spermatozoon.

We assume that the Reynolds number $Re \ll 1$ and therefore the Navier-Stokes equations reduce to the Stokes equation. As the latter is linear, a complete analytic treatment is possible.

Taking the divergence of the Stokes equation (8.38) and taking into account the solenoidal character of the velocity field, we get

$$\Delta p = 0 \ . \tag{8.134}$$

The pressure is thus a harmonic function for a Stokes flow.

Taking the curl of the Stokes equation (8.38), we obtain

$$\frac{\partial \boldsymbol{\omega}}{\partial t} = \nu \Delta \boldsymbol{\omega} \ , \tag{8.135}$$

where we have introduced the vorticity, $\boldsymbol{\omega} = \mathbf{curl}\, \boldsymbol{v}$ using equation (2.188). If the flow is stationary, the components of the vorticity are also harmonic functions.

### 8.8.1 Plane Creeping Flows

Consider a plane flow for which we have

$$\boldsymbol{v} = (v_1(x_1, x_2, t),\ v_2(x_1, x_2, t),\ 0); \quad p = p(x_1, x_2, t) \ . \tag{8.136}$$

In such a two-dimensional problem, incompressibility (8.16) is automatically satisfied by the introduction of a ***stream function*** $\psi$ so that

$$v_1 = \frac{\partial \psi}{\partial x_2}, \quad v_2 = -\frac{\partial \psi}{\partial x_1} \ . \tag{8.137}$$

As the vorticity reduces to a single component $\boldsymbol{\omega} = (0, 0, \omega)$, it follows that

$$\omega = \frac{\partial v_2}{\partial x_1} - \frac{\partial v_1}{\partial x_2} = -\Delta \psi \ , \tag{8.138}$$

and relation (8.135) becomes

$$\frac{\partial \Delta \psi}{\partial t} = \nu \Delta \Delta \psi \ . \tag{8.139}$$

For a stationary problem, we will have

$$\Delta \Delta \psi = 0 \ , \tag{8.140}$$

showing that in this case the stream function is a biharmonic function.

In polar coordinates $(r, \theta)$, the conservation of mass becomes

$$\frac{1}{r}\frac{\partial}{\partial r}(r v_r) + \frac{1}{r}\frac{\partial v_\theta}{\partial \theta} = 0 \ . \tag{8.141}$$

A stream function $\psi$ also exists such that

$$v_r = \frac{1}{r}\frac{\partial \psi}{\partial \theta}, \quad v_\theta = -\frac{\partial \psi}{\partial r} \ . \tag{8.142}$$

EXAMPLE 8.1

**Flow in a Corner**

Let us consider the flow in a corner as presented in figure 8.21. The lower wall is fixed while the wall inclined at an angle $\theta_0$ is in uniform translational motion at the constant velocity $U$ in the direction $x_1$. Near the origin, the velocity gradients are large; nonetheless, we expect the viscous forces to dominate in the neighborhood of the origin. To formulate the problem in steady state, we choose a coordinate system with the origin at the intersection of the two walls, in motion with the inclined wall. In this case, the boundary conditions are written as

$$\frac{1}{r}\frac{\partial\psi}{\partial\theta} = -U, \qquad \frac{\partial\psi}{\partial r} = 0 \quad \text{at} \quad \theta = 0 \tag{8.143}$$

$$\frac{1}{r}\frac{\partial\psi}{\partial\theta} = 0, \qquad \frac{\partial\psi}{\partial r} = 0 \quad \text{at} \quad \theta = \theta_0 \; . \tag{8.144}$$

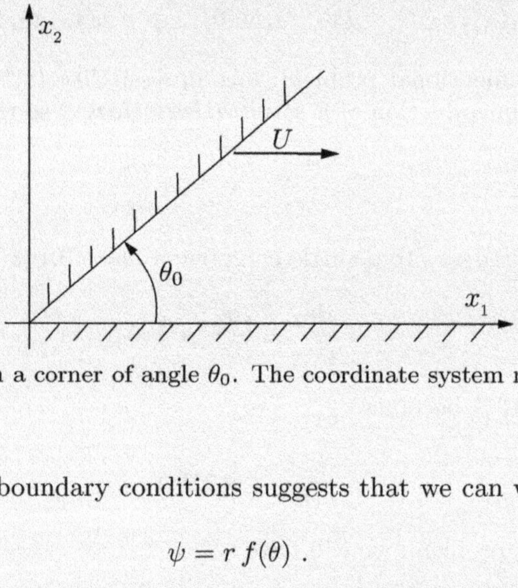


**Fig. 8.21** Flow in a corner of angle $\theta_0$. The coordinate system moves at the upper wall velocity $U$

The form of the boundary conditions suggests that we can write $\psi$ in the following form:

$$\psi = r\,f(\theta) \; . \tag{8.145}$$

Substituting (8.145) in the biharmonic equation (8.140), we find the relation

$$\frac{1}{r^3}\left(\frac{d^4 f}{d\theta^4} + 2\frac{d^2 f}{d\theta^2} + f\right) = 0 \; , \tag{8.146}$$

for which the solution is

$$f(\theta) = A\sin\theta + B\cos\theta + C\theta\sin\theta + D\theta\cos\theta \; . \tag{8.147}$$

(Recall that if $H$ is a harmonic function, $\theta H$ is a biharmonic function). The imposition of the boundary conditions (8.143) and (8.144) allows us to evaluate the constants which are

$$A, B, C, D = (-\theta_0^2, 0, \theta_0 - \sin\theta_0 \cos\theta_0, \sin^2\theta_0)\ \frac{U}{\theta_0^2 - \sin^2\theta_0}\ . \tag{8.148}$$

For the special case of a right angle, we have

$$\psi = \frac{rU}{(\frac{\pi}{2})^2 - 1}\left(-(\frac{\pi}{2})^2 \sin\theta + \frac{\pi}{2}\theta\sin\theta + \theta\cos\theta\right)\ , \tag{8.149}$$

from which we can easily obtain the velocity components

$$v_r = \frac{U}{(\frac{\pi}{2})^2 - 1}\left((1 - \frac{\pi^2}{4})\cos\theta + \frac{\pi}{2}(\sin\theta + \theta\cos\theta) - \theta\cos\theta\right) \tag{8.150}$$

$$v_\theta = -\frac{U}{(\frac{\pi}{2})^2 - 1}\left(-(\frac{\pi}{2})^2 \sin\theta + \frac{\pi}{2}\theta\sin\theta + \theta\cos\theta\right)\ . \tag{8.151}$$

In retrospect we can examine the correctness of the creeping flow assumption. We see that the acceleration components (A.11) and (A.12) evaluated with the preceding solution are proportional to $U^2/r$ with a factor that depends on $\theta$, that is, of the order of unity. As for the viscous effects, they are of the order of $\mu U/r^2$. Thus the creeping flow assumption is met when $\rho r U/\mu \ll 1$. This is the case in the region close to the origin such that $r \ll \nu U$. Further away, the solution will no longer be correct as the inertial forces rapidly become of the same order of magnitude as the viscous forces.

### 8.8.2 Parallel Flow Around a Sphere

A sphere of radius $R$ is in a viscous steady state flow for which the velocity at infinity upstream is $U$. We assume a creeping flow such that we can have a solution of the Stokes equation (8.38). We place the Cartesian coordinate system such that the axis $x_3$ is oriented in the direction of the flow incident on the sphere (fig. 8.22). The boundary conditions expressed in spherical coordinates (see fig. 1.7) are

$$\boldsymbol{v} = \boldsymbol{0} \quad \text{at} \quad r = R, \tag{8.152}$$

$$\boldsymbol{v} = U\boldsymbol{e}_3 \quad \text{at} \quad r = \infty\ . \tag{8.153}$$

The problem thus posed is symmetric about the axis $Ox_3$, and with respect to the longitude of the sphere. Consequently, $\partial(\bullet)/\partial\varphi \equiv 0$. Also then, $v_\varphi = 0$. Thus the mass conservation equation (B.30) reduces to

$$\frac{1}{r^2}\frac{\partial}{\partial r}(r^2 v_r) + \frac{1}{r\sin\theta}\frac{\partial}{\partial\theta}(v_\theta \sin\theta) = 0\ . \tag{8.154}$$

We deduce that a stream function $\psi$ exists such that

$$v_r = \frac{1}{r^2\sin\theta}\frac{\partial\psi}{\partial\theta}, \quad v_\theta = -\frac{1}{r\sin\theta}\frac{\partial\psi}{\partial r}\ . \tag{8.155}$$

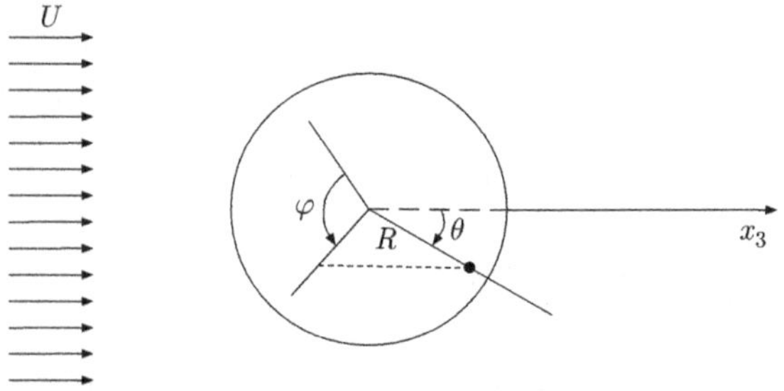


**Fig. 8.22** Flow around a sphere

Given the two-dimensional character of the flow, the vorticity will have a single component in the direction of the vector $\boldsymbol{e}_\varphi$ that we denote $\omega$. We can then write (see eqn. (B.5))

$$\omega(r,\theta) = -\frac{1}{r}\left[\frac{1}{\sin\theta}\frac{\partial^2\psi}{\partial r^2} + \frac{1}{r^2}\frac{\partial}{\partial\theta}\left(\frac{1}{\sin\theta}\frac{\partial\psi}{\partial\theta}\right)\right] . \tag{8.156}$$

In the case of the Stokes equation, vorticity is a harmonic function. We have (recall that the Laplacian of a vector is not equal to the Laplacian of its components, cf. (B.7))

$$\begin{aligned}\Delta\omega - \frac{\omega}{r^2\sin^2\theta} &= \frac{1}{r^2}\left(\frac{\partial}{\partial r}\left(r^2\frac{\partial\omega}{\partial r}\right) + \frac{1}{\sin\theta}\frac{\partial}{\partial\theta}\left(\sin\theta\frac{\partial\omega}{\partial\theta}\right) - \frac{\omega}{\sin^2\theta}\right)\\ &= \frac{1}{r}\frac{\partial^2}{\partial r^2}(r\omega) + \frac{1}{r^2}\frac{\partial}{\partial\theta}\left(\frac{1}{\sin\theta}\frac{\partial}{\partial\theta}(\omega\,\sin\theta)\right) .\end{aligned} \tag{8.157}$$

The combination of relations (8.156)–(8.157) gives the following biharmonic equation:

$$\left(\frac{\partial^2}{\partial r^2} + \frac{\sin\theta}{r^2}\frac{\partial}{\partial\theta}\left(\frac{1}{\sin\theta}\frac{\partial}{\partial\theta}\right)\right)^2\psi = 0 . \tag{8.158}$$

The boundary conditions (8.152) and (8.153), expressed in terms of the stream function, become

$$\begin{aligned}\frac{\partial\psi}{\partial\theta} &= \frac{\partial\psi}{\partial r} = 0, \quad \text{at} \quad r = R,\\ v_r &= U\cos\theta, \quad \frac{\partial\psi}{\partial\theta} = Ur^2\sin\theta\cos\theta, \quad \text{at} \quad r=\infty\\ v_\theta &= -U\sin\theta, \quad \frac{\partial\psi}{\partial r} = Ur\sin^2\theta.\end{aligned} \tag{8.159}$$

The condition at infinity can be easily integrated. It follows that

$$\psi_\infty = \frac{1}{2}Ur^2\sin^2\theta . \tag{8.160}$$

The form of this expression for $\psi$ suggests that the stream function can be written in the general form

$$\psi = \sin^2\theta\, f(r) . \tag{8.161}$$

Introducing (8.161) into (8.158), we find

$$\frac{d^4 f}{dr^4} - \frac{4}{r^2}\frac{d^2 f}{dr^2} + \frac{8}{r^3}\frac{df}{dr} - \frac{8}{r^4} f = 0 \,. \tag{8.162}$$

Seeking a solution as a power series in $r^n$, we obtain the characteristic polynomial

$$(n-2)(n-1)\left(n^2 - 3n - 4\right) = 0 \,,$$

whose roots are $n = -1, 1, 2$, and 4. The function $f(r)$ is thus

$$f = \frac{C_{-1}}{r} + C_1\, r + C_2\, r^2 + C_4\, r^4 \,. \tag{8.163}$$

The imposition of the boundary condition at infinity, (8.160), requires $C_4 = 0$ and $C_2 = \frac{1}{2}U$, while at the sphere, with $v_r = v_\theta = 0$, we can determine $C_{-1} = (1/4)U\,R^3, C_1 = -(3/4)U\,R$. The stream function is then

$$\psi = \frac{UR^2}{2}\sin^2\theta\left(\frac{R}{2r} - \frac{3}{2}\frac{r}{R} + (\frac{r}{R})^2\right) \,. \tag{8.164}$$

We can easily deduce the velocities from (8.155). The vorticity field is written (cf. (8.156))

$$\omega = -\frac{3}{2}UR(\frac{\sin\theta}{r^2}) \,. \tag{8.165}$$

The calculation of the pressure field can easily be accomplished by taking into account the vector identity (1.238) which leads to the Stokes equation

$$\boldsymbol{\nabla} p = -\mu\,\mathbf{curl}\,\boldsymbol{\omega} \,. \tag{8.166}$$

With (B.5), this leads to the system of equations

$$\frac{\partial p}{\partial r} = -\frac{\mu}{r\sin\theta}\frac{\partial}{\partial\theta}(\omega\,\sin\theta) = 3\mu\,U\,R\frac{\cos\theta}{r^3} \,, \tag{8.167}$$

$$\frac{1}{r}\frac{\partial p}{\partial\theta} = \frac{\mu}{r}\frac{\partial(r\omega)}{\partial r} = \frac{3\mu}{2}U\,R\frac{\sin\theta}{r^3} \,. \tag{8.168}$$

Integration of (8.167) yields

$$p = -\frac{3\mu}{2}\,U\,R\frac{\cos\theta}{r^2} + q(\theta) \,.$$

Inserting this result in (8.168), we have

$$\frac{3\mu}{2}\,U\,R\frac{\sin\theta}{r^3} + \frac{q'(\theta)}{r} = \frac{3\mu}{2}U\,R\frac{\sin\theta}{r^3} \,.$$

The pressure field is then finally given by

$$p = -\frac{3\mu}{2}\,U\,R\frac{\cos\theta}{r^2} + p_0 \,, \tag{8.169}$$

with $p_0$ a constant reference pressure.

The uniform velocity flow around a sphere will generate pressure and shear forces. To calculate the pressure force in direction $Ox_3$, we integrate the elementary forces over the surface of the sphere

$$dF_{3,p} = -(\frac{3\mu}{2} U \frac{\cos\theta}{R} + p_0)\cos\theta\,(2\pi R^2 \sin\theta)\,d\theta\ . \tag{8.170}$$

The factor $2\pi$ comes from the symmetry of the problem which permits us to take into account the longitudinal part of the integral. Integrating from $\theta = 0$ to $\theta = \pi$, we obtain

$$F_{3,p} = -2\pi\mu\, U\, R\ . \tag{8.171}$$

Friction drag is obtained by integration over the sphere of the shear stress that acts on it, that is, $\sigma_{r\theta}$ which is $-3\mu U \sin\theta/(2R)$ for $r = R$. This leads to

$$F_{3,\sigma} = -\int_{\theta=0}^{\theta=\pi} (\sigma_{r\theta}\,|_{r=R} \sin\theta)\,(2\pi R^2 \sin\theta)\,d\theta = -4\pi\mu\, U\, R\,. \tag{8.172}$$

Total drag, $F_3 = F_{3,p} + F_{3,\sigma}$, known as ***Stokes drag***, is the sum of the pressure force and the friction force

$$F_3 = -6\pi\mu\, U\, R\ . \tag{8.173}$$

If we define the drag coefficient by

$$C_D = \frac{F_3}{\frac{1}{2}\rho U^2\, \pi R^2}, \tag{8.174}$$

we obtain

$$C_D = \frac{24}{Re}, \tag{8.175}$$

where $Re = 2UR/\nu$. Note that the pressure drag represents a third of the total drag. Relation (8.175) is verified by experiments when $Re < 1$ which is valid in the neighborhood of the sphere. When we move away, the importance of the inertial terms grows and Stokes solution diverges from the exact solution. Note that the solution that we have obtained is not applicable to the case of a set of spherical particles, as the presence of a spherical obstacle in the flow has impact relatively far away since the velocity profiles decrease as $1/r$.

The solution for uniform flow around a fixed sphere can be transposed to the case of translation at uniform velocity $U$ of a sphere of radius $R$ in a fluid at rest at infinity. In this case, the coordinate system is still attached to the sphere and thus in translation at uniform velocity. This modifies the sign of $U$ to become $-U$ for the pressure and the vorticity. As for the velocity in the fluid, this is relative to the coordinate system, which leads to the following modifications: for the velocity and the stream function, $U$ becomes $-U$ and the uniform velocity field must also be subtracted from the corresponding relations.

**Oseen's Improvement**

The Stokes solution was improved by Oseen [37] who proposed the solution of the Navier-Stokes equations (8.17) as a sum of uniform velocity field and a perturbation such that

$$\boldsymbol{v} = U\,\boldsymbol{e}_3 + \boldsymbol{v}' \ . \tag{8.176}$$

In the case of the flow around a fixed sphere, the velocity $\boldsymbol{v}'$ then takes into account the perturbation caused by the sphere in a flow uniform at infinity. With (8.176), the stationary inertial term takes the form

$$\rho\frac{D\boldsymbol{v}}{Dt} = \rho\left(v_j'\frac{\partial v_i'}{\partial x_j} + U\frac{\partial v_i'}{\partial x_3}\right) \ . \tag{8.177}$$

Oseen's assumption amounts to neglecting the first term with respect to the second term on the right-hand side of equation (8.177). We obtain a linearized Navier-Stokes equation

$$\rho\,U\frac{\partial \boldsymbol{v}'}{\partial x_3} = -\boldsymbol{\nabla}\,p + \mu\,\Delta\boldsymbol{v}' + \rho\boldsymbol{b} \ . \tag{8.178}$$

The drag coefficient obtained with Oseen's solution is

$$C_D = \frac{24}{Re}(1 + \frac{3}{16}Re) \ . \tag{8.179}$$

Experimental results show that (8.179) is approximately valid for $Re < 5$. Using matched asymptotic expansions [28], the corrected coefficient becomes

$$C_D = \frac{24}{Re}(1 + \frac{3}{16}Re - \frac{19}{1280}Re^2 + O(Re^3)) \ . \tag{8.180}$$

## 8.9 Vorticity and Vortex Kinematics

In section 8.4, the Navier-Stokes equations were derived for a Newtonian viscous fluid in terms of primitive variables: velocity and pressure. The interactions observed in fluid flows have been interpreted by an equilibrium between the inertial forces, the pressure gradient, the volume forces such as gravity, and the viscous forces. In this section, we take a different point of view based on the concept of vorticity.

The presence of vorticity in a flow is an indication of the importance of the viscous effects, given that they are generated by viscous stresses. Therefore, under certain assumptions, vorticity possesses the following properties:

i) in the absence of viscosity, it is transported by the flow as an elementary material vector;

ii) in the presence of viscosity, it diffuses into the surrounding fluid while being continually produced at the solid walls that delimit the flow.

Thus the vorticity produced on a solid wall introduces the notion of a boundary layer for which we are led to modify certain conclusions coming from the theory of irrotational perfect fluids. In turbulence, flow dynamics are mostly the result of the stretching or shortening of vortex lines and their deformation.

**Kinematic Considerations**

The velocity gradient tensor $\boldsymbol{L}$ can be decomposed into the sum of a symmetric strain rate tensor $\boldsymbol{d}$ and an antisymmetric rotation rate tensor $\dot{\boldsymbol{\omega}}$ according to equation (2.184). The tensor $\boldsymbol{d}$ is given by (2.181) and $\dot{\boldsymbol{\omega}}$ by (2.183). Recall that the dual vector $\dot{\boldsymbol{\Omega}}$, corresponding to the rotation rate tensor, is the rotation rate vector introduced by (2.187).

In fluid mechanics, we classically introduce the vorticity vector $\boldsymbol{\omega}$, defining it as the curl of the velocity (2.188). To acquire familiarity with the concept of vorticity, we study the flow near a stagnation point at the origin. The velocity components are such that we have, with a constant $C$,

$$v_1 = Cx_1,\ v_2 = -Cx_2,\ v_3 = 0\ . \tag{8.181}$$

We easily calculate that for this flow $\boldsymbol{\omega} = \boldsymbol{0}$. A flow with zero vorticity is called ***irrotational***. Note that the corresponding stream function introduced in (8.137) is $\psi = Cx_1x_2$ which is represented by hyperbolas.

Now consider the plane Poiseuille flow in a channel of height $h$. If the coordinate system has its origin on the lower wall, the velocity profile, (8.63) with definition (8.68), is given by the relation

$$v_1 = 4v_{max}\frac{x_2}{h}(1-\frac{x_2}{h})\ , \tag{8.182}$$

with $v_{max}$ being the maximum velocity on the centerline of the channel at $x_2 = h/2$. The only component of the vorticity is $\omega_3$. It is perpendicular to the plane of the flow and its value is

$$\omega_3 = \varepsilon_{321}\frac{\partial v_1}{\partial x_2} = -\frac{4v_{max}}{h}(1-\frac{2x_2}{h})\ . \tag{8.183}$$

In this case, the absolute value of the vorticity attains a maximum at the two walls and goes to zero on the centerline of the channel.

From these examples we can conclude that the concept of vorticity has no relation to the curvature of the streamlines. In the first case, the streamlines are curved, but the vorticity is zero; while in the second example, the streamlines are straight lines and there is finite vorticity.

From the definition of vorticity, (2.188), and Stokes theorem, (1.229), we obtain the identity

$$I(S) = \int_S \boldsymbol{\omega}\cdot\boldsymbol{n}\,dS = \int_S \mathbf{curl}\,\boldsymbol{v}\cdot\boldsymbol{n}\,dS = \oint \boldsymbol{v}\cdot\boldsymbol{\tau}\,dl = \Gamma\ . \tag{8.184}$$

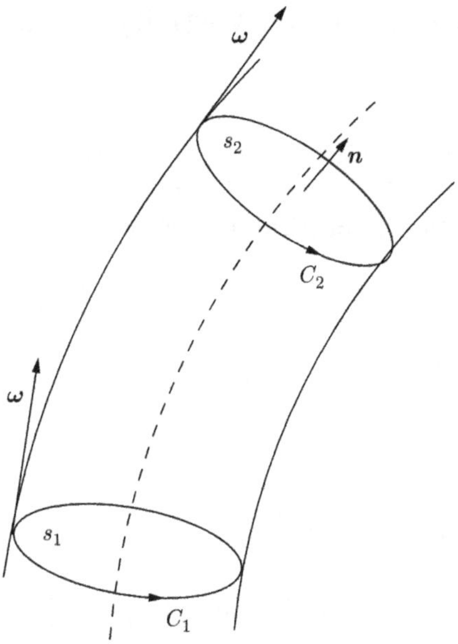


**Fig. 8.23** Vortex tube

The curvilinear integral in (8.184) defines the velocity circulation, Γ, along the closed curve $C$, of the unit tangent vector $\boldsymbol{\tau}$. It is thus equal to the vorticity vector flux through an arbitrary surface bounded by the curve. In the following, this property will permit us to systematically link the concept of circulation to an interpretation in terms of vorticity. Recall that a vortex line (fig. 8.23) is a line tangent at all its points to the vorticity vector, and that a vortex tube is a family of vortex lines circumscribed by a closed curve. The intensity of a vortex tube, for a surface $S$ defined by a closed line enclosing the vortex tube, is the flux $I(S)$ of vorticity through the surface.

**Helmholtz Theorem** (Vorticity properties)

*Helmholtz main theorems about vorticity are as follows:*

- *the vorticity flux through a closed surface is always zero;*
- *the intensity of a vortex tube does not depend on the transverse section considered;*
- *a vortex tube can only end connected to itself or extend to infinity unless it is cut by a wall.*

The proof of these theorems can be found in Panton's book [38].

## 8.10 Dynamic Vorticity Equation

### 8.10.1 General Equation

The formulation of the equation that governs vorticity dynamics requires the establishment of certain preliminary relations.

First, the acceleration term $\boldsymbol{a}$ can be written as follows:

$$\boldsymbol{a} = \frac{\partial \boldsymbol{v}}{\partial t} + \boldsymbol{\omega} \times \boldsymbol{v} + \boldsymbol{grad}\left(\frac{\boldsymbol{v}\cdot\boldsymbol{v}}{2}\right) , \tag{8.185}$$

from which we can deduce the relations

$$\begin{aligned} a_i &= \frac{\partial v_i}{\partial t} + \varepsilon_{ijk}\omega_j v_k + \frac{\partial}{\partial x_i}\left(\frac{v_j v_j}{2}\right) , \\ &= \frac{\partial v_i}{\partial t} + \varepsilon_{ijk}\varepsilon_{jlm}\left(\frac{\partial v_m}{\partial x_l}\right) v_k + v_j \frac{\partial v_j}{\partial x_i} , \\ &= \frac{\partial v_i}{\partial t} + (\delta_{kl}\delta_{im} - \delta_{km}\delta_{il})\left(\frac{\partial v_m}{\partial x_l} v_k\right) + v_j \frac{\partial v_j}{\partial x_i} , \end{aligned}$$

or

$$a_i = \frac{\partial v_i}{\partial t} + v_k \frac{\partial v_i}{\partial x_k} .$$

This last expression is none other than the definition of acceleration (2.33).

Thus, relation

$$\frac{1}{\rho}\,\mathbf{curl}\,\boldsymbol{a} = \frac{D}{Dt}\left(\frac{\boldsymbol{\omega}}{\rho}\right) - \frac{1}{\rho}(\boldsymbol{\omega}\cdot\boldsymbol{grad})\boldsymbol{v} \tag{8.186}$$

is an identity. As can be seen, applying the curl operator to relation (8.185) leads to

$$\mathbf{curl}\,\boldsymbol{a} = \frac{\partial}{\partial t}\mathbf{curl}\,\boldsymbol{v} + \mathbf{curl}(\boldsymbol{\omega}\times\boldsymbol{v}) + \mathbf{curl}\,\boldsymbol{grad}\left(\frac{\boldsymbol{v}\cdot\boldsymbol{v}}{2}\right)$$

or

$$\mathbf{curl}\,\boldsymbol{a} = \frac{\partial \boldsymbol{\omega}}{\partial t} + \mathbf{curl}(\boldsymbol{\omega}\times\boldsymbol{v}) . \tag{8.187}$$

The term $\mathbf{curl}(\boldsymbol{\omega}\times\boldsymbol{v})$ can be developed as follows:

$$\mathbf{curl}(\boldsymbol{\omega}\times\boldsymbol{v}) = \boldsymbol{v}\cdot\boldsymbol{grad}\,\boldsymbol{\omega} - (\boldsymbol{\nabla v})\,\boldsymbol{\omega} + \boldsymbol{\omega}\,\mathrm{div}\,\boldsymbol{v} - \boldsymbol{v}\,\mathrm{div}\,\boldsymbol{\omega} . \tag{8.188}$$

The last term of (8.188) is zero from (1.180). From (8.187) and (8.188), it follows that

$$\mathbf{curl}\,\boldsymbol{a} = \frac{D\boldsymbol{\omega}}{Dt} - (\boldsymbol{\nabla v})\,\boldsymbol{\omega} + \boldsymbol{\omega}\,\mathrm{div}\,\boldsymbol{v} .$$

From the mass conservation equation (3.41), we obtain the relation

$$\frac{1}{\rho}\,\mathbf{curl}\,\boldsymbol{a} = \frac{1}{\rho}\frac{D\boldsymbol{\omega}}{Dt} - (\boldsymbol{\nabla v})\left(\frac{\boldsymbol{\omega}}{\rho}\right) - \frac{1}{\rho^2}\frac{D\rho}{Dt}\boldsymbol{\omega} ,$$

which is equivalent to equation (8.186) and which can be written in the form

$$\frac{D}{Dt}\left(\frac{\boldsymbol{\omega}}{\rho}\right) = \left(\frac{\boldsymbol{\omega}}{\rho}\cdot \boldsymbol{grad}\right)\boldsymbol{v} + \frac{1}{\rho}\mathbf{curl}\,\boldsymbol{a}\ . \tag{8.189}$$

This relation constitutes a first description of the temporal evolution of vorticity. It is known as the ***Beltrami diffusion equation*** [46].

From the conservation of momentum (3.96), we write

$$\frac{D}{Dt}\left(\frac{\boldsymbol{\omega}}{\rho}\right) = (\frac{\boldsymbol{\omega}}{\rho}\cdot \boldsymbol{grad})\boldsymbol{v} + \frac{1}{\rho}\mathbf{curl}\,(\boldsymbol{b} + \frac{1}{\rho}\,\mathbf{div}\,\boldsymbol{\sigma})\ . \tag{8.190}$$

In order to separate the effects of pressure and viscosity, we use the constitutive equation (6.12) in (8.190). We have

$$\frac{D}{Dt}\left(\frac{\boldsymbol{\omega}}{\rho}\right) = \left(\frac{\boldsymbol{\omega}}{\rho}\cdot \boldsymbol{grad}\right)\boldsymbol{v} + \frac{1}{\rho}\mathbf{curl}\,\boldsymbol{b} + \frac{1}{\rho}\mathbf{curl}(-\frac{1}{\rho}\boldsymbol{\nabla}p) + \frac{1}{\rho}\mathbf{curl}(\frac{1}{\rho}\,\mathbf{div}\,\boldsymbol{T})\ , \tag{8.191}$$

with the deviatoric extra-stress tensor $\boldsymbol{T}$. Using relation (1.234), we write

$$\frac{D}{Dt}\left(\frac{\boldsymbol{\omega}}{\rho}\right) = \left(\frac{\boldsymbol{\omega}}{\rho}\cdot \boldsymbol{grad}\right)\boldsymbol{v} + \frac{1}{\rho}\mathbf{curl}\,\boldsymbol{b} + \frac{\boldsymbol{\nabla}p}{\rho}\times\boldsymbol{\nabla}(\frac{1}{\rho}) + \frac{1}{\rho}\mathbf{curl}(\frac{1}{\rho}\,\mathbf{div}\,\boldsymbol{T})\ . \tag{8.192}$$

The left-hand side of relation (8.192) contains the material derivative of the vorticity relative to the local density. On the right-hand side, we find two terms that describe the deformation (stretching and shrinking) and the curvature (bending-tilting) of the vortex lines, then the baroclinicity term (related to pressure) and, finally, the viscous diffusion of the vorticity.

If the volume force is conservative, it can be derived from a potential $\chi$, as is the case for gravity. Then we write

$$\boldsymbol{b} = -\boldsymbol{\nabla}\chi\ . \tag{8.193}$$

Consequently, $\mathbf{curl}\,\boldsymbol{b} = 0$, and this term disappears from (8.192). We adopt this hypothesis for the rest of this discussion.

### 8.10.2 Physical Interpretation of Vorticity Dynamics

#### Incompressible Perfect Fluid Case

For an incompressible fluid ($\boldsymbol{\nabla}\rho = 0$), that is inviscid ($\boldsymbol{T} = \mathbf{0}$), equation (8.192) simplifies to

$$\frac{D}{Dt}\boldsymbol{\omega} = (\boldsymbol{\nabla}\boldsymbol{v})\ \boldsymbol{\omega} = (\boldsymbol{\omega}\cdot\boldsymbol{\nabla})\boldsymbol{v}\ . \tag{8.194}$$

The term

$$(\boldsymbol{\omega}\cdot\boldsymbol{\nabla})\boldsymbol{v}$$

does not correspond to any term in the Navier-Stokes equations written with the primitive variables, velocity and pressure. Let us examine what it means for the flow from the physical point of view.

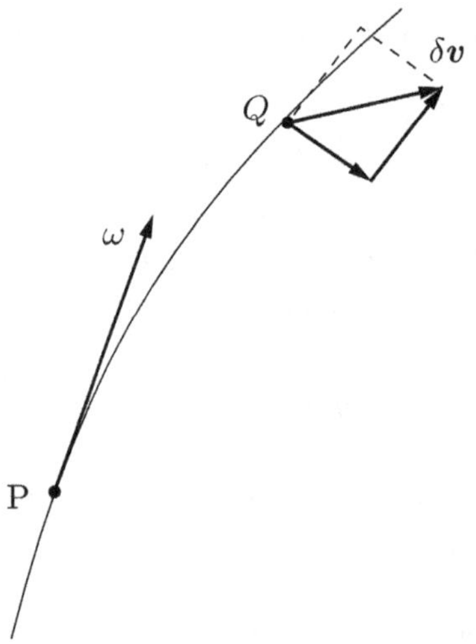


**Fig. 8.24** Portion of a vortex line

In figure 8.24, consider two neighboring points P and $Q$ on a vortex line. The points P and $Q$ also define a material line of length $dx = \| \boldsymbol{dx} \|$, and we can show the equality as

$$\frac{D(dx_i)}{Dt} = dv_i = \frac{\partial v_i}{\partial x_j} dx_j \text{ or, } \frac{D(\boldsymbol{dx})}{Dt} = \boldsymbol{dx} \cdot \boldsymbol{grad}\, \boldsymbol{v} \,. \tag{8.195}$$

This last equation simultaneously expresses the changes in length and direction of a material line element. Comparison of (8.194) and (8.195) shows that the vorticity vector $\boldsymbol{\omega}$ plays a role analogous to that of the vector $\boldsymbol{dx}$. Thus, $\boldsymbol{\omega}$ behaves *as if* it were a material line element instantaneously coinciding with a portion of the vortex line. Let $\delta \boldsymbol{v}$ be the relative velocity of the fluid at $Q$ with respect to P. In relation (8.194), we can make the substitution

$$(\boldsymbol{\nabla v})\, \boldsymbol{\omega} = \| \boldsymbol{\omega} \| \lim_{PQ \to 0} \frac{\delta \boldsymbol{v}}{PQ} \,.$$

One part of the change of $\boldsymbol{\omega}$ measured by (8.194) comes from the rigid body rotation of the material line element (from the component of $\delta \boldsymbol{v}$ normal to $\boldsymbol{\omega}$), and the other part is generated by the shrinking or stretching of the elementary line (from the component $\delta \boldsymbol{v}$ parallel to $\boldsymbol{\omega}$). Finally, equation (8.194) can be interpreted as follows: the vorticity is transported by the fluid particles, while being oriented and deformed *as if* it were an elementary material vector.

#### Compressible Perfect Fluid Case

The term $\boldsymbol{\nabla} \rho \neq 0$ is present in (8.192). The production of vorticity by baroclinicity occurs in flows where the constant pressure and constant density (isobaric and isopycnal) surfaces are not parallel. This can occur in some domains, including meteorology, oceanography, and astrophysics. In these cases, the center of gravity of the fluid does not coincide with the center of pressure, the latter being where the resulting vector for all the pressure forces is applied; the resulting force then acts as a torque, locally turning the fluid and creating circulation.

In the case of a barotropic fluid for which the density is only a function of the pressure (sec. 6.9), we have

$$\rho = \rho(p) \quad \text{or} \quad p = p(\rho) \,, \tag{8.196}$$

then the isobars and the isopycnal surfaces are parallel, and the baroclinic term is zero.

## 8.11 Vorticity Equation for a Newtonian Viscous Fluid

We assume now that the viscosities $\lambda$ and $\mu$ are invariable. With (6.13), we write

$$\frac{1}{\rho}\,\mathbf{div}\,\boldsymbol{T} = \frac{\lambda}{\rho}\boldsymbol{grad}\,(\operatorname{div}\boldsymbol{v}) + 2\nu\,\boldsymbol{grad}\,(\operatorname{div}\boldsymbol{v}) - \nu\,\mathbf{curl}\,\mathbf{curl}\,\boldsymbol{v} \tag{8.197}$$

or

$$\frac{1}{\rho}\,\mathbf{div}\,\boldsymbol{T} = \boldsymbol{grad}\left(\frac{\lambda+2\mu}{\rho}\operatorname{div}\boldsymbol{v}\right) - \boldsymbol{grad}\left(\frac{\lambda+2\mu}{\rho}\right)\operatorname{div}\boldsymbol{v} - \nu\,\mathbf{curl}\,\mathbf{curl}\boldsymbol{v} \,. \tag{8.198}$$

Taking the curl of (8.198) and accounting for (1.234), leads to

$$\mathbf{curl}(\frac{1}{\rho}\,\mathbf{div}\,\boldsymbol{T}) = \frac{\lambda+2\mu}{\rho^2}\boldsymbol{\nabla}\rho\times\boldsymbol{\nabla}(\operatorname{div}\boldsymbol{v}) - \mathbf{curl}(\nu\,\mathbf{curl}\,\boldsymbol{\omega}) \,. \tag{8.199}$$

The vorticity dynamics equation is obtained by combining (8.192) and (8.199):

$$\begin{aligned} \frac{D}{Dt}\left(\frac{\boldsymbol{\omega}}{\rho}\right) &= \left(\frac{\boldsymbol{\omega}}{\rho}\cdot\boldsymbol{grad}\right)\boldsymbol{v} + \frac{1}{\rho^3}\boldsymbol{\nabla}\rho\times\boldsymbol{\nabla}p \\ &+ \frac{\lambda+2\mu}{\rho^3}\boldsymbol{\nabla}\rho\times\boldsymbol{\nabla}(\operatorname{div}\boldsymbol{v}) - \frac{1}{\rho}\mathbf{curl}(\nu\,\mathbf{curl}\,\boldsymbol{\omega}) \,. \end{aligned} \tag{8.200}$$

This equation simplifies if the fluid is either compressible and barotropic (8.196) ($\nabla\rho$ is parallel to $\nabla p$) or incompressible ($\nabla\rho = 0$). We then have

$$\frac{D}{Dt}\left(\frac{\boldsymbol{\omega}}{\rho}\right) = \left(\frac{\boldsymbol{\omega}}{\rho}\cdot\boldsymbol{grad}\,\boldsymbol{v}\right) - \frac{1}{\rho}\,\mathbf{curl}\,(\nu\,\mathbf{curl}\,\boldsymbol{\omega}) \,. \tag{8.201}$$

**Special Case for Two-Dimensional Flow**

For an incompressible two-dimensional flow, equation (8.201) becomes, with the notation $\omega_3 = \omega$,

$$\frac{D\omega}{Dt} = \nu \, \Delta\omega \ , \tag{8.202}$$

because in this special case the term $(\boldsymbol{\omega} \cdot \boldsymbol{grad}\, \boldsymbol{v})$ is zero since $\boldsymbol{\omega}$ is orthogonal to the flow plane and thus to $\boldsymbol{grad}\, \boldsymbol{v}$. We note that equation (8.202) is analogous to that for heat conduction, with the kinematic viscosity replacing the thermal diffusivity. We also notice that equation (8.202) is satisfied for $\omega = 0$, that is, for an irrotational flow. However, that solution is inadequate. To understand why, we reason by analogy with the heat equation, which also allows an identically zero solution. We know from the study of heat flow, that any non-uniform distribution of temperature at the wall or non-zero heat flux will generate a variable temperature field in the material. Thus the analogy leads us to conclude that, in the case of a viscous fluid, the vorticity that is generated at the walls will diffuse out by shear and then be carried away by the flow. The creation of vorticity at the wall is the result of the shear stress on the wall. To obtain the value of vorticity at the wall, we resort to the classical method of Green's functions [51].

## 8.12 Circulation Equation

In the context of the hypotheses introduced in the previous section, we prove that for a material curve $C(t)$, along which the circulation of the velocity vector is $\Gamma(t)$, we can write the following:

$$\frac{d\Gamma}{dt} = -\oint_{C(t)} \nu(\mathbf{curl}\,\mathbf{curl}\,\boldsymbol{v}) \cdot d\boldsymbol{l} \ . \tag{8.203}$$

This relation expresses the fact that the variation of the circulation along the material curve is due to the viscosity which dampens the motion.

To obtain (8.203), we must first prove that for a material curve $C(t)$, we have the following identity:

$$\frac{d\Gamma}{dt} = \oint_{C(t)} \boldsymbol{a} \cdot d\boldsymbol{x} \ . \tag{8.204}$$

For that, we can write the equation

$$\frac{d\Gamma}{dt} = \frac{d}{dt}\oint_{C(t)} v_i dx_i = \frac{d}{dt}\oint_{C_0} V_i \frac{\partial x_i}{\partial X_j} dX_j \ ,$$

in which $C_0$ denotes the material curve $C(t)$ at the instant $t = t_0$ and $X_i$ are the associated Lagrangian coordinates. Denoting by $A_i$ and $V_i$ the Lagrangian representations of acceleration (2.32) and velocity (2.17), we have:

$$\frac{d}{dt}\oint_{C_0} V_i \frac{\partial x_i}{\partial X_j} dX_j = \oint_{C_0} \left( A_i \frac{\partial x_i}{\partial X_j} + V_i \frac{\partial V_i}{\partial X_j} \right) dX_j$$
$$= \oint_{C(t)} a_i dx_i + \oint_{C_0} \frac{\partial}{\partial X_j} \left( \frac{V_i V_i}{2} \right) dX_j \, .$$

The last term of the right-hand side of this equality is zero on a closed curve.

With relation (6.14), which we use in the motion equation (3.96), taking into account vector identity (1.238) and equation (8.10), we can write

$$\boldsymbol{a} = -\boldsymbol{grad}\,\chi - \frac{1}{\rho}\boldsymbol{grad}\,p + \left(\frac{\lambda}{\rho} + 2\nu\right) \boldsymbol{grad}\,(\mathrm{div}\,\boldsymbol{v}) - \nu\,\mathbf{curl}\,\mathbf{curl}\,\boldsymbol{v} \, .$$

And with the conservation of mass, it leads to

$$\boldsymbol{a} = -\boldsymbol{grad}\left(\frac{p}{\rho} + \chi - \frac{\lambda + 2\mu}{\rho}\mathrm{div}\,\boldsymbol{v}\right) - \frac{\lambda + 2\mu}{\rho^3}\frac{D\rho}{Dt}\boldsymbol{grad}\,\rho - \nu\,\mathbf{curl}\,\mathbf{curl}\boldsymbol{v} \, . \tag{8.205}$$

In the case of fluid that is weakly viscous or compressible, we consider the second term on the right-hand side to be of second order with respect to the term in the pressure gradient; we thus neglect it in the following. Inserting (8.205) in (8.204), we obtain (8.203).

## 8.13 Vorticity Equation for a Perfect Fluid

For an incompressible, perfect ($\nu = 0$), or barotropic fluid (especially in isentropic flow, $ds = 0$), the vorticity dynamics theorem (8.201) becomes

$$\frac{D}{Dt}\left(\frac{\boldsymbol{\omega}}{\rho}\right) = \left(\frac{\boldsymbol{\omega}}{\rho}\right) \cdot \boldsymbol{grad}\,\boldsymbol{v} \, . \tag{8.206}$$

In the two-dimensional case, $\boldsymbol{\omega}$ is orthogonal to $\boldsymbol{grad}\,\boldsymbol{v}$ and this relation reduces to

$$\frac{D}{Dt}\left(\frac{\omega}{\rho}\right) = 0 \, . \tag{8.207}$$

From equation (8.206), we deduce that, for a perfect barotropic or incompressible fluid, if the flow is irrotational at an instant, it remains so. In particular, an initially uniform flow will remain irrotational afterwards. This proposition, applied to isentropic flows of compressible, perfect fluids, is named Crocco's theorem [3].

In the case of a perfect fluid, equation (8.203) yields Kelvin's theorem [64]:

$$\frac{d\Gamma}{dt} = 0 \, . \tag{8.208}$$

The circulation of the velocity along a closed material line does not change, for an incompressible or barotropic perfect fluid (and particularly, in an isentropic flow).

## 8.14 Bernoulli's Equation

Bernoulli's equation is obtained from the Euler equation (8.36), for perfect fluids. Assume that the volume forces can be derived from a potential (8.193), then

$$\frac{D\boldsymbol{v}}{Dt} = -\frac{1}{\rho}\boldsymbol{\nabla}\, p - \boldsymbol{\nabla}\chi \,. \tag{8.209}$$

Using the vector identity

$$\boldsymbol{v}\cdot\boldsymbol{\nabla}\boldsymbol{v} = \boldsymbol{\omega}\times\boldsymbol{v} + \boldsymbol{\nabla}\left(\frac{\boldsymbol{v}\cdot\boldsymbol{v}}{2}\right) , \tag{8.210}$$

in the material derivative of the velocity, we obtain

$$\frac{\partial\boldsymbol{v}}{\partial t} = -\boldsymbol{\omega}\times\boldsymbol{v} - \frac{1}{\rho}\boldsymbol{\nabla}\, p - \boldsymbol{\nabla}\left(\frac{v^2}{2} + \chi\right) \,. \tag{8.211}$$

We also assume that the flow is irrotational, $\boldsymbol{\omega} = \mathbf{0}$. This assumption is strong, because real fluids produce rotational flows, such as those produced, for example, by the viscous effects near a wall. Thus, equation (8.211) can now be written as

$$\frac{\partial\boldsymbol{v}}{\partial t} = -\frac{1}{\rho}\boldsymbol{\nabla}\, p - \boldsymbol{\nabla}\left(\frac{v^2}{2} + \chi\right) \,. \tag{8.212}$$

Since the flow is irrotational, the velocity field can be derived from a potential, $\Phi$, such that

$$\boldsymbol{v} = \boldsymbol{\nabla}\Phi \,. \tag{8.213}$$

The Euler equation yields

$$\boldsymbol{\nabla}\left(\frac{\partial\Phi}{\partial t} + \frac{v^2}{2} + \chi\right) = -\frac{1}{\rho}\boldsymbol{\nabla}\, p. \tag{8.214}$$

As the left-hand side of (8.214) corresponds to the gradient of a scalar function, the same must be the case for the right-hand side. This is not possible unless the density $\rho$ is a function of $p$. This then amounts to requiring that the flow under consideration be that of a barotropic fluid according to relation (6.147). Consequently, equation (8.214) becomes

$$\boldsymbol{\nabla}\left(\frac{\partial\Phi}{\partial t} + \frac{v^2}{2} + \chi + \int\frac{dp}{\rho(p)}\right) = 0 \,. \tag{8.215}$$

We integrate this equation to obtain the general form of *Bernoulli's equation:*

$$\frac{\partial\Phi}{\partial t} + \int\frac{dp}{\rho(p)} + \frac{v^2}{2} + \chi = C(t) \,. \tag{8.216}$$

If the flow is stationary, then (8.216) yields the steady state form of Bernoulli's equation

$$\int\frac{dp}{\rho(p)} + \frac{v^2}{2} + \chi = \text{cnst}, \tag{8.217}$$

which, as is suggested by the second term, is an integral of the energy. Therefore, Bernoulli's equation is a first integral of the Euler equation for the case of a stationary, irrotational, perfect fluid.

If the flow is isentropic, the state relation (6.145) allows us to evaluate the pressure term in (8.217). We obtain

$$\int \frac{dp}{\rho(p)} = \frac{\gamma}{\gamma-1}\frac{p}{\rho} = \frac{a^2}{\gamma-1} , \tag{8.218}$$

where we have used (6.153) to deduce the last expression. With (8.218) and the definition $q^2 = v_i v_i$, the steady state Bernoulli's equation is written as

$$\frac{a^2}{\gamma-1} + \frac{q^2}{2} + \chi = \text{cnst} . \tag{8.219}$$

Assuming that the adiabatic flow of the ideal fluid is stationary and with no volume force, we have

$$\frac{a^2}{\gamma-1} + \frac{q^2}{2} = \text{cnst} . \tag{8.220}$$

Since $\gamma > 1$, we easily find from (6.145) and (6.153) that $a^2 = \gamma(p_0/\rho_0^\gamma)\,\rho^{\gamma-1} = \gamma C\,\rho^{\gamma-1}$. Thus, we obtain $a = 0$ when $\rho = 0$. Bernoulli's equation (8.220) becomes

$$\frac{a^2}{\gamma-1} + \frac{q^2}{2} = \frac{\gamma+1}{2(\gamma-1)}a_*^2 = \frac{q_{max}^2}{2} . \tag{8.221}$$

The two constants $a_*$ and $q_{max}$ denote the critical speed of sound and the maximum velocity on the streamline, respectively. It can easily be seen that if $q = a$, then $q = a = a_*$. This last equation defines the critical speed of sound. If $a = 0$, the velocity $q$ of the fluid is equal to the maximum velocity $q_{max}$. Equation (8.220) is especially used in aerodynamics.

If the flow is incompressible, then $\rho = \text{cnst}$ and Bernoulli's equation (8.216) yields

$$\frac{\partial \Phi}{\partial t} + \frac{p}{\rho} + \frac{v^2}{2} + \chi = C(t) . \tag{8.222}$$

For stationary flow of an incompressible, perfect fluid, Bernoulli's equation takes the well-known form

$$p + \frac{\rho v^2}{2} + \rho\chi = C , \tag{8.223}$$

where $C$ is a constant.

## 8.15 Acoustic Waves

Acoustic waves are generated by weak amplitude perturbations of the pressure or density and propagate at a certain velocity in a fluid flow. When the amplitudes are finite, a shock wave is produced.

We study a compressible, perfect fluid, in uniform isothermal flow, characterized by the variables $p_0, \rho_0$, and $\boldsymbol{v}_0$. The sound wave creates a perturbation $p', \rho', \boldsymbol{v}'$ such that the resulting flow is given by

$$p = p_0 + p', \quad \rho = \rho_0 + \rho', \quad \boldsymbol{v} = \boldsymbol{v}_0 + \boldsymbol{v}' \; . \tag{8.224}$$

As we assume that the perturbation is infinitesimal, we have the following estimations of magnitude:

$$\frac{|p'|}{p_0}, \quad \frac{|\rho'|}{\rho_0}, \quad \frac{\|\boldsymbol{v}'\|}{\sqrt{\frac{p_0}{\rho_0}}} \quad \ll 1 \; . \tag{8.225}$$

The equations for conservation of mass and momentum are

$$\frac{\partial \rho}{\partial t} + v_j \frac{\partial \rho}{\partial x_j} + \rho \frac{\partial v_i}{\partial x_i} = 0 \; , \tag{8.226}$$

$$\frac{\partial v_i}{\partial t} + v_j \frac{\partial v_i}{\partial x_j} + \frac{1}{\rho} \frac{\partial p}{\partial x_i} = 0 \; . \tag{8.227}$$

Inserting relations (8.224) in (8.226) and (8.227), we then linearize the equations, only retaining the first-order terms, and we assume that the underlying unperturbed flow satisfies the mass and momentum conversation equations. Then, it follows that

$$\frac{\partial \rho'}{\partial t} + v_{0j} \frac{\partial \rho'}{\partial x_j} + \rho_0 \frac{\partial v'_i}{\partial x_i} = 0 \; , \tag{8.228}$$

$$\frac{\partial v_{0i}}{\partial t} + v_{0j} \frac{\partial v'_i}{\partial x_j} + \frac{1}{\rho_0} \frac{\partial p'}{\partial x_i} = 0 \; . \tag{8.229}$$

Assuming, in addition, that the flow is globally isentropic ($ds = 0$), from (6.142), it follows for an ideal gas

$$\frac{dp'}{p_0 + p'} = \gamma \frac{d\rho'}{\rho_0 + \rho'} \; ,$$

or from (6.153) and (8.225)

$$\boldsymbol{\nabla} p' = \gamma \frac{p_0}{\rho_0} \boldsymbol{\nabla} \rho' \; . \tag{8.230}$$

In order to simplify the expression, we define the material derivative of the unperturbed flow

$$\frac{D_0(\bullet)}{Dt} = \frac{\partial(\bullet)}{\partial t} + v_{0j} \frac{\partial(\bullet)}{\partial x_j} \; . \tag{8.231}$$

The dynamic equations of the perturbation become

$$\frac{D_0\rho'}{Dt} + \rho_0 \operatorname{div} \boldsymbol{v}' = 0 \; , \tag{8.232}$$

$$\frac{D_0\boldsymbol{v}'}{Dt} + \frac{1}{\rho_0} \boldsymbol{\nabla} p' = 0 \; . \tag{8.233}$$

Inserting (8.230) in (8.232), we obtain

$$\frac{D_0\boldsymbol{v}'}{Dt} + \gamma\frac{p_0}{\rho_0^2}\boldsymbol{\nabla}\rho' = 0 \; . \tag{8.234}$$

Applying the divergence operator to (8.234) and using (1.189) leads to the relation

$$\rho_0 \operatorname{div}\frac{D_0\boldsymbol{v}'}{Dt} + \gamma\frac{p_0}{\rho_0}\Delta\rho' = 0 \; . \tag{8.235}$$

The material derivative of (8.232) yields

$$\frac{D_0^2\rho'}{Dt^2} + \rho_0 \operatorname{div}\frac{D_0\boldsymbol{v}'}{Dt} = 0 \; . \tag{8.236}$$

Combining (8.235) and (8.236), we produce the wave equation

$$\frac{D_0^2\rho'}{Dt^2} = \gamma\frac{p_0}{\rho_0}\Delta\rho' \; . \tag{8.237}$$

The perturbation propagates with respect to the uniform flow at a velocity given by

$$\sqrt{\gamma\frac{p_0}{\rho_0}} \; , \tag{8.238}$$

called the ***speed of sound***. In the general case of a non-uniform flow, we obtain the speed of sound

$$a = \sqrt{\gamma\frac{p}{\rho}} \; , \tag{8.239}$$

which is none other than expression (6.153). Note that with (8.230), we can write (8.237) in the form

$$\frac{D_0^2 p'}{Dt^2} = \gamma\frac{p_0}{\rho_0}\Delta p' \; . \tag{8.240}$$

Similarly, we have

$$\frac{D_0^2 v_i'}{Dt^2} = \gamma\frac{p_0}{\rho_0}\Delta v_i' \; . \tag{8.241}$$

Thus all the variables of the problem satisfy the wave equation.

Since the wave equation is linear with constant coefficients, we can consider that a Fourier harmonic is a solution, then

$$p'(\boldsymbol{k}, t) = \sum_{\boldsymbol{k}} \hat{p}_{\boldsymbol{k}}\, e^{i(\boldsymbol{k}\cdot\boldsymbol{x} - \omega t)} \; , \tag{8.242}$$

where we have the wave vector $\boldsymbol{k}$, the time frequency $\omega$, and the complex amplitude $\hat{p}_{\boldsymbol{k}}$ for the modes. The dispersion relation is obtained by introducing (8.242) in (8.240):

$$\omega^2 = a^2 k^2 \; . \tag{8.243}$$

The phase velocity of the wave is obtained by $|\omega|/|k| = a$ which shows that the speed of sound does not depend on the wave number; acoustic waves are not dispersive. In the case of air, with $T_0 = 288K$, we have $a = \sqrt{\gamma R T_0} = 340\,ms^{-1}$. For an acoustic wave with frequency $\omega/(2\pi) = 1000$ Hz, the wavelength, $\lambda = 2\pi/(\omega/a)$, is 0.34 m.

## 8.16 Stationary, Irrotational, Isentropic Flow of a Compressible Perfect Fluid

Consider the steady flow of a compressible perfect fluid for which we neglect the volume forces. We assume that the flow is adiabatic and thermodynamically reversible. This means that we exclude the presence of shocks. The flow is thus isentropic; the fluid is barotropic and its model corresponds to equation (6.147). Then, we assume that the flow is irrotational since this property is conserved in time for barotropic perfect fluids. With these hypotheses, the conservation equations simplify

$$\frac{\partial}{\partial x_i}(\rho v_i) = 0 , \tag{8.244}$$

$$v_j \frac{\partial v_i}{\partial x_j} = -\frac{1}{\rho}\frac{\partial p}{\partial x_i} . \tag{8.245}$$

For an isentropic flow, with (6.152), it follows that

$$\frac{\partial p}{\partial x_i} = \frac{\partial p}{\partial \rho}\frac{\partial \rho}{\partial x_i} = a^2 \frac{\partial \rho}{\partial x_i} . \tag{8.246}$$

Equation (8.245) becomes

$$v_j \frac{\partial v_i}{\partial x_j} + \frac{a^2}{\rho}\frac{\partial \rho}{\partial x_i} = 0 . \tag{8.247}$$

Multiplying (8.247) by $v_i$ and (8.244) by $a^2/\rho$ and then combining the two relations so obtained, we have

$$\frac{v_i v_j}{a^2}\frac{\partial v_i}{\partial x_j} = \frac{\partial v_i}{\partial x_i} . \tag{8.248}$$

Note that the speed of sound $a$ in (8.248) is a function of position. It is calculated from the fluid velocity with the ideal gas energy equation (8.221).

The irrotationality of the flow allows the introduction of a potential for the velocities (8.213). Inserting it in (8.248), we have

$$\frac{1}{a^2}\frac{\partial \Phi}{\partial x_i}\frac{\partial \Phi}{\partial x_j}\frac{\partial^2 \Phi}{\partial x_i \partial x_j} = \Delta \Phi . \tag{8.249}$$

Bernoulli's equation (8.221) yields

$$\frac{1}{2}\left(\frac{\partial \Phi}{\partial x_i}\right)^2 + \frac{a^2}{\gamma - 1} = \text{cnst} . \tag{8.250}$$

Since relations (8.249) and (8.250) are non-linear, we must make simplifications in order to find an analytic solution.

### 8.16.1 Small Perturbation Theory

We assume a uniform, parallel flow at velocity $U$ in direction $x_1$, perturbed by a thin obstacle of length $L$ as is shown in figure 8.25 or by walls slightly inclined with respect to the horizontal.

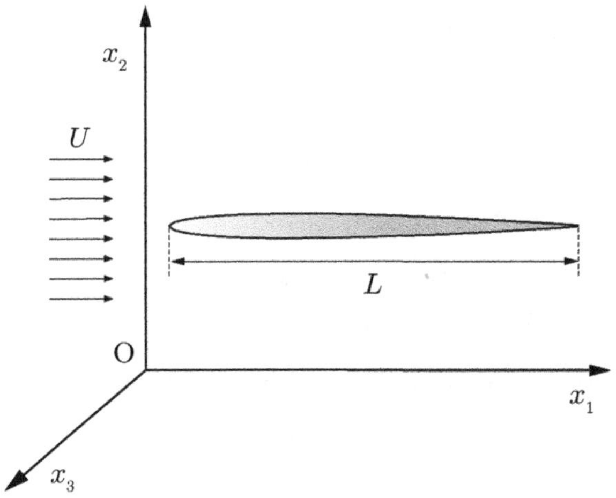


**Fig. 8.25** Thin obstacle in a uniform compressible flow

These geometries generate perturbations such that

$$v_i = U\boldsymbol{e}_1 + v_i' \,, \tag{8.251}$$

with the inequalities

$$\left|\frac{v_i'}{U}\right| \ll 1, \quad i = 1, 2, 3 \,. \tag{8.252}$$

The velocity field depends on the potential $\Phi(x_i)$ which we write in the form

$$\Phi(x_i) = Ux_1 + \varphi(x_i) \,, \tag{8.253}$$

where $\varphi$ denotes the perturbation potential.

We linearize equations (8.248) and (8.221), by neglecting all terms of order two or higher in the perturbation and retaining only those of first order. It follows for (8.248)

$$\frac{\partial v_i}{\partial x_i} = \frac{U^2}{a^2}\frac{\partial v_1}{\partial x_1} = \frac{U^2}{a^2}\frac{\partial v_1'}{\partial x_1} \,. \tag{8.254}$$

Note that (8.254) is not linear, as the local speed of sound $a$ depends on the perturbations $v_i'$.

Bernoulli's equation (8.221) can be written

$$\frac{q^2}{2} + \frac{a^2}{\gamma - 1} = \frac{U^2}{2} + \frac{a_\infty^2}{\gamma - 1} \,, \tag{8.255}$$

where $a_\infty$ is the speed of sound upstream at infinity. The linearization leads us to write

$$\begin{aligned} a^2 &= a_\infty^2 + \frac{\gamma-1}{2}\,(U^2 - q^2) \\ &= a_\infty^2 + \frac{\gamma-1}{2}\,\left[U^2 - (U+v_1')^2 - v_2'^2 - v_3'^2\right] \\ &= a_\infty^2 - (\gamma-1)Uv_1' + \dots \end{aligned} \tag{8.256}$$

Taking into account inequality (8.252), relation (8.256) is approximated by

$$a^2 \approx a_\infty^2 \; . \tag{8.257}$$

Combining relations (8.254)–(8.257), we obtain

$$\left(1 - \frac{U^2}{a_\infty^2}\right)\frac{\partial^2\varphi}{\partial x_1^2} + \frac{\partial^2\varphi}{\partial x_2^2} + \frac{\partial^2\varphi}{\partial x_3^2} = 0 \; . \tag{8.258}$$

With the definition of the Mach number (8.3), relation (8.258) becomes

$$(1 - M_\infty^2)\,\frac{\partial^2\varphi}{\partial x_1^2} + \frac{\partial^2\varphi}{\partial x_2^2} + \frac{\partial^2\varphi}{\partial x_3^2} = 0 \; . \tag{8.259}$$

When the Mach number is close to one, equation (8.259) degenerates and is no longer valid. We therefore cannot use it for the sonic (or transonic) case. Mathematically, equation (8.259) is elliptic for the subsonic case $M_\infty < 1$, and hyperbolic in the supersonic case $M_\infty > 1$. This difference in the mathematical nature also results in different physical behaviors.

The potential equation (8.259) is subject to boundary conditions that impose that the velocity vector $\boldsymbol{v}$ be parallel to the walls in their neighborhood. It is also necessary to verify that the perturbation method truly satisfies (8.252).

If the flow is delimited by the walls $F(x_1, x_2, x_3) = 0$, near them we have

$$\frac{v_2'}{U} \approx \frac{v_2'}{U+v_1'} = \left.\frac{dx_2}{dx_1}\right|_{x_3} = -\frac{(\partial F/\partial x_1)}{(\partial F/\partial x_2)} \tag{8.260}$$

$$\frac{v_3'}{U} \approx \frac{v_3'}{U+v_1'} = -\frac{(\partial F/\partial x_1)}{(\partial F/\partial x_3)} \; . \tag{8.261}$$

Thus the conditions

$$|\partial F/\partial x_1| \ll |\partial F/\partial x_2| \quad \text{and} \quad |\partial F/\partial x_1| \ll |\partial F/\partial x_3| \tag{8.262}$$

must be respected.

### 8.16.2 Two-Dimensional Flow of a Compressible Fluid in the Neighborhood of a Sinusoidal Wavy Wall

Consider the two-dimensional, stationary flow of a compressible perfect fluid in the neighborhood of a wavy wall of the form $x_2 = f(x_1) = \varepsilon \sin(\alpha x_1)$, for which

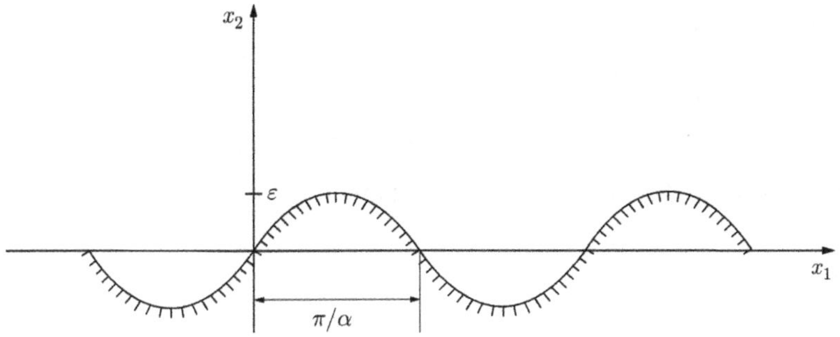


**Fig. 8.26** Wavy wall

the amplitude $\varepsilon$ is small compared to the wavelength $\lambda = \frac{2\pi}{\alpha}$, i.e., $\varepsilon\alpha \ll 2\pi$ as shown in figure 8.26.

We solve

$$\left(1 - M_\infty^2\right) \frac{\partial^2 \varphi}{\partial x_1^2} + \frac{\partial^2 \varphi}{\partial x_2^2} = 0 \tag{8.263}$$

with the condition at the wall $F(x_1, x_2) = x_2 - f(x_1) = 0$ (eqn. (8.260),

$$\frac{\partial f}{\partial x_1} = \frac{v_2'}{U + v_1'} \approx \left.\frac{v_2'}{U}\right|_{x_2 = f(x_1)} . \tag{8.264}$$

We can express the last term of (8.264) as

$$\frac{v_2'}{U}(x_1, \varepsilon \sin \alpha x_1) = \frac{1}{U}\left[v_2'(x_1, 0) + \left.\left(\frac{\partial v_2'}{\partial x_2}\right)\right|_{x_2=0} \varepsilon \sin \alpha x_1 + \ldots\right] . \tag{8.265}$$

The condition at the wall becomes, with the same level of approximation,

$$\frac{v_2'(x_1, 0)}{U} = \frac{\partial f}{\partial x_1} = \varepsilon\alpha \cos \alpha x_1 . \tag{8.266}$$

This amounts to imposing the condition on the average plane $x_2 = 0$; and thus the needed assumption $\varepsilon\alpha \ll 1 < 2\pi$. As $x_2 \to \infty$, for the subsonic case, we impose $v_1' = v_2' = 0$.

**Subsonic Flow**

We use the method of separation of variables, by setting in equation (8.263)

$$\varphi(x_1, x_2) = \varphi_1(x_1)\varphi_2(x_2) \tag{8.267}$$

which yields, with $n^2 = 1 - M_\infty^2$,

$$n^2 \frac{\varphi_1''}{\varphi_1} + \frac{\varphi_2''}{\varphi_2} = 0 , \tag{8.268}$$

where $\varphi_1''$ and $\varphi_2''$ are second-order derivatives. The boundary condition (8.266) becomes

$$\frac{v_2'(x_1, 0)}{U} = \frac{1}{U}\frac{\partial \varphi}{\partial x_2}(x_1, 0) = \varepsilon\alpha \cos \alpha x_1 , \tag{8.269}$$

from which we have

$$\varphi_1(x_1)\varphi_2'(0) = U\varepsilon\alpha\cos\alpha x_1 \ . \tag{8.270}$$

We deduce that

$$\varphi_1(x_1) = A\cos\alpha x_1 \ . \tag{8.271}$$

Now, putting (8.271) in (8.268), leads to

$$\frac{\varphi_2''}{\varphi_2} = n^2\alpha^2 \ . \tag{8.272}$$

Integrating (8.272), we obtain

$$\varphi_2(x_2) = Be^{-n\alpha x_2} + Ce^{n\alpha x_2} \ . \tag{8.273}$$

The condition for zero velocity at $x_2 \to \infty$ imposes $C = 0$, if we suppose that $n > 0$. With (8.270), we then have

$$\varphi_1(x_1) = -\frac{\varepsilon U}{Bn}\cos\alpha x_1 \tag{8.274}$$

and finally,

$$\varphi(x_1, x_2) = -\frac{\varepsilon U}{n}(\cos\alpha x_1)e^{-n\alpha x_2} \ . \tag{8.275}$$

We can easily verify that the condition $|v_1'/U| \ll 1$ implies

$$\frac{\varepsilon\alpha}{n} \ll 1 \ . \tag{8.276}$$

This is the case when $n \neq 0$, that is, when we are not too close to the transonic domain.

**Supersonic Flow**

Set $M_\infty^2 - 1 = m^2$. Equation (8.263) becomes

$$\frac{\partial^2\varphi}{\partial x_2^2} - \frac{1}{m^2}\frac{\partial^2\varphi}{\partial x_1^2} = 0 \ . \tag{8.277}$$

According to d'Alembert (sec. 7.5.3), the general solution of this wave equation is

$$\varphi = f(x_1 - mx_2) + g(x_1 + mx_2) \ . \tag{8.278}$$

The second term $g$ is neglected for the case where the flow is situated in the plane above the wall in the positive direction of the axis $x_1$. The boundary condition (8.269) yields

$$\frac{\partial\varphi}{\partial x_2}(x_1, 0) = U\varepsilon\alpha\cos\alpha x_1 = -mf'(x_1) \ . \tag{8.279}$$

Integration leads to

$$\varphi = f(x_1 - mx_2) = -\frac{U\varepsilon}{m}\sin\alpha(x_1 - mx_2) \ . \tag{8.280}$$

The small perturbation condition implies that $\varepsilon\alpha/m \ll 1$. Note that the perturbation potential retains the same values on lines inclined downstream with angular coefficient $dx_2/dx_1 = 1/m = \tan\alpha$, where the angle $\alpha$ is the Mach angle (8.4). The perturbation thus propagates downstream to infinity, whereas in the subsonic case, its amplitude diminishes rapidly as it moves away from the wall.

## 8.17 Exercises

**8.1** Starting from the conservation of momentum equation, (3.96), derive the Navier-Stokes for a Newtonian viscous fluid. What happens to these equations when the coefficients $\lambda$ and $\mu$ are constant? Look in particular at the incompressible fluid case (isochoric motion).

**8.2** Consider two-dimensional Couette-Poiseuille flow obtained by superimposing Couette flow induced by the constant velocity motion $U$ of the upper wall and Poiseuille flow resulting from a pressure gradient in direction $x_1$. Calculate the velocity profile, the shear stress, and the flow rate.

**8.3** Consider the helical flow of an incompressible viscous fluid between two rotating circular cylinders as in a Couette flow. The interior cylinder of radius $R_1$ and the exterior cylinder of radius $R_2$ have rotational velocities of $\omega_1$ and $\omega_2$, respectively. The fluid between the two cylinders is also subject to an axial pressure gradient. Calculate the non-zero velocity components $v_\theta$ and $v_z$. Calculate also the pressure field $p = p(r, z)$.

**8.4** A solid sphere of radius $R$ is immersed in an incompressible Newtonian viscous fluid that fills the space and is at rest at infinity. The sphere rotates about its diameter at a constant angular velocity $\Omega$. Assume that the Reynolds number is less than one and neglect the volume forces. The streamlines are circles centered on the rotation axis in planes perpendicular to this axis. Working in spherical coordinates, calculate the velocity profile.

**8.5** With the same hypotheses as in the preceding exercise, examine the flow of a fluid between two spheres of radii $R_1$ and $R_2$ such that $R_1 < R_2$, which rotate at the angular velocities $\Omega_1$ and $\Omega_2$ about a common, fixed axis. Calculate the velocity profile.

**8.6** By applying Bernoulli's theorem for perfect fluids (8.223), show that the velocity of a jet exiting an orifice in a wall at a distance $h$ from the free surface of the fluid is

$$v = \sqrt{2gh}\ . \tag{8.281}$$

**8.7** Consider the flow of a viscous fluid in a tube with an arbitrary section. With the hypothesis that the velocity field is of the form

$$v_1 = v_1(x_2, x_3), \quad v_2 = v_3 = 0 \ ,$$

show that the velocity field satisfies the equation

$$\frac{\partial^2 v_1}{\partial x_2^2} + \frac{\partial^2 v_1}{\partial x_3^2} = \frac{1}{\mu}\frac{dp}{dx_1} = C = \text{cnst} \ .$$

If the tube section is an ellipse with semi-axes $a$ and $b$ such that

$$\left(\frac{x_2}{a}\right)^2 + \left(\frac{x_3}{b}\right)^2 = 1 \ ,$$

the velocity field is written

$$v_1 = A\left[\left(\frac{x_2}{a}\right)^2 + \left(\frac{x_3}{b}\right)^2\right] + B \ .$$

Find the constants $A$ and $B$.

**8.8** A cylinder of radius $R_1$ moves parallel to its axis with a constant velocity $U$ inside a fixed, coaxial cylinder of radius $R_2$.

Calculate the velocity field of a viscous fluid which fills the space between the two cylinders. Find the friction force per unit length that acts on the moving cylinder.

Appendix A

# Cylindrical Coordinates

We list here some differential operators as well as the principal equations for a system of cylindrical coordinates. Shown in figure A.1 are the components of the stress tensor in the cylindrical coordinate system $(r, \theta, z)$.

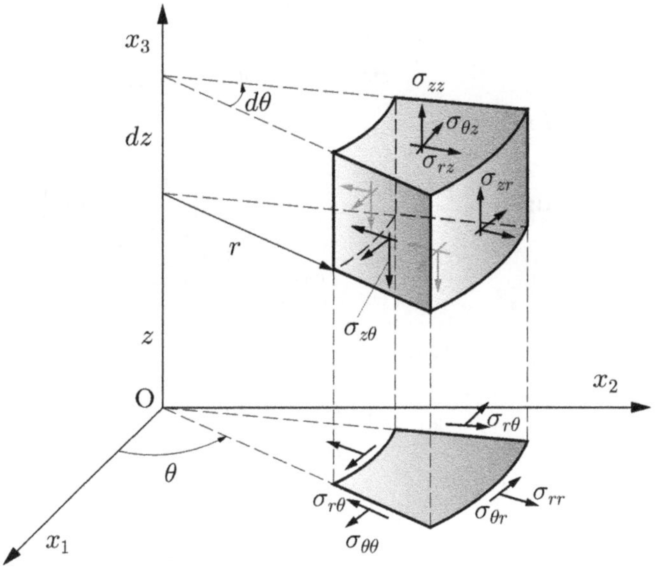


**Fig. A.1** Stress tensor components in a cylindrical coordinate system

*Divergence of a vector field* $\boldsymbol{v}(r, \theta, z)$*:*

$$\operatorname{div} \boldsymbol{v} = \frac{1}{r}\, v_r + \frac{\partial v_r}{\partial r} + \frac{1}{r}\frac{\partial v_\theta}{\partial \theta} + \frac{\partial v_z}{\partial z} \tag{A.1}$$

or

$$\operatorname{div} \boldsymbol{v} = \frac{1}{r}\frac{\partial}{\partial r}(r v_r) + \frac{1}{r}\frac{\partial v_\theta}{\partial \theta} + \frac{\partial v_z}{\partial z}\,. \tag{A.2}$$

*Divergence of a tensor field $\boldsymbol{\sigma}(r,\theta,z)$:*

$$\begin{aligned}\mathbf{div}\,\boldsymbol{\sigma} = &\left(\frac{\partial\sigma_{rr}}{\partial r}+\frac{1}{r}\frac{\partial\sigma_{r\theta}}{\partial\theta}+\frac{\partial\sigma_{rz}}{\partial z}+\frac{\sigma_{rr}-\sigma_{\theta\theta}}{r}\right)\boldsymbol{e}_r\\ &+\left(\frac{\partial\sigma_{r\theta}}{\partial r}+\frac{1}{r}\frac{\partial\sigma_{\theta\theta}}{\partial\theta}+\frac{\partial\sigma_{\theta z}}{\partial z}+\frac{2\sigma_{r\theta}}{r}\right)\boldsymbol{e}_\theta\\ &+\left(\frac{\partial\sigma_{rz}}{\partial r}+\frac{1}{r}\frac{\partial\sigma_{\theta z}}{\partial\theta}+\frac{\partial\sigma_{zz}}{\partial z}+\frac{\sigma_{rz}}{r}\right)\boldsymbol{e}_z\,.\end{aligned} \tag{A.3}$$

*Gradient of a scalar field $f(r,\theta,z)$:*

$$\boldsymbol{\nabla} f = \frac{\partial f}{\partial r}\boldsymbol{e}_r + \frac{1}{r}\frac{\partial f}{\partial\theta}\boldsymbol{e}_\theta + \frac{\partial f}{\partial z}\boldsymbol{e}_z\,. \tag{A.4}$$

*Gradient of a vector field $\boldsymbol{v}(r,\theta,z)$:*

$$\boldsymbol{\nabla}\boldsymbol{v} = \begin{pmatrix} \dfrac{\partial v_r}{\partial r} & \dfrac{1}{r}\dfrac{\partial v_r}{\partial\theta}-\dfrac{v_\theta}{r} & \dfrac{\partial v_r}{\partial z}\\ \dfrac{\partial v_\theta}{\partial r} & \dfrac{1}{r}\dfrac{\partial v_\theta}{\partial\theta}+\dfrac{v_r}{r} & \dfrac{\partial v_\theta}{\partial z}\\ \dfrac{\partial v_z}{\partial r} & \dfrac{1}{r}\dfrac{\partial v_z}{\partial\theta} & \dfrac{\partial v_z}{\partial z}\end{pmatrix}. \tag{A.5}$$

*Curl of a vector field $\boldsymbol{v}(r,\theta,z)$:*

$$\begin{aligned}\mathbf{curl}\,\boldsymbol{v} = &\left(\frac{1}{r}\frac{\partial v_z}{\partial\theta}-\frac{\partial v_\theta}{\partial z}\right)\boldsymbol{e}_r + \left(\frac{\partial v_r}{\partial z}-\frac{\partial v_z}{\partial r}\right)\boldsymbol{e}_\theta\\ &+\frac{1}{r}\left(\frac{\partial}{\partial r}(rv_\theta)-\frac{\partial v_r}{\partial\theta}\right)\boldsymbol{e}_z\,.\end{aligned} \tag{A.6}$$

*Laplacian of a scalar field $f(r,\theta,z)$:*

$$\Delta f = \nabla^2 f = \frac{1}{r}\frac{\partial f}{\partial r}+\frac{\partial^2 f}{\partial r^2}+\frac{1}{r^2}\frac{\partial^2 f}{\partial\theta^2}+\frac{\partial^2 f}{\partial z^2} \tag{A.7}$$

or

$$\nabla^2 f = \frac{1}{r}\frac{\partial}{\partial r}\left(r\frac{\partial f}{\partial r}\right)+\frac{1}{r^2}\frac{\partial^2 f}{\partial\theta^2}+\frac{\partial^2 f}{\partial z^2}\,. \tag{A.8}$$

*Laplacian of a vector field $\boldsymbol{v}(r,\theta,z)$:*

$$\begin{aligned}\boldsymbol{\nabla}^2\boldsymbol{v} = &\left(\nabla^2 v_r-\frac{2}{r^2}\frac{\partial v_\theta}{\partial\theta}-\frac{v_r}{r^2}\right)\boldsymbol{e}_r\\ &+\left(\nabla^2 v_\theta+\frac{2}{r^2}\frac{\partial v_r}{\partial\theta}-\frac{v_\theta}{r^2}\right)\boldsymbol{e}_\theta\\ &+(\nabla^2 v_z)\boldsymbol{e}_z\,.\end{aligned} \tag{A.9}$$

*Material derivative of a scalar field* $f(r,\theta,z)$:

$$\frac{Df}{Dt} = \frac{\partial f}{\partial t} + v_r \frac{\partial f}{\partial r} + \frac{v_\theta}{r}\frac{\partial f}{\partial \theta} + v_z \frac{\partial f}{\partial z}\,. \tag{A.10}$$

*Acceleration:*

$$\frac{Dv_r}{Dt} = \frac{\partial v_r}{\partial t} + v_r \frac{\partial v_r}{\partial r} + \frac{v_\theta}{r}\frac{\partial v_r}{\partial \theta} - \frac{v_\theta^2}{r} + v_z \frac{\partial v_r}{\partial z} \tag{A.11}$$

$$\frac{Dv_\theta}{Dt} = \frac{\partial v_\theta}{\partial t} + v_r \frac{\partial v_\theta}{\partial r} + \frac{v_\theta}{r}\frac{\partial v_\theta}{\partial \theta} + \frac{v_r v_\theta}{r} + v_z \frac{\partial v_\theta}{\partial z} \tag{A.12}$$

$$\frac{Dv_z}{Dt} = \frac{\partial v_z}{\partial t} + v_r \frac{\partial v_z}{\partial r} + \frac{v_\theta}{r}\frac{\partial v_z}{\partial \theta} + v_z \frac{\partial v_z}{\partial z}\,. \tag{A.13}$$

*Conservation of mass equation:*

$$\frac{\partial \rho}{\partial t} + \frac{\partial(\rho v_r)}{\partial r} + \frac{\rho v_r}{r} + \frac{1}{r}\frac{\partial(\rho v_\theta)}{\partial \theta} + \frac{\partial(\rho v_z)}{\partial z} = 0\,. \tag{A.14}$$

*Motion equations:*

$$\rho\left(\frac{\partial v_r}{\partial t} + v_r \frac{\partial v_r}{\partial r} + \frac{v_\theta}{r}\frac{\partial v_r}{\partial \theta} + v_z \frac{\partial v_r}{\partial z} - \frac{v_\theta^2}{r}\right) = \frac{\partial \sigma_{rr}}{\partial r} + \frac{1}{r}\frac{\partial \sigma_{r\theta}}{\partial \theta} + \frac{\partial \sigma_{rz}}{\partial z} + \frac{\sigma_{rr}-\sigma_{\theta\theta}}{r} + \rho b_r \tag{A.15}$$

$$\rho\left(\frac{\partial v_\theta}{\partial t} + v_r \frac{\partial v_\theta}{\partial r} + \frac{v_\theta}{r}\frac{\partial v_\theta}{\partial \theta} + v_z \frac{\partial v_\theta}{\partial z} + \frac{v_r v_\theta}{r}\right) = \frac{\partial \sigma_{r\theta}}{\partial r} + \frac{1}{r}\frac{\partial \sigma_{\theta\theta}}{\partial \theta} + \frac{\partial \sigma_{\theta z}}{\partial z} + \frac{2\sigma_{r\theta}}{r} + \rho b_\theta \tag{A.16}$$

$$\rho\left(\frac{\partial v_z}{\partial t} + v_r \frac{\partial v_z}{\partial r} + \frac{v_\theta}{r}\frac{\partial v_z}{\partial \theta} + v_z \frac{\partial v_z}{\partial z}\right) = \frac{\partial \sigma_{rz}}{\partial r} + \frac{1}{r}\frac{\partial \sigma_{\theta z}}{\partial \theta} + \frac{\partial \sigma_{zz}}{\partial z} + \frac{\sigma_{rz}}{r} + \rho b_z\,. \tag{A.17}$$

*Equilibrium equations:*

$$\frac{\partial \sigma_{rr}}{\partial r} + \frac{1}{r}\frac{\partial \sigma_{r\theta}}{\partial \theta} + \frac{\partial \sigma_{rz}}{\partial z} + \frac{\sigma_{rr}-\sigma_{\theta\theta}}{r} + \rho b_r = 0 \tag{A.18}$$

$$\frac{\partial \sigma_{r\theta}}{\partial r} + \frac{1}{r}\frac{\partial \sigma_{\theta\theta}}{\partial \theta} + \frac{\partial \sigma_{\theta z}}{\partial z} + \frac{2\sigma_{r\theta}}{r} + \rho b_\theta = 0 \tag{A.19}$$

$$\frac{\partial \sigma_{rz}}{\partial r} + \frac{1}{r}\frac{\partial \sigma_{\theta z}}{\partial \theta} + \frac{\partial \sigma_{zz}}{\partial z} + \frac{\sigma_{rz}}{r} + \rho b_z = 0\,. \tag{A.20}$$

*Navier's equations:*

$$\mu\left(\nabla^2 u_r - \frac{u_r}{r^2} - \frac{2}{r^2}\frac{\partial u_\theta}{\partial \theta}\right)$$
$$+ (\lambda+\mu)\frac{\partial}{\partial r}\left(\frac{1}{r}\frac{\partial(ru_r)}{\partial r} + \frac{1}{r}\frac{\partial u_\theta}{\partial \theta} + \frac{\partial u_z}{\partial z}\right) + f_r = 0 \qquad \text{(A.21)}$$

$$\mu\left(\nabla^2 u_\theta - \frac{u_\theta}{r^2} - \frac{2}{r^2}\frac{\partial u_r}{\partial \theta}\right)$$
$$+ (\lambda+\mu)\frac{1}{r}\frac{\partial}{\partial \theta}\left(\frac{1}{r}\frac{\partial(ru_r)}{\partial r} + \frac{1}{r}\frac{\partial u_\theta}{\partial \theta} + \frac{\partial u_z}{\partial z}\right) + f_\theta = 0 \qquad \text{(A.22)}$$

$$\mu\nabla^2 u_z + (\lambda+\mu)\frac{\partial}{\partial z}\left(\frac{1}{r}\frac{\partial(ru_r)}{\partial r} + \frac{1}{r}\frac{\partial u_\theta}{\partial \theta} + \frac{\partial u_z}{\partial z}\right) + f_z = 0\,. \qquad \text{(A.23)}$$

*Strain components as functions of displacement:*

$$\varepsilon_{rr} = \frac{\partial u_r}{\partial r} \qquad \varepsilon_{\theta\theta} = \frac{u_r}{r} + \frac{1}{r}\frac{\partial u_\theta}{\partial \theta} \qquad \varepsilon_{zz} = \frac{\partial u_z}{\partial z} \qquad \text{(A.24)}$$

$$\varepsilon_{zr} = \frac{1}{2}\left(\frac{\partial u_r}{\partial z} + \frac{\partial u_z}{\partial r}\right) \qquad \varepsilon_{r\theta} = \frac{1}{2}\left(\frac{1}{r}\frac{\partial u_r}{\partial \theta} + \frac{\partial u_\theta}{\partial r} - \frac{u_\theta}{r}\right) \qquad \text{(A.25)}$$

$$\varepsilon_{z\theta} = \frac{1}{2}\left(\frac{1}{r}\frac{\partial u_z}{\partial \theta} + \frac{\partial u_\theta}{\partial z}\right). \qquad \text{(A.26)}$$

Note that the constitutive equations are obtained by replacing the Cartesian components in (7.3) and (7.4) with the components above.

*Biharmonic equation:*

$$\nabla^4\Phi = \left(\frac{1}{r}\frac{\partial}{\partial r} + \frac{\partial^2}{\partial r^2} + \frac{1}{r^2}\frac{\partial^2}{\partial \theta^2}\right)\left(\frac{1}{r}\frac{\partial \Phi}{\partial r} + \frac{\partial^2\Phi}{\partial r^2} + \frac{1}{r^2}\frac{\partial^2\Phi}{\partial \theta^2}\right) = 0\,. \qquad \text{(A.27)}$$

*Stress components:*

Via the Airy stress function for a plane stress state

$$\sigma_{rr} = \frac{1}{r}\frac{\partial \Phi}{\partial r} + \frac{1}{r^2}\frac{\partial^2\Phi}{\partial \theta^2} \qquad \text{(A.28)}$$

$$\sigma_{\theta\theta} = \frac{\partial^2\Phi}{\partial r^2} \qquad \text{(A.29)}$$

$$\sigma_{r\theta} = \frac{1}{r^2}\frac{\partial \Phi}{\partial \theta} - \frac{1}{r}\frac{\partial^2\Phi}{\partial r\partial \theta}\,. \qquad \text{(A.30)}$$

*Incompressible Navier-Stokes equations:*

Conservation of mass equation

$$\frac{\partial v_r}{\partial r} + \frac{v_r}{r} + \frac{1}{r}\frac{\partial v_\theta}{\partial \theta} + \frac{\partial v_z}{\partial z} = 0 \ . \tag{A.31}$$

Motion equations

$$\rho(\frac{\partial v_r}{\partial t} + v_r\frac{\partial v_r}{\partial r} + \frac{v_\theta}{r}\frac{\partial v_r}{\partial \theta} + v_z\frac{\partial v_r}{\partial z} - \frac{v_\theta^2}{r}) = -\frac{\partial p}{\partial r} + \mu(\triangle v_r - \frac{v_r}{r^2} - \frac{2}{r^2}\frac{\partial v_\theta}{\partial \theta}) + \rho b_r \tag{A.32}$$

$$\rho(\frac{\partial v_\theta}{\partial t} + v_r\frac{\partial v_\theta}{\partial r} + \frac{v_\theta}{r}\frac{\partial v_\theta}{\partial \theta} + v_z\frac{\partial v_\theta}{\partial z} + \frac{v_r v_\theta}{r}) = -\frac{1}{r}\frac{\partial p}{\partial \theta} + \mu(\triangle v_\theta - \frac{v_\theta}{r^2} + \frac{2}{r^2}\frac{\partial v_r}{\partial \theta}) + \rho b_\theta \tag{A.33}$$

$$\rho(\frac{\partial v_z}{\partial t} + v_r\frac{\partial v_z}{\partial r} + \frac{v_\theta}{r}\frac{\partial v_z}{\partial \theta} + v_z\frac{\partial v_z}{\partial z}) = -\frac{\partial p}{\partial z} + \mu\Delta v_z + \rho b_z \tag{A.34}$$

with the Lapacian operator defined by (A.7) or (A.8).

Appendix B

# Spherical Coordinates

We list here some differential operators as well as the principal equations for a system of spherical coordinates. Shown in figure B.1 are the components of the stress tensor in the spherical coordinate system $r, \theta, \varphi$.

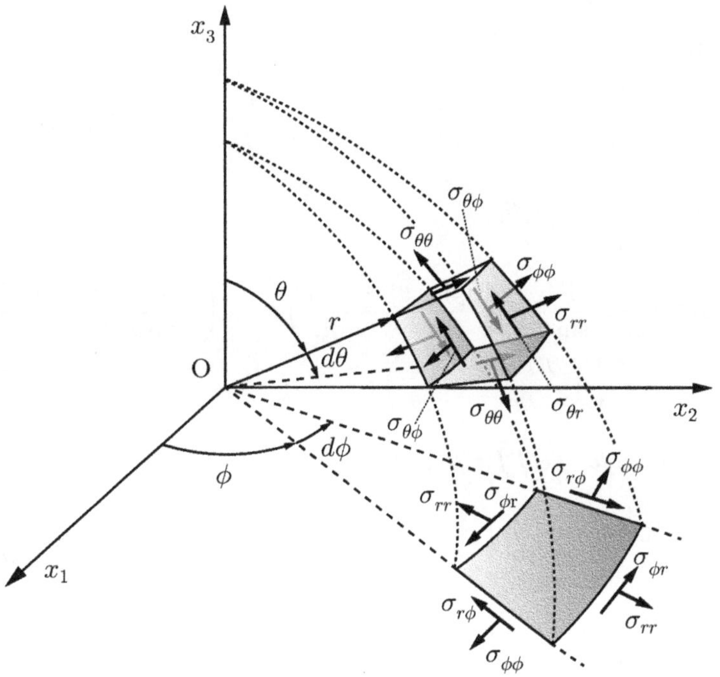


**Fig. B.1** Stress tensor components in a spherical coordinate system

*Divergence of a vector field* $\boldsymbol{v}(r, \theta, \varphi)$:

$$\operatorname{div} \boldsymbol{v} = \frac{1}{r^2}\frac{\partial}{\partial r}(r^2\, v_r) + \frac{1}{r\sin\theta}\frac{\partial}{\partial\theta}(v_\theta \sin\theta) + \frac{1}{r\sin\theta}\frac{\partial v_\varphi}{\partial\varphi} \tag{B.1}$$

*Divergence of a tensor field $\boldsymbol{\sigma}(r,\theta,\varphi)$:*

$$\begin{aligned}\mathbf{div}\,\boldsymbol{\sigma} = &\left(\frac{\partial\sigma_{rr}}{\partial r} + \frac{1}{r}\frac{\partial\sigma_{r\theta}}{\partial\theta} + \frac{1}{r\sin\theta}\frac{\partial\sigma_{r\varphi}}{\partial\varphi} + \frac{1}{r}\left[2\sigma_{rr} - \sigma_{\theta\theta} - \sigma_{\varphi\varphi} + \sigma_{r\theta}\cot\theta\right]\right)\boldsymbol{e}_r \\ &+ \left(\frac{\partial\sigma_{\theta r}}{\partial r} + \frac{1}{r}\frac{\partial\sigma_{\theta\theta}}{\partial\theta} + \frac{1}{r\sin\theta}\frac{\partial\sigma_{\theta\varphi}}{\partial\varphi} + \frac{1}{r}\left[3\sigma_{r\theta} + (\sigma_{\theta\theta} - \sigma_{\varphi\varphi})\cot\theta\right]\right)\boldsymbol{e}_\theta \\ &+ \left(\frac{\partial\sigma_{\varphi r}}{\partial r} + \frac{1}{r}\frac{\partial\sigma_{\varphi\theta}}{\partial\theta} + \frac{1}{r\sin\theta}\frac{\partial\sigma_{\varphi\varphi}}{\partial\varphi} + \frac{1}{r}\left[3\sigma_{r\varphi} + 2\sigma_{\theta\varphi}\cot\theta\right]\right)\boldsymbol{e}_\varphi\,.\end{aligned} \tag{B.2}$$

*Gradient of a scalar field $f(r,\theta,\varphi)$:*

$$\boldsymbol{\nabla} f = \frac{\partial f}{\partial r}\boldsymbol{e}_r + \frac{1}{r}\frac{\partial f}{\partial\theta}\boldsymbol{e}_\theta + \frac{1}{r\sin\theta}\frac{\partial f}{\partial\varphi}\boldsymbol{e}_\varphi\,. \tag{B.3}$$

*Gradient of a vector field $\boldsymbol{v}(r,\theta,\varphi)$:*

$$\boldsymbol{\nabla}\boldsymbol{v} = \begin{pmatrix} \dfrac{\partial v_r}{\partial r} & \dfrac{1}{r}\dfrac{\partial v_r}{\partial\theta} - \dfrac{v_\theta}{r} & \dfrac{1}{r\sin\theta}\dfrac{\partial v_r}{\partial\varphi} - \dfrac{v_\varphi}{r} \\ \dfrac{\partial v_\theta}{\partial r} & \dfrac{1}{r}\dfrac{\partial v_\theta}{\partial\theta} + \dfrac{v_r}{r} & \dfrac{1}{r\sin\theta}\dfrac{\partial v_\theta}{\partial\varphi} - \dfrac{v_\varphi}{r\tan\theta} \\ \dfrac{\partial v_\varphi}{\partial r} & \dfrac{1}{r}\dfrac{\partial v_\varphi}{\partial\theta} & \dfrac{1}{r\sin\theta}\left[\dfrac{\partial v_\varphi}{\partial\varphi} + v_r\sin\theta + v_\theta\cos\theta\right] \end{pmatrix}. \tag{B.4}$$

*Curl of a vector field $\boldsymbol{v}(r,\theta,\varphi)$:*

$$\begin{aligned}\mathbf{curl}\ \boldsymbol{v} = &\frac{1}{r\sin\theta}\left(\frac{\partial(v_\varphi\sin\theta)}{\partial\theta} - \frac{\partial v_\theta}{\partial\varphi}\right)\boldsymbol{e}_r + \left(\frac{1}{r\sin\theta}\frac{\partial v_r}{\partial\varphi} - \frac{1}{r}\frac{\partial(rv_\varphi)}{\partial r}\right)\boldsymbol{e}_\theta \\ &+ \frac{1}{r}\left(\frac{\partial(r\,v_\theta)}{\partial r} - \frac{\partial v_r}{\partial\theta}\right)\boldsymbol{e}_\varphi\,.\end{aligned} \tag{B.5}$$

*Laplacian of a scalar field $f(r,\theta,\varphi)$:*

$$\Delta f = \nabla^2 f = \frac{1}{r^2}\frac{\partial}{\partial r}\left(r^2\frac{\partial f}{\partial r}\right) + \frac{1}{r^2\sin\theta}\frac{\partial}{\partial\theta}\left(\sin\theta\frac{\partial f}{\partial\theta}\right) + \frac{1}{r^2\sin^2\theta}\frac{\partial^2 f}{\partial\varphi^2}\,. \tag{B.6}$$

*Laplacian of a vector field $\boldsymbol{v}(r,\theta,\varphi)$:*

$$\begin{aligned}\boldsymbol{\nabla}^2\boldsymbol{v} = &\left(\Delta v_r - \frac{2v_r}{r^2} - \frac{2}{r^2}\frac{\partial v_\theta}{\partial\theta} - \frac{2v_\theta\cot\theta}{r^2} - \frac{2}{r^2\sin\theta}\frac{\partial v_\varphi}{\partial\varphi}\right)\boldsymbol{e}_r \\ &+ \left(\Delta v_\theta - \frac{v_\theta}{r^2\sin^2\theta} + \frac{2}{r^2}\frac{\partial v_r}{\partial\theta} - \frac{2\cos\theta}{r^2\sin^2\theta}\frac{\partial v_\varphi}{\partial\varphi}\right)\boldsymbol{e}_\theta \\ &+ \left(\Delta v_\varphi - \frac{v_\varphi}{r^2\sin^2\theta} + \frac{2}{r^2\sin\theta}\frac{\partial v_r}{\partial\varphi} + \frac{2\cos\theta}{r^2\sin^2\theta}\frac{\partial v_\theta}{\partial\varphi}\right)\boldsymbol{e}_\varphi\,.\end{aligned} \tag{B.7}$$

*Material derivative of a scalar field $f(r,\theta,\varphi)$:*

$$\frac{Df}{Dt}=\frac{\partial f}{\partial t}+v_r\frac{\partial f}{\partial r}+\frac{v_\theta}{r}\frac{\partial f}{\partial\theta}+\frac{v_\varphi}{r\sin\theta}\frac{\partial f}{\partial\varphi}\ . \tag{B.8}$$

*Acceleration:*

$$a_r=\frac{Dv_r}{Dt}-\frac{v_\varphi^2+v_\theta^2}{r} \tag{B.9}$$

$$a_\theta=\frac{Dv_\theta}{Dt}+\frac{v_rv_\theta-v_\varphi^2\cot\theta}{r} \tag{B.10}$$

$$a_\varphi=\frac{Dv_\varphi}{Dt}+\frac{v_rv_\varphi+v_\varphi v_\theta\cot\theta}{r}\ . \tag{B.11}$$

*Conservation of mass equation:*

$$\frac{\partial\rho}{\partial t}+\frac{1}{r^2}\frac{\partial(r^2\,\rho\,v_r)}{\partial r}+\frac{1}{r\sin\theta}\frac{\partial(\rho\,v_\theta\sin\theta)}{\partial\theta}+\frac{1}{r\sin\theta}\frac{\partial(\rho\,v_\varphi)}{\partial\varphi}=0\ . \tag{B.12}$$

*Motion equations:*

$$\rho\left(\frac{Dv_r}{Dt}-\frac{v_\varphi^2+v_\theta^2}{r}\right)=\frac{\partial\sigma_{rr}}{\partial r}+\frac{1}{r}\frac{\partial\sigma_{r\theta}}{\partial\theta}+\frac{1}{r\sin\theta}\frac{\partial\sigma_{r\varphi}}{\partial\varphi}$$
$$+\frac{1}{r}(2\sigma_{rr}-\sigma_{\theta\theta}-\sigma_{\varphi\varphi}+\sigma_{r\theta}\cot\theta)+\rho\,b_r \tag{B.13}$$

$$\rho\left(\frac{Dv_\theta}{Dt}+\frac{v_rv_\theta-v_\varphi^2\cot\theta}{r}\right)=\frac{\partial\sigma_{r\theta}}{\partial r}+\frac{1}{r}\frac{\partial\sigma_{\theta\theta}}{\partial\theta}+\frac{1}{r\sin\theta}\frac{\partial\sigma_{\theta\varphi}}{\partial\varphi}$$
$$+\frac{1}{r}[(\sigma_{\theta\theta}-\sigma_{\varphi\varphi})\cot\theta+3\sigma_{r\theta}]+\rho\,b_\theta \tag{B.14}$$

$$\rho\left(\frac{Dv_\varphi}{Dt}+\frac{v_rv_\varphi+v_\varphi v_\theta\cot\theta}{r}\right)=\frac{\partial\sigma_{r\varphi}}{\partial r}+\frac{1}{r}\frac{\partial\sigma_{\theta\varphi}}{\partial\theta}+\frac{1}{r\sin\theta}\frac{\partial\sigma_{\varphi\varphi}}{\partial\varphi}$$
$$+\frac{1}{r}(3\sigma_{r\varphi}+2\sigma_{\theta\varphi}\cot\theta)+\rho\,b_\varphi\ . \tag{B.15}$$

*Equilibrium equations:*

$$\frac{\partial \sigma_{rr}}{\partial r} + \frac{1}{r}\frac{\partial \sigma_{r\theta}}{\partial \theta} + \frac{1}{r\sin\theta}\frac{\partial \sigma_{r\varphi}}{\partial \varphi} + \frac{1}{r}(2\sigma_{rr} - \sigma_{\theta\theta} - \sigma_{\varphi\varphi} + \sigma_{r\theta}\cot\theta) + \rho\, b_r = 0 \tag{B.16}$$

$$\frac{\partial \sigma_{r\theta}}{\partial r} + \frac{1}{r}\frac{\partial \sigma_{\theta\theta}}{\partial \theta} + \frac{1}{r\sin\theta}\frac{\partial \sigma_{\theta\varphi}}{\partial \varphi} + \frac{1}{r}[(\sigma_{\theta\theta} - \sigma_{\varphi\varphi})\cot\theta + 3\sigma_{r\theta}] + \rho\, b_\theta = 0 \tag{B.17}$$

$$\frac{\partial \sigma_{r\varphi}}{\partial r} + \frac{1}{r}\frac{\partial \sigma_{\theta\varphi}}{\partial \theta} + \frac{1}{r\sin\theta}\frac{\partial \sigma_{\varphi\varphi}}{\partial \varphi} + \frac{1}{r}(3\sigma_{r\varphi} + 2\sigma_{\theta\varphi}\cot\theta) + \rho\, b_\varphi = 0\,. \tag{B.18}$$

*Navier's equations:*

$$\mu\left(\nabla^2 u_r - \frac{2u_r}{r^2} - \frac{2}{r^2}\frac{\partial u_\theta}{\partial \theta} - \frac{2u_\theta\cot\theta}{r^2} - \frac{2\cot\theta}{r^2\sin\theta}\frac{\partial u_\varphi}{\partial \varphi}\right) + (\lambda+\mu)\frac{\partial}{\partial r}\left(\frac{1}{r^2}\frac{\partial (r^2 u_r)}{\partial r} + \frac{1}{r\sin\theta}\frac{\partial (u_\theta \sin\theta)}{\partial \theta} + \frac{1}{r\sin\theta}\frac{\partial u_\varphi}{\partial \varphi}\right) + f_r = 0 \tag{B.19}$$

$$\mu\left(\nabla^2 u_\theta + \frac{2}{r^2}\frac{\partial u_r}{\partial \theta} - \frac{u_\theta}{r^2\sin^2\theta} - \frac{2\cot\theta}{r^2\sin^2\theta}\frac{\partial u_\varphi}{\partial \varphi}\right) + (\lambda+\mu)\frac{1}{r}\frac{\partial}{\partial \theta}\left(\frac{1}{r^2}\frac{\partial (r^2 u_r)}{\partial r} + \frac{1}{r\sin\theta}\frac{\partial (u_\theta \sin\theta)}{\partial \theta} + \frac{1}{r\sin\theta}\frac{\partial u_\varphi}{\partial \varphi}\right) + f_\theta = 0 \tag{B.20}$$

$$\mu\left(\nabla^2 u_\varphi - \frac{u_\varphi}{r^2\sin^2\theta} + \frac{2}{r^2\sin^2\theta}\frac{\partial u_r}{\partial \varphi} + \frac{2\cot\theta}{r^2\sin^2\theta}\frac{\partial u_\theta}{\partial \varphi}\right) + (\lambda+\mu)\frac{1}{r\sin\theta}\frac{\partial}{\partial \varphi}\left(\frac{1}{r^2}\frac{\partial (r^2 u_r)}{\partial r} + \frac{1}{r\sin\theta}\frac{\partial (u_\theta \sin\theta)}{\partial \theta} + \frac{1}{r\sin\theta}\frac{\partial u_\varphi}{\partial \varphi}\right) + f_\varphi = 0\,. \tag{B.21}$$

*Strain components as functions of displacement:*

$$\varepsilon_{rr} = \frac{\partial u_r}{\partial r} \quad \varepsilon_{\theta\theta} = \frac{u_r}{r} + \frac{1}{r}\frac{\partial u_\theta}{\partial \theta} \quad \varepsilon_{\varphi\varphi} = \frac{1}{r\sin\theta}\frac{\partial u_\varphi}{\partial \varphi} + \frac{u_r}{r} + \frac{u_\theta}{r}\cot\theta \tag{B.22}$$

$$\varepsilon_{r\theta} = \frac{1}{2}\left(\frac{1}{r}\frac{\partial u_r}{\partial \theta} + \frac{\partial u_\theta}{\partial r} - \frac{u_\theta}{r}\right) \quad \varepsilon_{r\varphi} = \frac{1}{2}\left(\frac{1}{r\sin\theta}\frac{\partial u_r}{\partial \varphi} + \frac{\partial u_\varphi}{\partial r} - \frac{u_\theta}{r}\right) \tag{B.23}$$

$$\varepsilon_{\theta\varphi} = \frac{1}{2}\left(\frac{1}{r}\frac{\partial u_\varphi}{\partial \theta} + \frac{1}{r\sin\theta}\frac{\partial u_\theta}{\partial \varphi} - \frac{u_\varphi}{r}\cot\theta\right). \tag{B.24}$$

Note that the constitutive equations are obtained by replacing the Cartesian components in (7.3) and (7.4) with the components above.

*Strain rate components:*

$$d_{rr} = \frac{\partial v_r}{\partial r}, \quad d_{\theta\theta} = \frac{1}{r}\frac{\partial v_\theta}{\partial \theta} + \frac{v_r}{r}, \tag{B.25}$$

$$d_{\varphi\varphi} = \frac{1}{r\sin\theta}\frac{\partial v_\varphi}{\partial \varphi} + \frac{v_r}{r} + \frac{v_\theta \cot\theta}{r}, \tag{B.26}$$

$$d_{\varphi\theta} = \frac{1}{2}\left(\frac{1}{r\sin\theta}\frac{\partial v_\theta}{\partial \varphi} + \frac{1}{r}\frac{\partial v_\varphi}{\partial \theta} - \frac{v_\varphi \cot\theta}{r}\right), \tag{B.27}$$

$$d_{\varphi r} = \frac{1}{2}\left(\frac{\partial v_\varphi}{\partial r} + \frac{1}{r\sin\theta}\frac{\partial v_r}{\partial \varphi} - \frac{v_\varphi}{r}\right), \tag{B.28}$$

$$d_{r\theta} = \frac{1}{2}\left(\frac{1}{r}\frac{\partial v_r}{\partial \theta} + \frac{\partial v_\theta}{\partial r} - \frac{v_\theta}{r}\right). \tag{B.29}$$

*Incompressible Navier-Stokes equations:*

Conservation of mass equation

$$\frac{1}{r^2}\frac{\partial}{\partial r}(r^2\, v_r) + \frac{1}{r\sin\theta}\frac{\partial}{\partial \theta}(v_\theta \sin\theta) + \frac{1}{r\sin\theta}\frac{\partial v_\varphi}{\partial \varphi} = 0. \tag{B.30}$$

Motion equations

$$\rho\left(\frac{Dv_r}{Dt} - \frac{v_\varphi^2 + v_\theta^2}{r}\right) = -\frac{\partial p}{\partial r}$$
$$+\,\mu\left(\Delta v_r - \frac{2v_r}{r^2} - \frac{2}{r^2}\frac{\partial v_\theta}{\partial \theta} - \frac{2v_\theta \cot\theta}{r^2} - \frac{2}{r^2\sin\theta}\frac{\partial v_\varphi}{\partial \varphi}\right) + \rho b_r \tag{B.31}$$

$$\rho\left(\frac{Dv_\theta}{Dt} + \frac{v_r v_\theta - v_\varphi^2 \cot\theta}{r}\right) = -\frac{1}{r}\frac{\partial p}{\partial \theta}$$
$$+\,\mu\left(\Delta v_\theta - \frac{v_\theta}{r^2\sin^2\theta} + \frac{2}{r^2}\frac{\partial v_r}{\partial \theta} - \frac{2\cos\theta}{r^2\sin^2\theta}\frac{\partial v_\varphi}{\partial \varphi}\right) + \rho b_\theta \tag{B.32}$$

$$\rho\left(\frac{Dv_\varphi}{Dt} + \frac{v_r v_\varphi + v_\varphi v_\theta \cot\theta}{r}\right) = -\frac{1}{r\sin\theta}\frac{\partial p}{\partial \varphi}$$
$$+\,\mu\left(\Delta v_\varphi - \frac{v_\varphi}{r^2\sin^2\theta} + \frac{2}{r^2\sin\theta}\frac{\partial v_r}{\partial \varphi} + \frac{2\cos\theta}{r^2\sin^2\theta}\frac{\partial v_\theta}{\partial \varphi}\right) + \rho b_\varphi \tag{B.33}$$

with the Laplacian operator defined by (B.6).

# List of Symbols

This list of symbols gathers the principal notations used in this monograph. Those that are not shown here are defined where they appear and their context supplies all the necessary information for their understanding and use.

| Symbol | Description | Units |
|---|---|---|
| **Latin alphabet** | | |
| $a$ | speed of sound | $\mathrm{m\,s^{-1}}$ |
| $a_i$ | area of the face $s_i$ | $\mathrm{m^2}$ |
| $a_i$ | components of the acceleration | $\mathrm{m\,s^{-2}}$ |
| $\boldsymbol{a}$ | acceleration vector (spatial) | $\mathrm{m\,s^{-2}}$ |
| $\boldsymbol{\mathcal{A}}$ | tensor of order $n$ | – |
| $\mathcal{A}_{i_1 \cdots i_n}$ | components of the tensor of order $n$ | – |
| $A$ | area | $\mathrm{m^2}$ |
| $\boldsymbol{A}$ | acceleration vector (material) | $\mathrm{m\,s^{-2}}$ |
| $\boldsymbol{A}_i$ | unit material eigenvectors | – |
| $\boldsymbol{b}$ | volume force per unit mass | $\mathrm{N\,kg^{-1}}$ |
| $\boldsymbol{B}$ | volume force per unit mass | $\mathrm{N\,kg^{-1}}$ |
| $\boldsymbol{b}_i$ | unit spatial eigenvectors | – |
| $\mathcal{B}$ | body | – |
| $\mathbf{c}$ | left Cauchy-Green deformation tensor | – |
| $\boldsymbol{C}$ | right Cauchy-Green deformation tensor | – |
| $C_{ijk}$ | material parameters | – |
| $c_p$ | heat capacity at constant pressure | $\mathrm{J\,kg^{-1}K^{-1}}$ |
| $c_v$ | heat capacity at constant volume | $\mathrm{J\,kg^{-1}K^{-1}}$ |
| $C_D$ | drag coefficient | – |
| $\boldsymbol{d}$ | rate of deformation tensor | $\mathrm{s^{-1}}$ |
| $d_i$ | dual vector components | – |
| $e$ | thickness | m |
| $\boldsymbol{e}$ | Euler-Almansi strain tensor | – |
| $\boldsymbol{e}_i$ | basis vector | – |
| $E$ | Young's modulus | Pa |
| $E_k$ | kinetic energy | J |
| $E_{\text{int}}$ | internal energy | J |
| $\boldsymbol{E}$ | Green-Lagrange strain tensor | – |
| $E^3$ | vector space | – |
| $\boldsymbol{F}$ | deformation gradient tensor | – |
| $\boldsymbol{f}^b$ | volume force | N |
| $\boldsymbol{f}^c$ | contact force | N |
| $f$ | Helmholtz free energy | $\mathrm{J\,kg^{-1}}$ |
| $f$ | deflection | m |

| Symbol | Description | Units |
|---|---|---|
| $f$ | frequency | $s^{-1}$ |
| $\boldsymbol{f}(\boldsymbol{T})$ | tensor function of tensor $\boldsymbol{T}$ | – |
| $\boldsymbol{F}^b$ | volume force | N |
| Fr | Froude number | – |
| $\boldsymbol{g}_i$ | basis vectors in curvilinear coordinates | – |
| $g$ | acceleration of gravity | $m\,s^{-2}$ |
| $h$ | height | m |
| $h$ | enthalpy per unit mass | $J\,kg^{-1}$ |
| $I_i$ | scalar invariants | – |
| $I_3$ | moment of inertia with respect to $x_3$ | $m^4$ |
| $\boldsymbol{I}$ | identity tensor | – |
| $J$ | Jacobian | – |
| $k_B$ | Boltzmann's constant ($= 1.381\ 10^{-23}$) | $J\,K^{-1}$ |
| $k$ | coefficient of thermal conductivity | $W\,m^{-1}K^{-1}$ |
| $K$ | modulus of rigidity | Pa |
| $\boldsymbol{K}$ | second Piola-Kirchhoff stress vector | Pa |
| $l$ | length | m |
| $L$ | length | m |
| $\boldsymbol{L}$ | velocity gradient tensor | $s^{-1}$ |
| $m$ | mass | kg |
| $\overline{\boldsymbol{m}}$ | momentum | $kg\,m\,s^{-1}$ |
| $\widehat{\boldsymbol{m}}$ | angular momentum | $kg\,m^2\,s^{-1}$ |
| M | Mach number | – |
| $M$ | moment | N |
| $\boldsymbol{M}$ | deformation tensor | – |
| $\boldsymbol{N}$ | unit vector | – |
| $N_A$ | Avogadro number ($= 6\ 10^{23}$) | $mol^{-1}$ |
| $\boldsymbol{n}$ | unit outgoing vector of a domain | – |
| $\boldsymbol{n}_i$ | unit eigenvectors | – |
| $\boldsymbol{O}$ | orthogonal tensor | – |
| $\boldsymbol{O}$ | zero tensor | – |
| $p$ | pressure | Pa |
| $\boldsymbol{P}$ | first Piola-Kirchhoff stress tensor | Pa |
| $P_0$ | initial density | $kg\,m^{-3}$ |
| $P$ | current density in the material description | $kg\,m^{-3}$ |
| $P_i$ | internal pressure | Pa |
| $P_e$ | external pressure | Pa |
| $p_0$ | reference pressure | Pa |
| $p_i$ | pressure | Pa |
| $p_i$ | probability | – |
| Pr | Prandtl number | – |
| $Q$ | flow rate | $m^3\,s^{-1}$ |
| $\boldsymbol{Q}$ | orthogonal tensor | – |
| $\boldsymbol{q}$ | heat flux vector | $Wm^{-2}$ |
| $q$ | uniformly distributed load | $N\,m^{-2}$ |
| $\mathcal{R}$ | configuration | – |

| Symbol | Description | Units |
|---|---|---|
| $\boldsymbol{R}$ | orthogonal tensor | – |
| $\mathbb{R}^3$ | Euclidean space | – |
| $R$ | radius | m |
| $R$ | ideal gas constant | $\mathrm{J\,kg^{-1}\,K^{-1}}$ |
| Re | Reynolds number | – |
| $\mathrm{R} = (O, \boldsymbol{x}, t)$ | observer or reference frame | – |
| $r$ | volume heat production | $\mathrm{W\,m^{-3}}$ |
| $r_i$ | internal radius of a cylinder | m |
| $r_e$ | external radius of a cylinder | m |
| $\boldsymbol{S}$ | second Piola-Kirchhoff stress tensor | Pa |
| $S_u$ | surface | $\mathrm{m^2}$ |
| $S_t$ | surface | $\mathrm{m^2}$ |
| $s$ | entropy density | $\mathrm{J\,kg^{-1}\,K^{-1}}$ |
| $s$ | time | s |
| $dS$ | surface element | $\mathrm{m^2}$ |
| $ds$ | surface element | $\mathrm{m^2}$ |
| $s_i$ | face | $\mathrm{m^2}$ |
| $t$ | time | s |
| $T$ | temperature | K |
| $\boldsymbol{T}$ | deviatoric extra-stress tensor | Pa |
| $\boldsymbol{T}$ | first Piola-Kirchhoff stress vector | Pa |
| $\boldsymbol{t}$ | stress vector | Pa |
| $\boldsymbol{t}_{\boldsymbol{e}_i}$ | stress vector in direction $\boldsymbol{e}_i$ | Pa |
| $t_N$ | normal component of vector $\boldsymbol{t}$ | Pa |
| $t_T$ | tangential component of vector $\boldsymbol{t}$ | Pa |
| $\boldsymbol{u}$ | displacement vector (spatial) | m |
| $u$ | internal energy density (spatial) | $\mathrm{J\,kg^{-1}}$ |
| $\boldsymbol{U}$ | displacement vector (material) | m |
| $\boldsymbol{U}$ | second-order tensor | – |
| $U$ | internal energy density (material) | $\mathrm{J\,kg^{-1}}$ |
| $U$ | velocity | $\mathrm{m\,s^{-1}}$ |
| $\mathcal{V}$ | neighborhood | – |
| $\boldsymbol{v}$ | velocity vector (spatial) | $\mathrm{ms^{-1}}$ |
| $\boldsymbol{V}$ | velocity vector (material) | $\mathrm{m\,s^{-1}}$ |
| $V$ | potential function | Pa |
| $V$ | volume in material coordinates | $\mathrm{m^3}$ |
| $v$ | volume in spatial coordinates | $\mathrm{m^3}$ |
| $v_{\mathrm{moy}}$ | average velocity | $\mathrm{m\,s^{-1}}$ |
| $v_{\max}$ | maximum velocity | $\mathrm{m\,s^{-1}}$ |
| $\boldsymbol{W}$ | displacement vector | m |
| $W$ | strain energy function | $\mathrm{J\,m^{-3}}$ |
| $\mathcal{W}$ | scalar function of a tensor $\boldsymbol{T}$ | – |
| $\mathcal{W}, \widehat{\mathcal{W}}$ | strain energy function | $\mathrm{J\,m^{-3}}$ |
| $\boldsymbol{x}$ | position vector (spatial) | m |
| $\boldsymbol{X}$ | position vector (material) | m |
| $\mathcal{X}$ | physical system | – |

| Symbol | Description | Units |
|---|---|---|
| **Greek alphabet** | | |
| $\alpha$ | constant | – |
| $\alpha$ | angle | – |
| $\alpha$ | thermal expansion coefficient | $K^{-1}$ |
| $\alpha_i$ | material parameter | – |
| $\beta$ | angle | – |
| $\beta_i$ | scalar function | – |
| $\Gamma, \Gamma^-, \Gamma^+, \Gamma_1$ | surface | $m^2$ |
| $\gamma$ | heat capacity ratio | – |
| $\gamma$ | deformation | – |
| $\gamma_{12}$ | angle | – |
| $\gamma_0$ | thermal diffusivity | $m^2\,s^{-1}$ |
| $\boldsymbol{\delta}$ | Kronecker delta, components $\delta_{ij}$ | – |
| $\delta$ | growth | – |
| $\Delta$ | increment | |
| $\boldsymbol{\varepsilon}$ | infinitesimal strain tensor, components $\varepsilon_{ij}$ | – |
| $\varepsilon_{ijk}$ | permutation symbol | – |
| $\varepsilon$ | real number | – |
| $\theta$ | angle | – |
| $\Theta$ | angle | – |
| $\theta_i$ | curvilinear coordinates | – |
| $\lambda$ | eigenvalue | – |
| $\lambda$ | head loss | – |
| $\lambda_i$ | eigenvector | – |
| $\lambda_N$ | stretch ratio | – |
| $\lambda_i$ | principal stretches of the tensor $\boldsymbol{U}$ | – |
| $\lambda_i^2$ | principal stretches of the tensor $\boldsymbol{C}$ | – |
| $\lambda$ | Lamé coefficient | Pa |
| $\lambda$ | volume viscosity | Pa s |
| $\Lambda$ | thermal diffusivity | $m^2\,s^{-1}$ |
| $\mu$ | Lamé coefficient | Pa |
| $\mu$ | dynamic viscosity | Pa s |
| $\mu$ | shear modulus or modulus of rigidity | Pa |
| $\mu_i$ | parameter | – |
| $\nu$ | Poisson's coefficient | – |
| $\nu$ | kinematic viscosity ($= \mu/\rho$) | $m^2\,s^{-1}$ |
| $\Pi, \Pi^-, \Pi^+$ | portions of the body $\mathcal{B}$ | – |
| $\rho$ | current density | $kg\,m^{-3}$ |
| $\boldsymbol{\Sigma}$ | stress functional | Pa |
| $\boldsymbol{\Sigma}$ | tensor functional | – |
| $\boldsymbol{\sigma}$ | Cauchy stress tensor, components $\sigma_{ij}$ | Pa |
| $\sigma_i$ | principal stresses of $\boldsymbol{\sigma}$ | Pa |
| $\sigma$ | stress | Pa |
| $\sigma$ | surface tension coefficient | $N\,m^{-1}$ |
| $\sigma_0$ | hydrostatic stress | Pa |

| Symbol | Description | Units |
|---|---|---|
| $\tau$ | time | s |
| $\boldsymbol{\tau}$ | shear stress | Pa |
| $\varphi$ | function | – |
| $\phi$ | strain energy function | $\mathrm{J\,m^{-3}}$ |
| $\Phi$ | Airy function | $\mathrm{Pa\,m^2}$ |
| $\Phi$ | strain energy function | $\mathrm{J\,m^{-3}}$ |
| $\Phi$ | function | |
| $\boldsymbol{\chi}$ | vector motion function | m |
| $\dot{\Omega}$ | rotation rate tensor | $\mathrm{s^{-1}}$ |
| $\Omega$ | domain in the material representation | – |
| $\partial\Omega$ | surface of domain $\Omega$ | $\mathrm{m^2}$ |
| $\omega$ | domain in the spatial representation | – |
| $\omega_1, \omega_2$ | rates of angular rotation | $\mathrm{s^{-1}}$ |
| $\partial\omega$ | surface of the domain $\omega$ | $\mathrm{m^2}$ |
| $\dot{\boldsymbol{\omega}}$ | rotation rate tensor | $\mathrm{s^{-1}}$ |
| $\boldsymbol{\omega}$ | infinitesimal rotation tensor | – |

## Indices and exponents

| | | |
|---|---|---|
| $_{i,j,k}$ | indices of vectors and tensors, values 1, 2, or 3 | |
| $_i$ | index for a face, a physical value | |
| $_0$ | initial, reference or in the natural state | |
| $^*$ | with respect to the reference frame $\mathrm{R}^*$ or non-dimensional | |
| $\bar{\ }$ | imposed value | |
| $_{x,y,z}$ | Cartesian components | |
| $_{r,\theta,z}$ | components in cylindrical coordinates | |
| $_{r,\varphi,\theta}$ | components in spherical coordinates | |

## Notations

| | | |
|---|---|---|
| $[\cdot]$ | matrix | |
| $[\cdot]^T$ | transposed matrix | |
| $[\cdot]^{-1}$ | inverse matrix | |
| $\boldsymbol{v}, \boldsymbol{\sigma}$ | vector or tensor | |
| $\boldsymbol{L}^S$ | symmetric tensor | |
| $\boldsymbol{L}^A$ | antisymmetric tensor | |
| $\boldsymbol{L}^s$ | spherical tensor | |
| $\boldsymbol{L}^d$ | deviatoric tensor | |
| $O(\cdot)$ | order of term | |
| $\mathrm{o}(\cdot)$ | remainder of an expansion going to 0 when its argument goes to 0 | |

## Operators

| | | |
|---|---|---|
| $\cdot$ | scalar product | |
| $\otimes$ | tensor product of two vectors | |
| $\times$ | vector product | |
| $:$ | scalar product of two tensors | |
| $C^\infty$ | class of infinitely differentiable functions | |

| **Symbol** | **Description** | **Units** |
|---|---|---|
| det | determinant of a matrix | |
| $\mathrm{diag}(a, b, c)$ | diagonal matrix of components $a, b, c$ | |
| $\nabla$ | gradient | |
| div | divergence of a vector field | |
| **div** | divergence of a tensor field | |
| $\mathbf{curl} = \boldsymbol{\nabla}\times$ | curl of a vector field | |
| $\lVert \cdot \rVert$ | norm of a vector or tensor | |
| tr | trace of a tensor | |
| $\triangle$ | Laplacian | |
| $\triangle\triangle$ | biharmonic operator | |
| $\dfrac{D}{Dt}$ | material derivative | |
| $\dfrac{D_0}{D_0 t}$ | material derivative | |
| $\dfrac{\partial}{\partial t}$ | partial derivative with respect to time | |
| $\dfrac{\partial}{\partial x_i}$ | partial derivative with respect to coordinate $x_i$ | |
| $\sum_{i=1}^{n}$ | sum over $i$ from 1 to $n$ | |

# Suggestions for Solutions to the Exercises

## Chapter 1

**1.3** Follow the steps of example 1.2 (sec. 1.2.5).

**1.4** Follow the steps of example 1.2 (sec. 1.2.5).

**1.6** Follow the steps of examples in sections 1.4.6–8.

**1.7** Follow the steps of examples in sections 1.4.6–8.

**1.8** Follow the steps of examples in sections 1.4.6–8.

**1.15** Follow the steps of examples in section 1.3.8 for an antisymmetric tensor.

**1.16** Use relation (1.123) to express tensor $\boldsymbol{T}$ by its inverse and use it to replace $\boldsymbol{T}^2$ in relation (1.140).

## Chapter 2

**2.4** Eliminate the parameter $t$ from the motion equations.

**2.8** Use (2.77), (2.91), (2.167), and (2.181).

**2.11** Use (2.108), (2.111), (2.112), and (2.109).

**2.12** Introduce (2.147) in $(2.107)_1$ and eliminate the products with $O(\varepsilon)$.

**2.13** (a) Use (2.70) in (2.120) (in index form) and implement the approximations due to linearization.

**2.13** (b) Show first that $\boldsymbol{U}^{-1} \approx \boldsymbol{I} - \boldsymbol{\varepsilon}$ and use (2.73).

**2.14** Express the vectors $d\boldsymbol{x}$ and $d\boldsymbol{y}$ as functions of $\varepsilon_{ij}$ and use the approximation $(1+\alpha)^n \approx 1 + n\alpha$, when $\alpha \ll 1$.

**2.17** Use (2.81) when $e_{ij}$ is replaced by $\varepsilon_{ij}$ and use the following definitions $\cos\theta = \frac{dx_1}{dS}, \sin\theta = \frac{dx_2}{dS}$.

## Chapter 3

**3.2** Apply the conservation of mass equation to the given velocity field and deduce from it a relation for the density, $\rho$. Calculate the trajectories and combine the results to obtain the equality.

**3.4** (2) Find the normal vector on the plane and the sphere using the gradient and use (3.76).

**3.6** Consider the force and moment components as vector components. Express the equilibrium of the force and the moment on the surface and use the divergence theorem.

**3.9** With the divergence theorem, convert the surface integral to a volume integral. Then use the principle of conservation of momentum.

**3.11** Use (3.141) and (3.152); (3.149), (2.205), and (3.152).

## Chapter 4

**4.4** (1) For the perfect fluid, the term $\boldsymbol{\sigma} : \boldsymbol{d}$ becomes $-p\,tr\boldsymbol{d} = -p\boldsymbol{\nabla} \cdot \boldsymbol{v}$.

**4.4** (2) The relation to evaluate is written $\rho \frac{Dh}{Dt} = \frac{Dp}{Dt} - \frac{\partial q_i}{\partial x_i} + r$.

## Chapter 5

**5.2** Use relations (2.213), (2.181), (2.183), and (2.56).

**5.3** Note that $D/Dt^* = D/Dt$.

**5.4** Use the results of exercises 5.2 and 5.3.

**5.5** Relation (5.64) is that resulting from exercise 5.3 for $\boldsymbol{T} = \boldsymbol{d}$.

## Chapter 6

**6.1** Use (2.77) and (2.179).

**6.2** Insert (6.14) in (4.23).

**6.4** Use (2.88), (2.108), and (2.110); use (6.61).

**6.6** Express $\boldsymbol{C}^{-1}$ using (1.123) to modify (6.172); express $\boldsymbol{c}^2$ using (1.123) to modify (6.173).

**6.10** Use (6.78) in (6.86) for the state of plane stress and follow the steps of section 6.5.3.

**6.11** Simplify $(6.159)_1$ for simple traction and use $(6.110)_1$.

**6.14** (1) Introduce (6.175) and (6.176) in (6.174) and determine the trace of the resulting equation.

**6.14** (2) Use (1.109) and (6.175)–(6.177) to show that the principal axes of $\sigma_{ij}, \sigma_{ij}^d, \varepsilon_{ij}$ and $\varepsilon_{ij}^d$ coincide.

**6.14** (3) Introduce (6.176) in the first part of (6.178).

**6.15** (2) Introduce (6.182) in (6.104).

**6.15** (3) Introduce (6.183) in (6.106).

## Chapter 7

**7.1** Insert (7.18) in (7.21)–(7.23) and use (7.20).

**7.2** Insert (7.18) in (7.43) and use (7.20).

**7.3** Insert (7.317) in (7.7).

**7.4** (1) Use (1.190) to modify (7.7).

**7.4** (2) Use $(6.109)_2$ to modify (7.7).

**7.4** (3) Use (1.190) to modify (7.7).

**7.5** Take the divergence of (7.209) and use (1.191).

**7.6** Take the curl of (7.205) and use (1.237).

**7.7** Prove (7.38) and follow the steps of the example 7.3.

**7.8** Prove (7.38) and follow the steps of the example 7.3.

## Chapter 8

**8.2** Obtain the solution by combining two-dimensional Couette and Poiseuille flows.

**8.3** First show that $v_r$ is zero. To calculate the pressure, recall that the integration of

$$\frac{1}{\rho}\frac{\partial p}{\partial r} = \frac{v_\theta^2}{r}$$

yields

$$p = \rho \int_{R_1}^{r} \frac{v_\theta^2}{r'} dr' + f(z) ,$$

with $r'$ being a working variable.

**8.4** Show that in taking into account the problem symmetries, the only non-zero velocity component is $v_\varphi = v_\varphi(r, \theta)$. The condition imposed on this component on the sphere depends on the colatitude angle. Integrate relation (B.33) by separation of variables such that $v_\varphi = f(r)g(\theta)$. The boundary condition at infinity must also be taken into account.

**8.5** Same steps as for problem 8.4. In this case, a no-slip boundary condition is imposed on the outer sphere.

**8.7** First impose the condition at the boundary on the velocity component $v_1$ and take into account the relation that links the velocity field to the pressure gradient.

**8.8** Express the problem in cylindrical coordinates. Show that the only velocity component is $v_z = v_z(r)$ where the axis of the geometry is aligned with the coordinate $z$. To calculate the friction force, use the stress component $\sigma_{rz}$.

# Bibliography


[1] M. ABRAMOWITZ, I. A. STEGUN, *Handbook of Mathematical Functions with Formulas, Graphs, and Mathematical Tables*, Dover, New York, 1972.

[2] R. ARIS, *Vectors, Tensors, and the Basic Equations of Fluid Mechanics*, Dover, New York, 1989.

[3] G. K. BATCHELOR, *An Introduction to Fluid Dynamics*, Cambridge University Press, Cambridge, 1992.

[4] M. F. BEATTY, Topics in Finite Elasticity: Hyperelasticity of Rubber, Elastomers and Biological Tissues, With Examples, *Appl. Mech. Rev.*, 40, pp. 1699–1734, 1987.

[5] S. CHANDRASEKHAR, *Hydrodynamic and Hydromagnetic Stability*, Dover, New York, 1981.

[6] P. C. CHOU, N. J. PAGANO, *Elasticity: Tensor, Dyadic and Engineering Approaches*, Dover, New York, 1992.

[7] P. G. CIARLET, *Mathematical Elasticity: Three-Dimensional Elasticity*, Vol. I, North-Holland, Amsterdam, 1988.

[8] D. A. DANIELSON, *Vectors and Tensors in Engineering and Physics*, 2nd edition, Addison-Wesley, Reading, Massachusetts, 1997.

[9] M. O. DEVILLE, T. B. GATSKI, *Mathematical Modeling for Complex Fluids and Flows*, Springer, Berlin, 2012.

[10] P. G. DRAZIN, *Introduction to Hydrodynamic Stability*, Cambridge University Press, Cambridge, 2002.

[11] P. G. DRAZIN, W. H. REID, *Hydrodynamic Stability*, Cambridge University Press, Cambridge, 1981.

[12] A. C. ERINGEN, *Mechanics of Continua*, J. Wiley, New York, 1967.

[13] R. A. EUBANKS, E. STERNBERG, On the Completeness of the Boussinesq-Papkovich Stress Functions, *J. Rat. Mech. Anal.*, 5, pp. 735–746, 1956.

[14] L. EULER, *Lettres à une Princesse d'Allemagne*, Presses Polytechniques et Universitaires Romandes, Lausanne, 2003.

[15] Y. C. FUNG, P. TONG, *Classical and Computational Solid Mechanics*, World Scientific, New Jersey, 1994.

[16] K. F. GRAFF, *Wave Motion in Elastic Solids*, Dover, New York, 1991.

[17] M. E. GURTIN, *An Introduction to Continuum Mechanics*, Academic Press, New York, 1981.

[18] M. E. GURTIN, E . FRIED, L. ANAND, *The Mechanics and Thermodynamics of Continua*, Cambridge University Press, Cambridge, 2010.

[19] P. HAUPT, *Continuum Mechanics and Theory of Materials*, Springer, Berlin, 2000.

[20] G. A. HOLZAPFEL, *Nonlinear Solid Mechanics. A Continuum Approach for Engineering*, J. Wiley, Chichester, 2000.

[21] T. J. R. HUGHES, *Linear Static and Dynamic Finite Element Analysis*, Dover, New York, 2000.

[22] K. HUTTER, K. JÖHNK, *Continuum Methods of Physical Modeling*, Springer, Berlin, 2004.

[23] K. L. JOHNSON, *Contact Mechanics*, Cambridge University Press, Cambridge, 1985.

[24] W. M. LAI, D. RUBIN, E. KREMPL, *Introduction to Continuum Mechanics*, 4th edition, Butterworth-Heinemann, Oxford, 2010.

[25] SIR H. LAMB, *Hydrodynamics*, Cambridge University Press, Cambridge, 1995.

[26] L. LANDAU, E. M. LIFSHITZ, *Fluid Mechanics*, Butterworth-Heinemann, Oxford, 1997.

[27] L. LANDAU, E. M. LIFSHITZ, *Theory of elasticity*, Butterworth-Heinemann, Oxford, 1986.

[28] W. E. LANGLOIS, M. O. DEVILLE, *Slow Viscous Flow*, 2nd edition, Springer, 2014.

[29] C. C. LIN, L. A. SEGEL, *Mathematics Applied to Deterministic Problems in the Natural Sciences*, SIAM, Philadelphia, 1988.

[30] A. E. H. LOVE, *A Treatise on the Mathematical Theory of Elasticity*, 4th edition, Dover, New York, 1944.

[31] L. E. MALVERN, *Introduction to the Mechanics of a Continuous Medium*, Prentice-Hall, New Jersey, 1996.

[32] J. E. MARSDEN, T. J. R. HUGHES, *Mathematical Foundations of Elasticity*, Dover, New York, 1994.

[33] G. E. MASE, *Theory and Problems of Continuum Mechanics*, Schaum's Outline Series, McGraw-Hill, New York, 1970.

[34] G. T. MASE, G. E. MASE, *Continuum Mechanics for Engineers*, CRC Press, Boca Raton, 1999.

[35] M. J. MORAN, M. B. BAILEY, D. B. BOETTNER, H. N. SHAPIRO, *Principles of Engineering Thermodynamics, SI Version*, 8th edition, J. Wiley, Hoboken, 2015.

[36] R. W. OGDEN, *Non Linear Elastic Deformations*, Dover, New York, 1997.

[37] C. W. OSEEN, Über die Stokes'sche formel, und über eine verwandte Aufgabe in der Hydrodynamik, *Arkiv för matematik, astronomi och fysik*, vi, No. 29, 1910.

[38] R. L. PANTON, *Incompressible Flow*, J. Wiley, New York, 1984.

[39] S. B. POPE, *Turbulent Flows*, Cambridge University Press, Cambridge, 2000.

[40] M. Rappaz, M. Bellet, M. Deville, *Numerical Modeling in Materials Science and Engineering*, Springer, Berlin, 2003.

[41] R. S. Rivlin, Large Deformations of Isotropic Materials, IV. Further Developments of the General Theory, *Phil. Trans. Royal Soc. London*, A241, pp. 379–397, 1948.

[42] R. S. Rivlin, *Collected Papers*, Vol. I and II, Eds. G. I. Barenblatt and D. D. Joseph, Springer, New York, 1996.

[43] J. Salençon, *Handbook of Continuum Mechanics: General Concepts Thermoelasticity Volume I* , Springer, Berlin, 2001.

[44] P. J. Schmid, D. S. Henningson, *Stability and Transition in Shear Flows*, Springer, Berlin, 2001.

[45] L. A. Segel, *Mathematics Applied to Continuum Mechanics*, Dover, New York, 1987.

[46] J. Serrin, Mathematical Principles of Classical Fluid Mechanics, *Handbuch der Physik*, Vol. VIII/1, Ed. S. Flügge, Springer, Berlin, pp. 125–263, 1959.

[47] W. Slaughter, *The Linearized Theory of Elasticity*, Birkhäuser, Boston, 2002.

[48] R. W. Soutas-Little, *Elasticity*, Dover, New York, 1973.

[49] A. J. M. Spencer, *Continuum Mechanics*, Dover, New York, 2004.

[50] E. Sternberg, On Some Recent Developments in Linear Theory of Elasticity, Proc. 1st Symposium on Naval Structural Mechanics, Pergamon Press, New York, 1960.

[51] G. Strang, *Introduction to Applied Mathematics*, Wellesley-Cambridge Press, Wellesley, 1986.

[52] S. Succi, I. V. Karlin, H. Chen, Colloquium: Role of the $H$-Theorem in Lattice Boltzmann Hydrodynamic Simulations, *Rev. Modern Phys.*, 74, pp. 1203–1220, 2002.

[53] H. Tennekes, J. H. Lumley, *A First Course in Turbulence*, MIT Press, Boston, 1972.

[54] S. P. Timoshenko, J. N. Goodier, *Theory of Elasticity*, McGraw-Hill, Auckland, 1988.

[55] L. R. G. Treloar, *The Physics of Rubber Elasticity*, 3rd edition, Clarendon Press, Oxford, 1975.

[56] D. J. Tritton, *Physical Fluid Dynamics*, Van Nostrand Reinhold, New York, 1977.

[57] C. Truesdell, *A First Course in Rational Continuum Mechanics Volume 1 General Concepts*, Academic Press, New York, 1977.

[58] C. Truesdell, *Rational Thermodynamics*, Springer, 2nd edition, New York, 1984.

[59] C. Truesdell, K. R. Rajagopal, *An Introduction to the Mechanics of Fluids*, Birkhäuser, Boston, 2000.

[60] C. Truesdell, R. A. Toupin, The Classical Field Theories, In *Encyclopedia of Physics*, Vol. III/1, Ed. by S. Flügge, Springer, Berlin, pp. 226–793, 1960.

[61] A. C. UGURAL, S. K. FENSTER, *Advanced Strength and Applied Elasticity*, 3rd edition, Prentice Hall, New Jersey, 1995.

[62] M. VAN DYKE, *An Album of Fluid Motion*, Parabolic Press, Stanford, 1983.

[63] TH. VON KÁRMÁN, Supersonic Aerodynamics–Principles and Applications, *J. Aero. Sc.*, 14, pp. 373–402, 1947.

[64] R. VON MISES, *Mathematical Theory of Compressible Fluid Flow*, Dover, New York, 2004.

[65] M. WESTERGAARD, *Theory of Elasticity and Plasticity*, Harvard University Press, Cambridge, Mass., 1952 and Dover Publications, New York, 1952.

[66] C. H. K. WILLIAMSON, Vortex Dynamics in Cylinder Wake, *Annu. Rev. Fluid Mech.*, 28, pp. 477–539, 1996.

[67] C. R. WYLIE, *Advanced Engineering Mathematics*, McGraw-Hill, New York, 1966.

[68] H. ZIEGLER, *An Introduction to Thermomechanics*, North Holland, Amsterdam, 1977.

# Index


acceleration, 5, 66
  Coriolis, 99
affine transformation, 82
Airy stress function, 214
angle
  Mach, 273
angular momentum, 115
axis
  principal stress, 125

basis
  canonical, 4
  orthogonal, 4
beam
  long, 241
  long and thin, 239
  long and thin with uniform load $q$, 239
  short, 241
Beltrami-Michell compatibility equations, 209
Boltzmann's theory, 153
boundary conditions, 239
boundary layer, 270, 276, 308
Boussinesq's problem, 230

Cauchy
  deformation tensor, 97
  stress tensor, 121, 134
  tetrahedron, 120
Cauchy elastic material, 179
Cauchy stress tensor, 184
Cauchy's
  equation of motion, 123
  lemma, 118
  postulate, 114
  theorem, 119, 132, 144
Cauchy, Augustin Louis, 74
Cauchy-Green deformation tensor, 74–76, 97
Cayley-Hamilton theorem, 30
Cerruti's problem, 228
circulation of the velocity, 309, 314
Clausius, Rudolf, 155
Clausius-Duhem
  inequality, 156, 166, 196, 200
coefficient
  friction, 296
  Lamé, 188
  Lamé elasticity, 192
  thermal expansion, 202
compatibility equation, 90
compression, 122, 129
  uniform, 129
concept of fading memory, 171
condition
  boundary, 210
  free surface, 283
  incompressibility, 110, 185
configuration, 55, 77
conjugate parameter, 149
conservation
  of angular momentum, 117, 123, 149
  of energy, 141
  of mass, 107, 108, 149, 278
  of mechanical energy, 147
  of momentum, 116, 122, 149
continuity equation, 110
coordinate system, 6
coordinates
  Cartesian, 6
  curvilinear, 42, 43
  cylindrical, 43, 45
  spherical, 43–45
Couette
  two-dimensional flow, 290
creeping flow, 268
curl of a vector field, 39
cylinder
  hollow, under pressure, 223

d'Alembert
  solution, 249
decomposition
  polar, 32, 74
  spectral, of a tensor, 30
definition
  of a fluid, 170
  of a solid, 170
deformation, 56
  homogeneous, 81
  pure, 74
density, 5
  initial, 107
  spatial volume force, 113
derivative
  material, 64, 65
  material of the internal energy, 148
  material w.r.t. time, 104
  of a tensor function, 35
  of a vector function, 35
description
  Eulerian, 58
  Lagrangian, 58
  material, 58, 60
  spatial, 58, 60
determinant, 16
direction
  principal, 27, 184
  principal material, 79
direction cosine, 6
displacement, 57
  material, 61
  small, 85
  spatial, 61
displacement (small), 135
divergence
  of a tensor, 38, 39
  of a vector, 38
drag, 272
  Stokes, 306
Duhem, Pierre, 156

eigenvalue, 79
eigenvalue of a tensor, 26
eigenvector, 31
  normalized, 26
  unit, 78
elastic cord, 251
elasticity
  linear infinitesimal, 191
elastodynamic, 244
energy
  free, 180, 201
  Helmholtz specific free, 156
  internal, 142
  kinetic, 142
  strain, 181, 201, 202
  total, 142
  wave propagation, 254
enthalpy, 198
entropy of a system, 151
equation
  Beltrami diffusion, 311
  Beltrami-Michell compatibility, 209, 218, 233
  Bernoulli, 316
  biharmonic, 214, 238, 240
  Boltzmann's, 154
  characteristic, 27, 84
  compatibility, 90, 209
  constitutive, 160, 178, 182, 196, 208
  continuity, 110, 112
  deformation-displacement, 208
  equilibrium, 123, 128, 208
  Laplace, 41
  Liouville, 153
  motion, 133, 242
  motion in terms of potentials, 244
  motion, Cauchy's, 123
  Navier, 208, 218, 263
  Navier-Stokes, 154, 276, 280, 290
  Navier-Stokes, form reduced, 282
  Newton-Hamilton, 153
  Poisson, 41
  Saint-Venant-Kirchhoff, 192
  Stokes, 283
  wave, 319
  wave propagation, 253
Euler-Almansi strain tensor, 75, 76, 97
evolution
  adiabatic, 144
extension
  relative, 88
extra-stress, 175

fading memory, concept, 171
field
  irrotational, 39
  solenoidal, 38
first

Piola-Kirchhoff stress tensor, 132, 149
Piola-Kirchhoff stress vector, 132
principle of thermodynamics, 144, 149
flow
circular Couette, 293
hypersonic, 273
irrotational, 308
isentropic, 199
on an inclined plane, 289
subsonic, 272
supersonic, 273
two-dimensional Couette, 285
two-dimensional Couette for a compressible fluid, 290
two-dimensional Poiseuille, 286
flow rate, 288, 290, 296
fluid
barotropic, 199, 313
classical, 175
incompressible, 174, 175, 281
inviscid, 175
Newtonian viscous, 195
non-Newtonian, 175
perfect, 175
Rivlin-Ericksen, 175
simple, 173
flux, 48
heat, 143, 195, 291
force, 112
contact, 113, 148
Coriolis, 151
volume, 113, 129, 148
form
global material of conservation of mass, 109
local of the second principle, 155
material of conservation of mass, 107
of the strain energy function, 186
reduced Navier-Stokes equations, 282
spatial of conservation of mass, 110
formula
Nanson's, 78, 132
Fourier's law, 195
Fourier, Joseph, 195
frequency
circular, 254
wave, 255
function
Airy, 235
Airy stress, 214
displacement, 219
energy, 180, 183, 186
isotropic, 33
isotropic tensor, 179
Love's strain, 225
Maxwell, 232
Morera, 232
scalar, of a tensor, 34
strain energy, 186
stress, 232, 238
stress for plane strain problems, 214
stress for plane stress problems, 217

Galerkin vector, 224, 225, 228
Galilean
observer, 117
reference frame, 99
transformation, 99
gas dynamics, 272
Gauss' theorem, 49
Gauss, Carl Friedrich, 50
gradient
deformation, 97
of a scalar field, 36
of a scalar in orthogonal curvilinear coordinates, 46
of a scalar valued tensor function, 38
of a vector, 37
of a vector in orthogonal curvilinear coordinates, 46
of tensor valued tensor function, 38
velocity, 91, 92
Green elastic material, 180
Green, George, 75
Green-Lagrange strain tensor, 75, 76, 97

head loss, 296
heat
per unit mass, 202
produced, 143
received, 143
heat capacity ratio, 199
heat conduction, 195
heat flux, 144

Helmholtz free energy, 156
Helmholtz' theorem, 219
hollow cylinder under pressure, 236
Hooke's law, 202, 212
Hooke, Robert, 192
hyperelasticity, 180
hypothesis
  causality, 162
  determinism, 162
  Stokes, 278
  Valanis-Landel, 187

ideal gas, 198
ideal gas thermodynamics, 198
inequality
  Clausius-Duhem, 156, 166, 196, 200
infinite cord, 250
inflation of a balloon, 189
invariant
  principal, 126, 182
isochoric, 110

Jacobian, 73, 105

Kelvin
  problem, 226
kinematics, 55
  continuum, 55
Kirchhoff, Gustav, 132
Kronecker delta, 8, 14

Lagrange, Joseph-Louis de, 57
Lamé coefficient, 188, 192
Lamé, Gabriel, 188
laminar flow, 268
Laplacian
  of a scalar field, 40
  of a vector field, 41
law
  classical behavior, 173
  conservation of internal energy, 145
  Fourier, 195
  Hooke, 192, 202, 212
Leibnitz' theorem, 107
lemma
  Cauchy's, 118
line
  vortex, 309
linear elastic theory, 207
linear load, 234
linear static elasticity, 208
linearization
  kinematic, 87
  of the stress tensors, 135
Liouville equation, 153
local irreversibility, 198
Lord Rayleigh, 245

Mach number, 280, 281
mass, 107
material
  Cauchy elastic, 179
  Green elastic, 180
  homogeneous, 166
  incompressible, 185, 186
  isotropic, 166
  isotropic hyperelastic, 181
  linear elastic, 180
  Saint-Venant-Kirchhoff, 188
  simple, 166
matrix
  associated with a tensor, 16
  orthogonal, 8
maximum beam deflection, 241
mechanics
  Newtonian, 153
media
  continuous, 55
medium
  isothermal elastic, 177
method
  complex variables, 218
  finite elements, 218, 271
  inverse, 218
  potential, 218
  semi-inverse, 218, 234
  separation of variables, 253, 323
  variational, 218
model
  Mooney-Rivlin, 187, 189
  neo-Hookean, 186, 189
  Valanis-Landel, 187
modulus
  bulk, 194
  of rigidity, 194
  shear, 194
  Young's, 192
momentum, 115
Mooney-Rivlin model, 187, 189
motion, 56
  constant volume, 110

rigid body, 69, 71, 83, 98

Nanson's formula, 78, 132
Navier's equations, 208
Navier, Claude Louis, 209
Navier-Stokes equations, 276, 280, 282, 290
Newton-Hamilton equation, 153
norm
vector, 4
number
Avogadro, 1, 153
Froude, 282
Knudsen, 1
Mach, 272, 280, 281
Prandtl, 280
Reynolds, 268, 280
Strouhal, 270

objectivity
acceleration, 97
and rigid body motion, 98
notion, 98
of the conservation of energy, 150
of the kinematic quantities, 94
velocity, 97
observer, 56
operator
Laplacian, 40
linear, 16

Papkovich-Neuber presentation, 229
pathline, 67
penetration depth, 299, 300
period
wave, 255
permutation symbol, 12
physical components, 46
Piola-Kirchhoff
first stress tensor, 132
first stress vector, 132
second stress tensor, 134
second stress vector, 134
Piola-Kirchhoff stress tensor, 178, 184
plane
principal, 125
point
fixed, 70
material, 55
Poisson's ratio, 192
postulate
Cauchy's, 114
potential
Lamé strain, 220
power, 142
contact force, 145
mechanical, 149
Prandtl number, 280
presentation
Papkovich-Neuber, 229
pressure, 175, 176, 185
hydrostatic, 127, 130, 185
principal mode, 256
principle
admissibility, 166
conservation of angular momentum, 117, 123, 133
conservation of mass, 108
conservation of mechanical energy, 148
conservation of momentum, 116, 122
determinism, 168
equipresence, 168
local action, 162, 169
material invariance, 165, 178
memory, 166
objectivity, 163, 169, 174, 177, 192
regular memory, 170
Saint-Venant's, 210, 234
superposition, 211
thermodynamics, first, 144
thermodynamics, second, 154
problem
Boussinesq, 230
Cerruti, 228
Kelvin, 226
product
exterior, of two tensors , 19
scalar, 12
scalar, of two tensors, 24
scalar, of two vectors, 4, 11, 14
tensor, of two vectors, 16
vector, 12
vector, of two vectors, 13
pure bending, 130

quantity
materially objective tensor, 97
materially objective vector, 97
objective, 97
physical, 94

spatially objective tensor, 97
spatially objective vector, 97

ratio
Poisson, 192
Rayleigh surface wave, 244
reference frame, 56
relation
constitutive, 181
Reynolds
transport theorem, 106, 110, 116
Reynolds number, 280
Reynolds, Osborne, 106
rigid body
motion, 69, 71, 83
rotation, 70, 151
translation, 69
Rivlin-Ericksen theorem, 33
rotation, 82
rigid body, 72, 151
rigid body, around a fixed point, 70

Saint-Venant
compatibility equation, 90
Saint-Venant's Principle, 210
Saint-Venant-Kirchhoff equation, 192
Saint-Venant-Kirchhoff material, 188
scalar, 9
scalar product, 4, 11, 12, 14
second principle of thermodynamics, 154
shear, 194
simple, 83
shear angle, 89
shock, 274
SI, 5
simple material, 167
solution
d'Alembert, 249, 324
elastic cord, 258
self similar, 298
solution in linear elasticity, 217
sound barrier, 273
space
Euclidean, 3
vector, 4
speed
surface wave, 246, 247
waves in elastic solids, 262
speed of sound, 199, 319, 320
Sphere
hollow, under pressure, 221
state
of homogeneous stress, 129
one-dimensional stress, 126
plane strain, 211
plane stress, 215
pure hydrostatic stress, 127
stress, 121
three-dimensional stress, 125
two-dimensional stress, 125
Stokes
equation, 283
Stokes' hypothesis, 278
Stokes' theorem, 51
Stokes, George Gabriel, 51
strain potential
Lamé, 220
streakline, 67
stream function, 301
streamline, 67
stress
comparison, 128
material contact, 114
nominal contact, 114
normal, 122
octahedral, 128
plane, 131, 236, 239
principal, 125
shear, 122
uniform shear, 130
von Mises, 128
stretch
biaxial, 189
equibiaxial, 189
principal, 79, 187
uniaxial, 189
sum
of tensors, 20
summation convention, 7
surface tension, 284
symbol
permutation, 12

tension, 122, 129
uniform, 129
tensor, 3, 17
Green-Lagrange strain, 76
of order 2, 18
right Cauchy-Green deformation, 74

antisymmetric, 22, 72
Cauchy deformation, 97
Cauchy stress, 121, 134, 149, 184
Cauchy-Green deformation, 75, 97
components, 15
contraction, 19
deformation gradient, 73, 166
deviatoric, 23
deviatoric stress, 127
divergence of, 39
dual vector, 25
eigenvalue, 26
Euler-Almansi strain, 75, 76, 97
first Piola-Kirchhoff stress, 149, 184
Green-Lagrange strain, 75, 97
Green-Lagrange strain rate, 149
infinitesimal strain, 87, 191
invariants, 28
inverse, 22
left Cauchy-Green deformation and its inverse, 76
matrix, 15
of order $n$, 18
of order 2, 14
orthogonal, 24, 96
Piola-Kirchhoff stress, 178
positive definite, 30
product of two, 19
rate of deformation, 92
rate of rotation, 92
right Cauchy-Green deformation, 76
rotation, 71
rotation infinitesimal, 90
rotation rate, 92
scalar function of, 34
second Piola-Kirchhoff, 181
second Piola-Kirchhoff stress, 134, 149
singular, 16
spatially objective, 96
spectral decomposition, 30
strain rate, 92
stress, 119
symmetric, 22
trace of, 23
transpose, 21
unit, 15, 26
zero, 15, 19
tensors
interior product of two, 20
equivalent, 19
sum of, 20
tetrahedron
Cauchy, 120
theorem
Rivlin-Ericksen representation, 33
Cauchy's, 119, 132, 144
Cayley-Hamilton, 30
Crocco, 315
divergence, 49, 107, 122, 143
Gauss, 49
$H$, 154
Helmholtz, 219, 309
Kelvin, 315
kinetic energy, 145
Leibnitz', 107
localization, 109
polar decomposition, 32, 74, 167, 178
Reynolds transport, 106, 110, 116, 144
square root, 30
Stokes', 51
theory
Boltzmann's for hydrodynamics, 153
kinetic gas, 278
of elasticity, 179
thermal diffusivity, 280
thermoelasticity, 201
thin-walled container under pressure, 239
traction
simple, 193
transformation rule, 17
translation, 82
rigid body, 69
uniform, 150

uniform compression, 83
uniform expansion, 83

Valanis-Landel
hypothesis, 187
model, 187
value
principal, 184
variable
material, 60
spatial, 60

vector, 3, 10
    dual, 72, 151
    Galerkin, 224
    product, 8
    rotation rate, 93
    second Piola-Kirchhoff stress, 134
    spatial stress, 113
    spatially objective, 96
    stress, 125
    surface stress, 113
vector product, 13
velocity, 64
    average flux, 296
vibration of an elastic cord, 255
viscosity
    dynamic, 175
    kinematic, 176
    shear, 175
    volume, 175
volume flux, 288
von Helmholtz, Hermann, 156
vortex, 270, 307
vortex tube, 309
vorticity, 93, 310, 313, 315

wave
    compression, 243, 248
    dilatation, 242
    distortion, 243
    elastic plane, 248
    longitudinal in a beam, 261
    Rayleigh surface, 244
    shear, 242, 243, 248
    shock, 274
    stationary, 257
wavelength, 256
Weissenberg effect, 268

Young's modulus, 192

Achevé d'imprimer par Corlet Numéric, Z.A. Charles Tellier, 14110 Condé-en-Normandie
N° d'Imprimeur : 152388 - Dépôt légal : novembre 2018 - *Imprimé en France*